W0037035

ECOLOGY OF PLANT GALLS

MONOGRAPHIAE BIOLOGICAE

EDITORES

W. W. WEISBACH
Den Haag

P. VAN OYE
Gent

VOLUMEN XII

Springer-Science+Business Media, B.V. 1964

ECOLOGY OF PLANT GALLS

BY

M. S. Mani, M. A., D. Sc., F. L. S.,

Emeritus Professor of Zoölogy and Entomology, School of Entomology, St. John's College, Agra

Springer-Science+Business Media, B.V. 1964

ISBN 978-94-017-5801-7   ISBN 978-94-017-6230-4 (eBook)
DOI 10.1007/978-94-017-6230-4

Zuid-Nederlandsche Drukkerij N. V. — 's-Hertogenbosch

To

The Memory

of my

Parents who inculcated

early in me a Love

of

Nature

CONTENTS

PREFACE

This book presents a brief outline of the more outstanding features of the general ecology of plant galls. It is the outcome of over thirty-six years work by the author on the taxonomy and ecology of plant galls and the complex of organisms associated with them from different parts of the world. This is the first comprehensive book that attempts to cover all the recent advances on the fundamental principles of general cecidology. It supplies the urgent need for critical reviews and evaluations in a subject, in which there has been much new work from a variety of approaches in recent years.

Plant galls represent a unique and complex interspecific interaction and mutual adaptation between the plant and the gall-inducing organism. The nature and origin of the inter-relation, the rôle of the gall-inducing organism and the reaction of the plant, the cytological, histogenetic and morphogenetic processes involved in gall formation are essentially ecologic problems that need further investigations. This book differs from all other works on plant galls, which are also mostly out-of-date, mainly in its broad-based ecologic approach. Its outstanding merit consists in the synthesis of knowledge, gained in many different fields like not only botany, entomology, mycology and phytopathology, but also cytology, genetics, tissue culture, morphogenesis, cancer research, ecology, etc., into a unified ecologic problem. There is a reorientation of our ideas on the meaning of gall that marks a departure from the traditional teleological considerations, based on the doctrine of natural selection. The central idea in the book is that the development of gall represents a specific reaction of the plant, isolating the gall-inducing organism in space and time. The method of treatment is that of condensed summaries of the knowledge gained by the author in the course of his own studies and of the information from the extensive and much scattered literature on the subject. The scope of the work and the limitations of space naturally preclude any attempt at a detailed or an exhaustive treatment of any topic in such a diversified and large field as cecidology. Emphasis has on the other hand been laid throughout on a broad outline of the general and fundamental principles, which are likely to be of interest to the botanist and zoologist alike.

The book is divided into fifteen chapters. The first two chapters deal with a brief review of the current ideas on what constitutes plant galls, particularly in the light of recent advances in our knowledge of plant morphogenesis, the plants on which galls arise and the gall-inducing agencies. The third chapter is devoted to a brief outline of the fundamental principles of the morphology of galls in general. The salient structural, developmental and ecological features of galls on different parts of plants are briefly discussed in the next five chapters. Three more chapters are devoted to a consideration of the distinctive structure, development and ecology of zoocecidia, the interrelations of the diverse organisms associated with them, their life-cycles, etc. The characteristic features of galls caused by fungi, bacteria and viruses are summarized in two chapters. The problems of etiology, histogenesis, and the mechanism and factors in cecidogenesis and the morphogenesis of galls are dealt with in the penultimate chapter. This chapter is in a sense a general introduction to the problems of cancer analogy of plant galls. The last chapter presents a general and critical review of the recent advances in our knowledge of the crown-gall tumor in plants, etiology and nature of animal tumors and considerations on the fundamental similarity of plant and animal neoplasia. Although dogmatic statements are avoided, the general trend of current ideas is nevertheless sufficiently clearly indicated. There is at the end of the book a bibliography of 1300 titles, arranged alphabetically authorwise, and numbered serially, the numbers being quoted in the text. The bibliography does not claim to be exhaustive, but includes only the works considered relevant to the main theme of the book.

The book is illustrated by over one hundred and sixty text-figures, most of which are original and the sources of the others are suitably acknowledged in each case. The photographs reproduced in the plates are wholly original.

The author is grateful to his friends and pupils for specimens of plant galls and to Dr SANTOKH SINGH, Lecturer in Zoology, St. John's College, Agra, in particular for the preparation of microtome sections of galls. He is also indebted to his wife for her appreciation of his interests and for constant encouragement, and to the late Prof. em. Dr. W.W. WEISBACH, editor of the series Monographiae Biologiae, for his keen interest in the completion of this book.

M. S. MANI

CHAPTER I

INTRODUCTION

1. What are galls?

Most of us are familiar with plant galls or cecidia. Some galls, like the familiar oak-apple or the gall-nut produced by cynipids on *Quercus* Linn. (fig. 1A), have been well known in medicine and industry since time immemorial. Others like the bedeguar on *Rosa* Linn. (fig. 1E), the lenticular gall on leaf (fig. 84A), the gouty gall on stem and the roly-poly galls produced by cynipids on different species of *Quercus* Linn., the kammergalls of *Pontania* Costa on *Salix* Linn. (fig. 1D), the spirally-twisted covering gall on the petiole of *Populus pyramidalis* Linn. (fig. 1C) and the gladstone-bag-like pouch gall on *Pistacia* Linn. (fig. 1B) produced by aphids, the large woody galls on the branches of *Acacia* spp. produced by the fungus *Uromycladium* (Plate I, 5) or the spherical, solid, pellet-like galls on leaves of diverse plants (Plate VII), to mention only a few, though perhaps not so well known, are not, however, less common and are

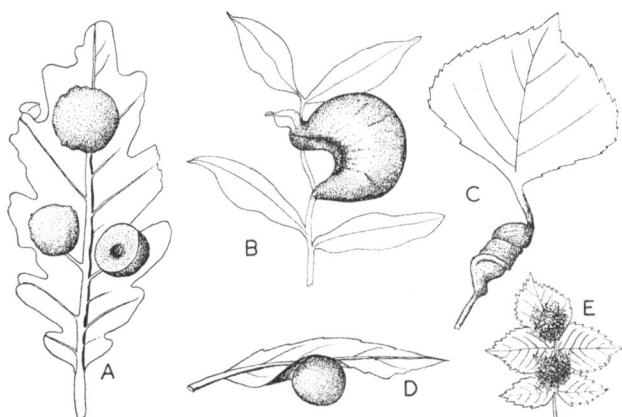

Fig. 1. Some common plant galls. A. The common oak-apple gall of *Diplolepis quercus-folii* (Linn.) on *Quercus pedunculata* Ehrh. B. The gladstone-bag-like gall of the aphid *Aploneura lentisci* Pass. on the leaflet of *Pistacia lentiscus* Linn. C. The spirally twisted covering gall of *Pemphigus spirothecae* Pass. on the petiole of *Populus pyramidalis* Linn. D. The kammergall of *Pontania* on leaf of *Salix*. E. The well known "bedeguar" gall of *Diplolepis (= Rhodites) rosae* (Linn.) on leaf of *Rosa* sp.

even more interesting for their complexity of structure, development and ecological relations.

Galls are pathologically developed cells, tissues or organs of plants that have risen mostly by hypertrophy (over-growth) and hyperplasy (cell proliferation) under the influence of parasitic organisms like bacteria, fungi, nematodes, mites or insects. They represent the growth reaction of plants to the attack of the parasite and are in some way related to the feeding activity and nutritional physiology of the parasite ([635, 639]). Some galls, like, for example, the so-called procecidia, are, however, known to arise even before the parasite starts feeding (fig. 2A-B). COSENS[221, 222] therefore considered as galls "any enlargement of plant cell, tissue or organ, induced by the stimulus of a parasitic organism as a regular incident in the life of the parasite."

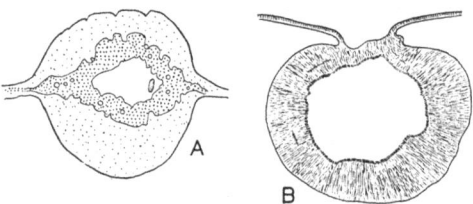

Fig. 2. Procecidia. A. Section through the kammergall of *Pontania capreae* Linn. on *Salix triandra* Linn. showing the initiation of cecidogenesis even before the hatching of the larva from the egg. B. L.S. through the kammergall of *Pontania viminalis* Linn. on *Salix purpurea* Linn. (Modified from Beijerinck).

In the biological association between the gall-inducing parasite or the gall-maker and the plant, it is the gall-maker that apparently derives all the benefit and the plant suffers loss of substance, deviations in the direction of growth, disturbances in sap flow, premature decay, increase of non-essential parts at the cost of essential and other injury. By producing the gall, the plant seems to have not only guaranteed free shelter but also free nourishment to the gall-maker and even to assure the dispersal of the species of gall-maker. Even when there is some benefit to the gall-bearing plant, as for example, in the fixation of atmospheric nitrogen in the root-nodule gall produced by *Bacterium radicicola* (BEYER.) on Leguminosae (fig. 28B) or in the cross pollination of flowers by the gall-forming fig-insects *(Blastophaga* GRAVENH.) in *Ficus* spp., it seems to be rather indirect and incidental. Although nearly all early definitions of gall have laid considerable stress on this teleological aspect, it has nevertheless been suspected that galls may be passive means of defence on the part of the plant against the parasite. COOK[214] believed, for example, that the formation of a gall is "probably an effort on the part of the plant to protect itself from an injury which is not sufficient to cause death." ZWEIGELT[1296-1299] has recently developed the idea that the formation of a gall is the result of a gradual process of mutual adaptation between the gall-inducing organism and the gall-bearing plant. The plant produces the gall primarily *not* for the benefit of the gall-inducing organism, as formerly assumed. The primary reaction of the plant to the gall-inducing organism is not in the interest of the organism, but represents a sort of "struggle against the attack of the parasite," the object of which is the neutralization of the toxins produced by it. In so far as the reaction of the plant favours the survival of the organ attacked by the gall-inducing organism, the primary advantage in the formation of the gall is to the plant. By producing the gall, the plant has in a sense localized the parasite in space and time and has forced it to extreme specialization. Considered from this point of view, a gall includes all structural deformities produced by plants under the influence of a foreign organism, which may either be a parasite or a symbiote.

Particular attention must be drawn to the fact that the abnormal structures which constitute galls are actively produced by the plant as a result of abnormal growth activity. Galls are essentially developmental and growth abnormalities. This peculiarity will therefore at once exclude a variety of abnormal structures, which are actively produced by diverse animals on plants, but are not the result of abnormal growth reaction on the part of the plant. The cigar-shaped or barrel-shaped leaf scrolls made by different species of weevils like *Attelabus* LINN., *Rhynchites* HERBST., etc., the rolling and webbing of the leaf blade by caterpillars of certain moths like *Gracilaria*, spiders, etc., for example do not represent the growth reaction of the plant to the attack of these species and are therefore not galls. In these cases the rôle of the plant is a purely passive one. We must also similarly exclude from our concept of gall the well known acarodomatia sometimes found on leaves, especially at the meeting point of some of the principal veins, as hairy outgrowths, in which mites like *Tarsonemus* live. These hairy outgrowths are normal features in the leaf structure, which are merely taken advantage of by the mites and do not likewise represent the growth reaction of the plant to the attack of the mite. A pathological structure, to be considered as gall, must thus represent the reaction of the plant in a specific manner, viz. by growth, to the attack of some foreign species. The gall is essentially the product of an interspecific association between a plant and another organism, characterized by the plant reacting with growth, which is abnormal in some respect.

While there is usually very little difficulty in recognizing a gall as such, one often comes across cases, which are, however, not so readily classed. The difficulty is partly due to the fact that we usually find every gradation between the apparently normal and the decidedly abnormal growths, so that it is often extremely difficult to be sure whether a given deformity is worthy of being called a gall. There is also no hard and fast rule to decide at what stage of abnormality a growth becomes a gall. Some workers include, for example, the mere curling of a developing leaf blade under the influence of thrips, aphids, etc. under the term gall, but others do not consider such deformities as galls, unless the curling is at the same time also accompanied by more or less pronounced swelling of the part. As growth is associated with cell hypertrophy, cell proliferation or both, wherever these processes are observed to arise under the influence of foreign organisms, we deal with galls. The fundamental character of a gall, as distinct from numerous other abnormal structures we find commonly on plants, is that the reaction of the plant to the attack of a foreign organism includes, without exception, these processes.

Even among the decidedly abnormal growths, one often comes across instances which are galls according to some workers, but not according to others. The mines in leaves produced by the larvae of certain insects illustrate some of the difficulty, we often find in determining galls. Leaf mines, in which the mesophyll surrounding the mine does not produce any new formation or growth, has no significance to our concept of gall. In these cases the plant remains more or less passive. In most cases, there is thus little doubt, but in the case of mines produced by certain species there is sometimes considerable cell proliferation, so that the mined leaf is also more or less conspicuously swollen, in addition to being passively crumpled and otherwise deformed. The active callus outgrowths from the uninjured neighbouring cells develop into the mined cavity generally as a tube-like or bladder-like thin-walled, plasma deficient but hyperhydric mass and may contain some chlorophyll or may lack this pigment altogether. The mined cavity seems in this case to particularly favour the growth of callus regenerative tissue. In the simplest case, the callus growth consists of typical, large branched cells. This is found, for example, in the leaf mines of *Pegomyia chenopodii* ROND. on *Chenopodium album* LINN. and *Chenopodium urbicum* LINN. and *Rhynchaenus quercus* LINN. on *Quercus*. The new tissue formed in these cases develops conspicuously near a vascular element and at some distance behind the mining larvae and is apparently not directly related to the feeding activity of the larvae. Abundant callus outgrowth fills the mine of *Phytomyza ilicis* CURT. in the leaf of *Ilex aquifolium* LINN. The larva feeds as a rule on the uppermost of the multi-layered palisade and the lowermost layer now exhibits cell proliferation and callus outgrowth in the form of a multi-layered, elongate tubular mass, so that even the epidermis becomes locally pushed upward to form a conspicuous bulge on the surface of the leaf blade. Such mines have a remarkable resemblance to what is usually known as parenchyma or pustule gall *(vide* Chapter VII). The epidermis is, for example, very conspicuously bulged outward in the mine of *Liriomyza strigata* MEIG.

In all these mines the neoplastic tissue is apparently the result of stimulus of mechanical injury and there is probably no direct specific relation to the mining larva. In the mines of other species, the larva at first mines in the midrib. From this main gallery the larva then produces some lateral galleries and often sometimes returns to the main gallery at a later stage to feed on the new regenerative tissue which has filled the gallery in the meanwhile. This callus regenerative tissue thus corresponds to the so-called nutritive tissue found in many typical galls (fig. 41). The larvae of the Microlepidoptera *Nepticula argyropeza* ZETT. and *Nepticula turbidella* ZETT. mine in the distal part of the petiole. The mined portion becomes swollen to about twice the normal thickness. Initially the mining larva is found in a gallery in parenchyma, with hypertrophied cells. VOIGT[1221] considers the mines of *Apion sedi* GERM. on leaf of *Sedum* spp. as galls on the basis of their histological peculiarities. Examples of leaf mines associated with such cell growth and renewed cell divisions and filling up of the lumen of mine with new tissue should be included among galls.

It should be pointed out that not all growth abnormalities arise at the seat of attack by the gall-maker, but sometimes involve distant organs also. The effect of the attack of the gall-maker does not seem to be strictly localized in such cases or perhaps the development of a gall in a distant part may be the secondary result of some as yet unrecognized effect at the seat of attack. In such cases the deformities do not generally show evidence of the presence of a parasite inside them. The inflorescence of many plants, for instance, may undergo anomalous development due to the attack of gall Nematodes and certain insects inside the root underground, far from the actual place of deformity. The gall midge *Rhopalomyia hypogaea* F. Löw develops in the stem of *Chrysanthemum leucanthemum* in many parts of Europe and causes the non-development of the rayflorets in the capitulum. MOLLIARD[810] has described several such growth abnormalities in flowers of *Scabiosa columbiaria* LINN. due to the attack of *Heterodera marioni* (CORNU) GOODEY within the root. Such growth abnormalities also constitute galls.

A number of galls are therefore essentially abnormally developed organs and are known as "organoid galls". The typical organoid galls include fasciations produced by mites, fungi and other organisms (fig. 3), abnormally elongated or abnormally stunted internodes and bunched leaves, abnormally shaped leaves, doubled and filled-up flowers, chloranthy (virescence) or greening of the petals of flowers, petalloidy, witches' brooms, etc. The abnormality is thus largely or also exclusively in the external form of the part, but the anatomical and histological structures do not differ fundamentally from those of the normal plant organ. The tissues of organoid galls are thus normal. Some of the organoid galls are often extensive

6

growths and involve the major part or the whole of the aerial portion of a plant (Plate II, 3). Organoid galls are predominantly produced by parasitic fungi (especially the Uredinaceae, Ustilaginaceae,

Fig. 3. Fasciation of the shoot axis of *Hibiscus canabinus* Linn.

Exoascaceae, etc.), mites and aphids and only exceptionally by Diptera and Hymenoptera. They develop also on all classes of plants.

Most galls are, on the other hand, far more complex in their structure and we cannot recognize in them any normal organ as such. These galls differ fundamentally in their anatomy and histology from the normal organ on which they arise and in contra-distinction to organoid galls, they are termed histioid galls.

2. The fundamental character of a gall

The essential character of a gall, however, is not really its cause – what gives rise to it – and its relation to the causative factor or even its structure, but its relation to the plant on which it arises – to the morphogenetic control of the plant body. A gall is essentially a *neoplastic* growth. Neoplastic growths are pathological structures, ranging from the nearly normal to the highly complex and abnormal outgrowths, characterized by cellular hypertrophy and hyperplasy. The cell growth and cell divisions in the gall are not coördinated with these processes in the normal organ. Abnormal cellular hypertrophy and hyperplasy are induced in plants as in animals, by diverse factors *(vide* Chapter XIV). Irrespective of the causative factor and complexity of the structure of a gall, there is in its development a more or less pronounced departure from the general morphogenetic restraint of the organ of the plant. Depending upon the degree of this departure, galls may be *limited* neoplasms or *unlimited* neoplasms. Neoplasms are described as "spontaneous" if internally conditioned and "induced" if externally conditioned. Most naturally occurring neoplasms on animals are usually presumed to be internally conditioned[321], but some animal neoplasia like the virus tumors, viz. the Shope papilloma[1044], Brown-Pearce tumor of rabbits[148], Rous sarcoma of fowl[985-987], the Bittner mammary tumor of mice[88] and the cysticercus tumor of rat[325] are considered as "induced". Tumors induced by the action of various carcinogens are externally conditioned. The line of demarcation between the so-called internally conditioned and the externally conditioned neoplastic growths is not, however, sharp and at any rate the distinction is purely arbitrary. The Shope papilloma, for example, ordinarily benign, becomes malignant as a result of the combination with agents, which are themselves only mildly carcinogenic in action[988]. Most of the tumors now classified as spontaneous will doubtless eventually be found to be induced on fuller investigation. Spontaneous neoplasms, so called, are generally rare on plants and are restricted, as at present known, to the relatively mild and delimited chimeras, fasciations, etc., due to somatic chromosomal mutations or hybrid incompatabilities as in the case of the genetic tumors on crosses of *Nicotiana glauca* with *Nicotiana langsdorfi (vide* Chapter XIV).

In limited neoplasms, the morphogenetic character of the galled tissue does not depart widely from that of the normal plant. The abnormal growth does not proceed beyond the strict limits in space and time set by the specific nature of the plant. Chimeras, fasciations; galls by mites, nematodes and insects (zoocecidia) *(vide* Chapter IX); galls by fungi (mycocecidia, *vide* Chapter XII), etc., are some of the better known examples of limited neoplasms on plants. Limited neoplasms may be non-biological or biological in origin. The non-biological limited neoplasms are galls induced by physical and chemical means.

The physical galls include the callus outgrowth at the margins of wounds. While mechanical injury is naturally the initial cause, a combination of derangements of water economy and respiratory rates, accumulations of the products of necrosis, all of which are essentially chemical changes, represents the immediate cause of the wound regeneration callus outgrowths. VÖCHTING[1217] showed that over-growth at the point of union of stock and scion in a graft results from the tendency of the individual cells of both the stock and scion to maintain their definitive polarities. Physiologically like cells of both animals and plants *(vide* Chapter XIV) exhibit repulsion, so that their distal ends tend to avoid the distal ends of adjacent cells and to approach the proximal ends. As the plant cells are not, however, capable of free movements to correct such unsuitable polarity orientations, such a correction can only come about by growth (fig. 149). Each new cell, as it is being formed, tends to be orientated to bring about adjustments between stock and scion. The result is an intumescence. The term intumescence was first applied by SORAUER to localized, pustule-like swellings of tissues due to the enlargement of cells on *Ribes, Eucalyptus rostrata* SCHLECHT., *Acacia pendula* A. CUNN., *Lavatera trimestris, Aloe grandiflora, Aphelandra porteana*, etc. The cortical cells often become radially elongated, especially on the sunny side, thus giving rise to loose swellings. These swellings become considerably large where the cells of the primary cortex are involved. Intumescences are also widely known on leaves and localized knob-like swellings arise either

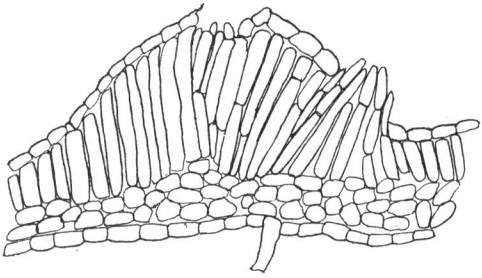

Fig. 4. T.S. through intumescence on leaf of *Cassia tomentosa* (Modified from Sorauer).

on the upper or also on the under side of the leaf blade in the form of green or white pustules of variable height and diameter. Some of these hypertrophied cells may also undergo division. This is, for instance, met with in the well known intumescence on leaf of *Ficus elastica* LINN., *Ruellia formosa* and *Cassia tomentosa* (fig. 4). The epidermis above the intumescent mesophyll becomes either simply bulged out as in *Epilobium hirsutum* or may also become in its turn hypertrophied, highly modified, with abnormally shaped stomata.

Although the galls referred to above are initiated by purely physical causes, it is, however, incorrect to assume that their development is not subsequently influenced by chemical factors. There is indeed no sharp dividing line between the so-called physical and chemical galls and in most cases both the initial and proximal causes are predominantly only chemical *(vide* Chapter XIV). It is known, for example, that even ordinary salt water and solutions of copper salts produce intumescence on leaves and stems[1036]. Organic acids like malic acid, acetic acid, formic acid, butyric acid, lactic acid, etc., are well known for their tumefacient effects on diverse plants[35, 97]. In 1932, DOLK & THIMANN[274] demonstrated the growth-promoting activity of indole-acetic acid. Formic acid has also been shown to be of some importance in the formation of certain insect galls. The wound hormones, considered to stimulate the formation of regeneration tissue[426], also appear to involve certain acid products. The heterauxins like indole-acetic acid, indole-butyric acid, naphthalene-acetic acid, etc., have well known tumefacient properties and may perhaps be in part least responsible for the formation of a many of our insect galls[146, 347, 629, 630]. Most of the chemical-induced intumescences are, with few exceptions, limited neoplasms and develop only so long as the external force of the chemical is available in the plant tissues.

Limited neoplastic galls of biological origin include the virus galls, bacterial galls, fungal galls, mite and insect galls, nematode galls, etc. Some viruses produce definite tissue distortions and the enations induced by certain strains of the tobacco mosaic virus are examples of this type ([90-92, 549];) (vide Chapter XIII). BLACK[90] has described a definitive virus tumor on a wide and varied range of plants. The gall-making virus is a systemic entity, but the gall arises only where local tissue proliferation is initiated by injury. Thus neither injury nor the virus alone seems to be capable of producing the gall, but when the cells invaded by the virus are injured, they are incited to rapid and disorganized growth and division. Though superficially non-limited in character, it is of course difficult to determine how the cells of the gall would behave in the absence of a chronic infection by the virus. Bacterial galls occur on a great variety of plants and some of them, like the root nodule galls of Leguminosae, are typically limited neoplastic galls (vide Chapter XIII). Fungal galls are too varied and numerous to be mentioned at this stage (vide Chapter XII) and there is at present no evidence that any mycocecidia ever become autonomous and non-limited. The zoocecidia are typically limited neoplastic galls. We must not, however, overlook the possibility that the limited character of the growth of these galls may actually be the result of nutritional or structural barriers, rather than intracellular physiological barriers, as may be generally assumed.

The galls, which we have so far considered, arise as a result of the presence of a foreign body of some kind and continue to develop only so long as this body or its immediate chemical products are present within the plant tissue and are thus in this sense strictly exogenous. In EWING's terminology[321] these limited neoplastic galls are "irritation teratoma". Although numerous attempts have been made to establish the similarity of plant galls to malignant growths on animals and man, it has been pointed out that vegetable neoplastic growths are mostly traceable to some recognizable and persistent stimuli, but the animal neoplasia are of unknown origin (vide Chapter XV). The discovery of virus and chemical tumors of animals has, however, weakened this objection. Certain galls are truly autonomous, at least in their final stages of development and are therefore typical non-limited neoplasms. In non-limited neoplasms the degree of abnormal growth is not controlled by the nature of the plant but only by the exciting factor. In 1912, SMITH and his collaborators[1064] described the development of galls on *Chrysanthemum frutescens*, inoculated with the bacterium *Phytomonas tumefaciens* (SM. & TOWNS.), as consisting of a tumor at the site of inoculation and secondary tumors at distances, often three or four internodes away from the site of inoculation. The secondary tumors arise in leaf axils, on midribs, petioles and seldom on the stem proper and are not connected with the primary tumor by continuous tumor tissue. As the secondary tumor does not develop in the stem tissue immediately adjacent to the primary tumor, it is not the result of direct diffusion of the bacterial metabolic products or even of the primary tumor. Also the bacterial sterility of the secondary tumors does not conform with the idea of "irritation teratoma". Though SMITH did not recognize the importance of his own observations, recent researches have demonstrated that the secondary tumors are truly autonomous: their origin is independent of the causative factor, once the initiation has been effected. JENSEN[548] demonstrated the transplantability of the secondary tumors. By grafting the bacteria-free secondary tumor tissue onto *Beta vulgaris* of different colours (for example, red tumor tissue grafted onto yellow healthy tissue and *vice versa*), JENSEN showed that there is no deep tissue invasion by the tumor tissue. The tumor corresponds to JENSEN'S rat sarcoma or artificial metastasis of animal tumors. BRAUN & WHITE later[132] successfully cultivated the bacteria-free fragments of the secondary tumor on *in vitro* media and established the characteristic independence of the tumor tissue. The tumor character was further proved by transplanting fragments of the tumor tissue cultures onto healthy tissues. The cause of the pathological condition is in the tumor cell itself and not in any accompanying organism. The tumor cells are endowed with the qualities of disorganized growth and division, both *in vitro* and *in vivo* and with the capacity for transplantability or artificial metastasis. The genetic tumors of *Nicotiana* hybrids, already mentioned, are also non-limited neoplastic galls. The cells of the non-limited neoplastic galls are generally assumed to be "permanently transformed" in some specific manner, but as has recently been pointed out by HOUGH[536], many cells of the so-called "cecidial galls", bacterial galls and virus galls also undergo irreversible transformation.

The term gall thus embraces a much wider range of morphogenetic abnormalities in plants than has hitherto been understood. A general synopsis of a morphogenetic classification of abnormal growths in plants, recently suggested by BLOCH[93], also favours this wider concept of galls. A fundamental difference thus exists between the earlier and the present concepts of galls. In our concept, there is a new emphasis on the ecological and morphogenetic aspects, as distinct from the earlier and rather narrowly morphological approach. Our approach to the problems of galls thus centres around the dynamic aspect rather than on the static structural. Whether we consider the term gall in a restricted sense or in its wider meaning, we must, however, recognize the fact that the concept is strictly ecological. A gall is essentially the product of mutual interaction of the plant cell and the gall-inducing agent. The development of a gall involves factors that release a cell from the general morphogenetic coördinating influences. For a correct understanding of the complex problems of gall formation, the inducing factors, the plant reactions, etc., an ecological approach is therefore a fundamental necessity.

CHAPTER II

GALL-BEARING PLANTS AND
GALL-INDUCING ORGANISMS

1. Plants that bear gall

Galls arise on nearly all the major classes of plants. Although no group of plants is entirely free from galls, even casual observation shows that galls are more common on some groups than among others. Relatively few galls are found, for example, on Cryptogams and Gymnosperms, but the Phanerogams, especially the Dicotyledonae, are remarkable for the abundance and great diversity of their galls (fig. 5).

Algae.

Although the Algae would at first sight appear to be unlikely to have galls, a number of interesting galls, caused by parasitic Algae, Bacteria, Fungi, Nematoda, Rotifers and Copepodes, are known on several species of Algae. Acarina and insects are not, however, known to give rise to galls on Algae. Some of the better known of these galls are found on the marine brown and blue-green Algae,

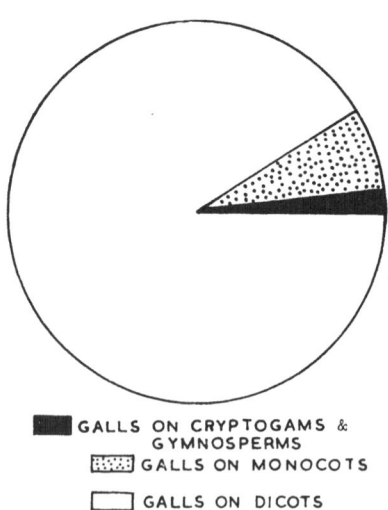

GALLS ON CRYPTOGAMS & GYMNOSPERMS
GALLS ON MONOCOTS
GALLS ON DICOTS

Fig. 5. Relative abundance of galls on the major classes of plants.

but some typical galls have also been described on a number of fresh-water Algae like *Batrachospermum, Urospora, Coleochaete, Vaucheria,* etc.[1287]. SCHMITZ[1030] has described nodular galls, caused by bacteria on the red Algae *Chondrus crispus, Pelesseria sanguinea*, etc. BRAND[118] has given an account of bacterial galls, in the form of irregular cellular outgrowths, on the filaments of *Chantransia*. Reference may also be made to CHEMIN[186] for an interesting account of several other bacterial galls on Algae like *Gracilaria confervoides, Cystoclonium, Saccorhiza bulbosa*, etc. Other galls on *Gracilaria confervoides* are caused by *Gloreocolex pachydermus* and *Gracilarophila oryzoides*. Small galls are reported by LAGER-HEIM[650] to be produced by *Sarcinastrum urosporae* LAGERH. on *Urospora mirabilis*. Several examples of galls produced by parasitic Algae on other species of Algae are also known. SAUVAGEAU[1010] has, for example, described a gall on *Cystoseira ericoides*, caused by the parasitic Alga *Ectocarpus valiantei*. VALIANTE[1200] reports *Streblonemapsis irritans* as causing nodular galls on *Cystoseira opuntioides*. Among the fungal galls on Algae, the best known is perhaps that of *Sphacelaria* and *Cladostephus*, produced by *Olpidium sphacelarum*. ROTHERT[984], TROTTER[1164] and BENKO[79] have given interesting accounts of the galls produced by the rotifer *Notommata werneckii* EHRENB. on *Vaucheria* spp. Swellings of the air-bladders of the marine Algae *Ascophyllum nodosum, Desemarestia aculeata* and *Rodymenia palmaza* produced by the Nematode *Tylenchus fuscicola*, were described by BARTON[58] in 1891 and by DE MAN[246] in 1892. BARTON[58] has also described the nematode galls on *Furcellaria* and *Chondrus*.

Fungi.

Galls are not really rare on Fungi, as may be supposed, but only a relatively small number of these have so far been studied in any detail. KÜSTER[639] refers to *Pleotrachelus fulgens* ZOPF. (Olpidaceae) as giving rise to irregular outgrowths on the mycelia of the fungus *Pilobolus crystallinus*. The so-called *Peronium aciculare*, formerly erroneously described as a distinctive species of fungus, is in fact *Saprolegnia*, galled by *Olpidiopsis saprolegniae*. Certain parasitic fungi are also known to give rise to galls on various Basidiomycetous fungi. VUILLEMIN[1224] refers to the deformations on *Boletus granulatus* caused by *Sepedonium chrysospermum*. According to him, *Mycogone rosea* gives rise to galls on the fructifications of *Clitocybe, Tricholoma*, etc. Although the fleshy fructifications of most higher Fungi are attacked by a great variety of insects, galls are, however, formed only exceptionally on these fungal organs. VOGLER[1220], RIEDEL[933] and RÜBSAAMEN[992] have reported interesting galls on the fructifications of different Basidiomycetous fungi. BOUDIER[115] and THOM[1133] have similarly described galls on certain well known Agaricaceae. According to DANGEARD[240], a Nematode produces small galls on *Ascobolus furfuraceus*.

The large fructifications of Polyporaceae often bear on the lower side numerous conical tubular galls. RÜBSAAMEN[992] believes that this tubular gall is produced by Diptera, which is perhaps *Scardia boleti* F.[981]. ULBRICH[1199] has recently described the galls on *Fomes salicinus* and *Ganoderma applanatum* from Europe.

Lichens.

The galls on lichens comprise the well known cephalodia and are generally caused by the "foreign" Cyanophyaceae, which penetrate into the thallus and give rise to horny or warty and often also other kinds of malformations. BEIJERINCK[73] and FORSSELL[339] were perhaps among the first to recognize the cephalodia as true galls. It is also of considerable ecologic importance to remark that lichens are the only Thallophyta on which Acarina are known to give rise to galls.

ZOPF[1289] has given accounts of several galls on lichens like *Ramalina kullensis*, *Ramalina scopulorum*, *Ramalina cuspidata*, etc. According to BACHMANN[44, 45, 47], a gall on *Ramalina fraxinea* is caused by the caterpillar of a moth. He found the thallus developing into a common gall chamber in *Ramalina kullensis*, but there are several globose chambers in the gall of *Ramalina fraxinea*. The gall on *Ramalina kullensis* is caused by Acarina. A number of different galls develop on the squamae and podetia of *Cladonia ochrochlora*. BACHMANN[40-43, 45-48] has also given accounts of the structure and development of several other interesting mycocecidia on the lichens *Cladonia ochrochlora*, *Cladonia fimbriata* forma *simplex minor*, *Cladonia cornutadiata*, *Cladonia fimbriata* forma *tubaeformis*, *Cladonia degerens phyllophora*, *Cladonia gracilis* and *Cladonia pityrea hololepis*. The mycocecidia on *Cladonia* are remarkable for the enormously swollen cells of the alga of these plants. The walls of the fungal cells of these lichens are also greatly thickened, with enlargement of the cell spaces. The gall on *Centraria glauca*, caused by the fungus *Abrothallus centrariae*, has the thallus irregularly bulged above, with a corresponding invagination below. The fungal galls on *Parmelia physodes* are globose tubercles on the upper surface. In species of *Ramalina*, the gall formation usually results in the branches of the stock being tubularly swollen the whole of their length or only basally.

Bryophyta.

A number of galls, like the swellings caused by *Nostoc lichenoides* on *Blasia pusilla*, are known on liverworts. MARCHAL[748] has described nematode galls on the growing tips of the mosses *Lophocolea bidentata* and *Cephalozia connivens laxa*.

The leafy mosses bear several zoocecidia, caused particularly by *Tylenchus* spp.[273, 762, 1018, 1019]. Most of these galls are typically artichoke-like swellings of the growing tips. GOODEY[413] has given an interesting account of another nematode *Anguillulina*, causing galls on the moss *Pottia*. Although different galls have been described on over thirty genera of leafy mosses from Europe alone, only those on *Dicranum longifolium* EHRENB., *Cephalozia* and *Lophozia* have been studied in detail (fig. 49D).

Pteridophyta.

A number of galls on ferns, caused by fungi, Acarina, Hymenoptera, Diptera, etc., have been described from different parts of the world. The gall of *Eriophyes pteridis* NALEPA on the fronds of *Pteridium aquilinum* KUHN (fig. 102D) is one of the best known of these. Some interesting Diptera galls on the stem and leaves of *Hymenophyllum* spp. are known from South America[535]. These galls are often remarkable for the dense covering of multicellular and branched hairs[975]. RÜBSAAMEN[997] has described an interesting leaf margin roll gall on the fronds of *Nephrolepis exaltata* SCHOTT, caused by *Eriophyes* from Malaya (fig. 102C). Bulbil-like galls on *Selanginella pentagona* are caused by Diptera[1111]. The gall-midge *Perrisia filicina* KIEFF. gives rise to leaf margin roll galls, with a curious resemblance to an abnormal indusium. GIESENHAGEN[389, 390] has also reported the finding of mycocecidia on ferns.

Gymnospermae.

As mentioned above, galls are more frequently met with on Gymnosperms than on Cryptogams. The best known of these is the ananas gall, caused by the aphids *Chermes (Adelges) abietis* LINN. and *Cnaphalodes (Adelges) strobilobius* KALT. on *Picea excelsa* (LAMK.) LINK. (fig. 52). Globose or fusiform galls on the stem of *Ephedra* spp., caused by Coleoptera, Lepidoptera and Diptera (fig. 47 D), are known from different parts of the world. Leaf galls

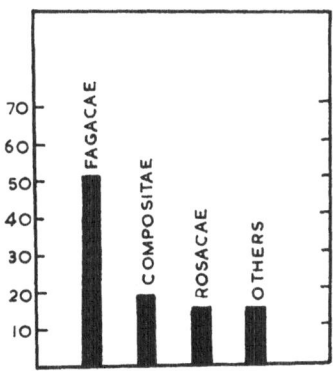

Fig. 6. Percentage frequency of galls on the dominant Natural Orders from the Holarctic.

on *Gnetum* are known from India and Java. Large galls on *Juniperus* spp. are caused by fungi of Gymnosporangiae. Bacterial galls occur on *Pinus* spp. A remarkable witches-broom gall on *Abies alba* is caused by the fungus *Melampsorella caryophyllacearum*. Different galls on *Juniperus* spp. are caused by *Eriophyes quadrisetus juniperinus* NAL., *Schmidtiella gemmarum* RÜBS., *Oligotrophus* spp., etc.

Angiospermae.

The Angiosperms bear not only the greatest bulk of all the known galls from the world, but they are also remarkable for the enormous diversity of their galls and the gall producing organisms associated with them. In 1911, KÜSTER[639] listed over 5500 galls on Angiosperms, mainly from Europe and North America. More recently ROSS & HEDICKE[982] reported 2900 galls on these plants from Central and North Europe alone. HOUARD[520] has listed nearly 3200 galls on the Angiosperms from Asia, Africa, Australia and Oceania and about 1300 galls from South America[535]. FELT[330] estimated that nearly 2000 galls are known on these plants from North America, north of Mexico. DOCTERS VAN LEEUWEN[307] described about 1500 galls on Angiosperms from Java, Sumatra and other parts of the East Indies. The total number of galls on Angiosperms so far known from all over the world is certainly not less than 14,750, representing almost 98% of all the known plant galls.

The greatest bulk, viz. over 93% of the galls on Angiosperms, are found on the Dicotyledonae. Although all Natural Orders of Dicotyledons bear galls, it is usual to find them more frequently on Cruciferae, Capparidaceae, Malvaceae, Sapindaceae, Anacardiaceae, Leguminosae, Rhamnaceae, Rosaceae, Geraniaceae, Vitaceae, Tiliaceae, Tamariscaceae, Myrtaceae, Combretaceae, Rubiaceae, Umbelliferae, Compositae, Convolvulaceae, Labiatae, Acanthaceae, Euphorbiaceae, Fagaceae, Salicaceae, Ulmaceae, Urticaceae, Moraceae, etc. than on the others. The frequency of gall on plants of different Natural Orders is naturally very different in different parts of the world. Galls on Gymnosperms, Salicaceae, Aceraceae, etc. are, for example, exceedingly rare in tropical regions. There is a very pronounced abundance of galls on Compositae and Fagaceae in North America and Europe. In these regions nearly 50% of the galls are found, for example, on Fagaceae, about 20% on Compositae and 15% on Rosaceae (Fig. 6). Galls on Leguminosae are particularly dominant in South America, parts of Africa and in India. More than half the known galls from Australia are found on Myrtaceae. Among the Monocotyledonae, the majority of galls occur on Gramineae and Cyperaceae. Some galls have also been described on Potamogetonaceae, Araceae, Liliaceae, Juncaceae, Orchidaceae, Amaryllidaceae and Musaceae. For further accounts of the frequency of galls on plant genera reference may be made to

HEIMHOFFEN[54] and for their frequency in different Natural Orders to KARNY[565] *(vide* Chapter IX).

2. Parts of plants galled

Galls are found on all the parts of plants, from the root tip to the growing point of shoot, both on the vegetative and on the reproductive organs. The aerial parts of plants, as may be expected, bear much larger numbers and greater varieties of galls than the underground parts. Galls on roots and other subterranean parts are generally less evident than those on aerial parts and are therefore often overlooked. In addition to the ordinary parts like roots, hypocotyl, stems, buds, petioles, leaf blade, flowers, fruits, etc., galls are often known to arise on other galls also. Not only the individual parts, but not infrequently the whole of the aerial portion of a plant may become galled in toto, giving rise to the familiar witches-broom or hexenbesen galls. This is for example the case with *Vaccinium vitis idaea* LINN., attacked by the fungus *Exobasidium vaccini* (FUCK.). Other common examples of galling of the whole of the aerial portion of a plant are frequently found on *Salix, Betula, Prunus, Syringa, Pinus, Ononis, Euphorbia cyparissias* LINN. and *Euphorbia gerardiana*. Some of these galls are caused by fungi, others by nematodes, Acarina and still others by insects. Not all witches-broom growths are the result of parasitic attack[1192].

Galls developing on other galls have been termed mixed galls and are often of great complexity. HEDICKE[450] distinguishes anacecidia from epicecidia, depending upon the degree of their complexity, particularly on the anatomical and histological modifications brought about by the second gall on the tissues of the first gall, on which it forms. One of the commonest anacecidia is caused by the aphid *Byrsocrypta gallarum* (GMELIN) on the galls of another species of aphid on *Ulmus*. The gall of *Diplolepis (= Rhodites) eglanteriae* (HTG.) on the gall of another cynipid *Diplolepis (= Rhodites) rosae* (LINN.) and the gall of *Iteomyia capreae* WINN. on the gall of *Pontania capreae* LINN. on *Salix* spp. are common examples of epicecidia. DOCTERS VAN LEEUWEN[308] has given an account of an interesting gall caused by the thrips *Gynaikothrips devriesii* KARNY on the gall of an unidentified midge on the leaves of *Elastostema sesquefolium* HASK. from the East Indies. The gall of *Andricus inflator* HTG. has been reported to develop on the bud gall of *Andricus globuli* HTG. on *Quercus*; not infrequently the gall of *Andricus fecundator* HTG. arises on the gall of *Andricus (collaris) curvator* HTG.[639].

Galls are not equally abundant on different parts. About 5% of the known galls produced by cynipids on *Quercus* from the world arise, for example on roots, 5% on buds, 22% on branches, 2% on flowers, 4% on acorns and about 63% on leaves (fig. 7). Over 80%

of the galls on Rosaceae are leaf galls. Nearly 70% of the South American galls are on leaves of diverse plants, 20% on branches and twigs, 7% on buds and only about 1–2% on flowers, fruits and roots. About 60% of the galls from the Oriental Realm are also leaf galls, about 15% are branch galls, 12% bud galls and the flower and fruit galls amount each to about 6%. Almost 55% of all the galls so far known from India arise on leaves; the stem galls represent about 25% and flower galls 10%. From the world, the cynipid galls on leaves amount to 80%, midge galls represent about 50%, Erio-phyid galls 80%, Homoptera galls 90%. Considering the galls caused by all agencies from the whole world, the largest number of galls are leaf galls, viz. 65%. Next in importance of abundance are the galls on twigs, branches and parts of the main axis, amounting to about 19%, with the bud galls standing at about 10% and the flower galls are about 4% and the root and fruit galls are each only 1% (fig. 8). The relative abundance of galls on different parts depends primarily on the species of the plant and the gall-maker and is also influenced by a variety of other environmental factors. It may also be observed that not all leaves, branches or buds of any given plant become galled, but the physiological conditions and the age of the organ seem to more or less greatly influence the development of galls. Furthermore, different gall-makers seem also to have become specialized to different degrees in their ability to induce galls on different parts. Certain types of galls are, for example, strictly found only on particular organs or even a part of an organ, such as leaf margin, the upper or the lower surface of the leaf blade, the petiole, etc. The development of a gall on a part presupposes the capacity of the part to react in a specific way to the influence of the gall-maker. As may be expected, all parts do not appear to be equal-ly responsive in this respect and their capacity also seems to vary with age and stage of development of the part. HOFMEISTER[479] was of the view that not only the normally actively growing, young and

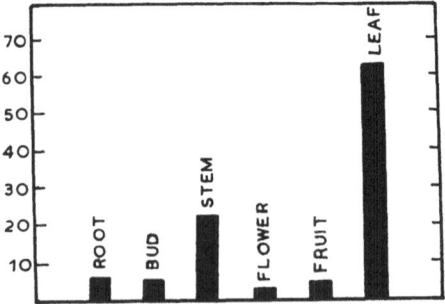

Fig. 7. Percentage frequency of galls caused by cynipids on different parts of *Quercus* from the world.

physiologically plastic structures, but also others that have ceased to grow and are more or less dormant and physiologically relatively inactive are capable of forming galls. THOMAS[1135] believed on the

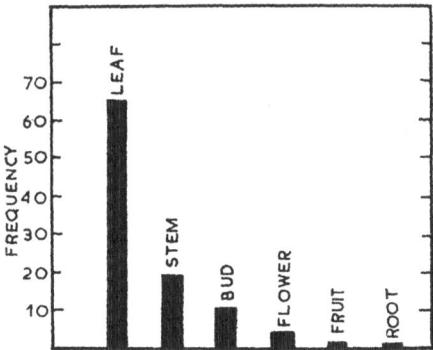

Fig. 8. Percentage frequency of galls caused by all agencies on different parts of higher plants from the world.

other hand that gall formation is only possible as long as the organ is still in a physiologically active state of development and growth. KÜSTER[639] has, however, cited numerous examples of galls by the aphid *Adelges fagi* HTG. on old bark layers of beech. ZWEIGELT[1297] recently reviewed the evidence on this important problem of the age of the gall-bearing organ and of the cecidozoan as factors in gall initiation and development. According to him, with increasing age of the plant part and thus with differentiation of normal tissues, the general complexity of the resulting gall decreases. In other words, the degree of organization of the gall is inversely proportional to the age of the gall-bearing organ. Old parts of plants do not show, however, any demonstrable loss of capacity to react to the gall-maker, but the latter only appear to be better adapted to the younger parts of plants than to the older ones.

3. Organisms which induce galls

The capacity to induce galls on plants is met with in widely unrelated groups of animals and plants, besides border organisms like bacteria and virus. Animals which give rise to gall are termed *cecidozoa* and the vegetable organisms that give rise to galls on other plants are *cecidophyta*. Not all species of a group cause the formation of galls and gall forming species are often found among genera and families, which are predominantly and typically non-gall making types and in some cases are not even phytophagous in their general habits.

i. Cecidozoa

The principal groups of cecidozoa are Trochelminthes, Nemathelminthes, Acarina and Insecta. With very few exceptions, the development of galls is closely bound up with the feeding activities of the
cecidozoa. The mode of feeding differs in different classes of cecidozoa and these differences are correlated with fundamental differences in the structure of their mouth-parts. Some cecidozoa, like
the larvae of Coleoptera, Lepidoptera and some Hymenoptera, have
typically biting and chewing mouth-parts and others like Homoptera,
Heteroptera, Nematoda and Acarina have piercing and sucking
mouth-parts. Still others like Thysanoptera have rasping and sucking mouth-parts. The larvae of the gall-midges (Itonididae: Diptera)
have very undeveloped mouth-parts and largely seem to absorb their
nourishment by suction. The salivary secretions of cecidozoa play
an important rôle not only in the feeding of the cecidozoa but also,
as we shall see later, in the initiation and development of the gall
itself *(vide* Chapter XIV). It is known, for example, that the salivary
glands of the gall-forming generation of the aphid *Chermes* are better
developed than in the non-gall-forming generation. Cecidozoa are
not capable of initiating the formation of galls throughout their life;
different groups of cecidozoa are able to influence the plant in such
a way as to produce galls at different stages in their life-history.
Most of them are able to do this only in their immature stages, but
some others are capable of initiating the development of a gall both
as larvae and as adults. It is also interesting to note that in some
species of cecidozoa, it is only one generation that is capable of
forming galls, while the succeeding generation in the life-cycle does
not produce any gall. Some of the outstanding characters of galls
caused by animals, viz. zoocecidia, are summarized in chapter IX.

Trochelminthes.

The only rotifer so far known to give rise to galls is *Notommata
werneckii* EHRENB., associated with unicellular galls on *Vaucheria*
spp., to which we have already referred. These galls are irregular,
unseptate, lateral outgrowths on the tubular green Algae[984], often
expanded and with 1–4 tubercles apically.

Nemathelminthes.

The gall Nematodes belong to the order Rhabditoidea, often also
called Anguilluloidea. They comprise a large and important assem-
blage of mostly moderately sized worms, with the cephalic sense
organs reduced to papillae and the amphids reduced to pockets. The
pharynx generally presents one to two bulbs, one of which is a
pseudobulb and the other is a valvulated bulb. The excretory system is usually asymmetrical. The rhabditoid Nematodes are terres-

trial, sarcophagous or coprophagous and by transition to partial or complete parasitism in plants and animals. The strictly gall-forming species belong to the family Tylenchidae. This family is recognized by its conspicuous buccal stylet, assymetrical form of the excretory system, with canals on one side. Exhaustive accounts of Tylenchids have been given by GREEF[416], GOODEY[404] and FILIPJEW & SCHUUR-MANS-STEKHOVEN[331]. The species pierce the plant tissues with the help of their buccal stylet and suck the plant sap. Their pharynx is characterized by a muscular median bulb and a posterior glandular dilatation. In the Aphelenchoid type the pharyngeal glandular dilatation projects backward as a lobe over the anterior portion of the intestine. All Tylenchoids are not, however, gall-formers. The gall species, often also called eelworms, belong to the genera *Tylen-chus*, *Aphelenchus* and *Heterodera*. Though parasitic in plants, most of these gall nematode worms are capable of leading a free life in soil for variable periods. The species usually give rise to galls on roots, but often also on stems, leaves, flowers and fruits. They give rise to galls predominantly on Dicotyledons, but sometimes also on Monocotyledons, especially the grasses and occasionally on mosses and marine algae. Most gall Nematodes are remarkably resistant to desiccation and remain viable in the dry state for several years. The dried Nematode galls on *Triticum vulgare* yield, for example, on soaking, thousands of the first stage juveniles of *Anguina tritici*. The dry cysts of *Heterodera* remain viable in soil for at least a whole year. STEINER *et al.*[1104] have described a remarkable case of *Tylenchus polyhypnus* remaining alive for almost thirty years on herbarium plants. Due to their capacity for active locomotion and also partly due to their high fecundity, gall Nematodes are generally distributed extremely rapidly. The genus *Heterodera* exhibits a high degree of parasitic adaptations, variations and sexual dimorphism (fig. 9). The second stage juvenile worm pierces a new host plant and sucks the sap from the cells. This leads to the formation of a gall at the site of attack. At the end of the third moult, the female becomes a plump, pear-shaped or gourd-shaped bag, containing 300–500 eggs. It then degenerates and finally leaves its cuticle as a protective cyst for the eggs. These cysts become free in the soil by the disinte-gration of the plant galls. The second stage juveniles develop within these cysts and escape into the soil and are able to live for several months without feeding, if a suitable host is not available. The males pass through an additional moult, but do not become pear-shaped bags. They escape from the decaying gall, seek a female and fertilize it. The detailed structure, habits and life-histories of several species have been investigated by a number of workers[199, 406, 478, 1101, 1113, 1154]. Most species are known to reproduce parthenogenetically [1196]. The species very often exist in several distinct biological races, structurally nearly indistinguishable from one another, but each

22

race attacks one set of host plants and does not breed on plants of the other groups [1101], [478], [406], [198]. Some species, like the common *Heterodera marioni* (CORNU) GOODEY, give rise to galls on the roots

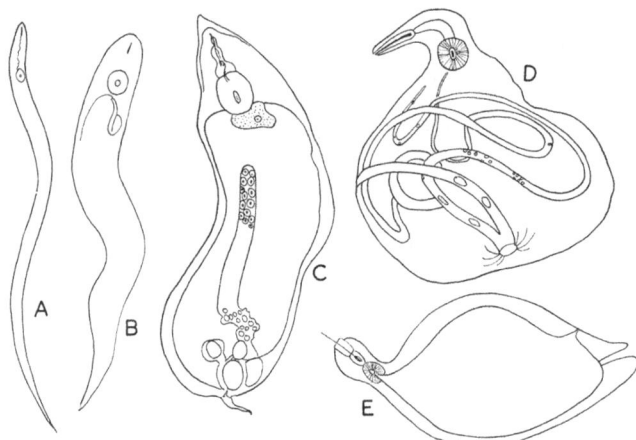

Fig. 9. Gall Nematodes. A–C. The second, third and fourth stage larvae of *Heterodera marioni* (Cornu) Goodey. D. Adult female of *Heterodera marioni* (Cornu) Goodey. E. Adult female of *Heterodera schachti* Schmidt.

of nearly 850 different species of plants. The family Allantonematidae, recognized by their degenerate pharyngeal musculature, is generally parasitic in diverse insects, but has a free-living phase in plants attacked by the insect. The family is remarkable for containing one species which gives rise to galls on *Eucalyptus* in Australia.

Acarina.

The gall mites belong to two families, viz. Tarsonemidae and Eriophyidae. The Tarsonemid mites are generally about 0.25 mm long and have four pairs of legs and a rounded body. The commonest species *Tarsonemus phragmitidis* SCHLECHT. is associated with peculiar galls on *Phragmites communis* TRIN. *Tarsonemus spirifex* MARCH. is associated with galls on grasses. Another related mite *Tenuipalpus geisenheyneri* RÜBS. (fig. 11 A) gives rise to small galls in the angles of leaf veins[996].

The Eriophyidae, an important group of cecidozoa, (fig. 10), are minute, elongate, cylindrical mites, with only two pairs of five-segmented legs and an annulated abdomen, ending in retractile caudal lobes. The genital organs, epiandrium and epigynium, are unpaired structures. Respiratory and circulatory organs and eyes are absent. The mite is usually 0.08 to 0.28 mm long. Long setae, found on different parts of the body, are of taxonomic importance.

Most eriophyid mites are white, yellowish-white, yellowish-red or brown. Relatively small numbers of eggs, each about 0.05 mm long, are deposited by the female. The eggs hatch in a short time and the

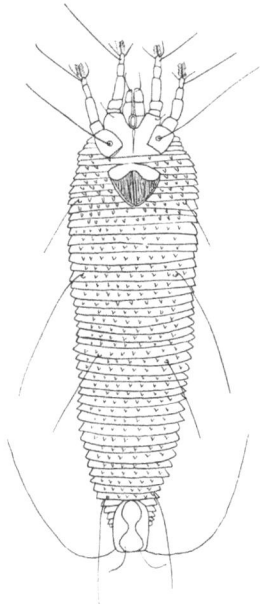

Fig. 10. *Eriophyes*, a typical gall mite. Ventral view of the adult.

newly hatched larvae resemble the adult mites very much, but they are much smaller, more weakly bristled and also lack the external genital apparatus. On moulting, the larvae transform into nymphs, which are larger and also have the external genital apparatus. The adult female mite is as a rule larger than the male adult. The mouth-parts of the eriophyid mites are adapted for boring into the plant tissues and sucking their contents out in liquid form. The mandibles are uniarticulate and needle-shaped, but them axillary palpi are free and have three segments. Owing to their small size, most erio-phyid mites can pierce single cells and suck its contents. NÈMEC[861] has given an interesting description of the mode of feeding by these mites. Eriophyids are remarkable for their relatively large salivary glands. The salivary secretion is poured into the plant cell through the wound at the time of sucking. Females are far more abundant than males and reproduce commonly by parthenogenesis. The greatest majority of gall mites hibernate on the plant, often below the protective scales of buds or in the angles of branches and buds and some species over-winter inside buds.

Not all species of eriophyids give rise to galls and some species live free on leaves, others occur in some abnormal growth on plants and still others are inquilines and commensals inside galls of other species. The structure, habits and classification of eriophyids have been placed on a sound basis by the monumental labours of NALE-PA[837, 839, 842]. Over 500 species of gall mites have so far been described, the greatest bulk of which belong to the genus *Eriophyes* (SIEB.) NAL. Other important gall mites are *Phyllocoptes* NAL., *Anthocoptes* NAL., *Oxypleurites* NAL. and *Epitremerus* NAL.

Insects

The principal gall insects belong to Thysanoptera, Heteroptera, Homoptera, Coleoptera, Hymenoptera, Lepidoptera and Diptera, but isolated examples of galls, caused by species of some other orders have also been described from time to time. *Lestes viridis* (Odonata), for example, is reported to cause galls by thrusting its eggs under the epidermis of various plants. The plant tissues immediately surrounding the egg become more or less conspicuously swollen, but as soon as the larvae hatch, undergo decay. Such galls, initiated by oviposition and not by the feeding activity of the cecidozoa, have been described as *procecidia* by THOMAS[1142] and PIERE[907]. All other insect galls are initiated when the larvae commence feeding. Some species of Thysanoptera, Heteroptera and Homoptera are capable of inducing the development of galls both in their larval and adult stages.

Thysanoptera.

Thysanoptera are minute insects, ranging from 0.5 to 10 mm in length, with a slender and depressed body and rasping and sucking mouth-parts. They are usually found on foliage, buds and flowers. They are readily recognized by their narrow, long-fringed membranous wings and by the swollen, bladder-like arolia between claws. Though Thysanoptera are known to give rise to galls on plants in nearly every floral region of the world, they occupy on the whole a minor place among the cecidozoa. The suborder Tubulifera contains many gall thrips. KARNY[561-564] has given excellent accounts of the thrips commonly found in galls. HOUARD[520] records over fifty species of gall thrips, belonging to about twenty-four genera, chiefly from the Indo-Malayan region and Africa, representing about 10% of the known species of cecidozoa from the area. In South America the gall thrips represent hardly 0.5% of the insects that induce galls. According to DOCTERS VAN LEEUWEN, over one hundred and twenty-five species of gall thrips are found in Java. The more important genera of gall thrips are *Cecidothrips* KIEFFER, *Cryptothrips* UZEL, *Dolerothrips* BAGN., *Gynaikothrips* ZIMMERMANN,

Mesothrips ZIMMERMANN, *Thrips* LINN., *Euthrips* TARG.-TOZ., *Neoheegeria* SCHMUTZ, etc.

Heteroptera.

A small number of Heteroptera, belonging to the family Tingidae, are remarkable for their characteristic galls on flowers of diverse plants. The species belong mostly to the genera *Copium* THUNB. (fig. 11 B), *Paracopium*, etc. LEACH & SMEE[654] have described the capsid bug *Helopeltis bergrothi* as giving rise to gnarled galls on branches of *Thea sinensis*.

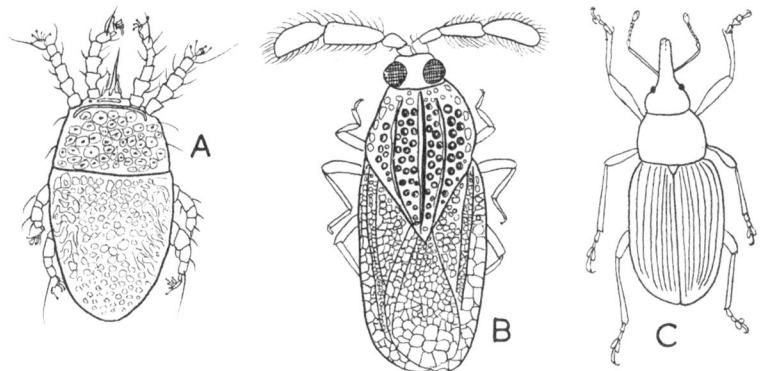

Fig. 11. Some typical cecidozoa. A. *Tenuipalpus geisenheyneri* Rübs. B. *Copium clavicorne* (Linn.). C. *Ceutorrhynchus pleurostigma* (Marsh.).

Homoptera

The order Homoptera are important gall insects nearly all over the world. Gall species are found in several families like Cercopidae, Jassidae, Chermidae (= Psyllidae), Aphidae, Phylloxeridae, Aleurodidae and Coccoidea. FRIEDERICHS[346] reports that the nymph of the Cercopid *Ptyelus (Aphrophora) spumarius* (LINN.) causes galls on certain plants. Procecidia are also produced by some Jassids like *Tettigoniella* JAC.

Chermidae.

Psyllids or jumping plant-lice are of considerable importance as gall insects in nearly all parts of the world. The adult psyllids have two pairs of wings, which are membranous and folded roof-like over the dorsum. The nymphs have a characteristically depressed body, usually coated with a waxy secretion and conspicuous wing-pads. They are all usually sluggish. Not all psyllids are gall formers. Gall forming species belong to *Psylla* GEOF., *Psyllopsis* Löw, *Trioza* Löw (Fig. 12 C), *Pauropsylla* RÜBS., *Megatrioza, Cecidotrioza, Neotrioza, Rhinocola* CRAWF., *Cecidopsylla, Dinopsylla, Livia, Ozo-*

trioza, Phacosema, etc. The number of species of gall-forming psyllids, so far described from the world, exceeds 350, of which almost half are from the Oriental Region and Africa and one-fifth from South America.

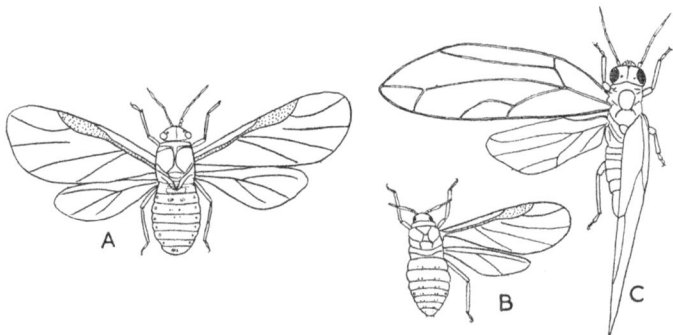

Fig. 12. Some typical cecidozoa. Homoptera. A. *Pemphigus bursarius* (Linn.), alate fundatrix. B. *Byrsocrypta gallarum* (Gmelin), alate sexupara. C. *Trioza alacris* Flor.

Aphidae.

Aphids are minute, winged or apterous, soft-bodied insects, with 3 ocelli in the alate form; antennae 3–6 segmented; fifth abdominal segment bearing a conspicuous tubular siphon. A very large number of aphids are known to give rise to galls, often of considerable complexity of structure and development. In some regions, especially in the temperate lands, the aphids appear to represent perhaps one of the most important gall insects. The gall aphids reproduce parthenogenetically and bisexually in alternate series of generations, in spring and summer respectively. The alternation of parthenogenetic and bisexual generations are often associated with changes in food plants and pronounced polymorphism. These cyclic changes are also associated with remarkable differences in the galls caused by the individuals of different generations. The female and male adults appear in autumn and after mating, the fertilized female lays a single egg. From this over-wintered egg an apterous female, called the fundatrix or stem mother, hatches in early spring. The fundatrix reproduces parthenogenetically and gives rise to several generations of apterous females, which likewise reproduce parthenogenetically and are thus virgins. Later in the season, alate females also appear and towards the next autumn once again males also appear. The galls produced by the same species of aphid on different parts of the same plant are frequently very different. The same species may also give rise to galls on different host plants in the course of its cyclic heterogenesis. Some of the more common gall aphids belong to *Adelges,*

Aphis, Byrsocrypta HALIDAY *(= Tetraneura* HTG.) (fig. 12 B), *Forda* HEYD., *Chermes* (LINN.) BÖRN., *Eriosoma* LEACH, *Hamamelistes, Pemphigus* HTG. (Fig. 12 A), *Dreyfusia* BÖRN., *Cnaphalodes* MACQ., *Pineus* SCHIM., *Peritymbia* WESTW. (= *Viteus* SHIM.), *Phylloxera,* etc. About 700 species of gall aphids are known so far from the world. Almost 50% of the species occur in the Holarctic and the gall aphids occupy a minor place in tropical countries. Important contributions to our knowledge of gall aphids may be found in MORD-WILKO[817, 818], INOUYE[539], ZWEIGELT[1292-1300] and others.

Coccoidea.

The female coccid is a degenerate sessile insect, which remains permanently fixed to a suitable place on the food plant and sucks copious quantities of sap. The males undergo a complete metamorphosis and have only a single pair of wings and non-functional mouth-parts. Although several thousand species of coccids are known to feed on various wild and cultivated plants all over the world, exceedingly few species seem to give rise to galls. Coccids thus occupy a minor place among gall insects in most faunal areas, but in Australia they appear to have reached their maximum development. Coccid galls are particularly abundant on Australian *Eucalyptus, Casuarina* and *Acacia. Amorphococcus, Aspidiotus* BOUCHÉ, *Asterolecanium* TARG., *Chionaspis* SIGN., *Fiorina* TARG. are some of the more important gall forming coccids. The Australian Pegtococcids of the family Apiomorphidae (= Brachyscelidae) like *Apiomorpha* RÜBS., *Ascelis* SCHR., *Cystococcus* FULL., *Opisthoscelis* SCHR. and *Cylindrococcus* MASK. are important gall makers; some of them are found in America also. Important contributions on gall coccids may be found in SCHRADER[1034], FROGGATT[348, 352], EHRHORN[317], etc.

Aleurodidae.

Aleurodids, popularly known as whiteflies, have two pairs of similar wings, coated with waxy bloom. The early larvae are free, but soon become sessile. There is a resting stage, within the larval skin before the emergence of the adult. One or two species give rise to pit galls on leaves. *Aleuromarginatus tephrosiae* CORB., from India, gives rise to epiphyllous, hemispherical pouch galls on leaflets of *Tephrosia purpurea* PERS. An unidentified species of *Bemisia* produces a remarkably fleshy, cup-like, beautifully pink coloured emergence gall from the epidermis on leaves of *Achyranthes aspera* LINN. from India[745].

Coleoptera.

Coleoptera are hard-bodied insects, with the fore wings modified into horny or shell-like elytra, which serve to protect the hind wings. The mouth-parts are biting and chewing. Development is holome-

28

tabolic. The larvae are variable and always have a distinct head. The phytophagous beetles feed on leaves, or they tunnel inside various plant parts and also mine inside leaves. Most gall-forming beetles are Curculionidae of the genera *Anthonomus* GERM., *Apion* HERBST., *Baris* GERM., *Ceutorrhynchus* GERM., (Fig. 11 C), *Cleonus* SCHÖNH., *Dorytomus* GERM., *Lixus* FAB., *Gymnetron* SCHÖNH., *Nanophyes* SCHÖNH., etc. Some species of Cerambycidae, Chrysomelidae, Buprestidae and Scolytidae have also been reported to give rise to galls. Altogether about 150 species of gall-forming Coleoptera are known from the world. Reference may be made to FROGGATT[349] for gall Buprestidae and to SCHMIDT[1026] for gall Curculionidae.

Hymenoptera

Hymenoptera have two pairs of membranous wings, with the fore wings much larger than the hind wings. Venation in both the wings is greatly reduced and often also wanting. Gall-forming species are met with among Tenthredinoidea, Cynipoidea and Chalcidoidea, which have all biting and chewing mouth-parts.

Tenthredinoidea or sawflies are remarkable for their saw-like ovipositor. Their larvae superficially resemble the caterpillars of Lepidoptera, but are distinguishable by the presence of 6–8 pairs of prolegs. Not all Tenthredinids, however, give rise to galls. The gall species belong to one of the following genera: *Pontania* COSTA (fig. 13 A), *Cryptocampus*, *Blennocampa* HTG., *Hoplocampa* HTG.,

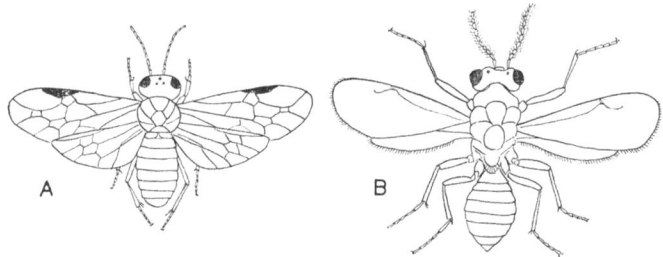

Fig. 13. Some typical cecidozoa. A. *Pontania viminalis* Linn. B. *Isthmosoma.*

Micronematus, *Monophadnus* HTG. and *Selandria*. The larvae occur in their galls, but pupation is often in soil. Most sawflies give rise to remarkably large, fleshy, hollow galls on leaves of various species of *Salix* and some species also give rise to procecidia on *Clematis*, *Helleborus*, *Pteridium*, etc.[319, 272].

Cynipoidea.

Cynipoidea are well known as gall insects in Europe and North America, where several hundred species give rise to often remarkably complex galls, especially on *Quercus* and *Rosa*. Cynipids are mostly small or minute, hard-bodied insects, with compressed abdomen, prothorax at the sides reaching back to tegulae, venation reduced and antennae simple (not geniculate). Their larvae are about 3–4 mm long, white, apparently apodous, but have distinct head and biting and chewing mouth-parts. Throughout its development, the larva, however, does not throw off any excrement, but at the time of pupation the fecal matter may be ejected. Pupation is always in the gall. Cynipids have well developed ovipositors, with the help of which the eggs are accurately positioned inside buds, leaves and other tissues. Many species reproduce parthenogenetically and have also alternations of parthenogenetic and bisexual generations. Some species of *Andricus* HTG. reproduce exclusively parthenogenetically. Very often the individuals of the two alternating generations are so markedly different that they have in the past been described as belonging to different species and even genera. In some species there is a change of the host plant associated with the alternation of generations. The galls induced by the unisexual and by the bisexual generations are usually fundamentally different. Some species do not give rise to galls, but occur as inquilines in the galls of other species and still other species are true parasites of various insects, especially Diptera. Important contributions to our knowledge of the structure, development, habits, ecology and classification of Cynipids are by MAYR[764-766], RIEDEL[932], ASHMEAD[32], DALLA TORRE & KIEFFER[237-238], KIEFFER[584], ADLER[1, 2], BEIJERINCK[75], FRÜHAUF[354], RAINER[923], KINSEY[594-597], WELD[1239-1242] and others.

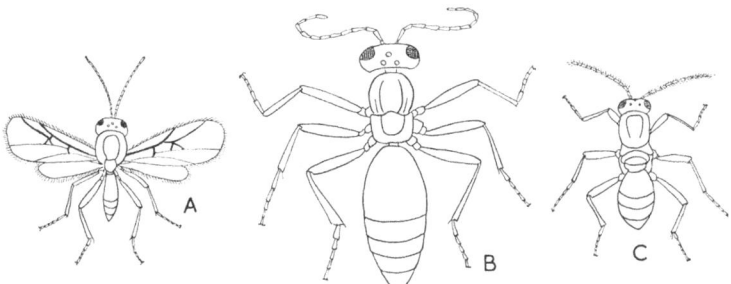

Fig. 14. Some common cecidozoa. A. *Biorrhiza pallida* (Ol.), male of the bisexual generation. B. The same, female of the unisexual generation. C. The same, female of the bisexual generation.

Chalcidoidea.

Chalcids are minute or small insects, with venation reduced to a single vein in the anterior part of the wing, pronotum at the sides not reaching back to tegulae, antennae elbowed at scape, ovipositor conspicuous and issuing some distance in front of the tip of abdomen.

Although predominantly entomophagous parasites, many gall-making species are met with among the families Agaontidae, Torymidae, Eurytomidae, Perilampidae, Pteromalidae, Encyrtidae and Tetrastichidae. The more common gall chalcids include *Blastophaga* GRAVENH., *Isthmosoma* (Fig. 13 B), *Pareunotus* GIR., *Coelocyba* ASHM., *Terobiella* ASHM., *Cecidoxenos* ASHM., *Eufroggatiana* GAHAN, *Espinosa* GAHAN, *Perilampella* GIR., *Pembertonia* GAHAN, *Lisseurytoma* CAMERON, *Brachyscelidiphaga* ASHM., *Systolomorpha* ASHM., *Trichilogaster* MAYR, etc. GAHAN[356, 357] has listed the phytophagous chalcids from the world, some of which are gall chalcids. GAHAN & FERRIÈRE[357] have given an account of some of the gall chalcids. CROSBY[233] described some interesting gall chalcids from Africa. The Agaontidae are the well known fig-insects, which breed in the figs of various species of *Ficus*, both wild and cultivated and turn the flowers into characteristic galls. Extensive studies of Agaontidae from the world have been made by GRANDI[415]. For the taxonomy of Chalcidoidea reference may be made to ASHMEAD[31] and SCHMIEDEKNECHT[1029].

Lepidoptera.

Lepidoptera are holometabolic insects, with two pairs of wings and siphoning mouth-parts, with the body and wings clothed with overlapping scales. Their larvae, the common caterpillars, have, however, biting and chewing mouth-parts and 8 pairs of legs, of which only the first three are true legs and the rest are prolegs on segments 6, 7, 8, 9 and 12.

About 100 species of gall-making Lepidoptera are known from the world. WILLE[1267] has given an interesting account of the Teneid *Cecidoses eremita* CURTIS, which forms a singularly interesting dehiscing gall on *Schinus dependens*. Certain species of *Ectoedemia* (Nepticuloidea: Nepticulidae a family of leaf-miners) are gall-makers in America. The South American *Ridiaschina* (Tineoidea) has vestigial mouth-parts and its larvae also give rise to galls. Other important gall-forming Lepidoptera are *Acella*, *Epiblema* HÜBN., *Gelechia* Z., *Laspeyresia* GERM., *Lobesia*, *Nepticula* Z., *Orneodes* LATR., *Phalonia* HÜBN., *Trochilium* SCOP., *Cecidolechia*, *Grapholitha*, *Amblypalpis*, *Gnorimoschema*, etc. For further account of galligenous Lepidoptera reference may be made to SORHAGEN[1073], SOLOWIOW[1072] and SCHMIDT[1027].

Diptera.

Diptera are the most important gall insects nearly all over the world. They are holometabolous insects, with only the fore wings; the hind wings are reduced to halteres. The most important gall-making Diptera are Itonididae (= Cecidomyiidae), Agromyzidae, Trypetidae and Anthomyiidae.

Itonididae or gall-midges (fig. 15) represent a large family of Nematocerous Diptera, with about 5000 species described so far from the world. They are readily recognized by their holoptic eyes, short coxae, absence of tibial spurs, reduced wing venation and by their characteristic antennae, with special sensoria called circumfila. Though some of the Itonididae are fungus feeders, debris feeders or are predaceous on various insects and mites, the bulk of the species are phytophagous. Some of the phytophagous midges do not really give rise to galls on their host plants, but the vast majority produce more or less complex galls on all parts of plants, particularly on Dicotyledonae. The adult midges usually never feed, but the larvae are apodous maggots, characterized by the presence of a peculiar sclerotic sternal spatula (fig. 15 C). The larvae have well-developed salivary glands. In some species the larvae have the ability to jerk and shoot up into the air by suddenly straightening the curved body. The larvae are also responsible for the galls. Pupation is sometimes in the gall, sometimes in small silken-lined chambers underground in the soil. Parthenogenesis and paedogenesis are known in many species. Gall-midges have been extensively studied by RÜBSAAMEN [996, 998, 999], KIEFFER[588], FELT[328], BARNES & MANI[744]. Some of the widely distributed gall-midges belong to the genera *Lasioptera* MEIG., *Neolasioptera* FELT, *Trotteria* KIEFF., *Misospatha* KIEFF., *Oligotrophus* LATR., *Dasyneura*, *Asphondylia* LÖW, *Schizomyia*, *Perrisia* ROND., *Rhabdophaga* WESTW., *Contarinia* ROND., *Mayetiola* KIEFF., *Itonida* MEIG., etc.

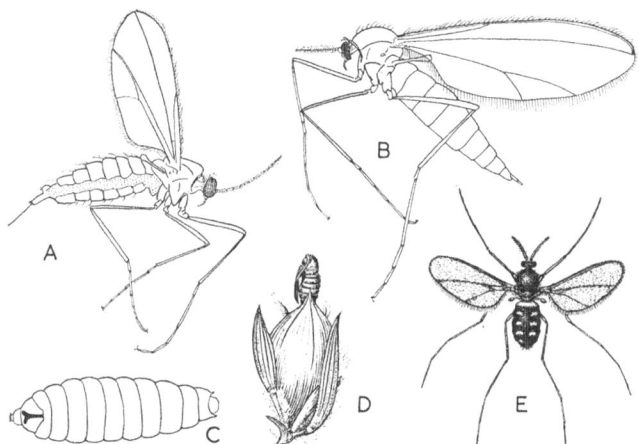

Fig. 15. Some typical cecidozoa. Itonididae (=Cecidomyiidae). A. *Asphondylia sarothamni* H. Löw. B. *Mikiola fagi* (Htg.) C. Larva of gall-midge, showing the characteristic sternal spatula. D. Pupa of gall-midge, sticking out of its gall on grass flower, just before the emergence of the adult. E. *Lasioptera falcata* Felt.

Among the other Diptera, Agromyzidae are of some importance as gall-makers. A few species are known to give rise to galls in stems or leaf veins of different plants. Some fruitflies like *Tephritis, Lipara, Dacus*, etc. occasionally give rise to galls in flowers and stems of most Compositae[315]. *Anthomyia* is also reported to give rise to galls occasionally.

Cecidophyta

The plant organisms that give rise to galls on other plants include the Bacteria, Algae, Fungi and Phanerogams. Though the cecidophyta are on the whole numerically inferior to the cecidozoa, some of them often give rise to nearly equally complex galls.

Bacteria.

The best known gall bacteria are undoubtedly *Rhizobium beyerinckii* and *Rhizobium radicicola* of the root nodule galls of Leguminosae. Bacterial galls are also known on a number of other groups of plants. *Phytomonas tumefaciens* (SM. & TOWN.) gives rise to the well known crown-gall *(vide* Chapters XIV and XV).

Fungi.

The fungi are the most important among the cecidophyta. They are remarkable for the great numbers of gall-forming species and the wide variety of plants on which they produce galls. The mycelia of the gall fungi may be intercellular or also intracellular in the host plant. Intercellular parasitic fungi produce haustoria for absorbing nourishment or their mycelia penetrate the cell wall in all directions. The gall fungi are chiefly Chytridiaceae, Ascomycetes, Uredinae, Ustilaginae, Exobasidiaceae and Peronosporinae. *Plasmodiophora brassicae* WORON. is perhaps the best known gall fungus. Some like the monadine *Aphelidium deformans* give rise to unicellular galls on *Coleochaete*[1297]. Of the class Basidiomycetes three orders are well known as gall fungi, viz. Exobasidinae, Uredinae and Ustilaginae. The Exobasidinae are characterized by free basidial layers, so that the spores cover the external surface of the host plant. They mostly attack the leaves of the host plant, which become more or less greatly swollen in consequence. *Exobasidium rhododendri* CRAM. gives rise to large spongy galls on leaves of *Rhododendron* spp. Other species of *Exobasidium* give rise to galls on *Vaccinium* spp. (fig. 141 B-C). The Uredinae[1079] produce different spore types, such as ascidia, uredospores and teleutospores. Very often all these three or at least two of these types of spores occur on the same host plant. The ascidiospores occur chiefly in spring and early summer. They are rounded, thin-walled, unicellular, rust-coloured, yellow or red spores that arise in chains, close together in beaker-shaped groups. The uredospores have an elongate stalk and are likwise unicellular.

The teleutospores are, however, thick-walled and retain their viability for much longer periods than the other two types (fig. 142 F). They are multicellular. Several fungi produce small, hemispherical or flat spermogonia or pycnidia, inside which small cells called spermato- or pycnospores arise. Some of the best known fungus galls are the witches-broom galls caused by the ascidial generation of *Melampsorella caryophyllacearum* SCHRÖT. on *Abies alba* LINN.; the teleutospores of this fungus develop on leaves of Caryophyllaceae. Species of *Puccinia* are known to give rise to some other common galls on plants like *Urtica, Origanum, Berberis, Jasminum, Carex*, etc. *Gymnosporangium* spp. give rise to extensive galling of the stem of *Juniperus. Uromyces* spp. give rise to galls on *Euphorbia* and Leguminosae. The teleutospore generation of *Calyptospora* is also known to cause galls and its acidial generation develops in the needles of *Abies*. Ustilaginae produce spores in the interior of the host plant tissue, such as the ovary of Gramineae, which thereby becomes more or less greatly swollen and contain ultimately a sooty powder. Very often Ustilagineae produce also pustular galls on leaves. The Ascomycetous fungi are also frequently gall makers. Their spores develop in the interior of elongate tubular cells called asci. *Taphrina* gives rise to a variety of galls. Even the Chytridiaceae, which are intracellular parasites, give rise to galls. The commonest fungus of this group belongs to *Synchytrium*. The white rust fungus *Albugo (Cystopus) candida* PERS. is associated with characteristic galls on Cruciferae, especially *Capsella bursa-pastoris*. Reference may also be made to *Uromycladium* that gives rise to galls on *Acacia* spp. [1002, 1003, 1159].

Algae.

Gall inducing species of Algae are met with among the green-algae, red-algae and brown-algae. *Phytophysa treubi* is reported to give rise to large galls on *Pilea* (Urticaceae). Many species of algae are also known to give rise to galls on other algae[242]. *Ectocarpus valiantei* gives rise to galls on *Cystoseira ericoides*. The alga *Rhodochytrium spilanthidis* gives rise to extensive galling on *Hibiscus sabdariffa*[887, 301].

Phanerogams.

Loranthaceae, especially *Viscum, Phoradendron, Loranthus, Cuscuta*, etc., parasitic on diverse other plants, give rise to large woody galls, with insignificant anatomical and histological abnormalities. For an account of the development of the haustoria of the parasite *Cuscuta* in the gall it produces reference may be made to THOMPSON [1149]. The haustoria arise underneath the epidermis of the parasite. The outer region of the modified epidermal cells of the parasite becomes closely applied to the epidermis of the host plant. Once this contact is established, the middle group of cells dissolve away their way through the host epidermis by the action of an enzyme and thus constitute the prehaustoria. The haustorium actually resembles a root in most respects but lacks a root cap. It follows the path prepared for it by the prehaustorium and it also dissolves away the host cells. The hyphae appear first from the haustorial dermis and undoubtedly represent modified root hairs. These hyphae often grow along the inner sides of the interfascicular cambium and come into contact with the xylem elements, but do not attack the phloem. The cells of the hyphae now become lignified, forming a string of tracheids from the host xylem to the haustoria. Phloem and sieve tissues do not arise in the haustorium. Cell proliferation in the host tissue provides abundant ready-made nutritive material. It is not therefore the phloem of the host, but the xylem that is the principal path of supply of water and carbohydrates to the parasite. MANGE-NOT[742] has recorded an extremely interesting case of the roots of *Thonningia coccina* (Balanophoraceae) coalescing with the roots of another tree in contact and forming large galls. with numerous absorbing strands.

CHAPTER III

STRUCTURE OF GALLS

Beginning with the early studies of MALPIGHI, an enormous amount of work has been done on the general external form, anatomy, histology and cytology of galls. The more important contributions to our knowledge of the general morphology of galls include those of THOMAS[1135-1143], BEIJERINCK[74-78], MAGNUS[726-733], KÜSTER [635-646], HOUARD[485-493], [496-501], COOK[212], BEGUINOT[72], TROTTER[1169], [1170, 1183], DOCTERS VAN LEEUWEN[291, 292, 297, 299, 304], BARGOGLI-PETRUCCI[55], APPEL[22], COSENS[221], WELLS[1242-1245], CAVADAS[178], ELCOCK[317], KOSTOFF & KENDALL[619, 620, 622], BOGGIO[108], PETERS-POSTUPOLSKA[900], AKAI[3-5], BUTLER[165], PHILLIPS & DICKE[906], ROBERTSON[963], HERING[461], KILCA[602], VERRIER[1208], MILOVIDOW[794-796], HYDE[538], RUGGERI[1000], GENEVES[377], GARRIGUES[361], BLUM[95], MEYER [778-790], WOLL[1272-1273], ZWEIGELT[1293, 1297], LERA[660], GOODEY[401, 411] and others listed in the bibliography. A general summary of current ideas on the fundamental morphology of galls may be found in KÜSTER[639, 641, 643] and ROSS[981]. It is therefore proposed to give here only a broad outline of the salient features of the general structure of galls, in so far as it is necessary for an understanding of the ecological problems associated with galls and the gall-inducing agents.

1. Gross morphology

As may be expected, the structure of galls depends primarily on the organ on which they arise and on the species of the plant and the gall-inducing organism. The importance of these factors may be readily seen from the fact that the galls induced by different species of gall-makers on the same organ of a given plant are structurally very different. Sometimes the different sexes of the gall-maker give rise to radically different types of galls (vide Chapter IX). The structure of galls induced by the same species on different organs of a plant is also distinctive of the organ. As may be expected, the structure of the organoid and histioid galls exhibit characteristic differences. Unlike the relatively simple organoid galls, the histioid galls are characterized by complex tissue abnormalities. Some of the histioid galls lack a definitive and constant external shape and size and constitute the *kataplasma* galls of KÜSTER *(loc.cit.)*. In sharp contrast to the kataplasma galls, KÜSTER recognizes the *prosoplasma* galls, characterized by their typical, definitive and constant external form, size, tissue differentiations, etc. The prosoplasma galls are structurally more highly specialized than the kataplasma galls and are believed by WELLS[1245] and TROTTER[1176] to be derived from them. The distinction into kataplasma and prosoplasma galls is not sharp however, since we often come across all intermediate stages of development in diverse galls. The prosoplasma galls present in general a most bewildering diversity of external form, colour, size, surface processes and internal structure that are on the whole characteristic of the gall. Most of these galls are spherical or subglobose, oval, cordate, pyriform, fusiform, cylindrical, conical, horn-shaped, clavate or sometimes also greatly flattened disc-like swellings. Some galls have a curious superficial resemblance to fruits and seeds of certain plants. Many galls are smooth, but others have variously rough or verrucose surfaces, which may also be ribbed, lobed, hairy, spiny or armed with other peculiar and more or less fleshy emergences. Some galls are sessile, but others are stalked, with the stalk sometimes slender and even much longer than the body of the gall itself (fig. 113 A). When a large number of galls develop closely crowded together, they generally tend to become fused into irregular agglomerate masses. The colour ranges mostly from green, yellowish-green, yellow to pink, violet or deep reddish-brown. The size varies from a fraction of a millimetre to several centimetres in diameter. Some galls are membranous bags, but numerous others are solid, fleshy, spongy or woody swellings and still others are capsule-like or nut-like growths. Most galls have one or more cavities, in which the gall-maker occurs, but in a number of remarkable galls, there are, in addition, a variable number of empty atria, surrounding the larval cavity (fig. 92 D). The gall cavity or the larval cell as it is also sometimes called may be basal, central or also peripheral and may communicate to the outside by a more or less narrow passage or may be completely closed.

Among the other major factors that influence the general structural pattern of galls, mention should be made of the position of the gall-maker on the gall-bearing organ and the mode of development of the gall. Galls developing at the growing tip of the main axis, viz. the *acrocecidia* of THOMAS[1138], arise by the arrest of the growth of the vegetative point and the neighbouring parts becoming deformed or by the tissues near the primary meristem turning into the swelling and the vegetative point itself becoming arrested in its growth and development. In contrast to the acrocecidia, THOMAS distinguishes the *pleurocecida*, which develop at a distance from the growing tip of the axis. The positioning of the gall-maker may be either external or internal. The following conditions are met with: i. gall-maker permanently external and exposed on the epidermis of the plant, ii. gall-maker external on the epidermis of the plant, but in course of development of the gall, becoming enclosed within the gall cavity, which, however, remains open to the outside throughout or at least in the early stages of development, iii. gall-maker initially external but becoming completely enclosed by over-growth of plant tissues, iv. gall-maker initially external, but sinks into the plant tissue, which then closes over it, thus completely enclosing it and v. gall-maker situated within the plant tissue from the beginning. The following structural types of galls are correlated with these differences: i. filzgall, ii. fold and roll galls, iii. krebsgall, iv. covering gall, v. pouch gall, vi. lysenchyme gall and vii. mark-gall.

Filzgalls.

While hairy outgrowths are common on galls, many galls comprise wholly or nearly wholly of such outgrowths; such galls are termed filzgalls. The filzgall is the simplest histioid gall, in which the gall-maker is mostly external. It is essentially hair-like outgrowths of epidermal cells and is generally predominantly caused by Acarina. By far the most preferred place for the development of filzgall is the leaf, especially the under side of the blade. Sometimes, however, filzgalls arise on the shoot axis, petiole and flowers also. The filzgall ranges from small obscure patches of abnormal hairs to extensive and dense hairy masses. The commonest examples of filzgalls are found on different Angiosperms like *Acer*, *Tilia*, *Nephilium*, *Alnus*, *Fagus*, *Vitis*, *Gossypium*, etc. The Acarina which give rise to filzgalls occur in among the hairs. Filzgalls are sometimes also caused by certain Fungi like *Synchytrium papillatum* FARL., which, however, occur within the hairy cells. The formation of excessive outgrowths is also

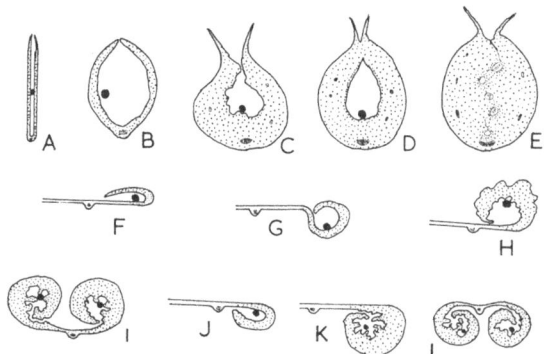

Fig. 16. Morphologic types of galls. A–B. Simple fold galls; mantel galls, with only slight swelling of the leaf blade. C. Fold gall, with pronounced swelling, but without tissue fusion. D. Fold gall, with very pronounced swelling and tissue fusion, enclosing a definitive gall cavity. E. Fold gall, developed into an enormously swollen solid mass. F. Simple epiphyllous marginal fold gall. G. Simple marginal roll gall. H. Epiphyllous leaf margin roll gall, with pronounced swelling. I. Epiphyllous double margin roll gall. J. Simple hypophyllous leaf margin roll gall. K. Hypophyllous margin roll gall. L. Hypophyllous double margin roll gall.

38

usually associated with other more or less pronounced abnormalities and modifications of the leaf, such as bulging and arching of the blade, rolling of the leaf margin and pouch-like evaginations or out-pocketings *(vide infra)*. The form and size of the hairs of the filzgall reveal characteristic variations *(vide* trichomes).

Fold and roll galls.

In many galls the margin of the leaf blade is folded or rolled more or less completely, either upward or downward, associated with greater or lesser swelling of the affected part. Such fold and roll galls arise either as a result of the failure of the leaves to properly unfold and unroll from the bud or also as the result of unequal surface growth in the course of development of the gall. Most fold and roll galls become enormously swollen, spongy, fleshy, woody and fibrous. Fold and roll galls are mostly caused by Acarina, Thysanoptera, Aphids, Psyllids, gall-midges and rarely by Hymenoptera and fungi. The folding and rolling processes have the result of enclosing the gall-maker.

Pouch galls.

The pouch gall arises from a localized intense out-arching of the leaf blade, producing a bulging on one side and a corresponding

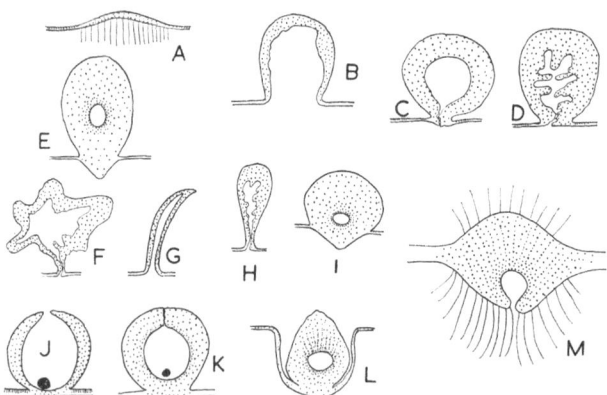

Fig. 17. Morphologic types of galls. A. Simple filzgall, with only slight swelling and upward arching of the leaf blade. B. Simple pouch gall, with wide open gall cavity. C. Pouch gall, with pronounced swelling and nearly complete obliteration of the ostiole. D. Pouch gall with fleshy emergences growing from the wall into the gall cavity. F–H. Lobed, ceratoneon and cephaloneon pouch galls. I. Pellet gall or "button gall". J. Simple covering gall, with the ostiole. K. Simple covering gall, with the ostiole obliterated. L. Pouch gall, with simultaneous downward curving of the blade. M. Midge gall on *Schoutenia ovata* Korth (after Houard) to show the dominant direction of cell division and cell elongation.

invagination on the other side. The arched part is usually also swollen. The cell proliferation leading to surface growth and to growth in thickness is strictly localized. The bulging and invagination take place in such a manner that the cecidozoa, always come to be enclosed within the invaginated cavity of the gall. As the name suggests, a pouch gall is typically hollow. The gall cavity contains the cecidozoa, but it may not be completely closed. In such cases the gall cavity communicates to the outside by a more or less narrow ostiole. The ostiole may, however, become secondarily obliterated by growth and fusion of tissue, especially when the gall is mature. It must be remarked that although the cecidozoan is within the gall, it is in reality outside the plant tissue. Pouch galls range from simple cup-like out-pocketings to oval, globose, cylindrical, horn-shaped, sessile or stalked, simple or lobed and branched growths. In nearly all pouch galls, the affected part is more or less greatly swollen, with considerable histological modifications. Pouch galls are clothed variously with hairy or other surface outgrowths. The cavity of the pouch gall is often filled with hairy processes, fleshy emergences, incomplete or complete septate outgrowths from the wall. These outgrowths often meet together and thus subdivide the gall cavity into compartments.

Krebsgalls.

The krebsgall is characterized primarily by its enormous growth in thickness. Cecidozoa like Coccids remain permanently fixed externally on the epidermis and repeated cell proliferation, localized in this place, results in the formation of irregular, globose or grape-bunch-like, solid outgrowths of tissues. Many mycocecidia are also typically krebsgalls.

Covering galls.

In the covering gall the externally situated cecidozoa induces the pathological tissue to grow around, so that eventually it comes to be enclosed within the gall. The tissue that thus grows around and over the cecidozoa usually also becomes fused together or a minute ostiole may also be left. Covering galls are typically caused by Acarina, Aphids, Coccids and gall-midges. The covering galls caused by Acarina are often morphologically distinguishable from those caused by others. The covering growth (Umwallung) often combines with other growth processes so as to give rise to complex gall types. In the gall of *Mikiola fagi* (HTG.) on the leaf of *Fagus silvatica* LINN., initial covering overgrowth is followed by pouch formation. The larva of the gall-midge is first enclosed by covering growth of tissue, the whole then becoming arched and bulged out. The spirally twisted gall of *Pemphigus spirothecae* PASS. on the petiole of *Populus* (fig. 1 C) is a typical covering gall. The well known ananas gall on *Picea excelsa* (LAMK.) LINK., caused by *Chermes abietis* LINN. (fig.

40

18), is also another common covering gall. The growth in thickness of several closely crowded parts results in the formation of sausage-shaped swellings, with numerous spaces in between the swollen parts, enclosing the aphids.

Lysenchyme galls.

WEIDEL[1234] and MAGNUS[731] applied the term lysenchyme gall to

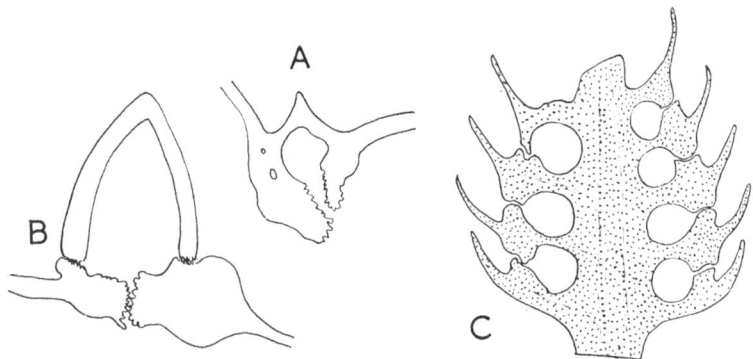

Fig. 18. Covering galls. A. L.S. through the initial covering growth, followed by the subsequent development of typical pouch gall. B. of *Mikiola fagi* (Htg.) on *Fagus silvatica* Linn. C. L.S. of the covering gall of *Chermes abietis* Linn. on *Picea excelsa* (Lamk.) Link.

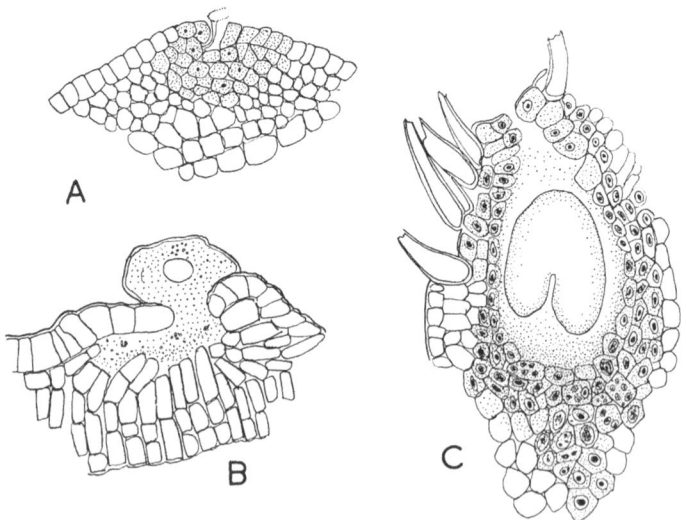

Fig. 19. Development of lysenchyme gall. A. The egg of *Diplolepis rosae* (Linn.) imbedded into the epidermis of petiole. B. The formation of primary larval cavity by lysigene action. C. Larva sunk within the tissue. (After Magnus, 1914).

the condition in which the plant tissue in contact with the cecidozoa dissolves, so as to form a hollow cavity, into which it now sinks. The aperture of this cavity eventually becomes obliterated through growth and fusion of tissue. The cells lining the lumen of the cavity now undergo proliferation and give rise to the swelling characteristic of the gall. Lysenchyme galls are caused by Cynipidae only, but are widely distributed. All lysenchyme galls are prosoplasma galls, with radially disposed tissues and often surpass most other galls in their diversity and complexity of external form and in the degree of tissue differentiation.

Mark-galls.

Mark-galls are characterized by the fact that the gall-maker is within the plant tissue from the very beginning of the development of the gall and the neoplastic tissue of the gall grows on all sides of it. The well known gall of *Pontania* COSTA on *Salix* spp. is a typical example of a mark-gall. The female of *Pontania* COSTA deposits the egg within the plant tissue and the larva becomes exposed only when it leaves the gall on maturity.

2. The cells of galls

The most striking feature of cells of galls is their conspicuously large size. The cells of not only the galls in which hypertrophy is the dominant histogenetic process but also of those in which hyperplasy plays a dominant rôle are usually much larger than those of the normal organ on which the gall arises. Sometimes the only difference from the normal organ is the larger size of the cells of the gall. It is only rarely that the gall cells are smaller than those of the normal organ.

Most gall cells are remarkable for their increased water content and there is also general enrichment of cytoplasm and increase of vacuoles. Among other cytoplasmic peculiarities of gall cells, mention may be made of disintegration of plastids and mitochondria, tendency of agglutination of chondriomes, fusion of plastids and mitochondria end to end and laterally to form reticular masses[54], arrest of formation of plastids and mitochondria[1208], inhibition of plastid divisions, etc. A number of other physical and chemical changes may be presumed to take place in the gall cell cytoplasm, as is evidenced by the nuclear changes[824].

A number of characteristic abnormalities may also be observed in the nucleus of gall cells. The position of the nucleus is usually not different from that of the normal cell of the gall-bearing organ. The most striking abnormality is a general depletion of chromatin material[849a]. GUTTENBERG[424] described the size of the nucleus of gall cells to range from 50 to 60 microns, but nuclear gigantism of gall

cells was first observed by MAGNUS[731] to represent an early phase
of development of the galls of *Diplolepis rosae* (LINN.) and *Pontania*
spp. The nucleus in the cells of galls caused by *Synchytrium* is
sometimes nearly 250 times larger than the nucleus of the normal
cell of the plant. Similar nuclear gigantism has also been observed
in some dipterocecidia and in the galls of other cynipids like *Aylax
glechomae* (LINN.). MEYER[780] has recently summarized the observations on the ratios of gall and normal nuclear sizes (fig. 20).

Table I.

The relative sizes of the nuclei of gall and normal cells

Cecidozoa	Plant	Normal cell	Gall cell
Perrisia urticae	*Urtica dioica*	1.8	3.0
Perrisia affinis	*Viola odorata*	1.2	2.1
Iteomyia capreae major	*Salix cinerea*	1.7	2.3
Rondaniella bursarius	*Glechoma hederacea*	1.0	2.0
Hartigiola annulipes	*Fagus silvatica*	0.8	1.9
Cystiphora sonchi	*Sonchus oleraceus*	1.7	2.3
Diplolepis rosae	*Rosa*-spp.	0.7	1.8
Neuroterus quercus-baccarum (bisexual)	*Quercus*	1.6	2.7
Peritymbia vitifolii	*Vitis vinifera*	0.7	2.7

It may therefore be observed that the size of the nucleus in gall
cells generally corresponds to that of the primary meristem cells of
the normal organ. There is considerable diversity of opinion regarding
the significance of the nuclear hypertrophy of gall cells. According
to some workers, nuclear hypertrophy precedes mitosis during cecidogenesis. Nuclear hypertrophy is exceedingly pronounced in the
cells of the nutritive tissue of galls, which are not, however, destined
to undergo proliferation. Others have therefore associated nuclear
gigantism of gall cells with increase in soluble glucides. MEYER *(op.
cit.)*, however, believes that in cecidogenesis nuclear hypertrophy
is not in any way related to the glucide metabolism, but is related
to the increase in ribose nucleic acid and proteogenesis. The abnormally large nucleus of gall cells contains also more than normal
nucleoli. Nuclei with three, four or even larger numbers of nucleoli
have often been observed in different galls. The form of the gall cell
nucleus is remarkably different from that of the normal cell. Many
of them are of the typical amoeboid type, with complicated lobes
and branches and sometimes also with a grape-bunch-like appearance in the galls caused by fungi and Nematodes. Similar abnormalities have also been described by different workers in the galls
of several Acarina and Homoptera. GUTTENBERG *(op. cit.)* has

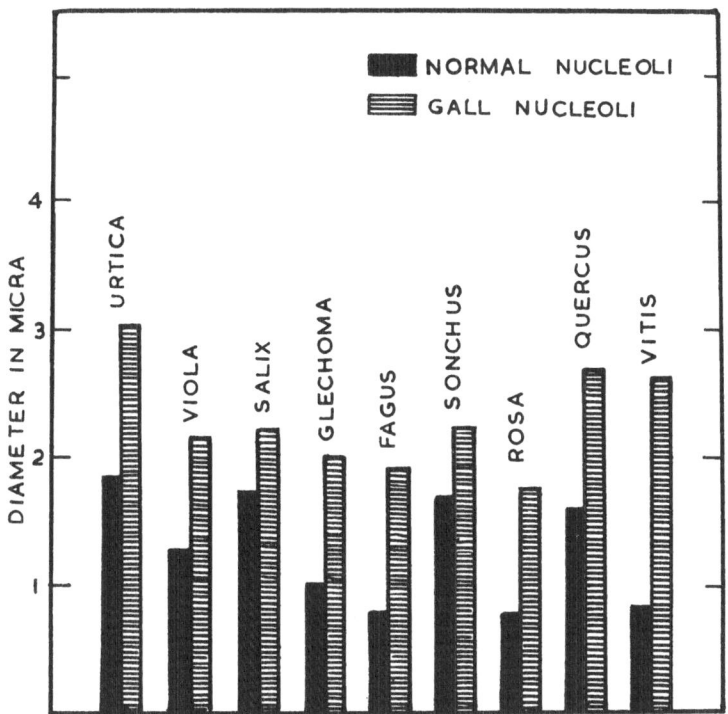

Fig. 20. Differences in the size of nuclei of the galls of *Perrisia urticae* (Perris) on *Urtica dioica* Linn., *Perrisia affinis* (Kieff.) on *Viola odorata, Iteomyia capreae* (Kieff.) on *Salix cinerea, Rondaniella bursarius* (Bremi) on *Glechoma hederacea, Hartigiola annulipes* (Htg.) on *Fagus silvatica* Linn., *Cystiphora sonchi* F. Löw on *Sonchus oleraceus, Diplolepis rosae* (Linn.) on *Rosa, Peritymbia vitifolii* Fitch on *Vitis vinifera* Linn. and the corresponding normal cells.

described an interesting modification of the gall cell nucleus, consisting of a series of canal-like structures, directed toward the gall-maker *Synchytrium mercurialis* LIBERT in its gall on *Mercurialis perennis*.

Another striking peculiarity of certain galls is the presence of multinucleate giant cells, particularly commonly found in the galls of *Eriophyes*, aphids, *Albugo*, bacteria, etc. Considerably large numbers of nuclei may be found in the erineal hairs of several mite galls. The best known example of the multinucleate giant cells is perhaps the root-knot galls caused by *Heterodera marioni* (CORNU) GOODEY [1157]. The number of nuclei in the giant cells of the nematode galls may be as many as 500. NÈMEC[859, 864] found this condition in the galls of *Heterodera* on the roots of *Vitis gongyloides*. Some of these numerous nuclei degenerate or they become fused together into groups or into a single giant nucleus. It is only rarely that the multi-nucleate giant cells become subdivided and the nuclei become parcelled out into the resulting daughter cells. TISCHLER[1157] found that in

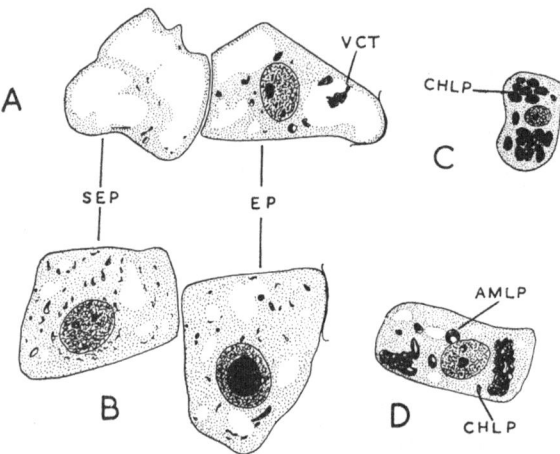

Fig. 21. Gall cells. A. Normal cells. B. Gall cells. EP epidermal and SEP subepidermal cortical cells; C. Normal cell D. Gall cells. CHLP chloroplast. AMLP amyloplast. VCT vacuolar tannin precipitate (After Meyer).

the course of the formation of the multinucleate giant cells in the gall of *Heterodera* on the roots of *Circea lutetiana* LINN., there is at first a karyokinetic division and later on rapid fragmentation. Such fragmentation is also described by GUTTENBERG[424] in the gall of *Albugo (Cystopus) candida* PERS. on *Capsella bursa-pastoris*. The multinucleate giant cells of these galls appear to elaborate some substance, which stimulates cell proliferation and cell polyploidy *(vide infra* chapter XIV).

A number of characteristic abnormalities in chromosomes have been repeatedly observed in gall cells. Some of these abnormalities include abnormal chromosome figures, massed chromosomes in lobed nucleus, hypoploid, heteroploid and polyploid condition, failure of chromosomes to reach the poles, etc. Gall cell polyploidy has been intensively studied by KOSTOFF in collaboration with KENDALL[622], WINGE[1270], LEVINE[675, 676] and others. The cells of many galls are regularly tetraploid, sometimes octoploid and occasionally even higher chromosome numbers are met with, especially in the crown-gall caused by the bacterium *Phytomonas tumefaciens* (SM. & TOWN.). In some galls, the cells in the peripheral zone are diploid and those of the deeper zone tetraploid or polyploid[676]. The tetraploid condition results from the diploid state, due to the nuclear division in the absence of cell division. Tetraploidy is thus the result of cytokinesis not following karyokinesis. WINGE believes *(op. cit.)* that the increased growth activity of the gall cells may be satisfactorily explained on the basis of tetraploidy. MOTTRAM[824] has expressed the view that the various abnormalities like endomitosis, polytene, polyploid and polynucleate conditions met with in galls are secondary to increased stiffness of the cytoplasm and are indeed no more than signs of such cytoplasmic changes.

The development of chlorophyll does not seem to be favoured in most histioid galls and indeed this pigment is extra-ordinarily sparse or even totally lacking in many of these galls. The chlorophyll may either disappear or it may not also form in the organs, in which this pigment is normally present. The development of a gall on an organ, in which chlorophyll is normally absent, is naturally accompanied by the lack of it in the gall cells. In other galls, which appear pale green or yellowish-green, some small-sized chlorophyll is often present. It is only exceptionally that galls are rich in chlorophyll. Some chlorophyll is present in the inner layers but absent in the outer layers of the gall of *Pontania* on *Salix*. The development of chlorophyll is not initiated in the gall even if the stimulus that has induced the development of the gall is removed.

3. Tissues of histioid galls

Tissue differentiation in galls reveals all gradations of complexity and in most galls there is some fundamental histological difference from the normal organ. In the simplest case, the gall is composed of nearly homogenous undifferentiated parenchyma and sometimes even the differentiation between epidermis and parenchyma is absent. In other galls, the tissues that are normally found in the plant do occur but in abnormal relations. A true epidermis may thus be present, but may be multi-layered.

In addition to the inhibition of differentiation of the normal tissues in the course of development of galls, we also know of several examples, in which there is an apparent recapitulation of tissues that are not formed in the normal organ. In the differentiation of characteristic gall tissues, the arrangement or the relative positions of the various tissues may be more or less comparable to that of the normal organ or fundamentally different and even inverted. A leaf gall may have, for example, an upper epidermis, a hypertrophied palisade tissue of several layers of cells, a similarly hypertrophied lower spongy tissue and a lower epidermis (fig. 63A). In other leaf galls there may be no such differentiation of the palisade and spongy tissues (fig. 63B). In most galls the differentiation of the gall tissues takes place radially around the larval chamber and not in reference to the organ on which the gall arises. The tissue differentiation in a gall is thus predominantly in relation to the position and nature of the gall-maker. The important gall tissues that deserve special mention are the epidermis, parenchyma, nutritive tissue, sclerotic tissue, primary and secondary tissues.

Epidermis

Differentiation of epidermis may or may not take place in the course of development of a gall. In some galls the development results in the formation of an abnormally large callus-like outgrowth (fig. 147), so that there is no true epidermis on the surface. In the gall of *Biorrhiza pallida* OLIV. on *Quercus*, for example, the epidermal tissue greatly resembles an ordinary callus growth. Even in the galls, in the development of which a typical dermatogen takes part, we may often fail to find a well defined epidermis, distinct from the subepidermal layer of cells. The gall then comprises homogenous tissue throughout, the upper layer of cells of which resembles more the callus tissue than an epidermis.

Most other galls have, however, a clearly defined epidermis, which may be derived typically from the dermatogen of the gall-bearing organ or may also arise from the parenchyma as endogenous new growth in the so-called free galls. The epidermal cells of galls sometimes resemble the neighbouring cells of the parenchyma in general size, shape and structure. Generally however, the gall epidermis is

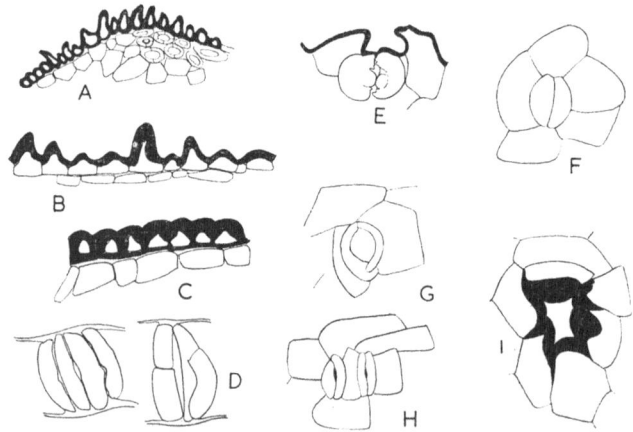

Fig. 22. Gall epidermis. A. Thick-walled epidermal cells of the gall of *Perrisia fraxini* (Kieff.) on leaf of *Fraxinus excelsior* Linn. B. Thickening of the outer wall of the epidermal cells of the gall of *Diplolepis longiventris* (Htg.) on *Quercus*. C. Thick-walled epidermal cells of the gall of *Neuroterus quercus-folii* (Linn.). D. Stomata in the gall of *Ustilago*. E. I. Stomata in the gall of *Trioza alacris* Flor. on *Laurus nobilis*, completely walled in in F.G. H. Stomata in gall of *Pontania proxima* Lepel on *Salix*.

characterized by its appreciably much larger cells than those of the normal epidermis. Diverse modifications of the epidermal cells of prosoplasmatic galls have been described [1234] and include thickened cell walls, thicker cuticle, lignified cells and other characteristically formed special cells. In a number of galls the epidermis becomes modified into elongate palisade-like cells. In some galls, like in the gall of *Copium teucrii* HOST. on *Teucrium capitulum* and in the gall of *Janetiella (= Oligotrophus) lemeei* (KIEFF.) on *Ulmus*, multicellular layer epidermis has been described [1140, 641].

The structure of the epidermis differs in different parts of a gall. In the gall, in the development of which both the lower and upper epidermis of a leaf take part, the structure of the gall epidermis differs very much in different parts. The striking difference that is seen in the epidermis of the upper and lower surfaces of the leaf blade is more or less completely obliterated. The histological character of the epidermis of the inner and outer surfaces of pouch galls shows fundamental differences (fig. 74). The epidermis on the outer surface of the gall tends usually in such cases to be composed of thick-walled cells, not rich in cytoplasm, but often also tending to form strong hairy outgrowths. The epidermis on the inner surface is most usually thin-walled, delicate, hyperhydric and rich in cytoplasm, relatively weakly cuticularized, sparsely or not at all hairy. The development of characteristically shaped hairs imparts a definitive appearance to the succulent epidermis of the inner surface of pouch galls (fig. 77). The specific character of the outer epidermis of pouch gall depends largely on the thickening of the cell walls (fig. 22) and in the deformation of stomata.

The thickening of the cell wall in the epidermis is usually very pronounced in the region of the ostiole in pouch galls and covering galls. KÜSTER has, for example, shown that in the gall of *Perrisia fraxini* KIEFF. on *Fraxinus excelsior* LINN. from Europe the epidermal cells near the ostiolar region are typically characterized by the presence of sclereids, thickened on all sides, resulting in a tooth-like growth. KÜSTER also found in the gall caused by an unknown midge on *Jacquinia* the cuticular epidermis described by DAMMS in his well-known work on the development and mechanical properties of the epidermis in Dicotyledons. The epidermal cells are very strongly cuticularized, strongly stretched with the continued swelling of the gall and are ultimately ruptured. The cuticularization occurs in the subepidermal layer of cells also. Typical but moderately cuticularized epidermal cells may be found in the galls of *Pemphigus bursarius* (LINN.) on the leaf of *Populus*. KÜSTER describes a definitive waxy coating on the epidermis in the gall of *Mikiola fagi* (HTG.) on *Fagus silvatica* LINN. Although the epidermal cells in most galls are more or less similar, it may be pointed out that in the galls of *Synchytrium* spp. it is individual single epidermal cells which are attacked by the fungal parasite and considerable heterogenity may therefore be met with. It may also be noticed that the epidermis in roll, fold and pouch galls usually lacks a definitive cuticular layer and the absence of this layer has considerable significance in altering the permeability of the cells.

Stomata

One of the striking features of the gall epidermis in general is that stomata are fewer than normal and in many cases totally absent. Even if some stomata are present, they are mostly functionless, due to the collapse of one or both the cells. The whole apparatus becomes often severely compressed by the pressure of enlarged neighbouring cells to mere membranous vestiges. The guard cells often become enormously swollen and fused and thus lose their capacity to close, so that the stomata remain permanently open pores (fig. 22). Total fusion of the two guard cells may be found commonly in the gall of *Diplolepis (= Dryophanta) quercus-folii* (LINN.) on *Quercus*. The guard cells in the gall of *Pemphigus bursarius* (LINN.) are so enormously swollen that the opening is reduced to a minute pore[641]. According to KÜSTER[641], in the stomata of the gall of *Pontania proxima* LEPEL on *Salix*, the two guard cells are torn apart from each other by the strong swelling of the parts from underneath. The neighbouring cells grow over and more or less completely cover the stomata in the gall of *Trioza alacris* FLOR on leaf of *Laurus nobilis*. Thus covered by epidermal papillae, the stomata become wholly non-functional. In some galls like that of *Pontania proxima* LEPEL on *Salix*, several stomata become closely aggregated together in irregular patches. One or both the guard cells are sometimes transversely divided, so that the stomata have four abnormally small-sized guard cells.

GERTZ[380, 381] has given interesting accounts of the different modifications of stomata found in galls, some of which are by no means specific to a gall, but may often be found quite irregularly in one and the same gall. We have an interesting record of the development of stomata both on the outer and inner surfaces of the gall of *Eriosoma (Schizoneura) lanuginosum* (HTG.) on the leaf of *Ulmus* from Europe. The outer surface of this gall corresponds to the upper surface of the normal leaf blade, where stomata are not present. GUTTENBERG[424] has also described stomata on the inner surface of the galls on fruit of *Capsella bursa-pastoris* caused by *Albugo candida* PERS. and of *Alnus incana* caused by *Exoascus amentorum*. Stomata are never found in this position in the normal organs of these plants.

Trichomes

Abnormally strongly and densely developed hairs are characteristic of many galls caused by Acarina, Diptera, Hymenoptera and occasionally also by aphids. The galls caused by some species of *Synchytrium* are also remarkable for their dense hairy covering. All aerial parts of plants seem to be capable of developing hairy outgrowths in the course of gall formation. Hairs develop even on the inner surfaces of galls on ovary and fruit of many plants.

Simple and clavate hairs develop in the gall on flowers of *Lonicera periclymenum*, caused by the aphid *Siphocoryne xylostei* (DEEG.)[270]. The development of hairs normally found on the plants is also occasionally inhibited in the course of gall formation. The ordinary hairs normally found on *Teucrium* are modified into glandular hairs in the gall of *Copium* on the plant[1140].

The hairs on galls resemble either those found normally on the plant or they are also in some way modified. In the former case, the hairy covering is usually far more dense than on the normal organ of the plant. In the latter case, the gall hairs differ more or less in size, shape and general structure from the normal hairs. Hairs also develop on galls borne on organs, which are normally glabrous, so that such gall hairs are new formations. In a number of galls, like the gall of *Wachtiella (Perrisia) persicariae* (LINN.) on the leaf of *Polygonum persicaria* LINN., the hairs develop haphazardly in a variety of modifications. In spite of the profound modifications

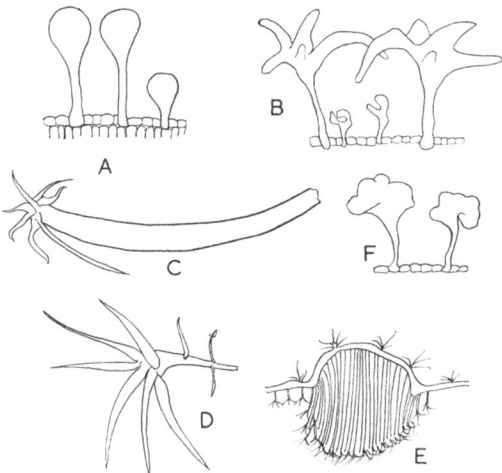

Fig. 23. Trichomes in gall. A. Clavate hairs in the gall of *Eriophyes macrorrhynchus ferruginus* Nal. on *Fagus silvatica* Linn. B. Lobed hairs of the gall of *Eriophyes brevitarsus typicus* Nal. on *Alnus glutinosa*. C.D.E. Trichomes of the gall of *Phlomis samia*: C. Galled hair, D. Normal hair, E. L.S. through the gall (After Rübsaamen). F. Lobed clavate hair of the gall of *Eriophyes macrorrhynchus aceribus* Nal. on *Acer pseudoplatanus*.

exhibited by gall hairs, their fundamental similarity to the hairs normally found in the species is usually evident.

The hairs that arise *de novo* on galls are as a rule characteristic for their remarkable diversity of size, shape and structure (fig. 23). They range from simple elongations of cells to large and complex branched and papillate growths. The epidermal cells in the gall of *Synchytrium myosotidis* become modified into stout hairs. The gall of *Synchytrium papillatum* on *Erodium cicutarium* consists of complex hairs, with slender stalk and greatly enlarged and lobed apex, usually thick-walled, but with a thin ring-like zone below, where the gall breaks off loose (fig. 140). Erineal galls with papillate hairs are not uncommon and in most cases even the simple tubular hairs which arise by hypertrophy of the epidermal cells are initially more or less papillate. Most erineal mats comprise slender, elongate, cylindrical hairs, nearly equally thick both basally and at the tip and usually also rounded apically. The erineal mat is produced by outgrowths from nearly all the epidermal cells. Non-cylindrical hairs are also sometimes found in erineal galls. Some hairs are slender basally and stout apically or shaped more or less like a miniature toadstool. Apical branching of the hair results in the formation of grape-bunch shaped formations in the erineal gall on *Alnus*. While the slender and cylindrical hairs arise as proliferation of large numbers of epidermal cells, the mushroom-shaped and clavate hairs develop usually scattered.

The erineal hairs of galls are usually rich in protoplasm, with large and numerous nuclei, thin wall and conspicuous vacuoles. They usually contain some chlorophyll and red pigment. The clavate hairs of some galls have been shown to contain starch granules, arranged like a mosaic and almost filling the lumen and oil globules have been found in other erineal hairs.

The thickening of the cell wall is usually very much less pronounced in the elongate and cylindrical hairs than in the clavate ones and in such cases it is usually the outer wall that is thickened. Thickening of the cell wall occurs also in places where the hairs come into contact with each other. The slender hairs of the erineal gall of *Tilia* have been described to become fused together at the points of their contact[342]. The hairy coverings of the outer and inner surfaces of pouch galls are strikingly different (fig. 103). The hairs on the outer surface may generally be acutely pointed, thick-walled and deficient in protoplasm. The hairs at the ostiolar entrance are remarkably stout, but the hairs on the inner surface are short, broad, thin-walled, apically rounded and rich in protoplasm. Some of these hairs on the inner surface have been termed nutritive hairs.

As already mentioned, not all hairs on galls are of epidermal origin. In the so-called free gall, not only hairs but also stomata arise as morphologically new formation from the ordinary parenchyma elements. In the gall of *Hartigiola (= Oligotrophus) annulipes* (HTG.) on *Fagus silvatica* LINN. (fig. 24, 76) the hairs do not, for example, arise from the dermatogen but from the palisade tissue of the leaf as

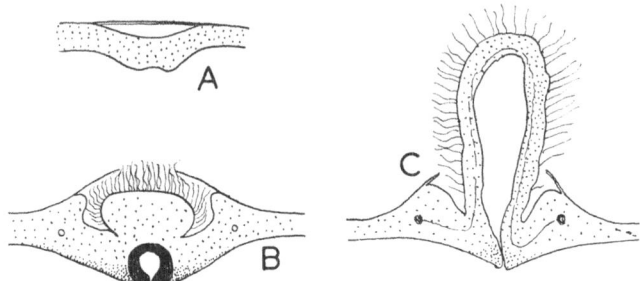

Fig. 24. Development of the pouch gall of *Hartigiola annulipes* (Htg.) on leaf of *Fagus silvatica* Linn.   A. The early stage of development, without the epidermis on the upper surface of the leaf taking part.   B. Later stage of development, with the gall breaking through the epidermis, trichomes developed not from the derma-togen but from the deeper mesophyll cells and with the nutritive zone differentiated round the larval cavity.   C. L.S. through the nearly mature gall.

elongate unicellular processes. The stellate hairs on the surface of the lenticular gall of *Neuroterus quercus-baccarum* LINN. *(Neuroterus lenticularis* OL.) on the leaf of *Quercus* from Europe (fig. 84) and the biramose hairs on the gall of *Neuroterus numismalis* FOURC. on *Quercus* are also *de novo* growths. Other examples of *de novo* hairs on galls are given in the following chapters. (vide fig. 77C).

Mechanical tissue

The mechanical tissue, consisting of thick-walled cells which give lignin reaction, is widely distributed in prosoplasmatic galls, but absent in filzgalls and pockengalls. Pouch galls with only predominant surface growth, many leaf roll and fold galls and the kammergall of *Pontania* also generally lack a mechanical tissue. With the possible exception of the gall of *Ustilago grewiae*, described by TROTTER[1169], mycocecidia also do not have mechanical tissue.

The development and arrangement of mechanical tissue in galls presents a remarkable complexity, characterized by dorsi-ventral and radial differences, particularly in relation to the larval chamber. Dorsi-ventral and separate plates of mechanical tissues arise in relation to the larval cavity in some galls like that of *Hartigiola (= Oligotrophus) annulipes* (HTG.) on the leaf of *Fagus silvatica* LINN. The mechanical tissue is also dorsi-ventral in the depressed gall of *Forda formicaria* HEYD. *(= Pemphigus semilunarius)* on *Pistacia*. In galls with central larval cavities and generally radially disposed tissues, the mechanical tissue develops as a continuous, woody, spherical or oval layer around the larval cavity, the shape of which corresponds more or less to that of the gall itself. The

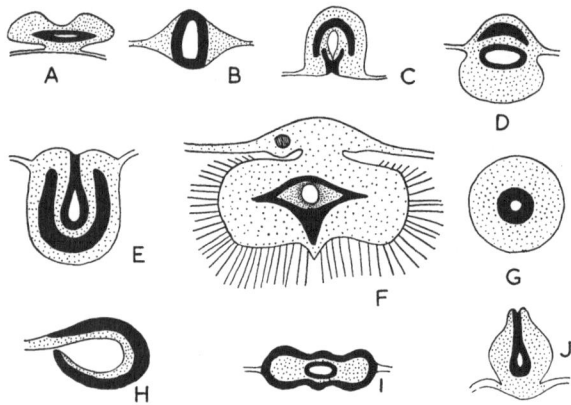

Fig. 25. Sclerenchyma in gall. A. Lenticular gall of *Neuroterus numismalis* Fourc. on *Quercus*. B. Gall of *Didymomyia reaumurianus* (F. Löw). C. *Harmandia globuli* Rübs. on *Populus tremula* Linn. D. *Arnoldia cerris* Kollar on *Quercus cerris*. E. *Perrisia fraxini* (Kieff.) on *Fraxinus excelsior* Linn. F. Midge gall on *Millettia sericea* W. & A. G. *Diplolepis quercus-folii* (Linn.) on *Quercus*. H. *Forda formicaria* Hey. I. *Banisteria*. J. *Craneiobia corni* (Gir.) on *Cornus*.

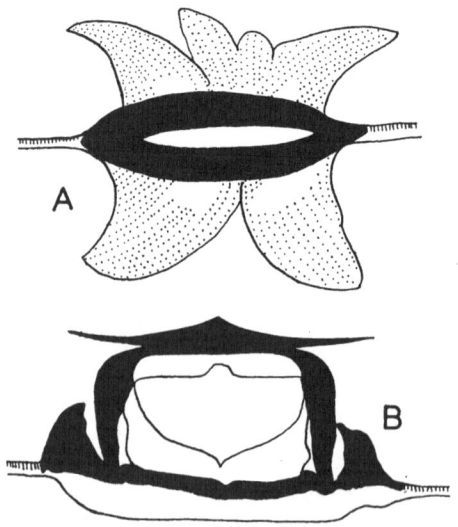

Fig. 26. Sclerenchyma in gall. A. Midge gall on *Dalbergia*. B. Midge gall on *Phialodiscus unifugalis* (After Rübsaamen).

mechanical tissue may arise either peripherally, separated from the outer surface of the gall by only a few layers of ordinary parenchyma cells or it may also be situated deeper and thus surrounded by several layers of parenchyma cells. There are sometimes two concentric layers of mechanical tissues surrounding the larval cavity. The mechanical tissue developing in the immediate neighbourhood of the larval cavity may not always form a continuous mantel, but occur in two separate pieces. Such sepatated pieces of mechanical tissues have significance in the dehiscence of the gall *(vide* chapter X). The unequal shrinkage of the mature gall, in which the mechanical tissue is less yielding than the softer parenchyma, leads to the spontaneous dehiscence at the site of the separation of the double mechanical tissue. An upper lid-like mechanical tissue piece becomes separated from a lower piece (fig. 26 B). The breaking loose of the gall of *Mikiola fagi* (HTG.) from its basal cup-like swelling on the leaf of *Fagus silvatica* LINN. is due to the presence of such a separation zone between the mechanical tissues at the base of the gall. Even in galls in which a continuous mechanical tissue layer is present, separation and free movements are possible on the ripening of the gall and the consequent differential drying and shrinking of tissues. In the gall of *Didymomyia (= Oligotrophus) reaumurianus* (F. Löw) on *Tilia*, for example, the hard inner core springs off from the softer tissues when the gall becomes mature.

The cells of the mechanical tissues of galls are derived from the general gall parenchyma. Their shape varies but very little and they are usually polygonal or short palisade-like cells. Where two separate zones of mechanical tissues are present, the outer zone is usually composed of larger, less thickened cells than the inner zone. The sclerotic cells of the mechanical tissues, especially in certain cynipid galls on *Quercus*, are frequently thickened more on one side than on the others, so that the lumen of these cells appears circular in cross sections. The sclerotic cells in the gall of *Neuroterus quercus-baccarum* LINN. on *Quercus*, for example, are more strongly thickened on the side turned away from the cynipid larva, but in the gall of *Diplolepis (Dryophanta) longiventris* (HTG.), also on *Quercus*, the thickening is more pronounced on the side of the larval cavity than on the other side. In the lenticular galls on *Quercus* the sclerenchyma cells undergo delignification at the later stages of development of the gall and become again thin-walled, but the other peripheral cells, which have so far remained thin-walled, become thickened to form sclerotic cells[1234].

Reserve material and nutritive tissues

The cells of these tissues are characterized by the abundance of proteins, fats or starch. The nutritive tissue forms most usually a continuous zone around the larval cavity. In some galls there may often be a second zone of reserve material tissue outside the zone of sclerenchyma; the inner zone then contains predominantly fats and proteids and the outer zone predominantly starch. The development and importance of the nutritive tissues of galls have recently been carefully investigated by MEYER[781, 783, 786, 788]. On the basis of their cytological characters, he has recognized the following principal types of nutritive tissues: 1. Nutritive tissue characterized by cytological differences, especially rich cytoplasm, abundant chondriome or plastidome and 2. nutritive tissue without special cytological characters, with large cells, peripheral cytoplasm, a large central vacuole and undifferentiated plastids. In the former type of nutritive tissue there may be a continous development of the tissue. Some of these are of the lipidase type, characterized by the presence of complex lipo-proteins of chondriomes and plastids. The nutritive tissue in the galls of *Perrisia urticae* (PERRIS) and *Hartigiola (= Oligotrophus) annulipes* (HTG.) and in the larval gall of *Chermes abietis* LINN. is mitochondrial. It is of filamentous chondriocontes type in the galls of the fundatrix of *Chermes abietis* LINN. *Diastrophus rubi* HTG. and *Perrisia affinis* KIEFF. In many other galls the nutritive tissue is plastid differentiated and many comprise chloroplast and amyloplast type as in the gall of *Iteomyia (= Oligotrophus) capreae* (KIEFF.). Nutritive tissue of free lipids is found in the gall of *Rondaniella (Oligotrophus) bursarius* (BREMI), *Cystiphora sonchi* (F. LÖW), etc. In the gall of *Eriophyes macrorrhynchus* NAL. on *Acer pseudoplatanus* LINN. the nutritive tissue arises by dedifferentiation. In some galls there is, however, no definitive nutritive tissue zone. This is for example the case with the gall of *Lasioptera rubi* HEEGER, in which the larva of the gall-midges derives its nourishment from the mycelia of an ambrosia fungus *(vide* chapter XI), introduced by midge itself.

Primary vascular bundles

The primary vascular bundles, which have been modified by gall-inducing influence, differ from the normally developed bundles in their characteristically weakly formed tracheids. The parenchyma portion, including also the medullary rays in between the bundles, is also strongly developed. The inhibition of the normal development of vascular bundles in galls on Dicotyledons results in the closed ring becoming pronouncedly wide meshed.

The vascular bundles that arise new in the gall are normally connected with the vascular bundles in the gall-bearing organ. Like the normal bundles, the new ones of the gall also arise from procambial cells, at least in so far as the tracheids are concerned, by direct modification of the thin-walled parenchyma cells into thickened tracheid elements. The new bundles are directed to the larval cavity and have been termed by HOUARD[485] as faisceaux d'irrigation. As shown by DOCTERS VAN LEEUWEN[295], the new bundles that arise in the gall and are directed to the larval cavity end only as phloem elements. The histological composition and the orientation of the vascular bundles in galls correspond usually to those found in the normal organ. The bundles are collateral, with the xylem elements on the inside and the phloem peripheral. The opposite condition is sometimes found, as in the gall of *Andricus nudus* ADLER ♀♀. In the hadrocentric concentrical bundles in the galls of *Trigonaspis megaptera* (PANZ.) and *Andricus albopunctatus* (SCHLECHT.) also the inverted condition has been described by BEIJERINCK. Inversion of the elements of the vascular bundles is found also in galls on some Monocotyledons, as for example, in the gall of *Lipara lucens* MEIGEN on *Phragmites*. In this gall new vascular bundles develop and the phloem is not found between two vessels, but toward the inner side [289]. Both excessive development and suppression of the phloem elements have been described in some galls. In the gall of *Mompha* on *Epilobium* the inner phloem is extremely richly developed[486]. In the gall of *Ustilago maydis* D.C. on *Zea mays* the vascular elements are composed of only phloem[424]. Doubled vascular bundles, due to complex folding and fusion of tissues, have been described by COURCHET[228] in the gall of *Baizongia pistaciae* (LINN.) *(= Pemphigus cornicularius)* on *Pistacia*.

Secondary tissue

The secondary tissue that takes part in the development of the gall is either a product of the normal cambium or of the newly formed gall meristem. The gall-inducing stimulus is also able to induce cell proliferation in the secondary tissue already present in the normal gall-bearing organ and the new cells thus formed now take on the character of the secondary tissue.

The strong activation of the cambium, lying between the xylem and phloem, in the course of development of the gall, results in the enormous production of parenchyma cells, far more than under normal conditions. In the course of gall development the production of xylem and phloem can be enormously increased. In the normal organ the xylem elements are ordinarily more densely developed than the phloem, but in the gall it is the phloem that is more massively developed than the xylem. This is especially the case if the cecidozoa is situated in the cortex of the stem. The phloem generally grows like a mantel near the larva of the cecidozoa. It is remarkable that wherever the larva lies, the xylem elements do not dominate over the phloem. The secondary tissue of mycocecidia also is characterized by the abnormally abundant formation of parenchyma elements. This increase in parenchyma is brought about by the young derivatives of cambium instead of becoming prosenchymatic xylem but developing into groups of parenchyma cells or by the cambium cells dividing transversely and by further subdivisions turning into parenchyma cells. Cambial cells may here and there modify, so that the medullary rays become enlarged or cell proliferation in the region results in a continuous mass of parenchyma.

Regeneration and renewal of meristem in the gall is brought about by the tissue proliferation in the medulla or by the medullary rays breaking up the ring of vascular bundles into isolated closed pieces. The cambium of these individual pieces becomes completed, so that each piece is by itself a complete closed ring and the gall has then a polystele structure. This condition is found, for example, in the gall of *Andricus inflator* HTG. on *Quercus*. The development of secondary mass around the primary phloem is described by HOUARD[485] in the gall of *Asterolecanium massalongianum* TARG.-TOZ. on *Hedera helix* LINN. Tissues, which can be compared with concentric leptocentric vascular bundles, are thus formed in this gall.

Other tissues

The secretory tissue of galls is on the whole not fundamentally different from that of the normal organ. Secretory organs are, however, developed often in abnormal abundance in galls. This is, for example, the case with the gall of *Baizongia pistaciae* (LINN.) *(Pemphigus cornicularius)* on *Pistacia*, characterized by the relatively large resin ducts. The development of secretory organs is often suppressed in many galls. Oil spaces, so abundant in *Eucalyptus*, are often absent in galls on its leaves. We have also record of the development of a gland cell in a gall on *Quercus* from America, although no gland normally occurs in the healthy organ. Gum canals in galls become blocked by proliferating cells[485]. Calcium oxalate crystal cells are less abundant in some galls. The distribution of these crystals is often characteristic in some galls, like that of *Perrisia ulmariae* (BREMI) on *Spiraea ulmariae*. It is the layer of cells nearest the innermost layer of epidermal cells which are richest in crystals in this gall. In the gall of *Mikiola fagi* (HTG.) on *Fagus silvatica* LINN. the crystal cells are situated between the small-celled inner and the large-celled outer tissues of the gall. Anthocyanin is often found in the epidermis and parenchyma cells of galls. Cork is not infrequently found in some galls.

CHAPTER IV

ROOT GALLS

1. Root galls in general

The study of root galls and the root gall communities belongs properly to the field of soil biology. The ecology of root galls is complicated by the fact that among the organisms, which induce galls on the roots of diverse plants, there are in addition to the strictly edaphon, many species of the aerial habitat, which have an edaphic phase in their life-cycle and cause galls during this phase. Some of the aerial species give rise to galls on roots in successive generations, but others give rise to galls on roots in alternate generations only and on different aerial organs in alternate generations by turn. Some of the outstanding features of the structure, development and ecology of root galls may be traced to these peculiarities. The edaphon that cause root galls include bacteria, fungi and Nematoda. To the second ecologic group belong aphids, Curculionidae, Buprestidae, Itonididae and certain species of Cynipidae with heterogony. Groups of cecidozoa like Acarina and Thysanoptera are excluded from the root gall community. Unlike in the case of aerial galls, it is only exception that the gall-maker is found externally on the surface of the galled root. However, as in

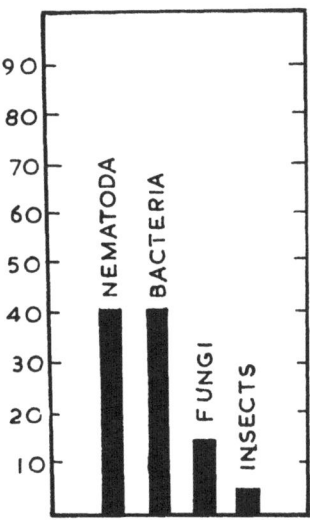

Fig. 27. Percentage frequency of root galls caused by different gall-inducing organisms from the world.

the case of galls on shoot axis, so also with root galls, the positioning of the gall-maker with reference to the region of active growth in the root has a profound influence on the structure and mode of development of the root gall. The bulk of the root galls have the gall-maker positioned in the cortex. Root galls in general lack the diversity and structural complexity of the galls on aerial organs and the ecologic inter-relations and communities in root galls are also very much simpler. The root galls caused by the two ecologic groups develop usually at definitive levels underground. The galls of true edaphon arise both in the surface layer and deeper soil, but the galls caused by aerial species arise generally in the upper levels. The relative frequencies of the root galls caused by different groups of gall-makers are shown in fig. 27. It may be observed that the galls of true edaphons are dominant over those of the aerial species.

2. Galls caused by edaphon

The root nodule galls caused by bacteria on Leguminosae are among the commonest and simplest of this ecologic group of root galls. The bacterial nodule galls on the roots of *Lupinus* and *Glycine hispida* are caused by *Bacterium beyerinckii* H. & ST., and those of other Leguminosae, especially Papilionaceae, are caused by *Bacterium radicicola* BEIJ. Similar bacterial root nodule galls are found commonly on *Alnus, Elaeagnus, Hippophaë, Myrica* and other non-leguminous species also.

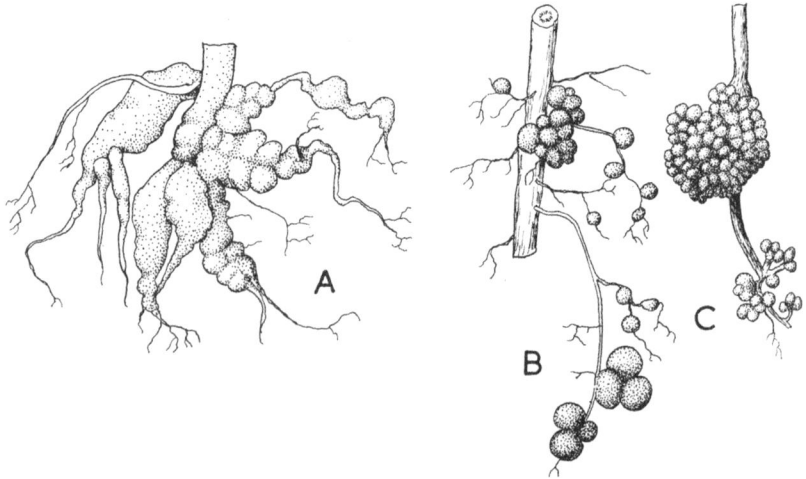

Fig. 28. Some common root galls of edaphon. A. Gall of *Plasmodiophora brassicae* Woron. on *Brassica*. B. Bacterial root nodule gall of Leguminosae. C. Bacterial root nodule gall on *Alnus glutinosa*.

Considerable literature exists on the bacterial root nodule galls of Leguminosae and other plants and useful summaries may be found in WORONIN[1276], VÖCHTING[1219], HILTNER & STÖRMER[473], SÜCHTING[1115], VIERMANN[1214], WENDEL[1246], SCHAEDEL[1021], HARRISON & BARLOW[438], DE ROSSI[262], RODELLA[967], GHEORGEVITCH[378], ZIPFEL[1286], KRÜGER[627], BARTHEL[56], DANGEARD[241], LÖHNIS[712], HANSEN & TANNER[436], KREBBER[626], LEWIS & MC COY[770], VIRTANEN, VAN HAUSEN & KARSTRÖM[1215], KORSAKOVA & LOPATINA[612], SCHEER-LINK[1016], ENGEL & ROBERG[318], PIETZ[909], SCHAEDE[1012] and others cited in the bibliography. WENDEL *(op. cit.)* has given a detailed account of the physiological anatomy of the nodule galls on the root of *Lupinus*. The general structure, origin and nature of the root nodule galls of Leguminosae are discussed by V. TIEGHEM & DOULIOT[1204].

The root nodule bacteria of Leguminosae enter the young root through the amyloid membrane spots of the primary root epidermis and multiply to begin with in this spot. The site of attack by the bacteria is marked by a bump-like swelling, produced due to rapid cell proliferation in the primary cortex of the principal root. The formation of this gall seems to stimulate a resting lateral root meristem to sprout, which pushes forward and expels the gall tissue, the primary seat of bacterial infection. The bacteria grow in lines, in the manner of plasmodesma, into the parenchyma cells newly formed by the meristem and multiply rapidly. The meristem of the sprouted lateral root primodrium now grows further in the flank and joins the meristematic zone of the primary infection to form bulblet meristem, which now multiplies in all directions. The bulblet meristem forms parenchyma cells, which separate into collenchyma, starch cells layer, bacterioidal tissue and vascular bundles. The connection with the cambium of the root is brought about by the meristematic connectives. The plerome arises from this parenchyma cell mass. The bacteria are not apparently able to follow the outgrowth of the meristem and are isolated by the formation of a starch cell layer in connective tissue traversed by isolated steles. In annuals the meristem disappears, but in perennials the meristem persists and produces additional bulblets in the growth next year. The bacterioidal cells have hypertrophied nucleoli and the nucleus becomes amoeboid. The nucleolus becomes fragmented and the fragments penetrate into nuclear pseudopodia. The starch cells form suberin lamella and thus occlude the bacterioidal tissue. The vascular bundles are cut off and decay with the empty bulblets.

Another root nodule gall, which has been somewhat intensively studied in recent years, is found to develop on *Alnus glutinosa* GAERTN., *Alnus cordata* and other species (fig. 29 C). According to some workers, this gall is caused by *Bacterium radicicola* BEIJ.

60

but according to others by a fungus *Actinomyces alni* (PEKLO) or by *Plasmodiophora alni* MÖLL. and still others believe the causative organism to be a polymorphic species[89,1075–1076, 1278]. A useful

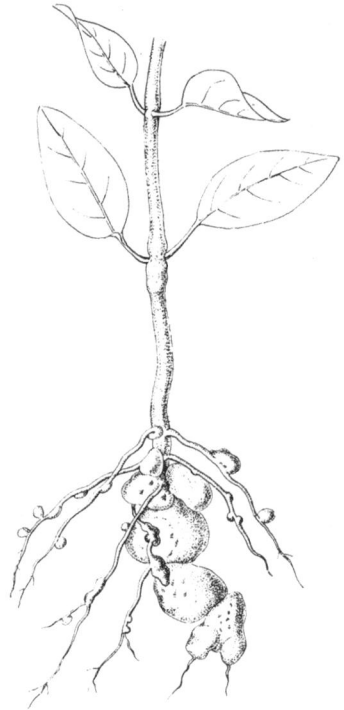

Fig. 29. Root gall of *Heterodera marioni* (Cornu) Goodey on a Dicot.

review of the important literature on these galls may be found in TROTTER[1175]. Nodular galls on the roots of other plants are known to be caused by fungi[29, 437]. In 1928, BARTLETT[57] described, for example, an interesting globose gall on the roots of cultivated Cruciferae, caused by the fungus *Olpidium radicicolum* DE WILLD., the resting stage of which is probably what is known under the name *Asterocystis radicis*. When exposed to direct sunlight above ground, some of these galls were found to turn green and even produce shoots with leaves in some cases. SCHAEDE[1013] has described the nodular gall on roots of *Podocarpus* spp., produced by a fungus, the hyphae of which are intercellular in the gall tissue[1077] (also *vide* Chapter XII). The best known of the root galls caused by fungus is that of *Plasmodiophora brassicae* WORON. on a variety of wild and cultivated Cruciferae (fig. 28 A), to which we have already referred. The gall is an irregular swelling[1277], mainly of the side roots, rarely of the main tap root and even of the subterranean portion of the main shoot axis. The gall arises as an elongately lobed, solid swelling, but turns into an irregular composite mass of considerable size by the fusion of several closely adjacent centres of swellings. The seat of proliferation is chiefly in the cortex, in which the parenchyma cells contain darker cytoplasm, in sharp contrast to the clear cytoplasm of the normal cortical cells. The fungus occurs in these cells in the form of naked plasma masses inside vacuoles. As the gall increases in size, the cambium in the vicinity becomes involved and leads to an enormous increase in the mass of parenchyma cells. The influence of the fungus seems to be exerted on distant cambial cells also. With the continued growth of the gall, the fungus also multiplies rapidly and the cytoplasmic contents of the gall cells become turbid and filled with fine granular masses of the parasite. In the second phase of development of the fungus spore formation occurs. The naked plasma fragments of the fungi probably fuse together to form a plasmodium, which then undergoes division to give rise to spores about 1.6 micron large. The spores remain in the gall cells, which have by now succumbed and can only be liberated by the rupture of these cells. The gall tissue now rapidly decays and the spores are thus liberated in the surrounding soil (*vide* Chapter XII). Reference may also be made here to the root nodule galls on *Casuarina equisetifolia* (FORT.), caused by bacteria, which occur in the gall cells[185, 891].

By far the commonest root galls are, as already mentioned, caused by the Nematode *Heterodera* (fig. 29). These galls arise either as large and irregular and agglomerated swellings or isolated cortical swellings from the root. A great deal is known about the anatomy, histology and physiology of the root galls caused by *Heterodera* spp.[805, 859, 1157]. The division of the pericycle gives rise to the formation of the so-called small-celled parenchyma, outgrowths from which often develop as lateral roots. Some of the innermost cells of the small-celled parenchyma may become differentiated into xylem and the phloem-like vessels that develop around the syncitial giant cells *(vide supra)* do not transport material to these cells but to the nematode.

3. Galls caused by aerial forms

Ecologically two distinct groups of cecidozoa are associated with this class of galls. In some species, though essentially aerial forms, there is a subterranean larval stage, associated with specific galls on roots. In these cecidozoa development is without heterogony, so that root galls are caused by successive generations. In the second ecologic group development is heterogony, with alternation of a generation associated with root galls and a generation associated with galls on some other aerial organ of the same or other plant.

A great many examples of galls of the first group are known. COCKERELL[205] has described, for instance, an interesting covering gall, about 5 mm in diameter, on the root of *Vitis vinifera* LINN. from the Cape Colony, Africa, caused by the coccid *Cryptinglisia lounsburyi* CKLL. A small cavity in the subglobose swelling contains the occid. Usually a number of galls become agglomerated into rugose, multilocular masses, about 20 mm long and 7 mm thick.

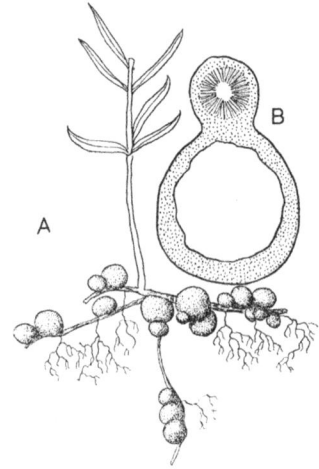

Fig. 30. Root galls of Coleoptera. The gall of *Gymnetron linariae* Payk. on the roots of *Linaria vulgaris* Mill.

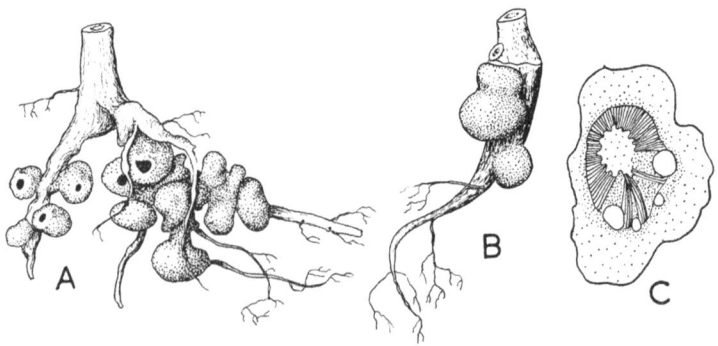

Fig. 31. Root galls of Coleoptera. A. Gall of *Ethon corpulentum* Boh. on *Dillwynia ericifolia* Sm. B. Gall of *Ceutorrhynchus pleurostigma* (Marsh.) on *Brassica*. C. T.S. through the gall of *Ceutorrhynchus pleurostigma* (Marsh.)

A number of root galls caused by Coleoptera belong also to the same group. QUINTARET[919, 920] has described the gall of *Gymnetron linariae* PAYK. on *Linaria striata* DC. A similar gall is caused on *Linaria vulgaris* MILL. from Europe (fig. 30). Perhaps the largest root gall is caused by the weevil *Liocleonus clathratus* OLIV. on *Tamarix nilotica* EHR. from Egypt and is described as being as large as the head of an infant[520]. HOUARD[520] has described the large woody swellings on the roots of the Egyptian *Echinops coeruleus* OWERIN, caused by the weevil *Larinus albolineatus* CAP. Other interesting root galls include those of the Buprestid beetles *Ethon corpulentum* BOH. on *Dillwynia ericifolia* SM. from Australia (fig. 31 A) and *Ethon marmoreum* LAP & GORY on the same plant, also from Australia[349]. These galls sometimes attain considerable sizes and arise as oval or subglobose and agglomerated hollow swellings. The best known root gall in this group is caused by *Ceutorrhynchus pleurostigma* MARSH. on Cruciferae like *Raphanus raphanistrum* LINN., *Brassica* spp., *Arabis albida* STEV. and others, (fig. 31 B-C) usually at the crown and upper part of the root. The gall develops as subglobose or irregular fleshy and multilocular swellings (*vide* Chapter IX).

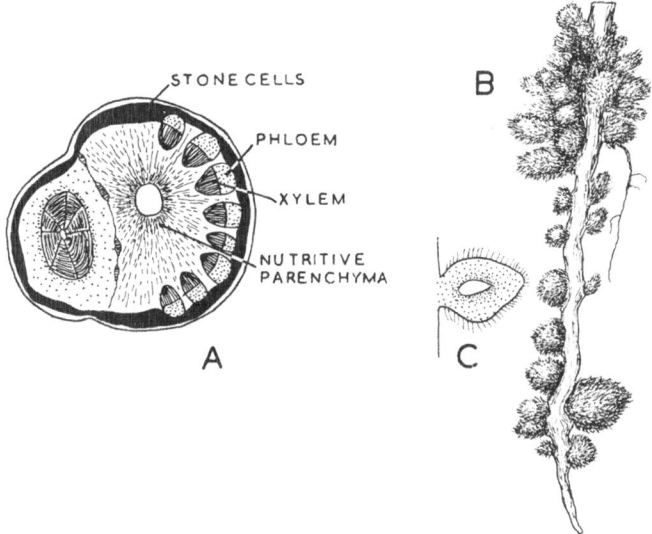

Fig. 32. Root galls of Hymenoptera and Diptera. A. Section through the Chalcid gall on the aerial root of *Ficus*, showing the larval cavity, the new vascular bundles. sclerenchyma and the direction of cell division and cell enlargement (Modified from Docters van Leeuwen-Reijnvaan). B. Gall of *Asphondylia strobilanthi* Felt on the root of *Strobilanthes cernuus*. C. The same, cut open. (After Docters van Leeuwen-Reijnvaan).

Other zoocecidia on root, belonging to the same ecologic group, include the interesting gall caused by an unknown Lepidoptera on *Helichrysum conglobatum* STEUD. from Tripoli, described by STE-

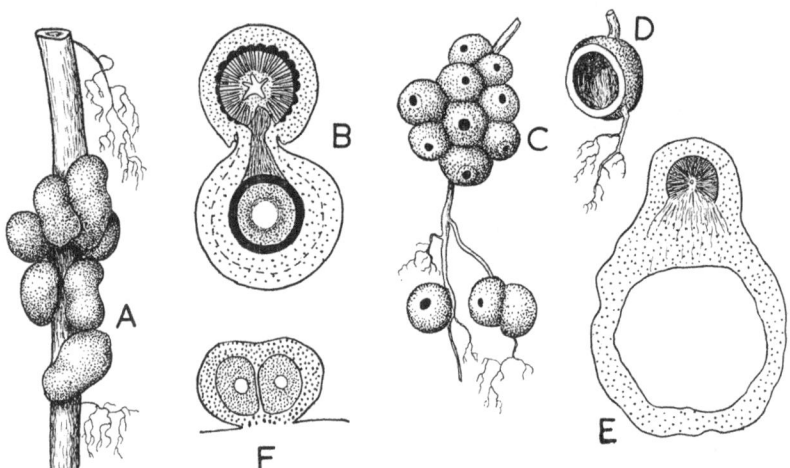

Fig. 33. Root galls of heterogonic Cynipids. A. *Biorrhiza pallida* (Ol.) unisexual generation. B–C. Sections through the same. D–E. Gall of *Pediaspis aceris* (Gmelin) on *Acer pseudoplatanus* Linn.

FANI-PEREZ[1095]. Each gall is a fusiform swelling, with rugose surface, fleshy and about 10 mm long and 6 mm thick, with a large larval cavity. Some root galls are also caused by gall-midges [551]. STEFANI-PEREZ[1096] has described a fusiform midge gall, about 22 mm long and 7 mm thick on the root of *Phagnolon rupestre* DC (Compositae) from Tripoli. According to TROTTER[1175], the gall-midge *Stefaniella* produces galls on the roots of *Atriplex halinus* LINN. in Tripoli; this gall is about 20 mm long and 12 mm thick, with a conspicuous projection on one side. Reference may also be made to the curiously hairy and pyriform rindengall, caused by the gall-midge *Asphondylia strobilanthi* FELT on the roots of *Strobilanthes cernuus* from Java (fig. 32 B—C.)

As a typical example of the root gall of the second ecologic group we may mention the galls of *Biorrhiza pallida* OL. and *Andricus quercus-radicis* FAB. on *Quercus* from Europe. Globose swellings, which often become agglomerated to form multilocular masses, develop sclerenchyma elements round the larval cavities[75, 247]. Another cynipid *Pediaspis aceris* FÖRST. is also well known for its remarkable root galls on *Acer pseudoplatanus* LINN., arising as bunches of spherical swellings (*vide* Chapter IX) (fig. 33). Among other cynipid galls on roots, common in North America, mention may be made of the galls of *Callirhytis ellipsoidea* WELD on the small fibrous roots of *Quercus bicolor,* developing in clusters just below the ground level; the galls *Callirhytis elliptica* WELD on *Quercus alba,* consisting of thin-walled, unilocular ellipsoid swellings of the fibrous roots; the globose agglomerate masses on the larger side roots of *Quercus stellata* caused by *Odontocynips nebulosa* KIEFF. and the fleshy fig-shaped clusters of galls caused by *Belenocnema treatae* MAYR on *Quercus geminata* (*vide* also Chapter X).

4. Galls on aerial roots

An account of root galls, to be complete, should mention some of the galls on aerial roots also. RÜBSAAMEN[994] has given an account of the root galls on *Listrostachys bidens* ROLFE caused by an unknown Coleoptera from Cameroun. According to DOCTERS VAN LEEUWEN[305], an unknown Lepidoptera causes galls on the aerial roots of *Vanda tricolor* LINDL. from Java. An undescribed gall-midge gives rise to fusiform galls, about 20 mm long and 5 mm thick, on the aerial roots of the Brazilian *Philodendron* sp.[533]. *Parellelodiplosis cattleyae* MOLL. causes globose or subglobose galls on the roots of *Cattleya* and *Laelia* from South America[533]. A number of interesting galls, caused by Chalcids, have also been described on the aerial roots of *Ficus pilosa* REINW. (fig. 32 A). The galls arise typically as lateral, unilocular swellings, but frequently becoming agglomerated[297].

CHAPTER V

SHOOT AXIS GALLS

1. General characters of galls on shoot axis

Structurally and ecologically, the galls on the shoot axis embrace a greater diversity and complexity of abnormal growths than on the root. Some of the essential characters of these galls may be traced to the fact that in the shoot axis we find both growth in length and growth in thickness. The development and structure of galls on the shoot axis are therefore profoundly influenced by their relation to the primary meristem and the cambial cylinder. Broadly speaking, two types of galls arise on the shoot axis, viz. the acrocecidia or the galls on the growing vegetative tip and the pleurocecidia or the galls that develop at a distance from the growing tip. The galls on the shoot axis are generally solid or hollow swellings, caused by

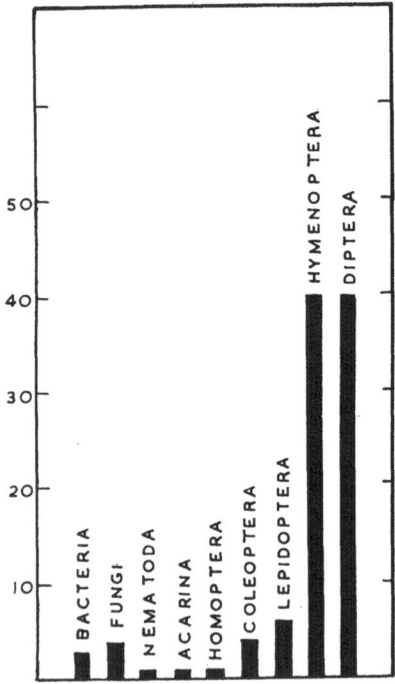

Fig. 34. Frequency of galls on shoot axis caused by different organisms from the world.

Bacteria, Fungi, some Nematodes, a few Acarina, Homoptera, Coleoptera, Lepidoptera, Hymenoptera and Diptera. No Thysanoptera are known to give rise to galls on the shoot axis. The relative abundance of galls caused by different organisms on the shoot axis is shown in fig. 34. In the simplest of the galls, the cecidozoa remain externally on the epidermis of the shoot. The detailed anatomy and histology of different galls on the shoot axis have been described by a number of workers, like HOUARD[485-489], TROTTER[1175], STEWART[1105], DOCTERS VAN LEEUWEN[298], MEYER[777], VERRIER[1208] and others.

2. Acrocecidia

The fundamental characters of acrocecidia have been beautifully illustrated by HOUARD[487, 489], with particular reference to several typical acrocecidia like those of *Eriophyes geranii* CAN. on *Geranium sanguineum*, of *Jaapiella* (= *Perrisia*) *genisticola* (F. LÖW) on *Genista tinctoria*, of *Taxomyia* (= *Oligotrophus*) *taxi* (INCHB.) on *Taxus baccata*, of *Isthmosoma graminicola* (GIRAUD) on *Agropyrum repens*, etc. In all these galls, there is typically an arrest of the normal development and growth of the growing tip, stunting of the stem, swelling of the growing tip, increase of cortex and medulla, increase of vascular elements which also become dissociated to a more or less conspicuous extent, arrest of differentiation of secondary tissue, absence of periderm, etc. The pronounced swelling of the growing tip and the consequent suppression of internodes generally result in a characteristic bunching of irregularly formed, stunted and malformed leafy outgrowths on the galls. This is, for example, the case with the gall on the growing point of *Sabicea venosa* BENTH. (Rubiaceae), caused by the larva of an unknown Lepidoptera from equatorial Africa[525]. An elongate-oval hollow swelling, about 20 mm long and 8 mm thick, is covered by narrow, stunted leaves (fig. 51 B). The gall of an unknown midge on the growing tip of *Strobilanthes involucratus* BLUME is a flattened swelling, covered by atrophied leaves and numerous larval cavities. The gall of *Euribia cardui* (LINN.) (Trypetidae) on *Cirsium arvense* (LINN.), from Europe, is a spherical or oval, multi-chambered swelling, about 15 mm long and surmounted by a tuft of stunted leaves and extending below to several internodes (fig. 49 B—C). The gall of *Lipara lucens* MEIG. on *Phragmites communis* TRIN., from Europe, is an acrocecidium, consisting of a tubular swelling, with crowded bunch of stunted leafy outgrowths on the surface (fig. 35). Bunched leafy outgrowths are found on the acrocecidia caused by *Oedaspis trotteriana* BEZZI on *Artemisia herba-alba* ASSO. and by *Asphondylia punica* MARCH. on *Atriplex halinus* LINN.

Some acrocecidia lack a dense outgrowth of bunched leafy processes, but may be covered by isolated stunted leaves. In the galls caused by *Dactylethra candida* (STAINT.) (Lepidoptera) on the growing point of *Tephrosia* spp. (Papilionaceae) from India and

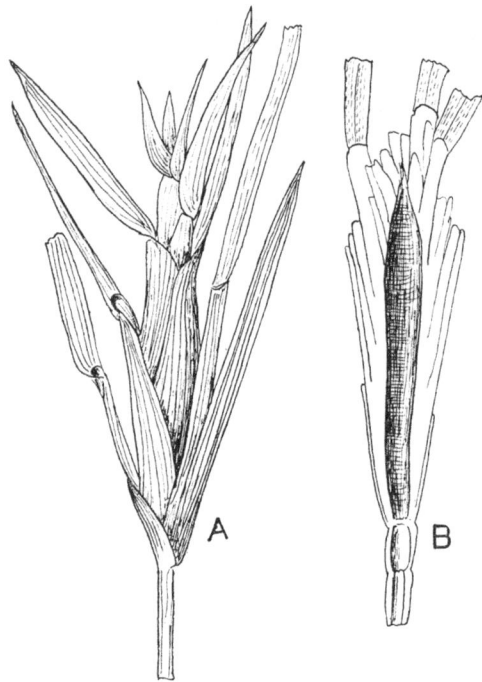

Fig. 35. Acrocecidia on Monocot.   A. Gall of *Lipara lucens* Meig. on the growing tip of *Phragmites communis* Trin., showing the characteristically stunted internodes and bunched leaves.   B. L.S. through the same, showing the spacious gall cavity.

parts of Africa, pyriform or flask-shaped, thick-walled, hollow swellings are surmounted by a few isolated and stunted leaves (plate I, 6). On the sides of the gall also we sometimes find dwarfed leaves, but the gall as a whole is not clothed densely with a bunch of leaf processes. This is also true of the gall of *Aylax* sp. on the growing tip of *Centaurea seridies maritima* LANGE from Algeria.

Other acrocecidia are naked. The gall of *Andricus championi* ASHM. on the growing tip of the South American *Quercus* is a spongy mass, about 75—100 mm long and 50—65 mm thick, with the surface mameleonated. The acrocecidia caused by *Amblardiella tamaricum* KIEFF. and *Amblylapis olivierella* RAGONOT on *Tamarix* spp. are also naked swellings. A chalcid *Trichilogaster* sp. produces a remarkable acrocecidium on *Acacia leucophloea* WILLD. from

India. Subglobose, hard, solid swellings of the growing tip have a flat, cup-shaped depression on the top. In this depressed flat area are situated 20—100 hemispherical, tubercular projections, representing the lids to the larval cavities immediately below, imbedded within the thin cortex. The bulk of the gall consists of the central truncated woody core (fig. 36). A remarkable acrocecidium, caused by an unknown Chalcid on *Prosopis* spp. from India, is a globose, solid, hard, often agglomerated mass of wood, about 25—75 mm in diameter (fig. 37). Over 200—500 peripheral and oval larval cavities are imbedded immediately beneath the bark, which is smooth and moderately thick. An unknown gall-midge gives rise to a most remarkable acrocecidium on *Terminalia paniculata* ROTH. from India. The gall comprises an agglomerated mass of over fifty hard, pyriform, woody, solid swellings, each measuring about 10 mm long and 7 mm thick basally, but the whole often attaining a diameter of 35 mm (Plate I, 4). The individual swellings

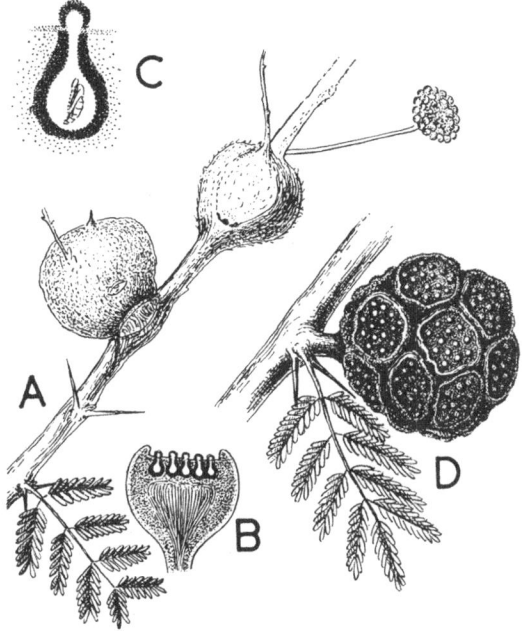

Fig. 36. Gall of the chalcid *Trichilogaster* on the growing tip of the shoot axis of *Acacia leucophloea* Willd. A. The gall in situ on branch. B. L.S. through a single gall, showing the arrested growing point and the numerous larval cavities arranged within a flattened, cup-like depression, peripherally in the cortex. C. A single larval cavity, enlarged to show the nutritive zone and the hemispherical lid of callus tissue. D. Agglomerated mass of several galls, with facetted flattened surfaces, containing the larval cavities.

of the agglomerated mass have a cylindrical, solid piston-like, dark brown callus tissue, projecting from a circular hole on the summit and leading into the axial larval cavities.

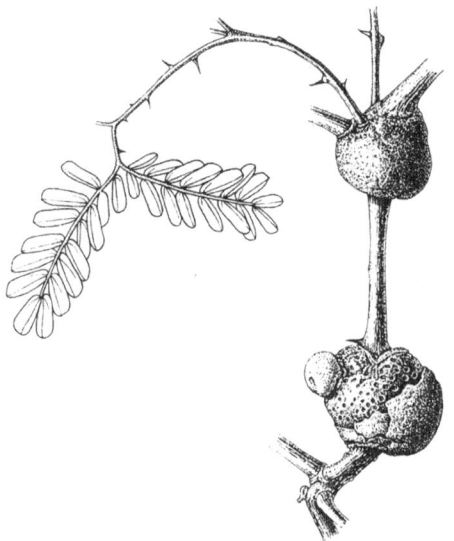

Fig. 37. Gall of an unknown Chalcid on the shoot axis of *Prosopis juliflora* Linn., with a part of the bark peeled off to show the numerous peripherally disposed larval cavities in one gall

In all the acrocecidia considered so far, the cecidozoa are situated within the gall cavity. There are, however, some remarkable acrocecidia, in which the cecidozoa are found permanently externally on the surface. *Eriophyes cernuus* MASSEE occurs, for example, externally, between the naked tuberculated parenchyma cells, on the surface of its gall on the growing tip of *Zizyphus jujuba* LAMK. from India and North Africa (fig. 38).

3. Pleurocecidia

In comparison to acrocecidia, pleurocecidia on the shoot axis are greatly diversified and have a highly complex structure and mode of development. The developmental and structural peculiarity of pleurocecidia depends largely on the positioning of the cecidozoa with reference to the vascular bundle cylinder and the number of larvae of the cecidozoa. As shown by HOUARD[485], the positioning of the cecidozoa may be external on the epidermis of the stem, within the cortex, in the region of the vascular bundles and within the medula. In accordance with these differences, HOUARD *(op. cit.)* recognizes four types of pleurocecidia on the stem: 1. cecidii caulinare laterale, with the larva of the cecidozoa situated externally on the epidermis, as in the gall of *Asterolecanium massalongianum* TARG.-TOZ. on *Hedera helix* LINN. (fig. 39 C), galls of coccids on *Potentilla hirta pedata* WILLD., the gall of *Perrisia fraxini* KIEFF. on *Fraxinus excelsior* LINN. and the gall of *Chermes abietis* LINN. on *Picea excelsa* (LAMK.) LINK. 2. cecidii caulinare laterale, with the larva of the cecidozoa situated within the cortex as in the gall of *Eriophyes pini* NAL. on *Pinus silvestris* LINN. 3. cecidii caulinare laterale, with the larva of the cecidozoa situated inside the formative zone of the secondary xylem-phloem tissue as in the galls of *Contarinia tiliarum* KIEFF. on *Tilia silvestris* DESF., *Harmandia petioli* KIEFF. on *Populus tremula* LINN., *Rhabdophaga salicis* SCHRANK on *Salix caprea* LINN., *Contarinia scoparii* RÜBS. on *Sarothamnus scoparius* KOCH, *Plagiotrochus fusifex* MAYR on *Quercus coccifera* LINN., *Lasioptera rubi* HAEGER on *Rubus fruticosa* LINN., *Ceutorrhynchus pleurostigma* MARSH. on *Brassica oleracea* LINN., *Aylax latreillei* KIEFF. on *Glechoma hederacea* LINN., *Agro-*

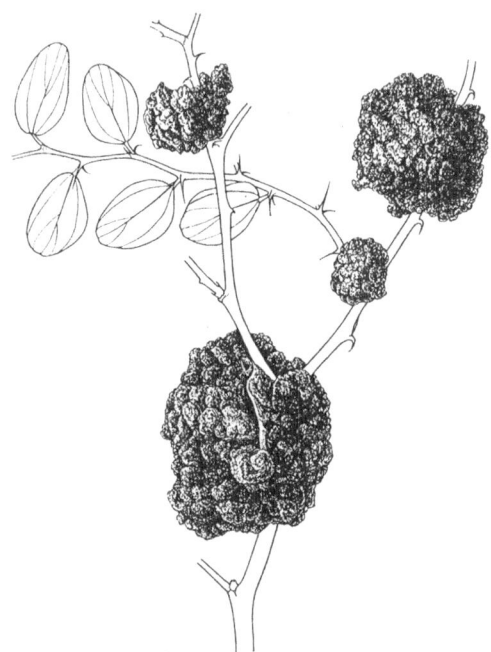

Fig. 38. The gall of *Eriophyes cernuus* Massee on the growing point of the branches of *Zizyphus jujuba* Lamarck.

72

myza kiefferi TAV. on *Cytisus albus* LAMK., *Agromyza pulicaria* MEIG. on *Sarothamnus scoparius* KOCH and *Andricus sieboldi* HTG. on *Quercus pedunculata* EHR. 4. cecidii caulinare, with the larva of

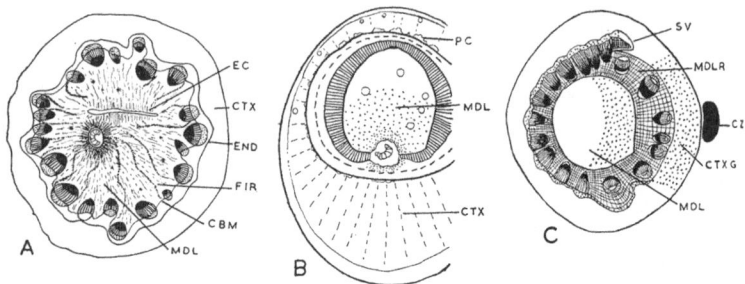

Fig. 39. Relation between the structure of galls on the shoot axis and the positioning of the cecidozoa. A. T.S. through the gall of *Aylax hieracii* (Bouché) on the stem of *Hieracium umbellatum* Linn. with the cecidozoa positioned in the medulla. B. Cecidozoa positioned in the zone of the vascular bundles. C. T.S. through the gall of *Asterolecanium massalongianum* Targ.-Toz. on the stem of *Hedera helix* Linn., with the cecidozoa positioned externally on the epidermis. CBM cambium. CTX cortex. CTXG galled part of cortex. CZ cecidozoa. EC original egg cavity. END endoderm. FIR faisceaux d'irrigation connecting the larval cavity with the vascular bundles. MDL medulla. MDLR medullary rays. PC pericycle. (After Houard).

the cecidozoa situated within the medulla, as in the galls of *Ceutorrhynchus atomus* BOH. on *Sisymbrium (Arabis) thalianum* GAY, *Xestophanes potentillae* VILL. on *Potentilla reptans* LINN., *Aylax hieracii* BOUCHÉ on *Hieracium umbellatum* LINN. (fig. 39 A), *Aylax hypochoeridis* KIEFF. on *Hypochoeris radicata* LINN., *Stefaniella trinacriae* (STEF.) and *Coleophora stefanii* JOANNIS on *Atriplex halinus* LINN., *Lasioptera eryngii* VALLOT on *Eryngium campestre* LINN., *Lasioptera carophila* F. LÖW on *Torilis anthriscus* GMELIN, *Nanophyes telephii* BEDEL on *Sedum telephium* LINN., *Apion scutellare* KIRBY on *Ulex europaeus* LINN., *Mompha decorella* STEPH. on *Epilobium montana* LINN. and *Epilobium tetragonum* LINN., *Gypsonoma aceriana* DUP. on *Populus alba* LINN. and *Evertia resinella* (LINN.) on *Pinus silvestris* LINN.

Galls with cecidozoa external

In this type of galls, the cecidozoa remain permanently exposed on the surface and never come to be enclosed within the plant tissue in the course of the growth of the gall. In the gall of the coccid *Asterolecanium variolosum* RATZB. on *Quercus robur* LINN. the coccid remains on the surface on one side and thrusts its piercing stylets into the stem. The gall is a unilateral swelling of the cortex, with cell proliferation localized on the side in which the coccid is fixed. The periderm shows an exaggerated development on this side. The sucking action of the coccid produces a general arrest of tissue differentiation, but there is an acceleration of ligninization. In the fusiform lateral gall of *Asterolecanium algeriense* (NEWST.) CKLL. on *Templetonia retusa* R. BR., we find an exaggerated development of the liber-ligneous elements in two characteristic longitudinal rows, on either side of the coccid. The vascular bundles immediately underneath the coccid are arrested in development, so that a depression arises and the coccid lies within the concavity. There is an abnormal development of periderm in the vicinity of the coccid in the gall of *Asterolecanium thesii* (DOUGLAS) CKLL. on *Pittosporum tobira* AITON. The cortex is hypertrophied and some dissociation of the vascular bundles may also be observed. In all these galls, the localization of cell proliferation below the coccid has the result of causing a pronounced asymmetry of the galled portion. On the distant side, away from the coccid, the vascular bundles are normally developed in a regular ring in cross section, but only somewhat hypertrophied. The dissociation of the bundles on the side of the coccid disturbs the ring arrangement and the vessels are arranged on this side in an arc, the alignment of which is not related to the ring of bundles on the opposite side. The cells of the cortex below the coccid are tangentially elongated and parenchyma cells are secondarily sclerotized. In some of these galls, although the cecidozoa may be permanently situated externally on the surface of the stem, their piercing stylets may penetrate relatively deep into the plant tissue and even reach down to the cambial zone. In such cases there is a great deal of similarity to the condition in which the cecidozoa are situated within the medullary rays in the zone of the vascular bundles. This is, for example, what we find in the gall of *Eriosoma lanigerum* (HAUSM.) on *Pirus malus* LINN. In place of normal wood elements, there arises a mass of thin-walled, elongate cells and the intensive cell proliferation in the deeper tissue results in the bulging out of the cortex, which also cracks and fissures irregularly. Mainly because of their irregular growth, these galls have been called "Krebsgallen" in German literature.

Galls with cecidozoa in the cortex

The positioning of the cecidozoa within the cortex of the stem results in the larval cavity being the centre of symmetry. Cell proliferation is naturally confined to the cortex, but a part of the vascular bundle ring near the larval cavity becomes also hypertrophied, due also to increase in the number of both xylem and phloem elements. The side opposite to the larval cavity is almost normally developed. In most cases, however, the entire gall tissue is derived from the cortex and the central stele is pratically unaffected. This condition is generally recognized as rindengall. Rindengalls are among the most diversified and complex of pleurocecidia on shoot axis.

74

The rindengalls caused by *Eriophyes pini* NAL. on *Pinus silvestris* LINN. and *Pinus montana* LINN. are unilateral, solitary or agglomerated subglobose swellings. The gall is covered by normal epider-

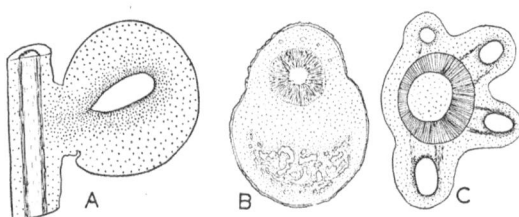

Fig. 40. Sections through typical rindengalls. A. L.S. through the stem and the rindengall of *Acroëctasis campanulata* Mani on *Sabia campanulata* Wall., to show the typical growth from superficial cortical layers and the zone of actively dividing cells surrounding the larval cavity. B. T.S. through the rindengall of *Eriophyes pini* Nal. on *Pinus silvestris* Linn., showing the completely enclosed subperipheral spaces in which the mites occur, with actively dividing cells concentrated in this zone. C. Multilocular rindengall of a gall-midge showing hypertrophy of vascular bundles ring on the side of the larval cavities and the faisceaux d'irrigation.

mis at first, but as it grows larger, the epidermis cracks off. In the young gall, the peripheral layers of cells of the cortex, which are under the influence of the mites on the surface of the plant, become enormously hyperthrophied and also soon undergo repeated divisions irregularly in all directions. In due course considerable intercellular spaces arise and thus give rise to a spongy swelling. Although an enormous swelling has developed, it may be observed that the vascular bundle, cambium, medullary rays and medulla are completely normal (fig. 40 B). Remarkably enough, the mites now occur in the intercellular spaces within the gall tissue. A number of typical rindengalls, often attaining large sizes, are caused by fungi. The well known Weisstannenkrebs on *Abies alba* MILL. is for example a rindengall caused by the ascidial generation of *Melampsorella caryophyllacearum* SCHRÖT. Another similar gall is found on *Juniperus* spp., caused by the teleutospore generation of *Gymnosporangium* spp., at least in the early stages of development. A subglobose, unilateral, hard, solid rindengall, about 25 mm thick, is caused by an unknown fungus on *Berberis lycium* ROYLE from India. The gall is composed of closely packed parenchyma cells, derived entirely from the sub-epidermal cortex (fig. 143).

Among the other rindengalls, mention should be made of the gall on *Combretum* sp., caused by an unknown gall-midge from West Africa (fig. 42 D—E)[505]. Subspherical, sessile swellings, about 10 mm in diameter, develop on branches and have nipple-like projections on the summit; these projections represent the lids closing the axial, cylindrical larval cavity. Through the localized swelling from the cortex, the gall is larger than even the thickness of the stem on which it arises, but the stele is completely unaffected. A remarkable rindengall is caused by the midge *Acroëctasis campanulata* MANI on the branches of *Sabia campanulata* WALL. from India[745]. This gall consists of regular, spherical, smooth, peppercorn-like, solid, fleshy swellings, about 5 mm in diameter, developing in truly enormous numbers all around the slender branches for several metres length, curiously giving the appearance of a gigantic pepper-corn spike (fig. 40 A). The gall is composed of simple parenchyma cells, with a distinct nutritive zone around the larval cavity, all derived from the sub-epidermal layer of cortex. Another Indian gall-midge *Daphnephila glandifex* KIEFF. gives rise to clusters of 5—20 subcylindrical, fleshy, sessile rindengalls on *Machilus gamblei* KING, each gall being about 25 mm long and 10 mm thick, with narrow, cylindrical axial larval cavity.

We have so far dealt with rindengalls that arise as sessile outgrowths, but a number of curiously pedunculated rindengalls are also known. HOUARD[530] has described, for example, a pyriform, pedunculated rindengall, about 20—25 mm thick and the stalk about 20 mm long, narrowed conspicuously both at the base and apex, on the branches of *Quercus* from Mexico. The gall is usually bunched together in clusters and has a central, spherical or ellipsoid larval cavity, about 8 mm long. The galls develop completely from the peripheral cortical cells of the branch.

Unlike the rindengalls mentioned above, with cell proliferation restricted to the subepidermal layers of cortex, a number of other

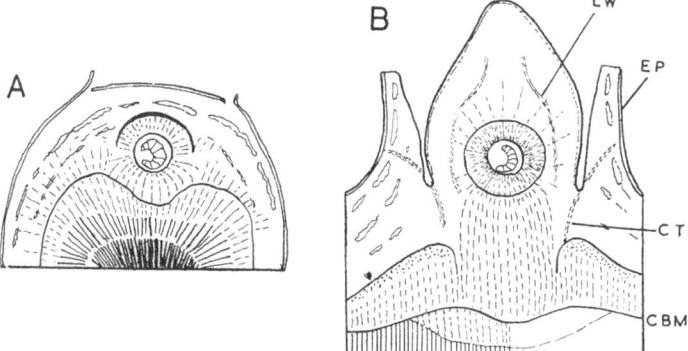

Fig. 41. Erupting rindengall of *Andricus testaceipes testaceipes* Htg., unisexual generation, on the stem of *Quercus pedunculata* Ehrh. A. Early stage of development of the gall, showing the localized cell proliferation in cortex and partial rupture of epidermis. B. Later stage of development of the gall, which has burst through the bark and erupted on the surface of the branch. CBM cambium. CT cicatrized tissue. LW liber-wood vessels. EP epidermis. The larval cavity is fully surrounded by nutritive tissue and connected by faisceaux d'irrigation with the central stele. (After Houard).

rindengalls develop from cell proliferation from the deeper layers of cortex. They form at first wholly within the epidermis of the stem and the epidermis does not take an active part in the gall develop-

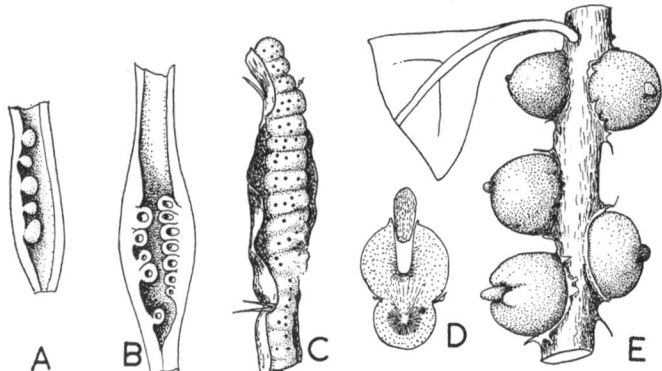

Fig. 42. Some remarkable rindengalls. A. The erupting rindengall of *Timaspis lusitanicus* Tav. on the stem of *Crepis taraxifolia* Th., growing from the inner cortex into the medullary space; the stem is in this region slightly swollen. B. The same with the galls cut open. C. Erupting rindengall of *Eschastocerus myriadeus* Kieff. & Jörg. on the stem of *Prosopis alpataco* Phil. The gall develops within the cortex, the epidermis cracks open due to excessive stretching, exposing the mature galls and thus facilitating the escape of the adult gall-midge. D–E. Rindengall, a pyxidiocecidium, caused by an unknown gall-midge on *Combretum* sp. The gall erupts through the epidermis. The larval cavity is an axial cylindrical space, plugged by a fusiform, solid stopper-like lid that becomes loosened and falls off. The section showing the nutritive zone, faisceaux d'irrigation and the loosened plug.

ment. As the development of the gall progresses, the sub-epidermal layer of cortex and the epidermis become greatly stretched and finally the gall breaks through these layers. The epidermis cracks off and peels away and we then have the so-called erupting galls. The simplest of these galls is found on *Holboelia latifolia* WALL. from India and is caused by an unknown midge. The gall of *Andricus testaceipes testaceipes* HTG.♀♀ (= *sieboldi* HTG.) on *Quercus* develops initially wholly within the deep cortex and finally breaks through as pyriform hollow swellings, pushing up the sub-epidermal cortex and epidermis like a cap (fig. 41). HOUARD[530] has described an interesting erupting rindengall on a Mexican species of *Quercus*. Spherical swellings, about 5 mm in diameter, arise under the epidermis and on maturing, appear through a fissure in the superficial layer of cortex and epidermis. A large series of such closely crowded galls arise in a linear row under the epidermis, through which they eventually break (fig. 42). The gall of the female coccid *Brachyscelis fletcheri* OLIFF. on *Eucalyptus* from Australia, is a large spherical swelling, 50—75 mm in diameter. The surface of the stem is irregularly crevassed, through which the galls are visible. The hemispherical gall of *Eschatocerus myriadeus* KIEFF. & JÖRG. from South America, develops in groups of 100—200 mm length under the outer bark of *Prosopis alpataco* PHIL. The epidermis breaks off under the pressure of the developing galls within. An unknown gall-midge causes on *Indigofera dosua* HAM. from the Himalaya an erupting rindengall, consisting of pyriform or oval hollow hard swellings, about 5 mm long and 3 mm thick, developing closely crowded together in large numbers along considerable lengths of the tender stem and initially strictly within the deep

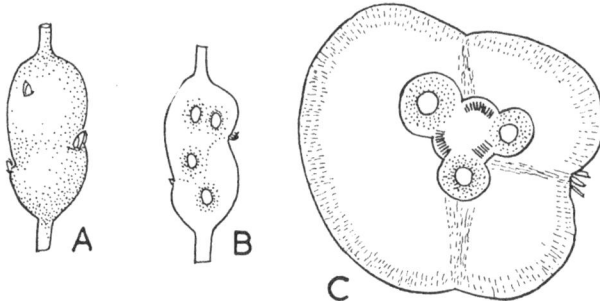

Fig. 43. Gall of *Plagiotrochus fusifex* Mayr on the stem of *Quercus coccifera* Linn. A. General appearance of the gall. B. Section through the gall. C. T.S. through the gall, showing the larval cavities situated within the zone of vascular bundles and surrounded by nutritive zone. (After Houard).

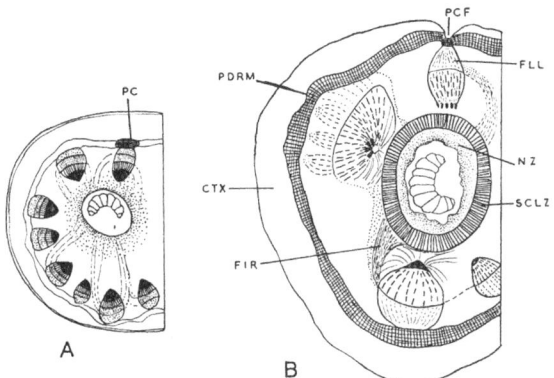

Fig. 44. Structure of the gall of *Xestophanes potentillae* Vill. on the stem of *Potentilla reptans* Linn. A. T.S. through young gall. B. T.S. through mature gall. CTX Cortex. FIR faisceaux d'irrigation. FLL liber vessels. NZ nutritive zone. PC pericycle. PCF pericycle fibres. PDRM periderm. SCLZ sclerenchyma zone. The larval cavity is situated in the medulla and is connected by the faisceaux d'irrigation to the vascular bundles. (After Houard).

cortex. The affected part of the stem is secondarily obscurely swollen, but as the galls mature, the epidermis and the peripheral cortical layer become ruptured longitudinally to expose the galls from within. The most remarkable of the erupting rindengalls is undoubtedly found on *Indigofera pulchella* ROXB. also from India. This midge gall comprises solid, parenchyma swellings, densely clothed with elongate, conspicuously red coloured, multicellular, straight, simple hair-like emergences. The gall and the emergences on its surface both develop completely from the deeper cortical cells and are wholly concealed under the epidermis at first (plate III, 6). When mature, the epidermis breaks loose and falls off in pieces, exposing the brilliant red hairy galls. It must be emphasized that the hairy outgrowths on the gall are not developed from the epidermis at all, but arise from the inner cortex.

Galls with cecidozoa in the zone of vascular bundles

In most of these galls the vascular bundle cylinder as a whole is much larger than in the normal stem. On the side of the larval cavity of the cecidozoa, the annular arrangement of the vascular bundles is much altered, particularly the bundles are greatly separated, due to hypertrophy of the medullary rays. The vascular

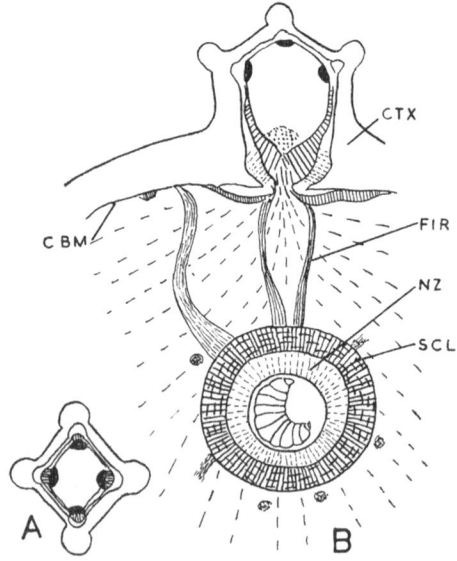

Fig. 45. Structure of the gall of *Aylax latreillei* Kieff. on the stem of *Glechoma hederacea* Linn. A. T.S. through normal stem. B. T.S. through galled part of the stem. CBM cambium. CTX cortex. FIR faisceaux d'irrigation. NZ nutritive zone. The larval cavity is situated within the cortex in one of the four rounded angles. (After Houard).

bundles are also often surrounded by lignified cells. The larval cavity is also lined by two zones of lignified cells; an inner zone is situated in the medulla and the outer zone in the cortex. The gall of *Aylax latreillei* KIEFF. on the rectangular stem of *Glechoma hederacea* LINN., studied by HOUARD *(op. cit.)* has many interesting features. One of the four rounded-ends of the rectangular stem (in cross section) is enormously swollen to form the gall, so that the remaining three other groups of vascular bundles in each of the other three rounded angles remain normal. The vascular bundles in the galled part are completely altered, with considerable hypertrophy of xylem. There is a double zone of continuous vascular bundle ring, united by irregular strands of faisceaux d'irrigation outside the larval cavity. The larval cavity is itself surrounded by a zone of nutritive cells and a protective zone of sclerenchyma cells (fig. 45).

Galls with cecidozoa in the medulla

The cell proliferation being predominantly localized in the medulla, these galls often attain considerable complexity. The medullary rays nearest the larval cavity show predominant cell proliferation, so that the vascular bundles become widely dissociated. The larval cell is surrounded by nutritive and sclerenchyma zones

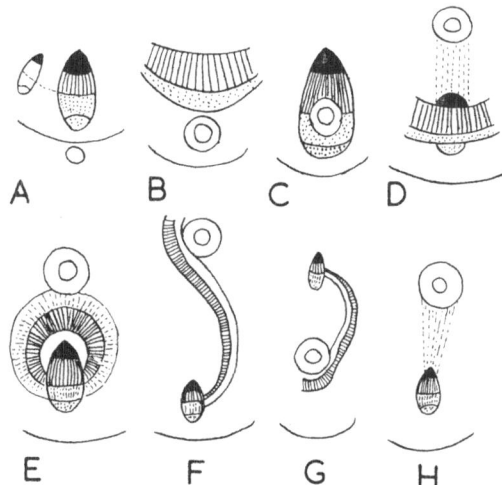

Fig. 46. Vascularization of the larval cavity in stem galls. A. Cecidozoa situated externally on the epidermis, with hypertrophy of vascular bundles. B. Cecidozoa positioned within the cortex, with hypertrophy of the vascular bundles on the side. C. Cecidozoa positioned within the zone of vascular bundles. D. Cecidozoa positioned in the medulla, with the nutritive zone connected to the vascular bundle by liber vessels. E. Cecidozoa deep within the cortex, in contact with the stele. F–H. Larval cavity connected by faisceaux d'irrigation to the vascular bundles. (After Houard).

and is connected by faisceaux d'irrigation with the hypertrophied vascular bundles. If within the medulla the larval cavity is situated in the middle, the hyperplastic nutritive zone is connected with the

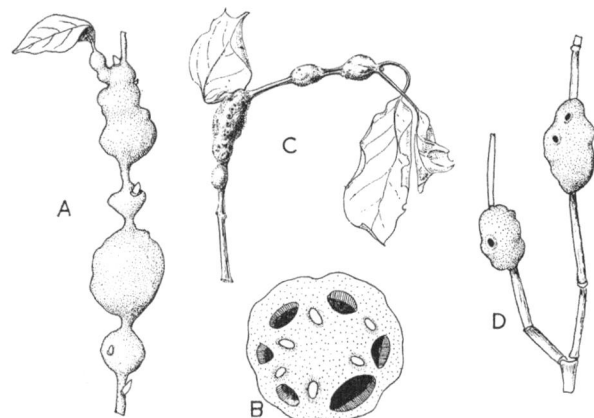

Fig. 47. Some typical stem galls. A. The gall of *Rhabdophaga salicis* Schrk. on *Salix*. B. T.S. through the same. C. The gall of *Thomasiniana salvadorae* Rao on *Salvadora oleoides* Dine. D. Midge gall on *Ephedra nephrodensis procera* Stapf.

vascular bundles through the liber elements. If, however, the larval cell is situated near the vascular bundles, the medullary rays undergo cell proliferation and the so-called faisceaux d'irrigation connect the larval cavity with the liber elements of the vascular bundles (fig. 42—45).

Galls with complex and irregular positioning of the gall-maker.

In a number of galls, the larval cavities are not situated wholly within a definite zone, but extend often through cortex and medulla. This has the effect of breaking up the stele irregularly and galls being more extensive than in other cases. This is, for example, the case with the galls of a number of midges like *Lasioptera* MEIG., *Neolasioptera* FELT, etc. on the stem of *Melothria, Bryonia, Momordica, Jasminum, Impatiens, Achyranthes, Galium*, etc. The same condition is also met with in the midge galls on *Salvadora* and *Dobera, Rhabdophaga oleiperda* DEL GUER. on *Olea chrysophylla*. The gall of the fungus *Uromycladium* on the shoot axis of *Acacia* spp. (Plate I, 5) and the gall of another fungus *Sphaeropsis tumefaciens* HEDG. on *Citrus medica acida* LINN. involve cell proliferation in cortex and medulla.

CHAPTER VI

BUD GALLS

1. General characters of bud galls

Bud galls are essentially acrocecidia, characterized by the localization of cell proliferation and other cecidogenetic alterations in the
rudiments of leaves, branches, etc. of the bud. The growing point
immediately below the bud may, however, be more or less enlarged
and flattened from above.

Bud galls range from simple enlargements of buds or the so-called
"big-buds" to solid or hollow, fleshy or hard and complex swellings,
with fleshy, spiny, filiform, imbricately scaly, foliaceous or cottony
surface processes, representing the bud leaves, or smooth and nut-
like, without trace of such rudiments. The size of bud galls varies
from a few millimetres to sometimes 150 mm. Most bud galls are
globose or pyriform, but some are also flask-shaped, clavate and
have other singular shapes. ZORIN[1290] has described a bud gall,
which greatly resembles a fruit superficially.

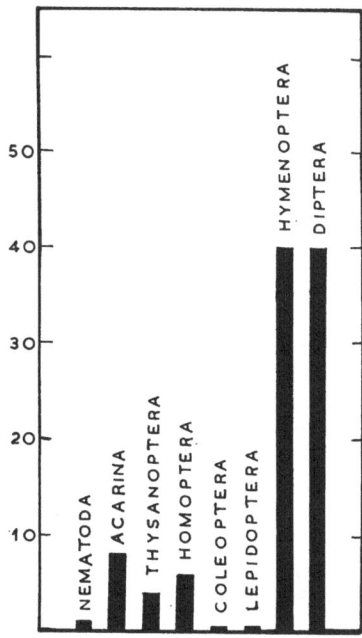

Fig. 48. Percentage frequency of bud galls from the world.

The outstanding features of the ecology of bud galls may be summarized as follows: Bud galls develop predominantly during spring or summer only and the greatest majority of them have also relatively short life-cycles. The relative abundance of bud galls caused by different groups of cecidozoa is shown in fig. 48. Some bud galls are caused by cecidozoa alternately with galls on other organs in cyclic heterogenesis (fig. 119, 121). The association of a larger complex of cecidozoa with bud galls than with galls on shoot axis in general is closely correlated with the fact that many species, which do not occur within the plant tissue, find, however, adequate shelter inside the bud. The difference is also partly attributable to the high concentration of nutritive material and intense protein synthesis in the unfolding bud, so that optimal conditions for the nutrition and development of a wider complex of cecidozoa may be expected. The proximity to the primary meristem should also explain, at least in part, the relative simplicity of bud galls; the intensity of cecido-genetic action seems largely to be counter-acted by the activity of the normal meristem. From the point of view of community stability, bud galls stand probably next to root galls and nearly in every case the communities are larger.

The following is a brief synopsis of the principal structural types of bud galls:

A. Bud galls in which the cecidozoa are not enclosed in a special gall cavity, but occur externally, often more or less exposed.

1. Simple enlargement of bud.

2. Cabbage-like enlargement of bud, with crumpling, crinkling and moderate swelling of the rudiments of leaves.

3. Swelling of bud, associated with more or less pronounced swelling of the growing point of the shoot axis below.

4. Fir-cone-like bud gall, with greatly enlarged, swollen and imbricating scaly processes.

5. Rosette galls, usually with cecidozoa exposed in part at least.

6. Swellings of only the basal portion of the leaf rudiments, but with the apices expanding more or less normally.

7. Fleshy swellings of the whole of the leaf rudiments of the bud, without fusion of the swollen parts.

8. Solid cauliflower-like, complex fleshy swellings, with lobed, convoluted fleshy masses and more or less fusion of the swollen parts.

B. Bud galls in which the cecidozoa are enclosed in a special gall cavity or they are within the plant tissue, at least never external and exposed.

1. Bud galls with central larval cavity; surface clothed with filiform, hairy, cottony, spiny or other fleshy emergences or foliaceous processes, representing the arrested rudiments of the component leaves of the bud.

2. Bud gall in which only the basal portion of the leaf rudiment is greatly swollen, the apices often forming imbricating scales on the surface of the gall.

3. Bud gall in which the whole of the leaf rudiments is swollen, but the swollen parts are not fused together.

4. Bladder-like or pouch-like and nut-like hollow swellings, without traces of rudimentary structures and surfaces processes.

5. Solid cauliflower-like fleshy swellings, with complex fleshy convoluted lobes, enclosing several larval cavities, with the swollen parts often also fused together.

2. Bud galls with external cecidozoa

The simplest bud gall, such as those on *Betula, Corylus, Ribes,* etc., caused by *Eriophyes* spp., is characterized by a general enlargement of the bud as a whole, without suppression of any of the rudiments and this enlargement occurs even before the healthy buds have opened in spring. This type of gall is essentially a case of gigantism of bud. Although the mites occur only between the altered bud rudiments, the development of the growing point of the shoot axis below is nearly always arrested. Scale leaves that remain small in the course of normal growth of the bud, become, however, enormously enlarged, swollen and green-coloured. While normal buds if prevented from unfolding, drop off, the bud gall persists on the shoot for a long time and is then readily distinguished from other healthy buds by its enormous size. In sections, we may recognize fleshy and parenchyma emergences, especially on the inner surface of the greatly enlarged and frequently swollen parts. The stunted parts are composed of thin-walled parenchyma cells and sometimes an epidermis may be differentiated, but there are no stomata and vascular bundles are only slightly differentiated (fig. 49).

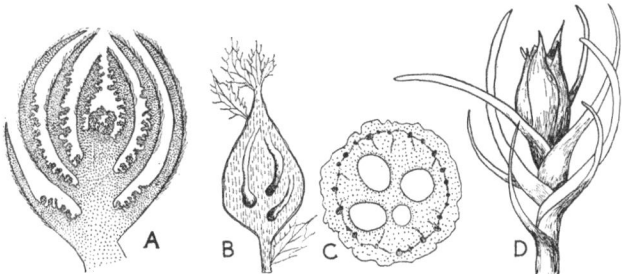

Fig. 49. Simple bud galls   A. L.S. through the bud gall of *Eriophyes* on *Corylus*, showing the irregular, fleshy emergences from the moderately swollen leaf rudiments. B. L.S. of the bud gall of *Euribia cardui* (Linn.) on the growing tip of *Cirsium arvense* Scop.   C. T.S. through the same.   D. Gall of *Tylenchus* on the growing tip of moss.

In the bud gall on *Syringa vulgaris* LINN., caused by *Eriophyes löwi* NAL. from Europe, the mites occur in large numbers inside the bud. With the advancing summer, the mites invade newer buds and

Fig. 50. Fir-cone-like bud gall of *Apsylla cistella* (Buckt.) on *Mangifera indica* Linn.

the fresh infestation results in the formation of a witches-broom-like cluster of bud galls, generally in the lower portion of the shoot. The mites over-winter in the galled bud. A curiously fir-cone-like gall is caused by *Apsylla cistella* (BUCKTON) on the bud of *Mangifera indica* LINN. from India, superficially resembling the imbricately scaled normal bud of *Rhododendron* spp. (fig. 50). Oval or pyriform swellings, about 20 mm long and 15 mm thick at base, are composed of green, thick, coriaceous, imbricating scaly outgrowths, arising from a flattened top of the growing point below the bud. More or less similar, fusiform or oval, often curiously beaked or rosette-shaped bud galls on *Salix* spp. are caused by *Rhabdophaga rigida* O.S., *Rhabdophaga coloradensis* FELT, *Rhabdophaga brassicoides* WALSH and other species from North America. Rosette gall is caused by *Rhabdophaga rosacea* FELT on the North American *Rosa* spp. and often attains a size of 30 mm. The rosette gall on the terminal bud of *Sageretia oppositifolia* BRONGN. (Rhamnaceae) from the Himalaya by an unknown gall-midge, is a globose, multi-chambered swelling, the leaf rudiments altered to irregularly crumpled, abnormally shaped, greenish-yellow fascicles of foliaceous processes, with fleshy emergences basally. A remarkably cabbage-shaped gall on the bud of *Acacia leucophloea* WILLD. from India is caused by *Thilakothrips babuli* RAMAKR. This gall is interesting for the enormously enlarged, mis-shaped, moderately coriaceously thickened, curled, rolled, crumpled leafy processes, in place of the normally developing minute leaflets of the bicompound leaf from the bud. The foliaceous processes do not show the typical anatomy of a normal leaf (fig. 51 C). The gall generally measures 25—30 mm in diameter. Large numbers of the gall thrips occur between the tightly folded foliaceous processes.

In the bud galls, which we have so far considered, the general appearance and character of a bud are not wholly obliterated. The predominant abnormality is due to excessive surface growth of the

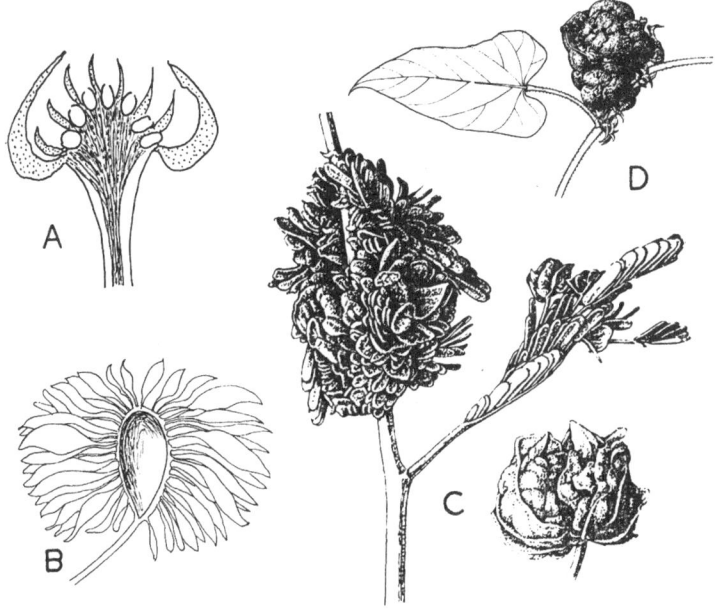

Fig. 51. Bud galls. A. Bud gall, with numerous larval cavities, surrounded by sclerenchyma cells, on the growing point, in between the bases of the stunted and swollen leaf rudiments. B. L.S. through the gall on the growing point of *Sabicea venosa* Benth., covered by tufts of stunted foliaceous growths. C. Miniature cabbage-like bud gall of *Thilakothrips babuli* Ramakr. on *Acacia leucophloea* Willd. D. Gall of *Eriophyes* sp. on *Ipomea scindica* Stock., involving buds, whole inflorescence, etc. to form a solid fleshy mass.

rudiments of organs inside the bud. We shall now turn our attention to bud galls, in which growth in thickness predominates and the bud gall does not also have the general appearance of a gigantic bud. In these galls the cecidozoa are not exposed, but also do not occur within the plant tissue. The rudiments of leaves in the bud become enormously swollen and fleshy, but the swollen parts do not fuse together. The bud gall of *Crataeva religiosa* FÖRST. caused by the gall-midge *Aschistonyx crataevae* (MANI) from India is a typical example of this type. This gall comprises oval or pyriform, solid, fleshy, succulent, greenish-yellow or pale yellow, rugose swellings, about 10 mm in diameter, with tortuous passages containing numerous larvae. The individual rudiments of the trifoliate leaf of the bud, though enormously swollen, remain distinct and enclose in between a cavity into which the tortuous passages open. The entire mass consists of undifferentiated small-celled parenchyma, without a trace of differentiation. Even the midrib is greatly hypertrophied. When axillary buds are galled, large semi-circular, concave, fleshy outgrowths arise from the superficial cortex cells of the branch adjoining the bud and together with the swollen part of the bud make the gall.

In the gall of *Eriophyes populi* NAL. on the bud of the common European *Populus tremula* LINN. we find all the rudiments in the bud, including even the outer protective scales, swollen into irregularly lobed fleshy masses, bearing fleshy emergences. The mites occur in large numbers in the tortuous passages between the emergences and lobes. The gall as a whole has a curious superficial resemblance to a miniature cauliflower. Adventitious buds arise near the galled bud and become in turn galled and eventually all the closely crowded galls become agglomerated to a complex fleshy mass, often 10 cm in diameter. Similar cauliflower-like swellings of the buds of *Sarothamnus*, *Genista*, etc. have also been described from Europe. The most remarkable of these bud galls is caused by *Eriophyes hoheriae* LAMB on the Australian *Hoheria populnea* A. CUNN. (Plate II, 2). This is a solid, hard but fleshy, globose swelling, 15—20 mm in diameter, with complex fleshy lobes, large tubercles and convolutions, superficially resembling in general appearance a human brain, but with finely granulated surface, without a true epidermis, trace of leaf or other rudiments of the bud. The mites crowd in the intricate passages between the fleshy lobes. An eriophyid bud gall on the Brazilian *Inga fagifolia* WILLD. is a spherical, solid, verrucose swelling, about 35 mm thick and greatly resembling the bud gall on *Populus* caused by *Eriophyes populi* NAL. from Europe.

3. Bud galls with cecidozoa in a special larval cavity

This group of bud galls is structurally and ecologically more specialized than the bud galls, in which the cecidozoa are external. There is, however, considerable parallelism in these two groups. Fir-cone-like and rosette galls are met with, for example, among the members of this group also. The larval cavity may be single or numerous and may arise as a result of covering growth or may also form around the cecidozoa situated within the plant tissue from the commencement of gall development.

One of the simplest of the bud galls of this group is the well-known ananas gall caused by *Chermes (= Adelges) abietis* LINN. (fig. 52) and by *Cnaphalodes strobilobius* KALT. on *Picea excelsa* (LAMK.) LINK. As mentioned earlier, the part of the shoot axis immediately below the bud becomes swollen, the needles become expanded and swollen basally into peculiar disc-shaped structures, which thus enclose between them nearly spherical cavities. The aphid is found in these cavities. Further growth of the affected parts results in the fusion of the swollen and expanded bases of the needles. The larval cavity is formed by covering growth.

In other bud galls the larval cavity is differently formed. We have, for example, large numbers of bud galls caused by cynipids on *Quercus* spp., in some of which the general bud-like appearance is retained, but the swollen parts are usually clothed with tubercles, spines, rugosities and other appendages and the larval cavity is situated variously in the middle or bottom part. The gall of *Cynips tomentosa* TROTT. on *Quercus pubescens* WILLD. is a curious mitre-shaped swelling, 12—18 mm long and 15—18 mm thick at base, densely clothed with short hairs. A slight constriction near the tip separates

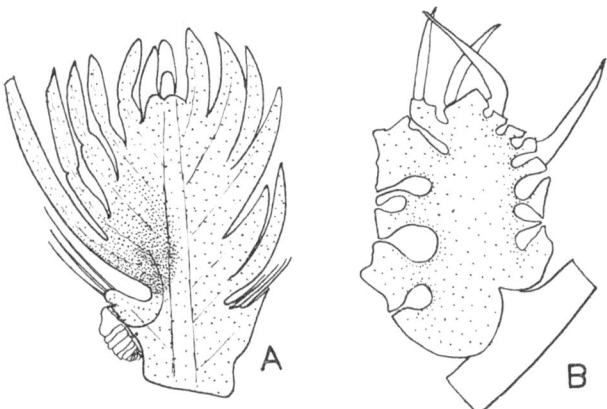

Fig. 52. The "ananas gall" of *Chermes abietis* Linn. on the bud of *Picea excelsa* (Lamk.) Link. A. L.S. through the winter bud, showing the fundatrix, with its piercing stylets inserted into the plant tissue and thus inducing cell division (shaded area). (After Börner). B. L.S. through mature gall, with covering growth formed by swollen and flattened bases of needles and enclosing gall cavities.

an apical, somewhat spherical part from the rest and this portion is a hollow, rimmed, cup-like top of the gall. The larval cavity is large and central. The bud gall caused by *Andricus fecundator* HTG. on

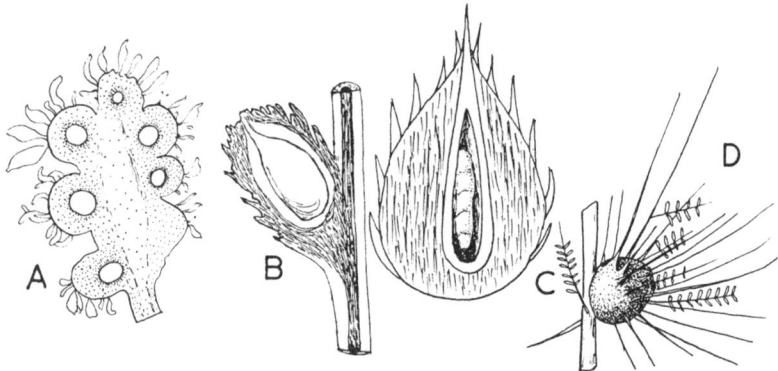

Fig. 53. Bud and growing tip galls. A. L.S. through the gall of *Asphondylia punica* M. on the growing tip of *Atriplex halinus*. B. L.S. through the gall of a Trypetid on the axillary bud of *Berberis lycium* Royle, showing the imbricating fleshy scaly outgrowths on the surface. C. L.S. through the gall of *Cylindrococcus casuarinae* Mask. on *Casuarina*. D. Gall of *Eriophyes* on the bud of *Gourliea decorticans* Gill. (After Houard).

Quercus is about 2—3 cm long and initially green, but turns later yellow-brown or brown. It is an artichoke-like swelling, popularly called "Eichenrosen". The bulk of the gall is composed of imbricating scaly structures, in the middle of which is the oval larval cavity, pointed apically and about 9 mm long. Not only the true scales but also the leaf rudiments are altered into scaly structures. Pine-cone-like bud galls are caused by Hymenoptera on *Casuarina quadrivalvis* LABIL. from Australia[280]. These solid and woody swellings, covered with imbricating scales, contain numerous larval cavities in the middle. In the pine-cone-like gall on bud of *Lissanthe strigosa* R. BR., caused by a midge from Australia[280], the growing point of the shoot axis immediately below the bud is conspicuously swollen, with the larval cavity situated in the middle; the surface of the gall is also clothed densely with imbricating scaly outgrowths. In the bud gall of the cynipid *Dryocosmus floridensis* BEUTENM. on the North American *Quercus cinerea*, we find a central mass of 3—4 larval cavities, covered by short aborted leaves. The leafy processes on the rosette gall on the Australian *Pultenaea gunnii* BENTH. are, unlike in most rosette galls, unusually long, indeed five or six times longer than even the normally developed leaves of the plant[280].

RÜBSAAMEN[993] has described an interesting example of a bud gall, in which the tendency of the leaf rudiments to become modified into long, filiform and hairy processes is best observed. In the gall on *Inga stringillosa* BENTH. spherical swellings of buds, about 25 mm thick, consist of dense tufts of hair-like and foliaceous outgrowths 10 mm long and usually straight. *Opisthoscelis prosopidis* KIEFF. & JÖRG., from Argentina, causes a bud gall on *Prosopis adamsioides* GRIESB., in which the larval cavity is covered by short, stunted, narrow, filiform leafy processes, 20 mm long. On another Argentinian *Prosopis alpataco* PHIL. the midge *Misospatha prosopidis* KIEFF. & JÖRG. gives rise to a globose bud gall, curiously resembling the bedeguar on *Rosa*, 10—25 mm in diameter and covered with dense rays of elongate, serrated hairy outgrowths from a central, fleshy, subglobose, multilocular core, enclosing the numerous larvae of the gall-midge[533]. The gall of the midge *Panteliola bedeguaris* KIEFF. & JÖRG. on the South American *Lycium chilense* BERT. (Solanaceae) is also remarkable for its curious resemblanche to the bedeguar on *Rosa*, due to the presence of dense hairy or filiform growths, often branched and hairy. The larval cavity, about 2 mm in diameter, lies in the middle of these processes. Cottony galls are caused by the gall-midge *Misospatha giraldii* KIEFF. & JÖRG. on *Artemisia vulgaris* LINN. (Plate II, 5). A similar cottony bud gall about 3—4 cm in diameter, is caused by an unknown gall-midge on *Achyranthes aspera* LINN. in Ethiopia[1187]. The hard larval cavity is situated in the centre of the cottony mass. A somewhat similar cottony bud gall on *Achyranthes bidentata* is recorded from Java[520]. In many bud galls the rudiments are modified into more or less spiny processes. Numerous long, straight, fleshy, spine-like emergences cover the globose bud gall caused by *Allodiplosis crassa* KIEFF. & JÖRG. on the South American *Gourliea decorticans* GILL. (fig. 54 A) [533]. The best known of spiny bud galls is perhaps the gall

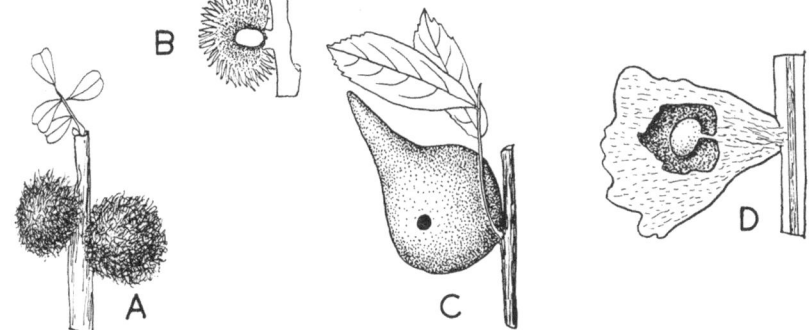

Fig. 54. Bud galls. A–B. Miniature sea-urchin-like gall on the axillary bud of *Gourliea decorticans* Gill., caused by *Allodiplosis crassa* Kieff. & Jörg. C. Beaked bud gall of cynipid on *Quercus*. D. Bud gall of *Cynips hungarica* on *Quercus*, showing the inner gall.

of *Hamamelistes spinosus* SCHIM. on *Hamamelis* in North America. This is a hollow, thick-walled swelling, with numerous, stout, curved, acutely pointed, spine-like, fleshy emergences.

A somewhat similar spiny bud gall, caused by the gall-midge *Oxasphondylia echinata* MANI on *Indigofera gerardiana* WALL. is recorded from the Himalaya[745]. *Schizomyia scheppigi* RÜBS. gives rise to spherical echinate gall on the bud of the South African *Stoebe cinerea* THUNB. A spherical, woody bud gall on the South American *Gourliea decorticans* GILL., caused by *Eriophyes* sp. (fig. 53 D), is also remarkable for the greatly elongated spiny processes, bearing stunted leaflets and arrested branches; the mites occur within the gall cavity. In some bud galls, like the midge-gall on *Sinapis pubescens* LINN. from Algeria[520], hollow oval swellings of the bases of the petioles of the leaf rudiments, about 8 mm long and 5—8 mm thick, enclose the larval cavity and the surface of the gall bears aborted leaf blades. A bud gall on *Pisonia* sp., described from Brazil[533], is a large swollen mass, covered with small stunted leafy processes, and enclosing a number of oval larval cavities in the basal swollen portion.

The gall of *Cynips polycera marchali* KIEFF. on *Quercus lusitanica mirbecki* is a curiously shaped, elongate, fleshy, solid, cylindrical, rayed swelling of the bud, from the flattened top of which rises a short, truncated cylindrical process, containing the larval cavity. The midge-gall on the bud of the Brazilian *Myrciaria* sp. is a spherical fleshy swelling, about 20 mm in diameter, with a small tuft of undeveloped leaves apically, the whole swelling being concealed partly within the greatly enlarged and inflated scale. A number of small larval cavities are contained in the fleshy mass[533]. An undescribed gall-midge causes an irregular solid, hard bud gall on *Indigofera pulchella* ROXB. from India[745]. The surface of the gall is densely covered with a silvery-white pubescence, stunted fleshy vestiges of leaves and branches. Four or five larval cavities are irregularly scattered in the fleshy mass. The whole swelling often attains a size of 25 mm in diameter. In a number of other bud galls the rudiments of the organs are reduced to obscure verrucosities on the surface. This is for example the case with the solid, fleshy, globose bud gall on *Potentilla*, caused by *Gonaspis potentillae* BASS., from North America. The rudiments of the bud leaves are represented by surface rugosities in the gall of *Misospatha tricyclae* KIEFF. & JÖRG. on the South American *Tricycla spinosa* CAV. (Nyctaginaceae)[533].

A number of bud galls are characterized by the complete absence of all traces of the bud rudiments, so that we find hollow or solid swellings. A familiar example of this type is the gall of *Biorrhiza pallida* OL. on *Quercus* from Europe (fig. 55). This is typically an irregularly globose mass, about 4 cm in diameter, with a number of spherical lobes and numerous larval cavities (vide Chapter IX).

Another common bud gall is that of *Pemphigus* on *Populus ciliata* WALL. from India. TROTTER[1187] has described a regular, globose, unilocular bud gall, about 5—7 mm in diameter, on *Dalbergia melanoxylon* GUILL. & PAT. from Eritrea.

An unknown Lepidoptera gives rise to subglobose, hollow and longitudinally striated swellings of the axillary buds of *Randia longifolia* LAMARCK from Java. The gall is apically produced into a hooked appendage, about 10 mm long. The most interesting of the smooth and nut-like bud galls is caused by *Cecidoses eremita* CURTIS on the South American *Schinus* spp. (fig. 123 C). A midge gall on the buds of *Bridelia tomentosa* BLUME from Cochin-China is a pyriform, solid, fleshy, beaked swelling, about 10 mm in diameter. The beak is about 5 mm long and curved like a hook. The surface of both these galls is smooth, without traces of leaves, branches or other rudiments of the bud organs. The bud gall of the Chalcid *Trichilogaster pendulae* MAYR on *Acacia pendula* A. CUNN. *(vide* Chapter IX) from Australia is a spherical swelling, like the bud gall of *Cynips lignicola* HTG. on *Quercus*, with the surface ridged and covered with short scales. Of the paired larval cavities, one is larger and contains the female and the other is a smaller excentric cavity that contains the male.

Many bud galls are remarkable for the large cavities, which they enclose. The bud gall of *Rhopalomyia millefolii* H. Löw on *Achillea millefolium*, has an oval or flask-shaped fleshy swelling of the bases of the leaf-sheaths of the bud. The mouth of the flask forms the ostiole, leading by a narrow cylindrical passage to the spacious larval cavity. The ostiolar passage is covered by downwardly directed multicellular, interlacing erect hairs. As the gall matures, the ostiole becomes enlarged and the rim of the ostiole curls outward, forming a sort of fringe around the ostiole. Interesting bud galls, caused by gall-midges on *Berberis* spp. from Brazil, are globose fleshy swellings, with imbricating, swollen, scaly outgrowths, 8—12 mm long. The

Fig. 55. Bud gall of *Biorrhiza pallida* (Ol.) bisexual generation on *Quercus*.

scales arise from a central, hollow, fleshy, cylindrical or spherical mass, formed by the fused and tumescent bases of the leaf rudiments. The tips of these rudiments are altered to form the imbricating scales. A Trypetid causes a similar gall in the axillary bud of *Berberis lycium* ROYLE from Himalaya (fig. 53 B). An oval or globose, fleshy, unilocular, thick-walled swelling of the basal portions of the leaf rudiments has the tips of the rudiments modified as the imbricating scales. Irregular fleshy emergences project from the wall in the gall cavity basally.

4. Some highly specialized bud galls

Some of the highly specialized bud galls are curiously coriaceous bags, often enormously inflated, with greatly convoluted and lobed cavities. The fungus *Exobasidium symploci* ELLIS & MARTIN gives rise to an utricular swelling of a North American *Symplocos*, measuring about 25—50 mm in diameter. A remarkably large utricular bud gall on the Indian climber *Calycopteryx floribunda* LAMARCK is caused by *Austrothrips cochin-chinensis* KARNY. The gall attains a size of 4 cm in diameter (Plate III, 2). A similar bladder-like coriaceous bud gall is caused by an unknown aphid on *Vaccinium leschanaulti* WT. from the Kodaikanal Hills, South India (Plate I, 1).

Some very unique bud galls, caused by different species of *Brachyscelis* on *Eucalyptus*, are known from Australia. The female of *Brachyscelis pomiformis* FROGGATT gives rise to an apple-shaped, globose, smooth gall, about 75 mm in diameter, with a circular depression 6 mm deep on the summit, in the centre of which is the ostiole, leading to a spacious gall cavity. A sessile, globose smooth gall, about 57 mm long and 43 mm thick, narrowed and lobed apically, is caused by the female of *Brachyscelis variabilis* FROGGATT. The lobes surround a fusiform axial gall cavity, containing the coccid. An ovoid, somewhat curved hollow gall, about 23—35 mm long and 15—20 mm thick, is caused by the female of *Brachyscelis ovicola* SCHRADER. An elongate, subcylindrical gall, about 65 mm long and 13 mm thick, reduced both apically and basally, is associated with the female of *Brachyscelis sloanei* FROGGATT. The gall of *Brachyscelis pedunculata* OLLIFF. (fig. 113 A—C) is a curiously elongate-clavate, strongly curved, woody swelling, 75—100 mm long and 10—13 mm thick, with a minute ostiole at the truncated apex. The gall cavity extends down into the basal pedunculate portion also. Variously bottle-shaped or jar-shaped hollow galls are caused by other species of *Brachyscelis*. For example, the gall of *Brachyscelis calycina* TEPPER is a remarkably vasiform swelling, about 15—25 mm long, 4—8 mm thick basally and 9—15 mm thick apically, with the mouth of the vase rimmed dentately. A trumpet-shaped gall, with the ostiolar rim expanded, is caused by the male of *Brachyscelis neumanii* TEPPER. The female of *Brachyscelis crispa* OLLIFF. gives rise to a sessile, subglobose, verrucose swelling, distally truncated and crateriform, 17—25 mm in diameter and the surface covered with pyramidal processes. A cupuliform gall, with numerous erect, bracteate appendages apically and an elongate axial larval cavity is caused by *Brachyscelis* sp. The gall of *Brachyscelis duplex* SCHRADER (female) consists of a large basal elongate bag 50—90 mm long and 30—50 mm thick and paired, apical, horny, flattened re-curved appendages often 150 mm long and tucked inside the basal larger utricular part. A small transverse slit-like aperture at the axil of these horns is the ostiole, leading into the spacious gall cavity. A remarkable tetrahedral, subcylindrical swelling, 20 mm long and 19 mm thick apically, caused by the female of *Brachyscelis tricornis* FROGGATT, bears 3 long, curved, slender horny appendages 65—75 mm long. In the gall of *Brachyscelis munita* SCHRADER (female) there are four such horny appendages, each measuring 50 mm long.

CHAPTER VII

LEAF GALLS

1. General characters

The great dominance and complexity of galls on leaves, the association of so many diverse groups of gall organisms and numerous other peculiarities of leaf galls are related to the distinctive structural and functional characters of the leaf as organ of the plant. The high synthetic activity, rapid growth and the peculiarities of morphogenetic patterns of leaf development are other important contributory factors to the wealth of leaf galls. Cecidozoa find in the leaf a much wider choice of ecological optima (fig. 57). The conditions for development of galls on different parts of the leaf, such as for example the leaf margin, the general surface of the blade, the midrib and the larger veins, the upper side or the under side of the blade, the petiole, etc. differ within wide limits. The fact that cecidozoa may become enclosed more readily in a leaf fold or roll than in any other organ should not also be overlooked. Some of the outstanding problems in the structure, differentiation and morphogenesis of leaf galls depend largely on the fact that surface growth predominates in the leaf. Striking differences exist, however, in the cell polarities, directions of cell elongations and cell divisions be-

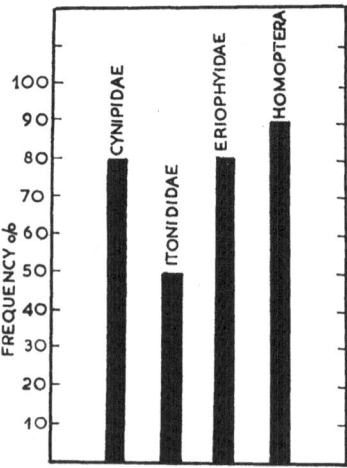

Fig. 56. Percentage frequency of leaf galls, caused by dominant cecidozoa from the world.

tween the epidermis of the upper and lower sides of the leaf blade, palisade and spongy tissues. Galls thus arise wholly from one or the other of these leaf tissues, combine growth in thickness, more or less

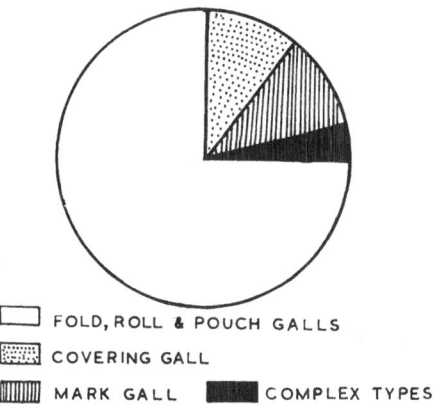

FOLD, ROLL & POUCH GALLS

COVERING GALL

MARK GALL   COMPLEX TYPES

Fig. 57. Relative abundance of the structural types of leaf galls from the world.

completely obliterate surface growth and thus exhibit considerable complexity of developmental patterns and structure. A number of interesting galls develop, for example, entirely by hyperplasy of the mesophyll, while the epidermis and the sub-epidermal layer of mesophyll do not take any part, so that eventually the gall breaks through these tissues. The communities associated with leaf galls are generally characterized by their large size, greater complexity of inter-relationships than in the case of other galls, abundance of pioneer associations and presence of both phyletically generalized and higly specialized groups of gall-makers. Phenologically leaf galls are by no means so restricted as galls on the vegetative buds or roots. The majority of leaf galls are mantel-galls and arise by folding, rolling or out-pocketing of the leaf (fig. 57). In a considerable number of cases the cecidozoa are hypophyllous.

The number of organoid galls on leaves is considerable. They range from an abnormally shaped leaf blade or lobed leaves to complex adventitious foliaceous outgrowths from the leaf blade[639]. The adventitious development of branches has also been described in the gall of *Taphrina laurencia* on the leaf of *Pteris quadriaurita*[639]. The bulk of leaf galls are, however, histioid galls. We may recognize two broad groups of leaf galls, on structural and ecologic grounds, viz. galls on petiole, midrib or larger veins and galls on the blade proper. The following is a synopsis of the fundamental types of leaf galls:

A. Galls on stipules, petioles and veins
    1. Galls on stipules
    2. Galls on petioles
       i. Simple swellings of petioles
       ii. Covering galls on petioles
    3. Midrib and vein galls
       i. Simple swellings of midrib or other larger veins
       ii. Covering galls on veins
       iii. Bedeguar gall
B. Galls on leaf blade
    1. Emergence galls on veins and blade
       i. Filzgalls
       ii. Fleshy emergence galls
       iii. Parenchyma galls
       iv. Epidermal galls
    2. Pockengall and leaf mine gall
    3. Kammergall
    4. Swellings of entire leaf blade
    5. Fold galls
       i. Fold gall without significant swelling of the blade
       ii. Fold gall with more or less pronounced swelling of the folded part of the blade, but without fusion of the folded parts.
       iii. Fold gall with considerable swelling and fusion of the swollen parts
    6. Roll gall
       i. Entire leaf rolled, without fusion of the rolled part; blade only slightly swollen
       ii. Roll gall, with pronounced swelling and fusion of the rolled and swollen parts
       iii. Margin roll gall
       iv. Roll and fold gall, with fusion and swelling of the blade
    7. Pouch gall
       i. Pit gall
       ii. Pocket gall
       iii. Pit or pocket gall, with swelling of the blade
       iv. Pouch gall with covering growth
       v. Complex and special pouch galls
    8. Mark-gall
    9. Oak-apple gall
    10. Lenticular gall
    11. Other specialized types of leaf galls
    12. Galls formed by the fusion of more than one leaf

The literature on the structure, development and ethology of leaf galls is very extensive, but the reader will find valuable information in HOUARD[506], GERHARDT[379], TROTTER[1183], BAIS[50], COSENS[221] and others.

2. Galls on stipules, petioles and veins

Galls on stipules

Among the galls on stipules, special mention should be made of the often enormously inflated, solid, fleshy or hollow swellings of the stipular thorns of *Acacia* spp., caused by gall-midges from parts of Africa and India. In some cases only the bases of the paired thorns are swollen to a globose basal mass, with the tips of the thorns projecting like curved horns apically (fig. 58). HOUARD[520] has described an interesting fusiform hollow gall, formed by the swelling of the middle part of the stipular thorns of *Acacia bussei* HARMS. Ecologically galls on stipules of *Acacia* spp. are remarkable for the community of secondary organisms, especially ants, regularly associated with them.

Galls on petioles

Structurally and ecologically, galls on petioles are greatly related to those on the shoot axis and likewise show differences dependent on the positioning of the cecidozoa. Specializations arise, however, from differences in the mechanical and anatomical conditions in the petiole and shoot axis, the relative size of petiole and leaf blade and a number of other factors, which do not operate in the development of galls on the shoot axis. Some petiole galls extend also to the midrib of the leaf above and are thus continuous with swelling of the latter. It is also remarkable that most petiole galls seem to develop on the

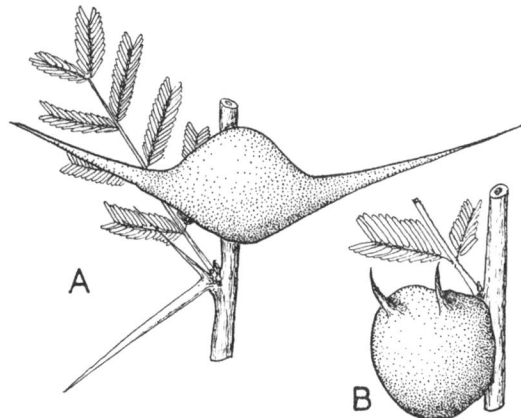

Fig. 58. Galls on stipules. A. Gall of an unknown gall-midge on the base of the paired stipular thorns of *Acacia leucophloea* Willd. B. Gall of an unknown gall-midge, with only the tips of thorns remaining as paired horn-like spines, on the same plant.

epiphyllous side. The developmental types of petiole galls include filzgalls, rindengalls with cecidozoa external, rindengalls with cecidozoa internal, rindengalls with cecidozoa initially external but later becoming enclosed, galls with cecidozoa situated in the zone of vascular bundles, galls with cecidozoa in the medulla, covering galls and kammergalls. Some of the more common galls on petioles include the gall of *Arnoldia nervicola* KIEFF. on *Quercus cerris*[1183], *Atrichosema aceris* KIEFF. on *Acer campestris* LINN. from Europe, the ribbed petiole gall of *Ectoaedemia populella* BUSCK on *Populus* from North America, etc. HOUARD[520] has described an interesting gall on the petiole of *Dioscorea praehensilis* BENTH. from Liberia, consisting of an irregular, solid, striated mass, 10—15 mm long and 10 mm thick, with the entire petiole swollen. The aphid *Schlechtendalia chinensis* BELL causes a curious petiole gall on *Rhus semialata osbecki* DC from China and Java. This is a curious coriaceous, utricular hollow outgrowth, formed partly from the foliaceous expansions on the petiole and partly from the leaf blade, with irregular, finger-shaped bulges externally. The gall on the petiole of *Fraxinus excelsior* LINN., caused by *Perrisia fraxini* KIEFF. from Europe[485], is remarkable for the enormous swelling of the two low, ridge-like fleshy wings on the epiphyllous side of the petiole, enclosing between them a deep groove (fig. 59 B). This swollen part is thicker than the body of the petiole proper. The larval cavity is surrounded by a typical nutritive zone and sclerenchyma zone, connected by a patch of sclerenchyma cells with the vascular bundle semi-ring, within the body of the petiole. In the gall on the petiole of *Tilia silvestris* DESF., caused by *Contarinia tiliarum* KIEFF., also from Europe, the

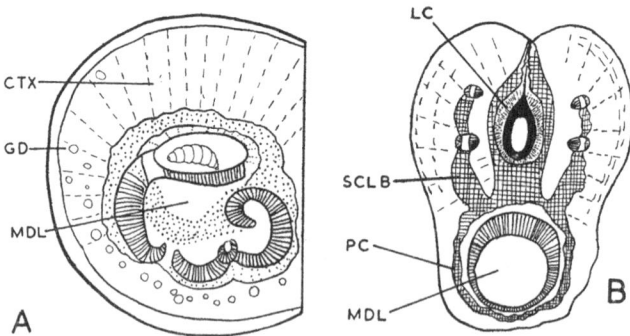

Fig. 59. Structure of galls on petiole. A. T.S. through the gall of *Contarinia tiliarum* Kieff. on the petiole of *Tilia*, with the larval cavity situated in the zone of vascular bundles. B. T.S. through the gall of *Perrisia fraxini* Kieff. on the petiole of *Fraxinus excelsior* Linn., with the larval cavity in the cortex and enclosed between the paired fleshy ridges above, surrounded by nutritive zone (black) and sclerenchyma band (strippled). CTX cortex. GD gum ducts. LC. larval cavity. MDL medulla. PC pericycle. SCLB sclerenchyma band. (After Houard).

midge larva is situated in the zone of the vascular bundles (fig. 59 A). The gall develops at the distal end of the petiole, just below the base of the leaf blade. The cells of the cortex undergo very pronoun-

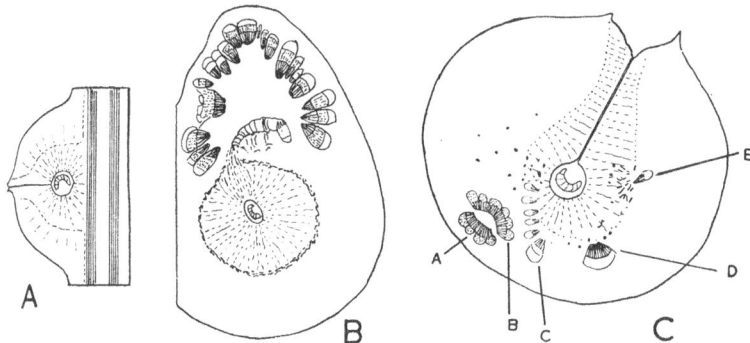

Fig. 60. Structure of the gall of *Harmandia petioli* Kieff. on the petiole of *Populus tremula* Linn. A. L.S. through the young gall, showing the covering growth, larval cavity and the radially dividing and elongating cells. B. T.S. through the same gall, with the larval cavity in cortex and connected with the nearest vascular bundle, C. T.S. through mature gall, with the larval cavity situated in the zone of the vascular bundles, dissociated from the stele. A-E vascular bundles (After Houard).

ced proliferation and hypertrophy. The vascular bundles in the neighbourhood of the larva are also hypertrophied, but lack lignification in the secondary tissue. Sclerotic cells appear in the cortex and medulla. The petiole gall of *Harmandia petioli* KIEFF. on *Populus tremula* LINN. is remarkable for the fact that one or more larval cavities are situated in the zone of vascular bundles, with abundance of hypertrophied liber and wood elements, producing abnormal secondary tissue, without lignification (fig. 60). In the gall on the petiole of *Potentilla reptans* LINN., caused by *Xestophanes potentillae* VILLERS, the larval cavity is situated within the medulla.

The most remarkable petiole gall is undoubtedly the "Spirallockengall" of *Pemphigus spirothecae* PASS. on *Populus* from Europe (fig. 1 C). The subglobose or pyriform, hollow gall arises by the ribbon-like flattening, swelling and spiral twisting of the middle part of the petiole, so as to enclose a spacious cavity, containing the aphid. To begin with, a slight elevation of tissue arises all around the fundatrix of the aphid that settles down on the petiole of the unfolding leaf. The aphid thus comes to lie in a small, flat, crater-like depression. The petiole undergoes then at this point a sharp downward torsion through 180°. As a result of this twisting, the leaf blade comes to be pointed upward and the aphid now lies within the angle of the bend in the petiole. Continued twisting of the petiole upward brings the leaf blade back to its original position and the second turning of the petiole results in the formation of a spirally twisted, ribbon-shaped structure, with the twisted margins of the swollen ribbon coming into accurate contact with each other, but not becoming fused. The aphid is thus completely enclosed within the spherical, coriaceous bag. The spiral twisting of the ribbon-shaped swelling of the petiole may be clock-wise or also counterclock-wise. Two spirally wound, distinct fleshy carinate processes, conspicuous externally on the gall, indicate vascular bundles in the gall tissue. The expansion of the petiole into a swollen, ribbon-shaped structure is chiefly due to cell proliferation in the peripheral parenchyma of the cortex. The collenchyma that is typically found in the normal petiole of *Populus* spp. is, however, completely absent in the gall tissue. The gall tissue is also made of thin-walled cells, so that the gall as a whole remains largely fleshy. The epidermis lining the gall cavity bears short, unicellular hairs, often papillate or clavate. In between these papillae there may also be found some elongate, multicellular and acutely pointed hairs. It will be seen from its mode of development that this is typically a covering gall.

Galls on midrib and larger veins

The galls on midrib and the larger secondary veins are, to some extent, related to those on the petiole and are often, as already pointed out, continuous with them. In most of them the blade on either side of a galled vein is not involved, but many interesting galls include swellings of the tissues of vein continuously with part of the mesophyll of the blade on either side. Galls on midrib and larger veins are generally more or less localized or extensive, globose, fusiform or diffuse, mostly hypophyllous, solid, fleshy or hard swellings and are popularly called gouty-vein galls in North America. The commonest gouty-vein gall from North America is, for example, caused by *Dasyneura communis* FELT on *Acer rubrum*. The midrib gall of *Arnoldia nervicola* KIEFF. on *Quercus cerris* LINN., from Europe, consists of diffuse and extensive swellings of midribs, with the cell proliferation localized mostly in the cortex[1183]. Even the Nematode *Tylenchus* sp. causes a gall on the midrib of *Crepis leontodontoides* ALL. from Europe [1183]. Extensive, fusiform, solid and woody swellings of midribs and several of the larger lateral veins close to the midrib are caused by the midge *Pipaldiplosis pipaldiplosis* MANI on *Ficus religiosa* LINN. from India [745]. In this gall the cell proliferation is not localized strictly within the cortex but embraces medullary rays and medulla, so that there is considerable dissociation of the vascular bundles. The surface is marked by longitudinal and irregular cracks, due to the fact that cell proliferation is deep-seated and the peripheral cells do not take an active part in the development of the gall. The gall of *Perrisia fraxini* KIEFF. on the midrib of *Fraxinus excelsior* LINN. from Europe is a nearly hypophyllous swelling that appears on the upper surface of the leaf as deep, elongate and pouch-like invaginations. A part of the blade on either side of the midrib is also swollen. The epiphyllous, slit-like (fig. 59 B) aperture is almost completely closed by interlacing, papillar outgrowths from the epidermis. Subcylindrical and nodular, multilocular swelling of the midrib of *Fragaria*, caused by *Diastrophus fragariae* BEUTEN. is common in several parts of North America. An undescribed Chalcid gives rise to oval or fusiform, solid, hard, unilocular and often moniliform swellings of nearly all the veins on the under side of the leaf blade in *Bassia latifolia* ROXB. from India (Plate VI, 3—4). Solid, fleshy, globose, yellowish-brown galls, about 20 mm in diameter, are caused by the midge *Odinadiplosis odinae* MANI on the rachis and midribs of *Lannea coromandelica* (HOUT.) (= *Odina wodier* ROXB.) from India (Plate V, 3).

A number of vein galls are caused by cynipids. The gall of *Diplolepis douglasi* ASHM. on *Quercus lobata* NEE from North America is a solid, top-shaped, hard, hypophyllous swelling, inserted by the apex of the top to the main side vein or to the midrib. This gall is about 7 mm long and 7 mm wide at the flattened top, the edges of which are armed with stout, fleshy, acute, spinous processes. An ellipsoidal larval cavity opens to the outside by a lateral passage, ending in the exit hole on the side of the conical gall (fig. 61 A—B). Among the interesting leaf-vein galls caused by fungi, mention may

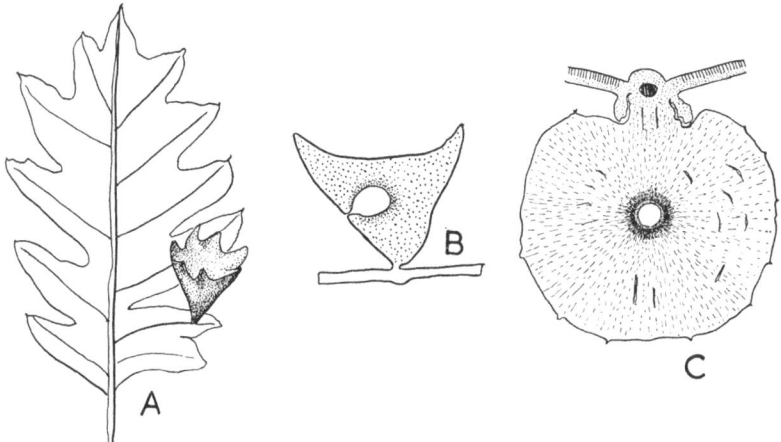

Fig. 61. Leaf vein galls of cynipids. A. Turbiniform gall of *Diplolepis douglasi* Ashm. on *Quercus lobata* Nee. B. L.S. through the same gall, showing the larval cavity situated on one side. C. L.S. through the oak-apple gall of *Diplolepis quercus-folii* Linn. unisexual generation on *Quercus* showing the radial zonation of tissues, attachment of the gall to the leaf vein and the eruption of the gall through the cortex and epidermis of the vein.

be made of the gall of *Urocystis violae* Sow., which is remarkable for its solid, fusiform and spirally twisted structure[888].

3. Epidermal and emergence galls on leaf blade

It is of considerable interest to remark that a great many galls arise from single epidermal cells, attacked by gall-inducing organisms: the mesophyll and the neighbouring epidermal cells remain apparently unaffected. The proliferation of a single epidermal cell gives rise to a minute vesicular blister on the surface of the leaf blade. The best known perhaps of this type of gall is caused by the fungus *Synchytrium* (fig. 140). Usually however, a number of adjacent epidermal cells proliferate or grow out to give rise to hairy or fleshy emergences on the surface of the blade. The cecidozoa are external and feed on the epidermis from the outside. The simplest of the emergence galls is the filzgall, already referred to. Most filzgalls are caused by *Eriophyes* spp. The hairy outgrowths in the filzgall are generally abnormally enlarged epidermal cells. Intense cell proliferation, strictly localized to a narrow patch of epidermis, has the result of an irregular fleshy, parenchymatous emergence. The emergence galls comprise multicellular, solid, often branched growths. The gall of *Phyllocoptes populi* NAL. on the leaf of *Populus tremula*

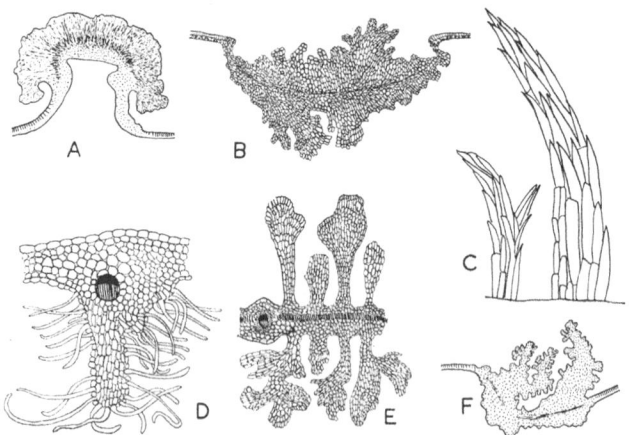

Fig. 62. Emergence galls of *Eriophyes*. A. *Commiphora campestris* Engl., showing the persistence of the direction of growth and cell division in the palisade tissue even in the emergence growth. B. Irregular parenchyma emergence outgrowth from both sides of the leaf blade of *Rumex nervosa usambarensis* Engl. C. Elongate fleshy needle-shaped emergence outgrowths on *Ossea* (After Rübsaamen). D. Fleshy emergence and erineal hairs on *Juglans regia* Linn., caused by *Eriophyes juglandium* Pers. E. Irregularly lobed, clavate fleshy emergence gall of *Phyllocoptes populi* Nal. on *Populus tremula* Linn. F. Irregular emergence gall on *Ipomea cairica* Sw.

LINN. from Europe is a typical emergence outgrowth (fig. 62 E). A stout, clavately cylindrical, fleshy emergence on the under side of the leaf of *Juglans regia* LINN. is an emergence gall, in which the epidermis has completely disappeared; there is often a slight hypertrophy of sub-epidermal cells also (fig. 62 D). Remarkably beautiful, bright red or pink coloured emergence galls, consisting of irregular and extensive, granular fleshy epidermal outgrowths on the underside of the leaves of *Ruellia, Asystasia* and *Achyranthes aspera* LINN. from India are caused by Aleurodidae. A somewhat similar, spongy, solid emergence gall of bright carmine-red epidermal outgrowths is described by RÜBSAAMEN[997] on both sides of the leaf of *Rumex nervosa usambarensis* ENGL. from Africa (fig. 62 B). Irregular, parenchymatous and spongy emergences cover more or less extensive areas of the lower surface of the leaves of *Ficus sycamorus* LINN.[997]. To RÜBSAAMEN[997] we owe again the record of an interesting emergence gall on the leaves of the African *Commiphora campestris* ENGL., caused by *Eriophyes* sp. (fig. 62 A), consisting of irregular, fleshy, solid out-growths on the upper surface. In this gall the deeper mesophyll is also greatly altered and the blade is curled up to form a hollow depression.

4. Pockengalls and Kammergalls

Pockengalls

The pockengall is typically a biconvex, mostly solid swelling of the leaf blade, equally developed on both sides and usually not exceeding a few millimetres in size. The Eriophyid mites that give rise to these galls occur inside the gall, in among the cells and seem to gain entry into the leaf tissue through stomata. The differentiation of the palisade tissue is greatly suppressed. Chlorophyll is abnormally developed or altogether absent, so that the pockengalls are generally bright yellow, but may also be tinted red or violet, especially when exposed to the direct sunrays during their development. One of the commonest pockengalls is caused by *Eriophyes piri typicus* NAL. on the leaves of *Pirus communis* LINN. (fig. 63 A). In some pockengalls a narrow canaliculous ostiole is found (fig. 63 B). In the pockengall caused by *Eriophyes tristriatus typicus* NAL. on leaf of *Juglans regia* LINN. there is a relatively large central hollow space, communicating to the outside by an hypophyllous ostiole. Pockengalls are also caused by *Tylenchus millefolii* F. Löw and *Aphelenchus olesistus* RITZ. The pockengall caused by the midge *Cystiphora pilosellae* KIEFF. on the leaf of *Hieracium* spp. is generally classed as a parenchyma gall. The leaf mine galls of *Phytomyza ilicis* CURT. on *Ilex aquifolium* LINN., arising from callus-like outgrowth within the mine and resulting in the formation of a yellow, bladdery bulging of the leaf blade are also a type of parenchyma gall.

Kammergalls

Kammergalls are completely closed, hollow fleshy swellings on leaves of *Salix, Quercus, Rosa, Acer* and other plants, usually caused by *Pontania*. The typical kammergall of *Pontania* on *Salix* is a

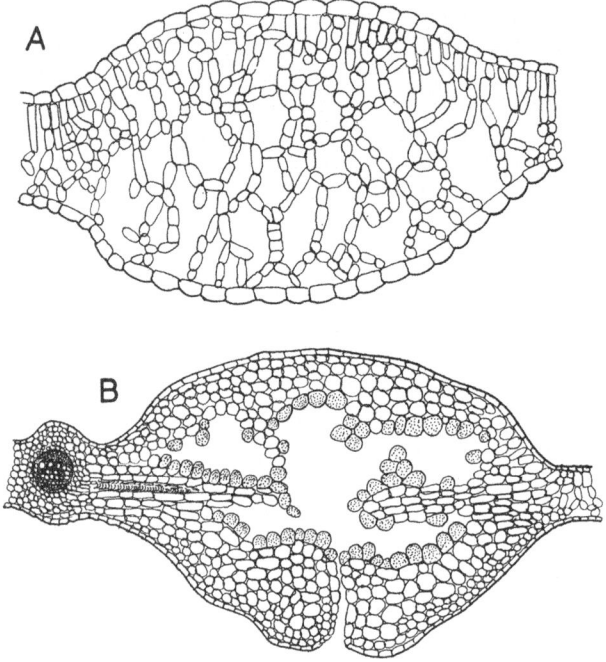

Fig. 63. Pockengalls. A. Pockengall of *Eriophyes typicus* Nal. on leaf of *Pirus commnnis* Linn. B. Pockengall of *Eriophyes tristriatus typicus* Nal. on leaf of *Juglans regia* Linn.

smooth, regular, spherical or oval, yellow or brown, thick-walled, fleshy, succulent swelling, inserted narrowly on the under side of the blade (fig. 1 D). The gall epidermis is composed of small-sized cells, with thick cuticle and lacking in stomata, although lentices arise in older galls. The bulk of the gall tissue is composed of uniform, thin-walled parenchyma cells, with numerous irregularly branching and anastomosing vascular bundles. The innermost layer of cells, lining the gall cavity, is less closely packed and is also rich in reserve food materials.

The pockengalls and kammergalls are more specialized than the emergence galls in that the cecidozoa are within the plant tissue and are not exposed on the surface of the plant. The cell proliferation is also far more extensive and the inhibition of normal tissue differentiation is generally very pronounced. While the emergence galls often cover extensive areas of the leaf blade, the pocken- and kammergalls are more restricted.

5. Fold galls

Fold galls arise predominantly from the failure of the leaf to unfold normally when it opens, so that the fold galls start with the natural folding in the bud and the cecidozoa are incapable of actively folding the leaf blade. The mode of development and the general form of the fold gall depend also largely on the differences in the position of the leaf rudiment and its folding within the bud. The cecidozoa deposit their eggs between the folds of the leaf rudiment in the bud and the larva feeds from this position, so that the folding characteristic of the bud persists in the gall. Arrest of normal development has the effect of failure of unfolding of the rudiments. Most of these galls are therefore "Hemmungsgallen" of the German literature.

In the simplest case, the folded parts of the leaf rudiments fail to open, but are not conspicuously swollen or otherwise malformed

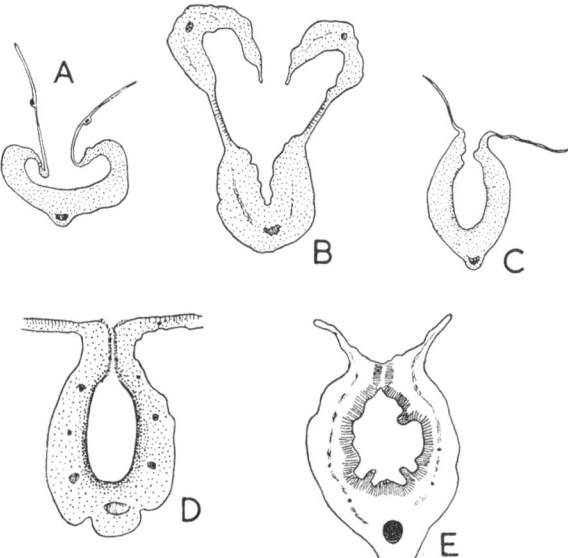

Fig. 64. Leaf fold galls   A. Gall of *Gynaikothrips crassipes* Zimmermann on leaf of *Piper*.   B. Gall of *Gynaikothrips intorquens* Karny on leaf of *Smilax*.   C. Gall of *Gynaikothrips chavicae* Zimmermann on *Piper*.   D. Gall of *Perrisia fraxini* Kieff. on leaf of *Fraxinus excelsior* Linn.   E. Gall of *Tychius crassirostris* Kirsch on leaf of *Melilotus alba*. There is no tissue fusion in A–D, but in E the folded halves of the blade are fused together.

externally. Such a simple leaf fold gall is caused by *Cecidomyia duttai* MANI on *Aegle marmelos* CORR. from India[745]. Each of the three young leaflets of the trifoliate leaf is folded upward along the midrib while inside the bud and this folding is retained in the course of development of the gall, but the folded leaflets become inflated to form an oval, biconvex, pod-like, coriaceous bag, in which the margins at the fold do not fuse together. The outer surface of the gall corresponds to the lower side of the leaf. The leaflet is only very slightly swollen, due largely to hypertrophy of the cells, but differentiation of the characteristic tissues of the leaf is also lacking. In the fold gall of *Tychius crassirostris* KIRSCH on the leaf of *Melilotus alba* from Europe (fig. 64 E) the folded halves of the blade on either side of the midrib are moderately swollen and form thick parenchyma tissue. There is also a partial fusing together of the swollen halves, so that the gall cavity is completely enclosed[976]. The entire blade of the leaflets of *Pistacia lentiscus* LINN. is turned into a gladstone-bag-like, coriaceous, hollow, fold gall (fig. 1 B). The fold gall of *Baizongia pistaciae* (LINN.) on *Pistacia terebinthus* LINN. is an enormously elongated, horn-shaped, contorted, cylindrical and hollow structure, about 20 cm long and 30—35 mm thick. A pyriform, hollow, utricular gall of the whole leaf blade, 125 mm thick and apically pointed is recorded on *Distyllum racemosum* SIEB. & ZUCC. from Java[520].

Many fold galls are remarkable for the very pronounced swelling and complete fusion of the tissues of the folded parts of the leaf blade. Some of them are globose or biconvex-oval swellings, often 30 mm long and 20 mm thick. The fold gall of *Schizomyia meruae* FELT on the leaf of *Maerua arenaria* HOOK. & BL. from India is, for example, an enormously swollen, solid, spongy, pod-shaped, smooth, yellowish-green mass, derived from the basal portions of the blade on either side of the midrib folded upward, with only the narrow marginal fringe expanding flat and normal (Plate II, 4, 7, fig. 65 B). The outer surface of the gall corresponds to the under side of the leaf. The gall epidermis is characterized by the pronounced hypertrophy of the cells, with relatively thick cuticle and malformed or totally absent stomata. The mass of the gall is composed of large-sized, thin-walled, undifferentiated parenchyma cells, with considerable intercellular spaces. The upper epidermis of the folded halves of the leaf rudiments is totally undifferentiated, as a result of the tissue fusion between the folded halves. The larval cavities are small and irregular spaces in the gall parenchyma, lined by a narrow zone of small, elongate or branched, irregularly lobed interlacing cells, greatly resembling fungal hyphae, but really derived from the general gall parenchyma. This cell layer represents no doubt the nutritive zone. A somewhat similar but usually larger and more spongy leaf fold gall is caused by *Asphondylia riveae* MANI on

Rivea hypocrateriformis CHOISY from India (Plate VII, 2, fig. 65 A). This is a solid, spherical or oval-biconvex, smooth or pubescent, yellowish-green spongy swelling, about 75 mm long and 25 mm thick.

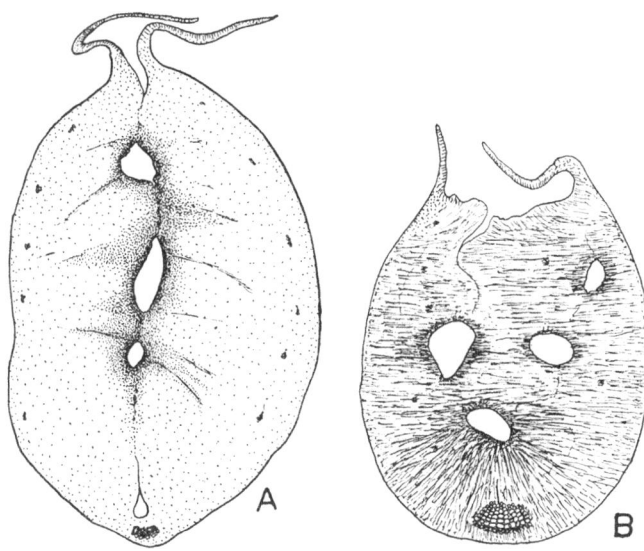

Fig. 65. Leaf fold galls. A. T.S. through the gall of *Asphondylia riveae* Mani on leaf of *Rivea hypocrateriformis* Choisy, showing the enormous swelling of the folded halves of the blade and extensive tissue fusion. B. T.S. through the gall of *Schizomyia meruae* Felt on the leaf of *Maerua arenaria* Hook. & Bl.

It arises by the tissue fusion of the enormously swollen folded halves of the blade, with only a narrow marginal fringe of the blade remaining as foliaceous wings on one side of the gall. The gall epidermis, derived from the epidermis of the under side of the leaf rudiment, is characterized by hypertrophied cells, thick cuticle and lacks stomata. The great mass of parenchyma of the gall is composed of large-sized, thin-walled cells, with considerable intercellular spaces. The vascular bundles irregularly scattered in the gall parenchyma represent the degenerated veins. The larval cavities are irregularly scattered in the middle of the gall parenchyma.

6. Roll galls

The leaf roll galls are on the whole more specialized than fold galls. The rolling of the leaf blade, representing the dominant cecidogenetic alteration, is caused by the cecidozoa and it is a part of the active growth reaction of the plant. In this reaction the bud folding or position is more or less obliterated. In leaf roll galls the blade may curl either upward or downward, more or less completely, so that one of the leaf margins is covered and the cecidozoa become enclosed in a tubular space. The rolling of the blade is usually also accompanied by swelling, crinkling and other malformations. The entire leaf blade may thus become rolled and the roll also at the same time may be spirally twisted like a cigar. This is, for example, the case with the roll gall of *Eothrips aswamukha* RAMAKR. on *Jasminum pubescens* WILLD. from India. The gall is a cylindrical or elongate fusiform, spirally twisted, verrucose and rugose leaf roll, about 75 mm long and 25—40 mm thick. The rolling of the blade commences obliquely from one margin at the base and extends right across beyond the midrib to the other margin. The rolling takes places on the upper side of the leaf, so that the outer surface of the gall represents the under side of the leaf. The spacious gall cavity is remarkable for the labyrinthine passages, in which numerous thrips occur. The rolled leaf blade is moderately swollen and lacks tissue differentiation in the mesophyll. The larvae of *Dasyneura mali* KIEFF. and *Dasyneura piri* BOUCHÉ from Europe roll the two halves of the leaf blade to the midrib. In the roll galls on *Rosa*, *Trifolium*, *Fraxinus*, etc. caused by different gall-midges from Europe the leaf roll is combined with partial folding upward and tissue fusion. The roll galls caused by gall-midges on leaves of *Acer* and *Alnus* consist of rolling on the under side, with the veins greatly swollen. The gall of *Eriophyes lacticinctus* NAL. on the leaf of *Lysimachia vulgaris* LINN. from Europe is an irregularly shaped folding, spiral twisting and rolling of the blade (fig. 66 B), with simultaneous crinckling, swelling, abnormal fleshy emergences and hairy outgrowths. The margins of the blade are rolled inward toward the midrib on the under side. The hairs on the normal leaf are cylindrical, unbranched and only 2—5 cells long, but in the gall they are branched and often 25 cells long. The lower epidermis that lines the

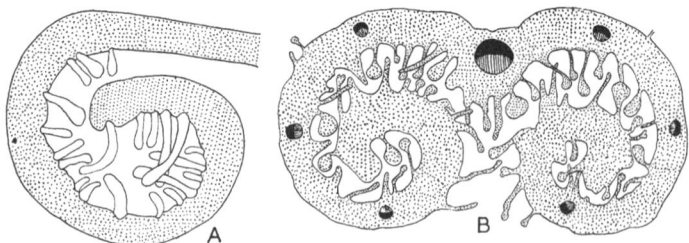

Fig. 66. Leaf roll galls.   A. T.S. of the leaf margin roll gall of *Eriophyes*, showing enormously enlarged, clavate, nutritive hairs.   B. T.S. through the double margin roll gall of *Eriophyes lacticinctus* Nal. on *Lysimachia vulgaris* Linn., with multicellular, clavate emergences.

gall cavity is highly abnormal, with pronounced cell proliferation, taking place first tangentially and later irregularly in all directions, giving rise to numerous, small-sized, thin-walled, plasma rich nutritive cells. In the roll gall of *Exobasidium* on *Vaccinium vitis idaea* LINN. the rolled part of the blade is greatly swollen, without tissue differentiation, with abnormally formed vascular elements, without trace of lignification.

7. Leaf margin roll galls

The leaf margin roll gall is a specialized leaf roll gall, in which the rolling and gall formation are confined to a narrow marginal area of the leaf blade. Some of the margin roll galls become complex due to combination of folding, evagination and out-pocketing growths. The margin roll gall may be epiphyllous or sometimes also hypophyllous and these differences are largely dependent upon the nature of folding and arrangement of the leaf rudiments within the bud. The margin roll gall may extend the whole length of the leaf or may also be of short length, confined often within two veins. Cecidozoa which cause margin roll galls are generally Acarina, Thysanoptera, Psyllids, Itonididae and some Tenthredinidae. Margin roll galls are generally

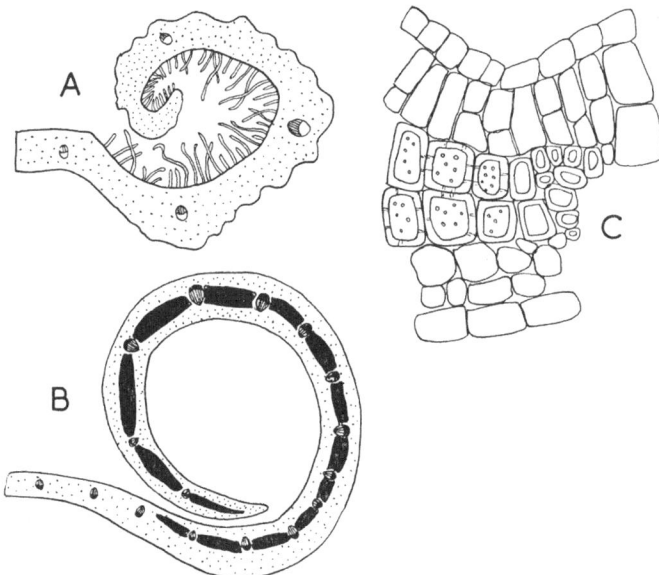

Fig. 67. Leaf margin roll galls. A. Gall of *Eriophyes tetrastichus* Nal. on *Tilia silvestris* Desf. (After Houard). B. Gall of *Perrisia tiliamvolvens* Rübs. on *Tilia silvestris* Desf., showing the pitted, lignified cells between veins. C. The pitted lignified cells of the same highly magnified.

110

characterized by complete absence of normal tissue differentiation, but consists largely of small-sized, thin-walled parenchyma cells, usually with very little intercellular space. The thickness of the

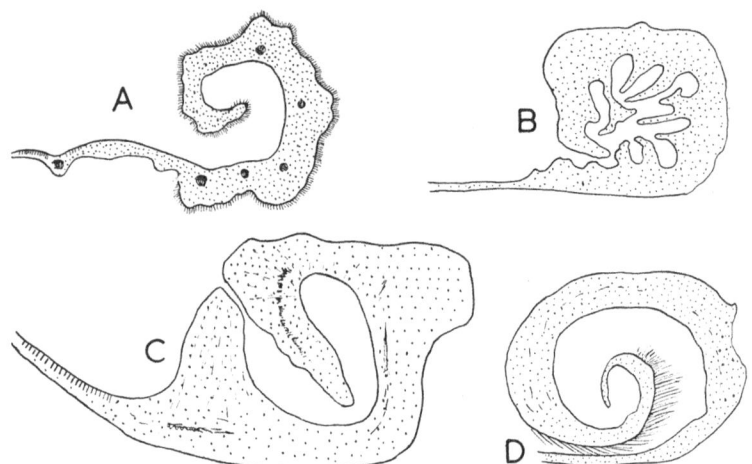

Fig. 68. Leaf margin roll galls. A. Gall of *Trichochermes walkeri* (Först.) on *Rhamnus cathartica* Linn., showing the pronounced thickening of the blade. B. Gall of *Eriophyes sp.* on an unknown Indian plant, showing the fleshy emergences from the greatly swollen sides. C. Midge gall of *Symmeria*, with swelling of the rolled part and covering growth on the left (After Rübsaamen). D. Midge gall on an unknown Brazilian plant, showing the elongate hairy outgrowths (After Rübsaamen).

rolled part is increased considerably (fig. 68). The epidermal cells are generally hypertrophied, lack stomata, or have malformed and non-functional stomata. The epidermis lining the gall cavity usually lacks a cuticular layer[506]. In many margin roll galls fleshy papillae or hairy outgrowths develop, especially on the inner surface. The cecidozoa are positioned on the upper surface of the leaf blade and the leaf margin is naturally rolled in upwards.

The margin roll galls caused by Itonididae are relatively strongly developed, typically coriaceous or fleshy. In the gall of *Dasyneura marginemtorquens* WINN. on the leaf of *Salix viminalis* LINN. from Europe, the midge larvae hatch from eggs deposited in spring between the bud scales. The larvae feed on the underside of the leaf rudiments along the margin, proceeding upward to the leaf tip. The leaf margin curls downward around the larva, thus enclosing it. The rolling is chiefly the result of cell hypertrophy and very little cell proliferation may be observed. A larger and more extensive upward rolling of the leaf margin, with coriaceous and abnormal swelling and development of hairs is met with in the gall of *Dasyneura tiliamvolvens* RÜBS. on *Tilia*. The margin roll gall of *Wach-*

tiella (= *Dasyneura) persicariae* (LINN.) on *Polygonum* spp. is larger and epiphyllous, with a characteristic spongy texture and four to five times thicker than the normal blade. The margin roll gall of *Macrodiplosis dryobia* F. Löw on the leaf of *Quercus* spp., about 10 mm long, consists of a part of the leaf lobe near the margin being sharply folded and fixed on the under side of the blade (fig. 69 B). The differentiation of palisade tissue is not always inhibited in this gall. The roll gall of *Macrodiplosis volvens* KIEFF. on *Quercus* from Europe is a tubular growth confined between two leaf lobes on the upper side of the leaf (fig. 69 A). Some roll galls, such as that of *Psyllopsis fraxini* LINN. on *Fraxinus* from Europe is a loose utricular structure on the upper surface of the leaf. A fleshy-coriaceous margin roll gall is caused by *Trichochermes (= Trioza) walkeri* (FÖRST.) on the leaf of *Rhamnus cathartica* LINN. from Europe (fig. 68 A, 70 A). This is largely a localized roll of a small section of the leaf margin, on the upper side, abnormally hairy and often swollen nearly ten times the normal thickness of the leaf blade. The vascular elements are greatly swollen and anastomosed in a complex manner. The rolling is usually confined in between two adjacent lateral veins, which appear to act as a kind of barrier to an extended rolling. A remarkable margin roll gall from North America, caused by *Diplosis silvestrii* TROTTER on *Quercus*, appears wholly on the under side of the leaf, but is in reality an epiphyllous rolling, accompanied by simultaneous curling downward of the rolled margin. The rolled-in margin of the blade comes to lie in contact with the edge of the downwardly curled part (fig. 16G)[1172]. In the margin roll gall of *Cryptothrips intorquens* KARNY on the leaf of *Smilax* spp. from Java[520] the leaf is folded into two along the midrib on the

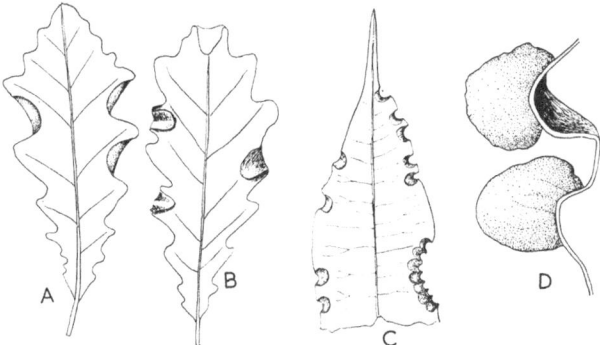

Fig. 69. Leaf margin fold and roll galls. A. Margin fold gall of *Macrodiplosis volvens* Kieff. on *Quercus robur* Linn. B. Margin fold gall of *Macrodiplosis dryobia* F. Löw on leaf of *Quercus robur* Linn. C–D. Leaf margin fold and out-pocketing galls on unknown Brazilian plants (After Rübsaamen).

112

upper side and the margins of the folded halves are rolled towards the midrib. The rolled part of the blade is generally swollen 5—6 times the normal leaf and the whole gall is also spirally contorted,

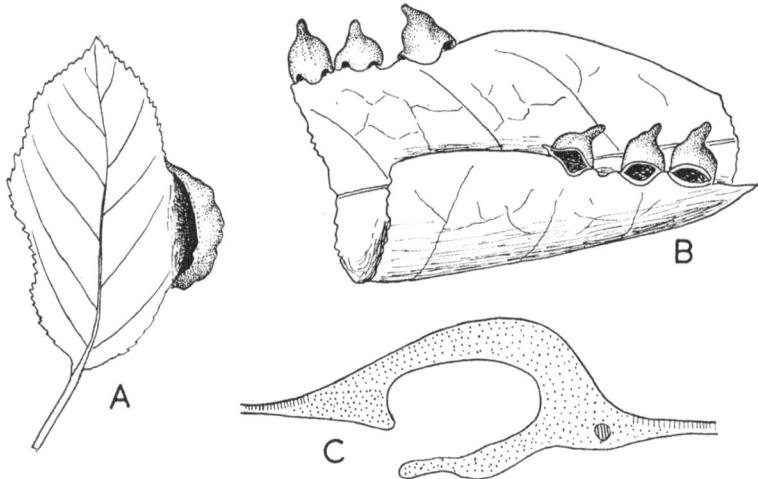

Fig. 70. Leaf margin out-pocketing galls. A. Leaf margin roll and out-pocketing gall of *Trichochermes walkeri* (Först.) on *Rhamnus cathartica* Linn. B. Psyllid gall on *Ficus nervosa* Hey. (After Trotter). C. Psyllid gall on *Ocota* (After Rübsaamen).

with the surface rugose and with tuberosities (fig. 64 B). The margin roll gall caused by *Gynaikothrips chavicae* ZIMMERMANN on *Piper nigrum* LINN. from India extends the whole of the length of the leaf (Plate VII, 1, fig. 64 C). A marginal fold-roll gall on *Ficus nervosa* HEY. caused by an unknown Psyllid from Formosa, has a curious conical pouchlike fold, projecting outward horizontally from the leaf margin, with the ostiole turned inwards to the midrib (fig. 70 B).

8. Pouch galls

Leaf galls reach their maximum specialization in structure and ecologic relations in the pouch galls, which are also the dominant type (fig. 57). In the simplest case, the pouch gall is merely a pit gall—a more or less deep invagination on the under side of the leaf blade (fig. 104C), caused by fungi, psyllids, aphid, aleurodids, etc. The invagination is also usually accompanied by more or less pronounced swelling, a conspicuous bulging on the upper surface of the blade and other abnormalities. As an example of a typical pit gall, we may mention the gall of *Trioza* sp. on the leaf of *Ficus religiosa* LINN. from India[745]. We have in this case a hemispherical, epiphyllous bulge, about 2 mm in diameter. Similar pit galls are also caused by *Eriophyes* sp. on *Quercus incana* ROXB. from India[745]. Other common pit galls include those of *Aleuromarginatus tephrosiae* CORBETT on the leaflets of *Tephrosia purpurea* PERS. and by other aleurodids on *Neolitsea odoratissima* NEES and *Phoebe lanceolata* NEES from India[745]. All these pit galls are about 2 mm in diameter, with an hypophyllous pit and subconical epiphyllous bulging-out of the blade. The pit-like depression of the blade is only as deep as the thickness of the aleurodid larva, so that the dorsum oɪ the latter lies flush with the general surface of the leaf.

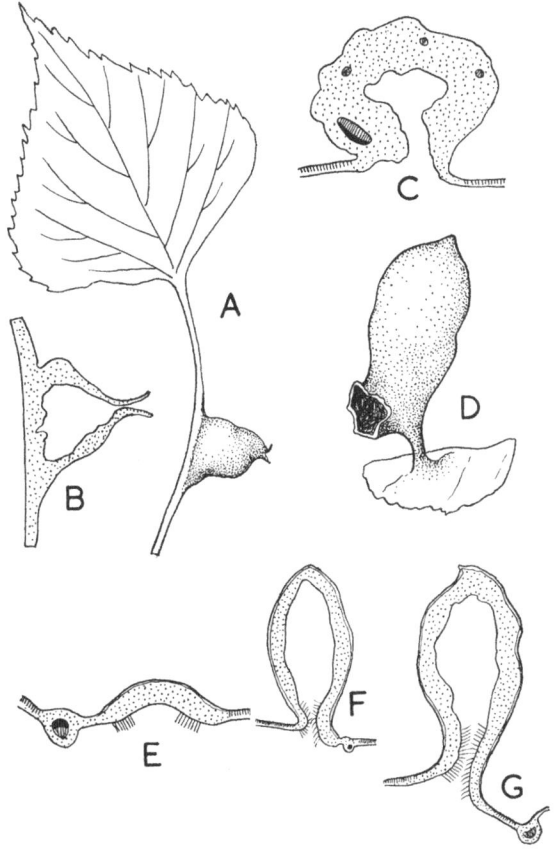

Fig. 71. Covering and pouch galls by aphids. A. Covering gall of *Pemphigus bursarius* (Linn.) on *Populus tremula* Linn. B. L.S. through the same. C. L.S. through the gall of *Pemphigus filaginis* Fonsc. on *Populus*. D. Gall of *Byrsocrypta gallarum* (Gmelin) on *Ulmus*. E–G. T.S. through successive stages in the development of the same gall.

A somewhat more complex type of pouch gall consists of more or less foliaceous or coriaceous, utricular and widely open out-pocketings of the leaf, sometimes covered by erineal outgrowths on the inner surface, but without conspicuous swelling of the blade. Such "Ausstülpungsgallen" of KERNER[578] are caused by *Eriophyes laevis inangulis* NAL. on *Alnus glutinosa* GAERTN., *Eriophyes tiliae exilis* NAL. on *Tilia platyphyllops* SCOP. from Europe, *Eriophyes cordiae* NAL. on *Cordia myxa* LINN. from the Orient, *Eriophyes salvadorae* NAL. on *Salvadora persica* LINN. (Plate IV, 1), *Eriophyes* spp. on *Sarcococa brevifolia* STAPF., *Holoptelea integrifolia* PLANCH (Plate IV, 3) and by *Trioza fletcheri* CRAWF. on *Trewia nudiflora* LINN. from India[745]. Such out-pocketings arise sometimes only in the angles of veins, as in the case of the gall of *Eriophyes pulchellum* SCHLECHTEN. on *Carpinus betulus* LINN. from Europe[982]. While most of the Ausstülpungsgalls are sac-like, TROTTER[1185] has described an interesting pouch gall, caused by an unknown gall-midge on the leaf of *Boscia* sp. from Italy, consisting of a series of flattenend, elliptical, hollow, squamiform outgrowths along the midrib and larger side veins on the under side of the blade.

The greatest majority of pouch galls develop on the upper surface of the leaf and the cecidozoa are naturally situated on the underside. A wide range of shapes and external characters are met with in pouch galls, some of which are diagrammatically shown in fig. 17. Most pouch galls are fleshy or hard, with thick walls and often have the ostiole more or less completely obliterated. Several of them have also fleshy or other outgrowths projecting from the wall into the cavity. Most pouch galls are remarkable for the total inhibition of differentiation of the normal tissues. The gall epidermis on the outer surface is continuous with that of the upper surface of the leaf and has often a conspicuously thicker cuticle than on the normal leaf (fig. 22). The outermost layer of cells of the gall tissue is sometimes characterized by distinct thickening of cell walls (fig. 74), but the cells are generally closely packed together. Chlorophyll is only sparingly present in these cells, particularly in younger galls. The deeper cells generally tend to increase in size, so that the bulk of the gall tissue is composed of large-sized, thin-walled parenchyma cells, with considerable intercellular spaces or closely packed. The innermost layer of cells, lining the gall cavity are small-sized and closely packed. Numerous vascular bundles, scattered in the gall tissue, are connected with those of veins. These vessels have distinct xylem and phloem, the latter being directed to the inner side of the gall, in other words to the gall cavity. Latex and other secretory canals occur in pouch galls on plants in which these are normally found. The epidermis lining the gall cavity is very profoundly modified and is typically composed of rather large cells, with thin cuticle or without cuticle (fig. 103A).

Stomata are naturally never developed. Thick-walled erect or curved hairs are commonly found, especially near the ostiolar opening. Larger, plasma-rich, thin-walled, stumpy and spirally twisted,

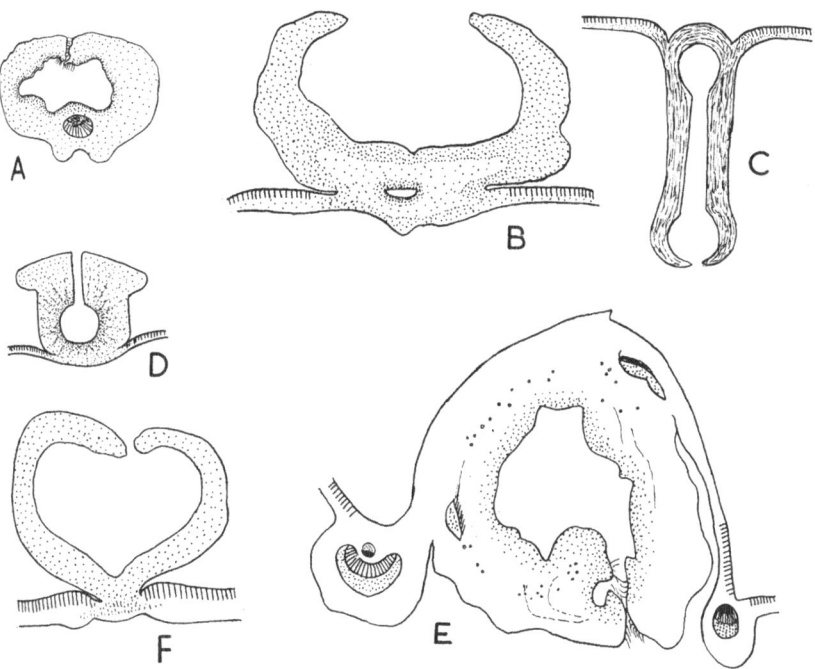

Fig. 72. Covering galls and pouch galls in L.S.   A. Gall of *Geocrypta galii* H. Löw on *Galium verum*, with narrow ostiole.   B. Midge gall on *Machaerium*, with wide open cavity.   C. Cylindrical pouch gall on *Erisma uncinatum* Warm., with elongate atrium.   D. Covering gall on *Erisma*.   E. Gall of *Perrisia urticae* (Perris) on *Urtica dioica* Linn., initially a covering gall, but just developing into an out-pocketed pouch (After Meyer).   F. Midge gall on *Tetrapteris*.

nutritive hairs are found in the cavity of many pouch galls (fig. 103).

The outstanding characters of pouch galls, caused by Acarina, are dealt with in Chapter IX. These pouch galls are generally of two basic types, viz. the elongate and horn-shaped or ceratoneon and the globose or the cephaloneon pouch galls. Eriophyid pouch galls are mostly remarkable for the fleshy emergences and hairy outgrowths and persistent ostiole. Some of them, like the gall of *Eriophyes cheriani* Massee on the leaf of *Pongamia glabra* Vent. from the Oriental Region, are curiously lop-sided (fig. 106). Others like the eriophyid gall on *Antidesma moritzi* Müll. from Saigon[527], are erect, cylindrical and longitudinally striated. A cylindrical or

116

conical, erect, stout and sessile pouch gall is caused by *Eriophyes* sp.
on *Cinnamomum* (Plate V, 4). A cylindrical pouch gall on the leaf of
Viburnum cotinifolium DON. caused by *Eriophyes* sp. from the Himalaya, is curiously curved apically, like the air-vent tube of ships.
The eriophyid pouch gall on the leaf of *Grewia* from the Orient is
enormously expanded apically and lobed and branched in a complex manner, but is inserted on the blade by a relatively short and
narrow neck-like stalk.

Lop-sided, but sac-like large pouch galls are caused by aphids
and are characterized by a persistent ostiole, which is, however, too
small to permit the escape of the adult aphids. Most of them develop, especially when approaching maturity, an asymmetrical bulge on
one side near the base. A lacerated crack appears in the bulge, thus

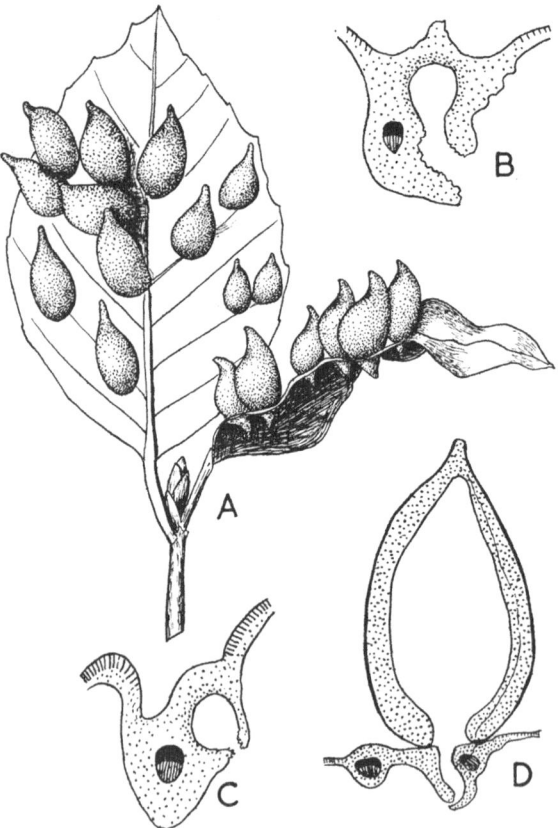

Fig. 73. Pouch gall of *Mikiola fagi* (Htg.) on *Fagus silvatica* Linn.   A. Galls in situ
on leaf.   B. Early covering growth.   C. Initiation of out-pocketing.   D. Mature
gall to show the typical separation layer below at base.

facilitating the escape of the aphids. This is, for example, the case with the gall of *Byrsocrypta gallarum* (GMELIN) on the leaf of *Ulmus campestris* LINN. and sometimes also of *Ulmus laevis* PALL. (fig.

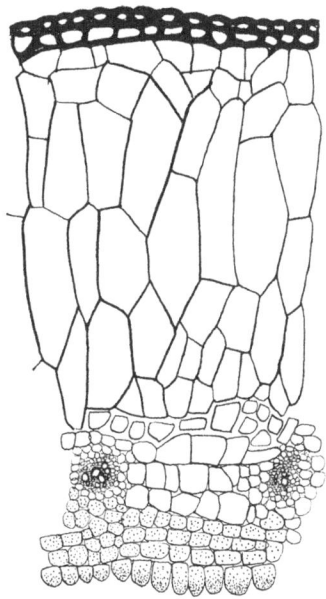

Fig. 74. Pouch gall of *Mikiola fagi* (Htg.) Part of the T.S., showing the nutritive cells on the left and the sclerenchyma element peripherally.

71 D). As the aphid always feeds at the top of the pouch gall, the growth is also confined to this region, so that the lower part remains small and narrow, but the upper portion expands and grows out irregularly large. In immature specimens of this gall the epidermis on the outer surface is exposed to the direct sunrays and often develops anthocyanin. Stomata are not developed on this epidermis in the normal leaf of *Ulmus*, but in the gall these appear in small groups and the epidermal cells are also elongated. Below the gall epidermis, there is a layer of closely packed subepidermal cells, with relatively little chlorophyll. The innermost layer of cells are stretched in all directions. On the inner epidermis, lining the gall cavity, we often find abnormally developed and non-functional stomata. Other common pouch galls of this type include the gall of *Eriosoma lanuginosum* HTG. on *Ulmus montana*, of *Byrsocrypta coerulescens* PASS. on *Ulmus montana* and *Ulmus campestris*, of *Pemphigus filaginis* FONSC. on *Populus*, etc. from Europe.

A number of pouch galls are hard and nut-like structures, with the ostiole completely closed. This is, for example, the case with the pouch gall on *Rhus semialata roxburghi* DC, caused by an unknown psyllid from Java and Indo-China. This gall is a subcylindrical, oval, hollow, hard, capsule-like swelling, with a beak-like apex and inserted in the middle of a short, circular, cylindrical, hollow cup-like swelling on the upper surface of the blade. The pouch gall of another psyllid *Phacopteron lentiginosum* BUCKTON on *Garuga pinnata* ROXB. from India (Plate II, 7) is a subspherical or subcylindrical, somewhat compressed, nut-like, yellowish-green, violet or reddish-brown, epiphyllous outgrowth, inserted in the middle of a conspicuous cup-like intumescence of the blade, basally near the midrib. Usually half a dozen galls develop close together in a cluster. The gall cavity is large and the ostiole is completely obliterated.

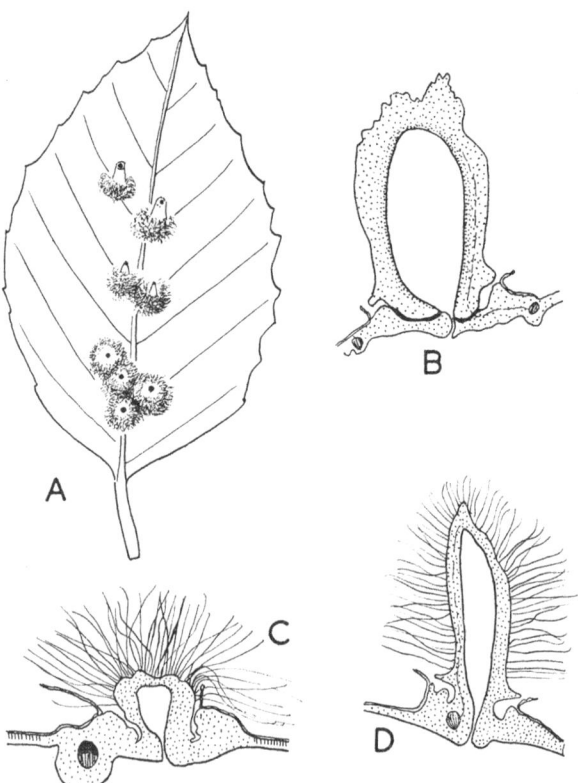

Fig. 75. Pouch gall of *Hartigiola annulipes* (Htg.) on *Fagus silvatica* Linn.   A. Gall on leaf.   B. T.S. through mature gall, showing the separation layer.   C. Early stage of development, showing the gall erupting through the upper epidermis and the hairs arising from the deeper mesophyll cells.   D. Nearly mature gall.

Very often pouch galls do not arise as a result of localized out-pocketing, but develop as elongate, fold-like bags. A unique pouch gall, caused by an undescribed species of *Eriophyes* on the leaf of

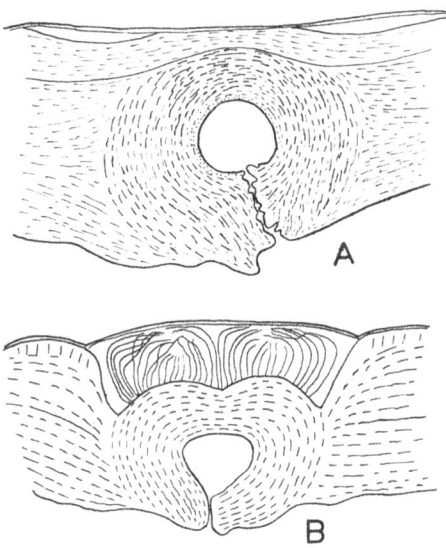

Fig. 76. Development of the pouch gall of *Hartigiola annulipes* (Htg.) on *Fagus silvatica* Linn. A. Cell proliferation around the larval cavity wholly localized within the lower mesophyll cells, the epidermis not taking any part. B. Hairs develop not from the epidermis but from the mesophyll on the internal gall, just before the rupture of the upper epidermis.

Millettia sp. (?) from India (Plate VIII, 1), is a hypophyllous, elongate and irregularly sinuate, hollow, worm-like ridge, 10—25 mm long and 10—15 mm high, but only 3—5 mm thick and running parallel submarginally on the blade[745]. The cavity is filled with typically developed erineal hairs. On the upper surface of the leaf, the site of the gall is indicated by a more or less deep and narrow groove. The gall of *Pemphigus imaicus* CHOLOD. on the Himalayan *Populus ciliata* WALL. is also a specialized pouch gall (Plate VI, 1), in which the blade on the sides of the midrib is out-pocketed, so as to give rise to a hypophyllous, cylindrical, stout, thick-walled swelling, with a superficial resemblance to a naked caterpillar.

In some pouch galls the out-pocketing of the blade on the upper surface is also associated with a partial downward invagination of the part, so that the pouch gall, when full grown, comes to lie in a sort of trough and though really epiphyllous, appears to be largely hypophyllous. The evaginated part becomes enormously swollen and fused below at the sides with the edge of the invaginated por-

120

tion. This is, for example, the condition in the gall on *Sterculia* sp., caused by an unknown gall-midge from Togo[520]. This is a globose-conical, nearly solid swelling, 5 mm in diameter, with a solid, epiphyllous, conical and a larger hemispherical hypophyllous part, containing a flattened oval larval cavity (fig. 79 A).

Pouch galls often arise initially as covering galls, but subsequently follow the usual course of out-pocketing. This is, for example, what we find in the gall of *Perrisia urticae* (PERRIS) on *Urtica dioica* LINN. from Europe[777]. This is a typical pouch gall that develops initially as a covering gall (fig. 72 E). The pouch galls on *Rhus semialata roxburghi* DC and *Garuga pinnata* ROXB., referred to above also belong to this type; the cup-like basal swellings in which these galls are inserted are indeed what are left of the initial covering growths. The best example of a pouch gall arising at first as a covering gall is found in the gall of *Mikiola fagi* (HTG.) on the leaf of *Fagus silvatica* LINN. from Europe (fig. 73). This is typically an oval, epiphyllous gall, about 10 mm long, pointed apically like a beak and strongly constricted basally and a minute hypophyllous ostiole is situated on a short conical process. The gall tissue consists of an outer sclerenchyma zone and an inner thin-walled nutritive zone (fig. 74). The

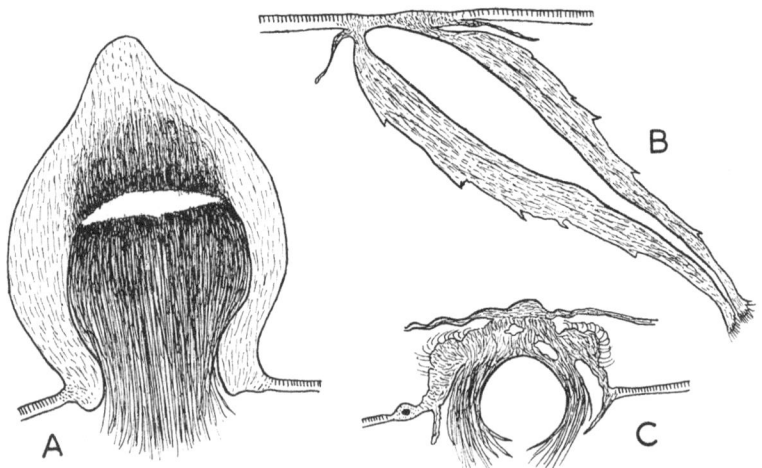

Fig. 77. Covering and pouch galls.   A. Midge gall on leaf of *Vitex*, with the larval cavity plugged by a tight tuft of hairy processes, formed as a result of cell division and cell elongation at right angle to the surface of the blade. The hairy tuft falls off when the gall matures. B. Covering gall of an unknown gall-midge on *Psidium* (After Rübsaamen), developing wholly from under the lower epidermis and finally breaking through it.   C. Pouch gall of an unknown Psyllid on *Trichilia* sp., developing wholly within the mesophyll and later breaking through and carrying the upper epidermis like a flat hood above.

outer epidermis lacks stomata and is made of flat, thick-walled cells. Immediately below the epidermis is a layer of smaller, thick-walled cells, which become gradually larger inward. Then follows the scleren-

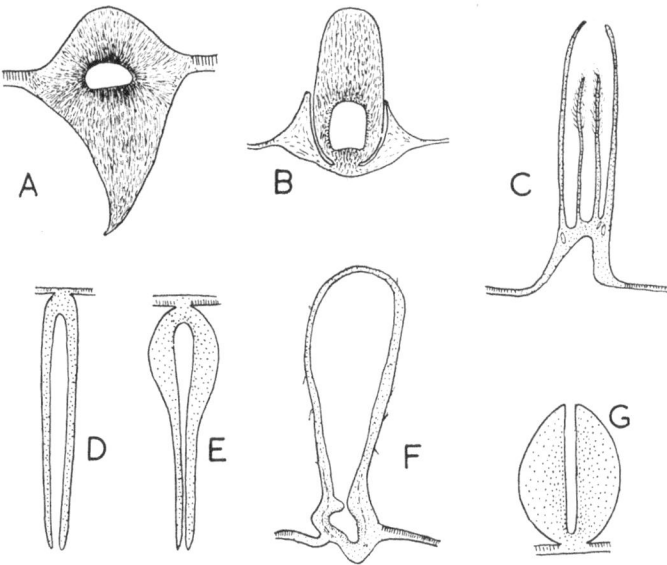

Fig. 78. Some remarkable covering and pouch galls of gall-midges on leaves. A. Gall on *Ficus benjamina* Linn., showing the actively dividing cells near the larval cavity and the general direction of cell division and cell elongation. B. Gall of *Cecidomyia tubicola*. C. Double-tubed gall on *Serjania*. D. Gall on *Antidesma montanum* Bl. E,G. Gall on *Machilus* sp. F. Gall on *Brownea*, with a small larval cavity below and a relatively enormous empty atrium above.

chyma zone of radially elongated and partially lignified cells. Between this sclerenchyma and the inner nutritive zones we find some collateral vascular bundles. The vascular elements of the gall are in communication with the vascular bundles in the veins. As shown by Ross[981], cell proliferation does not take place precisely where the midge larva feeds, but at some distance away. Around the larva a zone of about 15—20 cells is thus formed. The cells of the lowest part of the mesophyll become greatly hypertrophied, elongated and then undergo rapid proliferation. The larva becomes soon surrounded and closed in, the tissues gradually meeting together, but leaving only a minute aperture (fig. 73 C). The whole growth is at this stage visible as a bulge on the under side of the leaf. Hereafter the typical pouch-like growth follows. A very similar pouch gall, arising initially as a covering gall, is caused by an unknown gall-midge on the leaf of *Quercus incana* Roxb. from India.

Reference should also be made to pouch galls that develop initially wholly within the mesophyll, without the upper epidermis taking a part in cecidogenesis. In the interesting pouch gall on the leaf of *Trichilia* sp., caused by an undescribed psyllid from Africa[997], the epidermis of the upper surface of the leaf is carried like a flat hood on the top. The gall is a short, cylindrical, hairy swelling, formed by the upward bulging growth from within, breaking through and carrying the epidermis above (fig. 77 C). From the hypophyllous concave side grow elongate hairs, filling the space between the cylinder and the bulged part of the blade. The cylindrical part of the gall arises primarily wholly under the epidermis, which does not take part in cell proliferation, so that the gall breaks through the epidermis and carries it like a flat hood. On the under side and deeply sunk in a thickly swollen infolding of the blade, arises the hypophyllous spherical portion, the end of which bears numerous long, multicellular hairy emergences, curving inward to cover the ostiole. The pouch gall caused by an unknown gall-midge on the leaf of the Brazilian *Nectandra* sp., consists of a basal, globose, hollow, thin-walled, shortly stalked swelling on the under side of the blade and surmounted by a circular, flat, thin disc, larger than the spherical base and densely clothed with hairs. This gall also arises from below the epidermis.

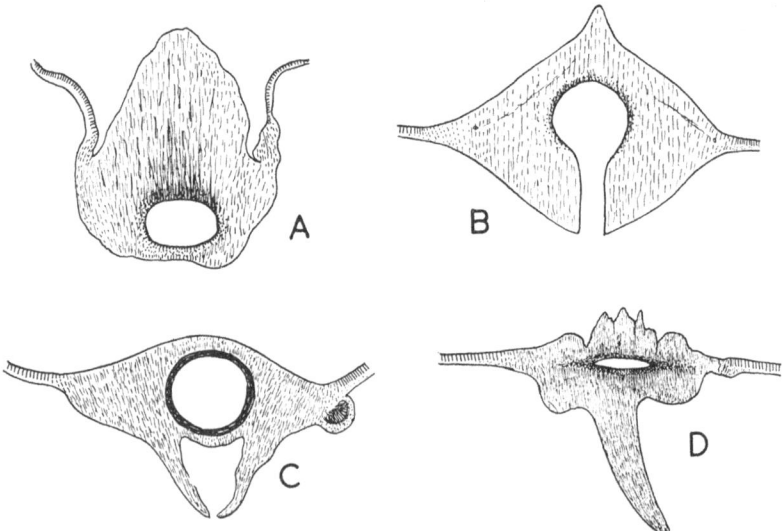

Fig. 79. Some remarkable leaf galls. A. Midge gall on *Sterculia,* showing the dominant direction of cell division and cell elongation. B. Gall on *Baccaurea racemosa* Müll.-Ar. C. Midge gall on *Melhania futteporensis* Munro, with sclerenchyma zone round the larval cavity and atrium below. D. Midge gall on *Hydnocarpus laurifolia* (Dennst) Steun.

Some pouch galls arise in between the folded halves of the leaf blade and bind these halves together, although the rest of the blade is normally developed. *Asphondylia ipomaeae* FELT causes on

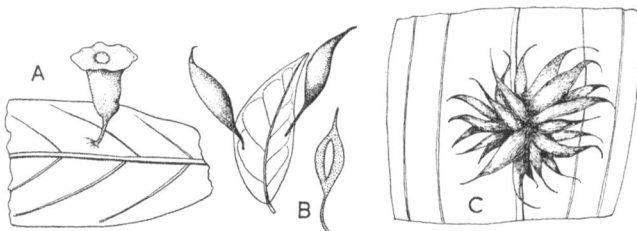

Fig. 80. Some remarkable leaf galls. A. Midge gall on *Machilus* (After Trotter). B. Beaked and pedicelled gall on *Hymenaea courbaril* Linn. (After Houard). C. Agglomerate bunch of galls of an unknown Lepidoptera on leaf of *Tibouchina* sp. (After Tavares).

Ipomea staphylina ROEM. & SCH. from India a smooth, spherical or pyriform, yellowish-green, thick-walled, fleshy pouch gall, lying wholly within the leaf folded along the midrib upwards and connected by short, slender, neck-like prolongations at the two ends transversely to the two halves of the blade. The gall develops usually in a linear series of 6—8, immediately above the midrib on the upper surface of the leaf.

We may also now refer to some of the other singular types of pouch galls. The pouch gall of the gall-midge *Daphnephila haasi* KIEFF. on the leaf of *Machilus gamblei* KING from India is an elongately fusiform, basally stout and narrowly stalked swelling, 12—18 mm long and 3—5 mm thick sub-basally (fig. 78E, G). *Neolasioptera* sp. causes on the leaves of *Machilus odoratissima* NEES from India an oval or sub-pyriform, solid, shortly stalked pouch gall, about 15 mm long and 8 mm thick, with narrow, cylindrical axial larval cavity (Plate IV, 4). A curiously horned pouch gall is caused by an unknown gall-midge on the leaf of *Hydnocarpus laurifolia* (DENNST) STEUN. *(Hydnocarpus wightiana)* from India (fig. 79D). This is a flat, almost disc-like, circular swelling, with a conical, slightly curved, solid horn about 2 mm long on the epiphyllous side. A curiously conical epiphyllous pouch gall on the leaf of *Vitex grandiflora* GÜRKE, caused by an unknown midge, from East Africa, is a hard, thick-walled out-pocketing, about 9 mm high, with the rim of the bulged part wide. The cavity thus enclosed is practically completely filled with a closely packed, solid mass of straight hairy outgrowths, projecting below as a conspicuous sub-spherical mass (fig. 77 A). The larval cavity is a narrow lenticular space within the out-pocketed portion, above the large hairy bundle. An elongately fusiform pouch gall, inserted by a long stalk on the lower side of the leaf of the South American *Hymenaea courbaril* LINN. (fig. 80 B), is a hard, woody swelling, about 35 mm long and only 6—8 mm thick, with an apical, pointed filiform process and narrow axial larval cavity[521]. A pouch gall on the leaf of the South American *Rubus* sp. is a hypophyllous, fleshy, globose swelling, 3 mm in diameter, with a large, central larval cavity and an atrium of somewhat flattened shape, opening to the outside. The greatest part of the pouch gall on *Brownea* sp. from Brazil [533] is an atrium, but the larval cavity is within the hollow swelling of the midrib or one of the larger side veins. The pouch gall of *Pachypsylla mamma* RILEY on the leaf of *Celtis* sp. from North

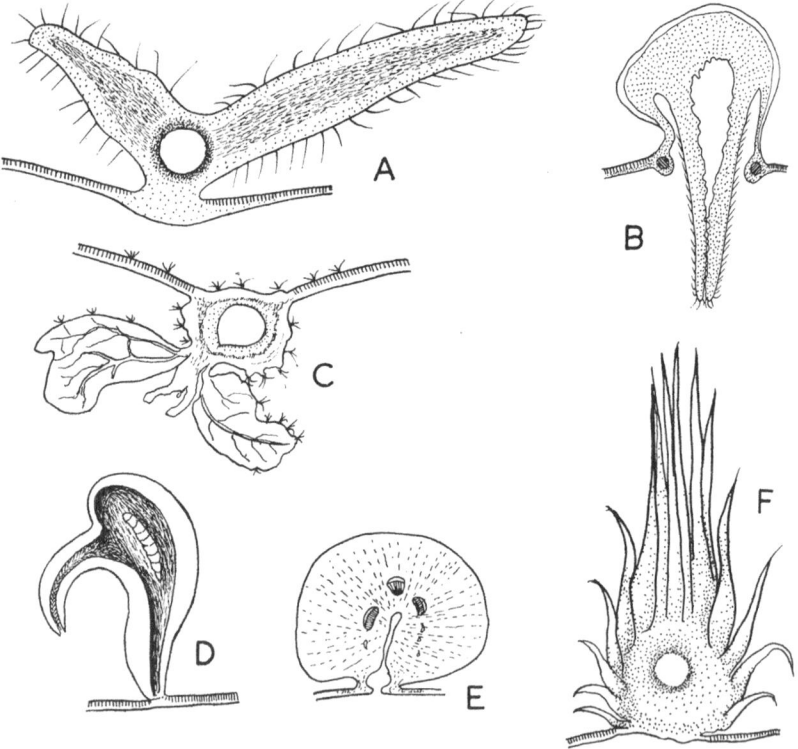

Fig. 81. Some remarkable midge galls on leaf. A. Gall on a Brazilian species of *Swartzia*, with unequal, fleshy horn-like outgrowths. B. Pouch gall of *Dasyneura ulmariae* Bremi on *Ulmaria pentapetala*, with a peculiar covering growth below. C. Midge gall on an unknown Brazilian plant (Tiliaceae) with foliaceous outgrowths (After Rübsaamen). D. Covering gall, with lateral recurved exit tube. E. Pouch gall (button or pellet gall) of *Cecidomyia pellex*. F. Covering gall on leaf of *Piptadenia communis* Benth., with elongate imbricating outgrowths (After Rübsaamen).

America, is a curiously thimble-shaped, stout cylindrical, truncated, sessile, solid outpocketing, with the base deeply excavated below. The bulk of the gall is solid, except for a narrow, miniscus-shaped larval cavity in the middle.

9. Nodular galls

Nodular galls are specialized pouch galls, hemispherical on the upper surface of the blade and more or less conically developed on the lower side and nearly also solid. The galls of *Eriophyes ulmicola punctatus* NAL. on *Ulmus campestris* LINN., *Eriophyes laevis lionotus* NAL. on *Betula verrucosa* EHR. from Europe, *Eriophyes prosopidis* SAKSENA on *Prosopis spicigera* LINN. from India (Plate V, 5, 6) are some of the common examples of nodular galls. These galls do not generally exceed a diameter of 2 mm. A nodular gall is also caused by *Tylenchus nivalis* KÜHN. on the leaf of *Leontopodium alpinum* CASS.[1183]. Bacterial nodule galls are dealt with elsewhere[537].

10. Lenticular galls

Lenticular galls are characteristically thin, flattened, disc-shaped swellings of the blade or outgrowths on the blade, with a

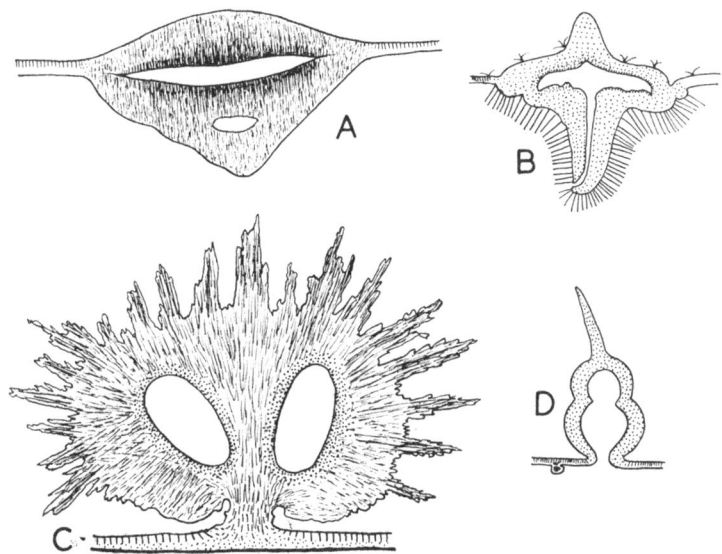

Fig. 82. Midge galls on leaf in sections. A. *Vitex* sp. (After Rübsaamen). B. *Arnoldia homocera* F. Löw on *Quercus cerris*. C. Echinate gall of *Amradiplosis echinogalliperda* Mani on *Mangifera indica* Linn. D. Horned pouch gall on *Berlinia acuminata* Sol.

small central larval cavity; they are typically mark-galls. A number of typical lenticular galls are caused by cynipids on *Quercus*. The structure and development of lenticular galls present a number of peculiarities. The reader will find useful information on lenticular galls in the works of LACAZE-DUTHIERS[648], FOCKEU[336], HIERONY-MUS[462], PRILLIEUX[917], SAJO[1006], WEIDEL[1234], MAGNUS[731], KÜSTEN-MACHER[634], GREVILLIUS & NIESSEN[419] and others.

The general features of lenticular galls may be illustrated by reference to some of the typical lenticular galls on *Quercus*, caused by the cynipid *Neuroterus*. The lenticular gall of the bisexual generation of *Neuroterus quercus-baccarum* LINN. (= *Neuroterus lenticularis* OL.) is a flat, circular disc, fixed by a short stalk to the lower surface of the leaf blade (fig. 84 A). The disc is usually 6 mm in diameter, but never more than 2 mm thick in the centre. The margin of the disc is flat and the upper surface is generally greenish-yellow or also tinted reddish and clothed with a few stellate hairs. The gall arises from cell proliferation of the cambiform tissue of the vascular bundles that happen to lie closest to the eggs of the cynipid, deposited within the leaf tissue. The young gall breaks through the layers of the cells above, including the spongy parenchyma and subepidermal cells and pushes its way through to the surface of the leaf as hemispherical growth. From now onwards, the growth of the gall is horizontal. The relatively small larval cavity is central and is nearly filled by the larva. The larval chamber is surrounded by a uniform layer of nutritive cells, but all other tissues are typically flattened and disc-shaped. The disc is itself composed of several of simple parenchyma cells. Around and outside of the nutritive layer

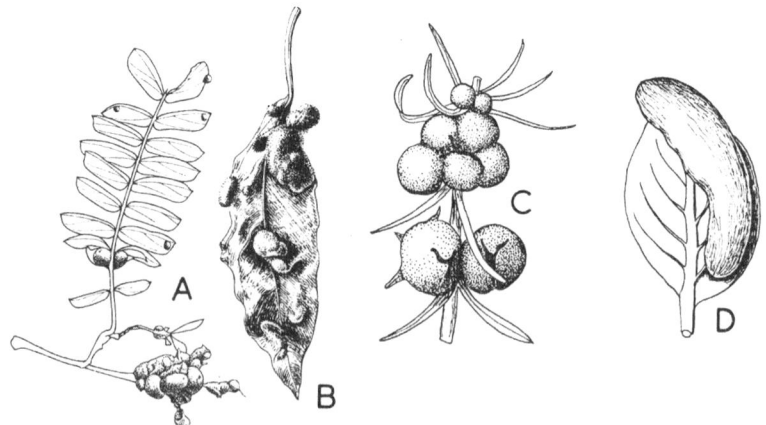

Fig. 83. Some remarkable leaf galls. A. Pouch gall of *Eriophyes prosopidis* Saksena on *Prosopis*. B. Pouch gall of *Eriophyes* on *Salvadora persica* Linn. C. Covering gall of *Geocrypta galii* H. Löw on *Galium verum*. D. Midge gall on *Sambucus*.

are one to three layers of sclerenchyma cells, which are absent on the side of the stalk, where only thin-walled parenchyma cells are found. The thickening of the walls of the sclerenchyma cells is very pronounced, but often only on the outer side of the gall. Vascular bundles radiate from the stalk into the gall parenchyma and are connected with the vessels of the blade. The epidermal cells of the gall in the centre of the disc are generally thickened all around and sometimes even the subepidermal cells are thickened in this area. It may also be noted the gall epidermis is a *de novo* tissue, derived not from the epidermis of the leaf, but from the mesophyll cells.

The lenticular gall of *Neuroterus numismalis* FOURC. is associated with the unisexual generation of the cynipid and develops often in enormous numbers, as many as 1000 galls on a single leaf. BAUDYS[65] observed, for example, no less than 1342 galls on a single leaf. This lenticular gall (fig. 84 C) is also a hypophyllous, circular disc, about 3 mm in diameter and only 1 mm high, flat at first, but developing later a stout cylindrical rim all around the margin, so that the upper surface of the gall becomes a shallow flat basin. The lenticular gall of the unisexual generation of *Neuroterus tricolor* HTG. has also the margin of the disc raised upward to form a shallow cup-like depression on the upper surface. In the centre of this cup there is a characteristic hemispherical tubercle (fig. 84 J). One of the commonest lenticular galls in North America is formed by *Andricus discalis*

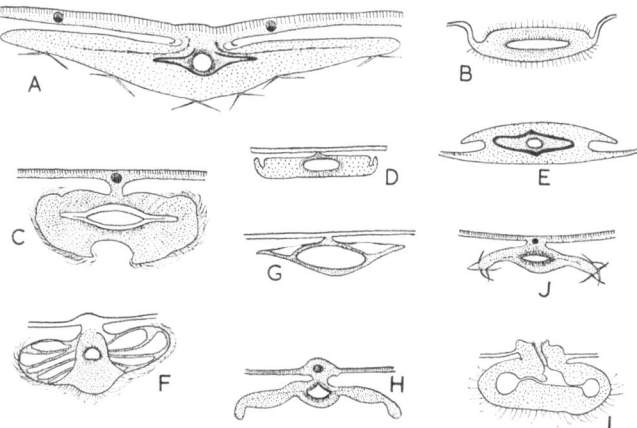

Fig. 84. Lenticular galls in sections. A. *Neuroterus quercus-baccarum* Linn. B. Midge gall on *Zizyphus horsfieldii* Nig. C. *Neuroterus numismalis* Fourc. asexual generation. D. *Andricus discularis* Weld on *Quercus garryana*. E. Midge gall on *Ardisia attenuata* Wall. F. *Neuroterus lanuginosus* Gir. on *Quercus cerris*. G. *Andricus discalis* Weld on *Quercus subturbinalis*. H. *Neuroterus albipes* Sch. unisexual generation. I. *Bryomyia circinans* Gir. on *Quercus cerris*. J. *Neuroterus tricolor* Htg. asexual generation.

WELD on the leaf of *Quercus subturbinella* (fig. 84 G). This is a circular disc, somewhat convex above and almost flat below, about 3 mm in diameter, with a large, flat-oval larval chamber in the middle. Another lenticular gall from North America is caused by *Andricus discularis* WELD on *Quercus garryana*. This gall is a uniformly thick, flat, button-like, circular disc, about 6—8 mm in diameter but only 1—1.5 mm thick, flat above and below and with a circular submarginal groove on the under side (fig. 84 D). It is inserted on the blade by an extremely short, point-like stalk. The larval cavity is a flat-oval, central space.

Lenticular galls, caused by gall-midges, have also been described from different parts of the world. We have, for example, a lenticular gall on *Millettia sericea* W. & A., from Java, in the form of a circular disc 5 mm in diameter and 1 mm thick, inserted by a short, stout stalk. The surface of the gall is clothed with simple, elongate, unicellular, acutely pointed hairs. The depressed central larval cavity is surrounded completely by nutritive and sclerenchyma zones. There is a characteristic conical projection in the centre of the disc on the surface of the gall; the sclerenchyma zone extends into the tissue of the conical projection (fig. 25 F)[302]. A lenticular swelling of the leaf of *Zizyphus horsfieldii* NIG., caused by a gall-midge from Java[302], is a disc-like swelling in a depressed fold of the blade, 5 mm in diameter and 1.5 mm thick (fig. 84 B). The larval chamber is an extremely narrow flattened space in the centre of the gall. An epiphyllous lenticular gall on the leaf of *Ardisia attenuata* WALL., also caused by an unknown gall-midge from Java, is an extremely short, flattened-cylindrical outgrowth, with a circular, flat larger rim all around, 4—5 mm in diameter. The central larval cavity is surrounded by a depressed oval zone of sclerenchyma cells. An interesting lenticular gall, caused by an unknown midge from Africa on *Psychotria* sp., is a flat-cylindrical disc, 05.—0.7 mm thick, inserted on a vein on the under side of the blade [520].

11. Mark-galls

Although most of the lenticular galls dealt with above are also mark-galls, the best known mark-galls on leaves are perhaps the oak-apple galls on *Quercus* spp. These galls have usually a hard larval cavity, with spongy or fibrous material around and are generally caused by species of *Amphibolips, Cynips, Callirhytis, Disholcaspis*, etc. The gall of *Andricus amphora* WELD is a cylindrical outgrowth, inserted by a short stalk on the under side of the leaf. The larval cavity is at the base of a vase-shaped gall cavity. The roly-poly galls from America are also oak-apple galls, in which the central larval cavity becomes free from the surrounding gall tissue and rolls about in the space around. They are

usually spherical or conical, hollow and succulent galls, some
of which are extraordinarily delicately and loosely inserted on
the leaf blade and others are rather imbedded in the leaf. They

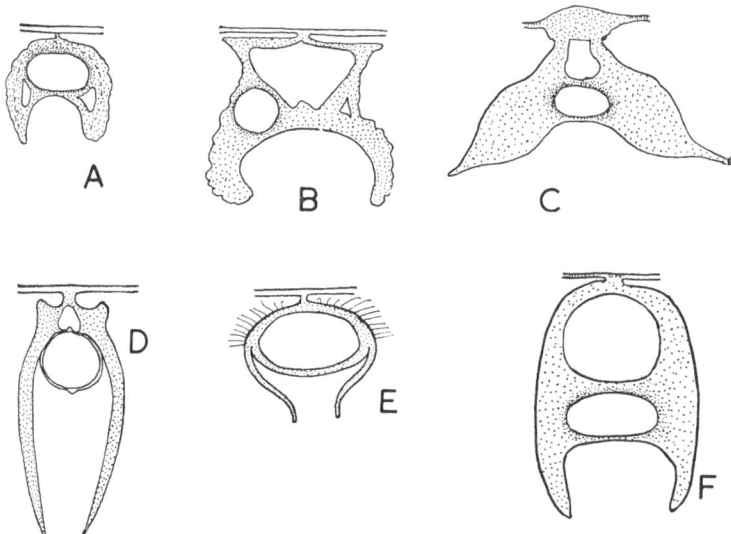

Fig. 85. Cynipid galls on leaf of *Quercus* in sections, showing atria. A. *Trigonaspis
cupella* Weld on *Quercus grisea*. B. *Andricus sessilis* Weld on *Quercus arizonica*.
C. *Andricus petalloides* Trotter. D. *Andricus amphora* Weld on *Quercus undulata*.
E. *Cynips caepula* Weld on *Quercus undulata*. F. *Trigonaspis vaccinioides* Trotter.

are caused by *Andricus palustris* O.S., *Neuroterus verrucosum*
O.S., *Amphibolips nubilipennis* HARR., etc. Some of the roly-poly
galls have facetted surfaces, or as in the sea-urchin galls (fig. 86 C) of
Acraspis echinus and *Andricus stellaris*, have spiny outgrowths.
Some of the galls on *Quercus* are curiously multi-chambered, but
only one of these numerous gall cavities is the real larval cavity and
the others are atria. In the gall of *Andricus sulfureus* WELD on
Quercus from North America we have a hypophyllous, cylindrical-
conical hairy outgrowth, in which the larval cavity is the bottom-
most of the three cavities (fig. 86 F). In the barrel-shaped hypophyllous gall of *Andricus sessilis* WELD from North America the
gall cavity is a large space in the wide base. Above this lies a broad,
deep, open, cup-shaped cavity and the real larval cavity is a small,
somewhat laterally placed hollow space between these two (fig. 85 B).
A tubular gall, about 25 mm long and 8 mm thick, slender basally
and widened apically and clothed with long, acutely pointed fleshy,
spine-like emergences, is caused by *Xanthoteras tubifaciens* WELD
on the leaf of *Quercus garryana* from North America (fig. 86 D).

This gall is remarkable for its three separate cavities, of which only the middle one is the actual larval cavity. The basal cavity is a conical empty space and the apical one is a vase-shaped open space. A clavate, short, wedge-shaped, hypophyllous gall, caused by the unisexual generation of *Cynips insolens* WELD on *Quercus chrysolepis* from North America, is about 12 mm long and 5 mm thick. It has a flat bottom, but is inserted on the leaf by a short, slender stalk. The bulk of the gall is occupied by an axial, clavate-shaped hollow space, but the larval cavity is an extremely small, flat oval cell, just beneath the tip of the gall. A curiously drumstick-shaped pubescent gall is caused by *Xanthoteras teres* WELD on the veins on the under side of the leaf of *Quercus garryana* from North America (fig. 86 B). The small, rounded, pinhead-like larval cavity is borne at the tip of a long, slender, straight stalk, about 6 mm long and 2 mm thick. The gall of *Dryophanta pedunculata* BASS. from North America is also a drumstick-like growth, with a pyriform, beaked larval cavity, borne at the tip of the long, slender stalk, 5—6 mm long. An extraordinarily elongated, slender, pedunculate gall is caused by *Andricus chinquapin* FITCH on *Quercus stellata* from North America. At the

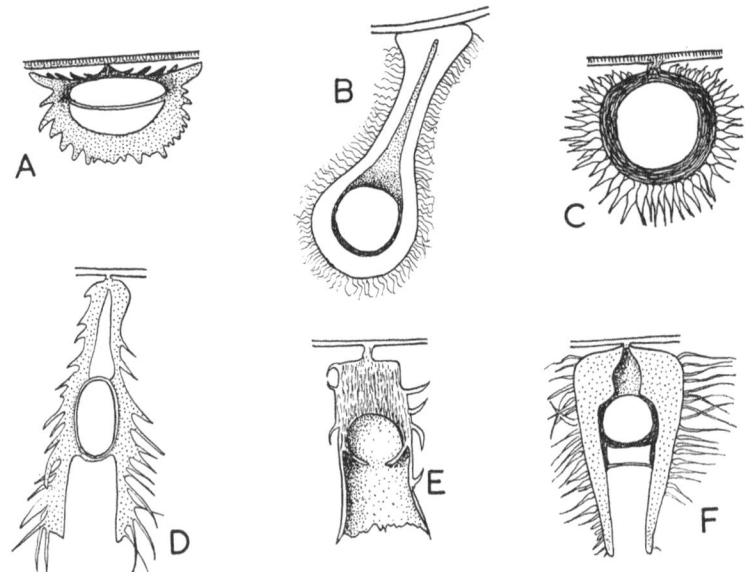

Fig. 86. Cynipid galls on leaf of *Quercus* in sections. A. *Andricus stellaris* Weld on *Quercus garryana*, with larval cavity above and empty cell below. B. *Xanthoteras teres* Weld on *Quercus garryana*. C. Sea-urchin gall of *Acraspis acraspiformis* Weld on *Quercus grisea*. D. *Xanthoteras tubifaciens* Weld on *Quercus garryana*. E. *Diplolepis splendens* Weld on *Quercus grisea*. F. *Andricus sulfureus* Weld on *Quercus oblongifolia*.

end of a slender, long, straight, cylindrical stalk is a diffuse, elongate-fusiform swelling, produced into a slender style apically. A finger-bowl-shaped gall is caused by *Cynips petalloides* WELD on

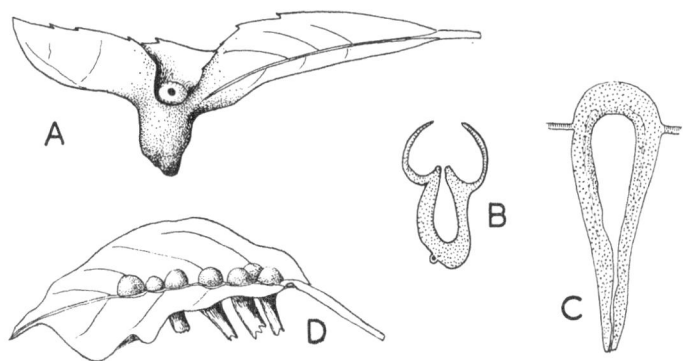

Fig. 87. Some remarkable leaf galls. A. Pouch gall of an unknown psyllid on *Quercus mexicana* Humb. B. Section through the same. C. Pouch gall of *Craneiobia corni* (Gir.) on *Cornus sanguinea* Linn. D. L. S. through the same.

Quercus chrysolepis from North America. The upper surface of this gall is a deep, cup-shaped depression. The oval larval cavity is situated in the centre, below this depression and beneath it is an empty oval atrium. The globose or elongate multilocular yellow or reddish gall, caused by *Acraspis prionoides* WELD (unisexual generation) on *Quercus alba* from North America, is the common hedgehog gall, characterized by its dense covering of acutely pointed spines. Cylindrical spiny galls are caused by *Diplolepis splendens* WELD on the Mexican oak *Quercus grisea* (fig. 86 E). These are solid basally, with a central spherical larval cavity and an apical, truncated, wide-open atrium[528]. The gall of *Trigonaspis vaccinioides* TROTTER on the leaf of the Californian oak[1172] is a hypophyllous, subcylindrical outgrowth, fixed by a short stalk to a vein, 7 mm long and 5 mm thick, truncated apically and with a deep cup-like depression, forming a wide-open atrium. The larval cavity is situated below this atrium and at the base of the gall is another atrium.

Among the numerous galls on the leaf of *Mangifera indica* LINN. mention may be made of the conical-cylindrical, hard, epiphyllous gall, caused by an undescribed gall-midge from India (Plate VI, 5). An echinate sea-urchin gall on the leaf is caused by *Amradiplosis echinogalliperda* MANI (Plate VII, 4, fig. 82 C). This is an epiphyllous, sessile, reddish-brown swelling, about 5—7 mm in diameter and with numerous fleshy, spiny emergences on the surface. The larval cavities are generally paired and surrounded by a thick nutritive

132

and sclerenchyma zones. This gall is derived by cell proliferation of
the upper epidermis and subepidermal cells. The mass of the gall and
the emergences are composed of simple parenchyma cells. The gall
of *Haplopalpus serjaniae* RÜBS. on the Peruvian *Serjania* sp.
(Sapindaceae) is remarkable for its double-walled, cylindrical-tubu-
lar structure, swollen basally. The outer wall is much longer than
the inner wall (fig. 78 C), which is also thicker and encloses an axial
larval cavity. A curiously bell-shaped, solid, parenchymatous gall,
inserted by the handle of the bell on the under side of the leaf of
Machilus sp., is caused by an unknown gall-midge from Formosa
(fig. 80 A)[1181].

12. Some complex leaf galls

All the galls, which we have so far considered, belong to single
leaf blades. We may now refer to the complex galls, in the development of which more than one blade is involved. The simplest type
of this gall is found on *Dichrostachys cinerea* W. & A. from India.
Eriophyes dichrostachia TUCKER gives rise on this plant to biconvex,
spongy swellings between two adjacent leaflets of the same side of
the pinna of the bicompound leaf and the leaflets are in consequence
bound together (fig. 88 E, G). The gall tissue is derived partly from
one leaflet and partly from the other and the tissues from the two
sources are indistinguishably fused together. Very often 2—3 galls

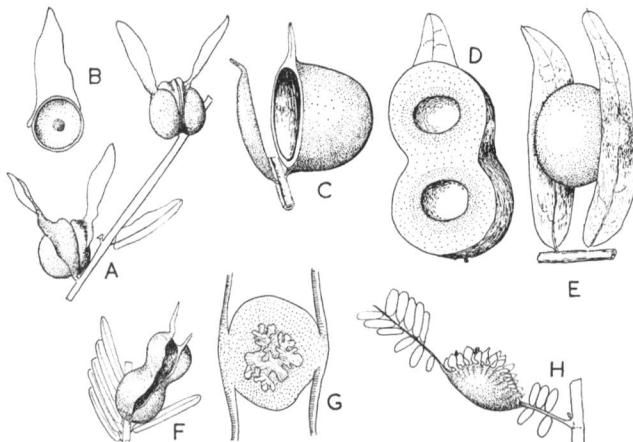

Fig. 88. Complex galls on leaf. A. Bivalve gall on leaflets of *Acacia venosa*
Hochst caused by an unknown gall-midge. B. The same gall cut open. C. Bivalve
gall of *Lobopteromyia bivalviae* (Rao) on the leaflets of *Acacia catechu* Willd. D,F
Lomentum-shaped midge gall on leaflets of *Acacia leucophloea* Willd. E,G.
Pockengall between two leaflets of *Dichrostachys cinerea* by *Eriophyes dichrostachia*
Tucker. H. *Parkia biglobosa* Benth.

arise between two adjacent leaflets. The galls are brilliant red, violet or pale yellow coloured. TROTTER[1178] has described a midge gall on *Acacia venosa* HOCHST from Eritrea (fig. 88 A), formed from a pair of leaflets. Subglobose swellings, about 3—5 mm thick, comprise two separable, symmetrical halves, each of which is derived from one leaflet, respectively one above the other, as basal cup-like swellings. These halves fit each other accurately by their rims, but do not become fused together. One of these cups is an epiphyllous growth from one leaflet and fits into a corresponding hypophyllous cup from the next leaflet immediately above. We may also refer to the utricular hemispherical, sessile galls, caused by *Eriophyes acaciae* NAL. on *Acacia leucophloea* WILLD. from India (Plate VIII, 3). Each gall is formed by subtriangular, coriaceous outgrowths from the cortex of the secondary rachis of the pinna and by similar swellings of the basal halves of the adjacent pairs of leaflets. These coriaceous outgrowths curve toward each other and meet in a cross-shaped slit, so as to enclose a spherical or hemispherical hollow cavity, containing numerous mites. The tips of the affected leaflets project as green, foliaceous wings, spread outwards on the summit of the gall. The gall measures only 2 mm in diameter, but a large number of them arise in a continuous series on every leaflet of every pinna of nearly all the leaves, causing a very conspicuous curling of the entire main rachis.

A singular looking, miniature lomentum-like green, solid gall, formed of a pair of fused leaflets of *Acacia leucophloea* Willd. is formed by an unknown gall-midge from India (fig. 88 D, F). Each gall is 2—2.5 mm long, 1 mm thick. The larval cavities are usually 2—3, one above the other, with a conspicuous nutritive cell zone. The outer surface of the gall represents the upper surface of one leaflet and the under side of another leaflet. *Schizomyia* sp. gives rise to subglobose, pubescent, solid swellings on the leaf of *Acacia leucophloea* WILLD. (Plate VII, 3)[745]. Each gall is externally bilobed, the lobes indicating the two swollen leaflets, the tissues of which have fused together to give rise to the solid structure. The apices of these leaflets remain, however, normal as minute wing-like growths on the summit of the gall. A densely villous and reddish-brown, elongate oval, solid gall, caused by *Asphondylia trichocecidarum* MANI on the same plant from India is also the result of tissue fusion of the enormously swollen basal parts of paired leaflets, a proximal and a distal leaflet on the same side of the pinna of the bicompound leaf. In both these galls there is no differentiation of the normal tissues and the gall consists of only a greatly altered epidermis, parenchyma and typical nutritive zone round the central larval cavities.

There are three other extra-ordinary galls on the leaves of *Acacia* spp. from India. The bivalve gall (Plate VIII, 2, fig. 88 C), described by MANI[745] is caused by *Lobopteromyia bivalviae* (RAO) on *Acacia catechu* WILLD. This gall consists of two unequal and asymmetrical, cup-shaped, thick, hard, basal swellings of two adjacent leaflets on the same side of the pinna. These cup-shaped valves fit each other accurately by their rims, without tissue fusion. The smaller valve is formed by the proximal leaflet and the larger valve by the terminal leaflet. The gall cavity is spacious. The gall is externally a curious spherical nut-like growth, armed with foliaceous wings apically and arranged serially one above the other on the leaf. Each gall is about 3.5 mm in diameter. The most remarkable of leaf galls from the world is the cylinder-piston gall caused by *Lobopteromyia* sp. on the leaflets of *Acacia suma* HAM.-BUCH. from India (Pl. VI, 2). This is a hypophyllous, sessile, elongate, slender, cylindrical, smooth, solid, yellowish-green, brown, red or vivid violet coloured outgrowth. Each gall consists (fig. 89 A) of a hollow, cylindrical, hypophyllous out-pocketing from one leaflet, into which fits accurately a solid, cylindrical, piston-like, epiphyllous, parenchymatous outgrowth from the leaflet immediately above. The solid piston fits into the hollow cylinder, filling out its cavity nearly completely, except for a short narrow nipple-like larval cavity at the very tip of the cylinder. The outer hollow cylinder is basally provided with a short, stout, circular hypophyllous rim, projecting conspicuously beneath on the

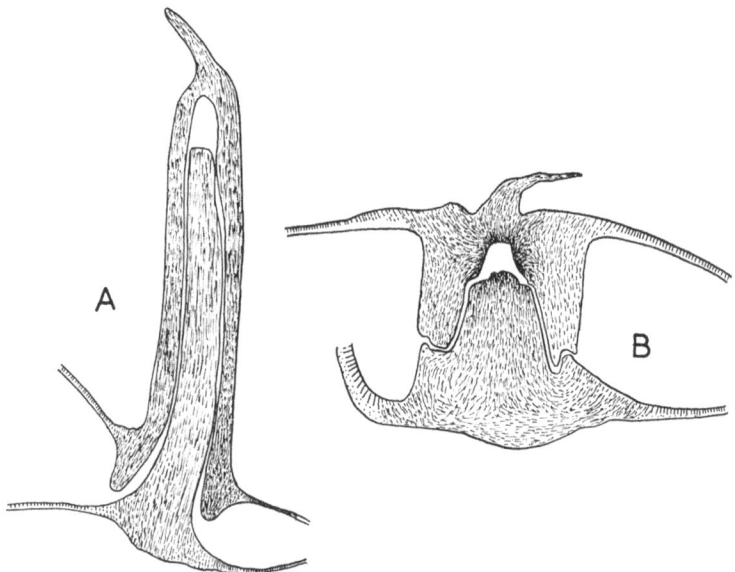

Fig. 89. Complex midge galls on leaf.   A. Cylinder-piston gall of *Lobopteromyia* sp. and   B. Barrel gall of *Lobopteromyia ramachandrani* Mani on leaflets of *Acacia suma* Ham.-Buchn.

other side of the leaflet and girdling the solid piston. The site of the piston on the other side of its leaflet is indicated by a discolourized and somewhat depressed area. The gall develops usually in an inter-locked series; one leaflet producing the cylinder on one side for one gall and the piston on the other side for another gall, so that all the leaflets of the pinna are curiously interlocked by their galls. Each gall is about 10 mm long, 1—1.5 mm thick and is thus actually much longer than even the leaflet on which it had risen. It may also be observed that there is nowhere tissue fusion between the solid piston and the hollow cylinder, although they fit each other extra-ordi-narily correctly and tightly. The piston is composed of closely pack-ed, thin-walled, greatly axially elongated parenchyma cells, lacking pigments. The cylinder is composed of nearly similar parenchyma cells, which contain however pigment in great abundance. A single larva of the gall-midge is found in the larval cavity, which it fills. A barrel gall is caused by *Lobopteromyia ramachandrani* MANI on the leaflets of the same plant from India (Plate VIII, 4). Each gall is a stout, hollow, thick-walled swelling from the upper surface of one leaflet, into which accurately fits a short, stout, stumpy-cylindrical, pestle-like plug, developed from a swollen disc-like lower side of the next leaflet immediately above (fig. 89 B). This leaflet in its turn bears the barrel for the next gall. Each barrel is prolonged into a long, curved, fleshy process from the free conical end. The gall is 3.5—4.0 mm long and 2.5 mm thick. The surface is somewhat rugose and reddish-brown. The plug reaches down to the bottom of the barrel, where the small larval cavity is situated. The apices and the narrow margins of the leaflets, bearing the serially interlocked galls, remain completely normal. The gall tissue consists of greatly hypertrophied and elongate parenchyma cells. The epidermal cells of the gall are subcolumnar, thick-walled and have a thick crenated cuticle.

CHAPTER VIII

GALLS ON FLOWER AND FRUIT

1. General characters of flower galls

Although flowers are highly specialized organs of plants, flower galls are on the whole relatively simple structures. Structurally and ecologically, they show some affinity to galls on vegetative buds (Chapter VI), but do not surpass them in complexity of structure and development. Inspite of this relative simplicity and low incidence, flower galls exhibit certain ecologic specializations, not found among galls on other organs. The morphogenesis of flower galls is in general more or less complicated by the morphogenetic patterns of the individual floral parts and often embraces an abnormal accentuation of these normal patterns. Though flower galls are caused by Fungi, Nematodes, Acarina, Thysanoptera, Heteroptera, Psyllidae, Aphidae, Brachyscelidae, Coleoptera, Cynipidae and Itonididae, the communities associated with these galls are perhaps the smallest of plant gall communities in general. The number of inquilines, parasites and other ecologic groups is exceptionally small in flower galls. As most flower galls are typically deciduous, successori (vide Chapter XI) may be said to be practically non-existent. Some of the other peculiarities of galls on flowers are related to the restricted seasonal appearance of flowers and the short duration of their development. Flower galls are thus phenologically far more restricted than galls on vegetative buds, so that in many cases the gall of the same species on flowers alternates with the gall on some other organ. The high proportion of galls in which the cecidozoa are situated not within the plant tissue but lie externally may perhaps be traced to the fact that a flower bud provides natural cover and other optimal conditions for groups like Acarina, Thysanoptera, Heteroptera, Psyllidae, etc. It seems remarkable that even in a high proportion of galls caused by Itonididae, the midge larvae remain external, between the different floral envelopes. The relative abundance of flower galls caused by different cecidogenetic organisms is shown in fig. 90. In the greatest majority of cases, the cecidogenetic attack is initiated in the bud stage or at least it synchronizes with the opening of the flower. The flower being a complex organ, not only the individual parts may be galled separately, but the flower as a whole and often also the entire inflorescence may develop into a gall. The cecidogenetic development of either the calyx or the corolla usually affects the development of the other, but the localization of cecidogenetic growth in the perianth whorls does not influence the normal development of the androecium and gynoecium. This localization does not, however, favour the normal development and opening of the flower. While the mode of folding of the rudiments inside the vegetative bud has a profound influence

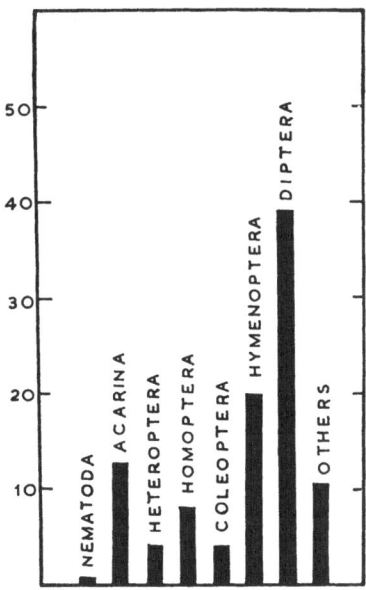

Fig. 90. Percentage frequency of galls on flowers caused by different organisms from the world.

on the development and structure of the bud gall and often also the gall on leaf, the arrangement and folding of the floral envelopes inside the flower bud have no significant relation to the resulting flower gall.

The simplest flower gall involves arrest of normal development and eventual failure of the flower opening, so that the gall is usually more or less bud-like in general external appearance. In some of these galls, either the calyx or the corolla or both and even the stamens and pistil turn into leaf-like growths, resulting in the condition known as phyllomany *(Verlaubung)*. In some cases, the growing point in the flower axis gives rise to numerous new leafy organs, reduced leaves, stunted branches, etc.[902]. The condition in which the floral envelopes merely turn green is known as chloranthy *(Vergrünung)*. Some flower galls have the effect of morphological sex reversal (vide Chapter XI). In the gall of *Ustilago violacea* PERS. on the pistillate flower of *Melandrium album* GARCKE stamens arise as abnormal growths and similarly in the gall of *Cintractia* on the staminate flower of *Carex* the pistil develops. Other abnormalities include swelling of the parts, inhibition of normal differentiation of parts and tissues, tissue fusion, etc.

As shown by GAMBIER[360], the general structure of a flower gall is determined to some extent by the position of the cecidozoa (fig. 91). We have already referred to the galls on flowers induced by cecidozoa, positioned inside the root or other distant organs[810] (vide Chapter I). When the cecidozoa occur within the flower gall, they may be situated externally on the corolla between the sepals and petals, inside the corolla between the petals and stamens, at the base of the stamens, inside the whorl of stamens or they may also be within the plant tissue and inside the ovary or become enclosed by covering growth from the different parts.

The outstanding features of flower galls are described by GIARD[385-387], SCHLECHTENDAL[1021], MAGNIN[724, 725], HIERONYMUS[462], MOLLIARD[799, 800, 802, 804, 810], VUILLEMIN[1225], GUÉGEN & HEIM[423], DIELS[270], TROTTER[1174], HOUARD[513], GAMBIER[360], GOEBEL[397], FISCHER & GÄUMANN[334], HAUMESSER[443], LEMESLE[658, 659], MOYSE[824] and others.

2. Cecidogenetic reaction localized

In a number of flower galls the predominant cecidogenetic reaction is localized more or less in the perianth whorls or in the androecium and gynoecium. The cell proliferation induced by the cecidozoa is thus restricted to one of these floral parts, but the others remain stunted or their normal growth is arrested. The individual parts, often profoundly altered and abnormally formed, remain, however, distinct and separate. Though inside the gall, the cecidozoa are in these cases outside the plant tissue and not enclosed in special larval cavities. The floral envelopes, especially the calyx and corolla and sometimes also the bracts, are swollen, enlarged, elongated or inflated particularly basally, but remain separate. The stamens and pistil, though arrested in normal development, are distinct. The filament may, however, be somewhat swollen. Most of the abnormalities observed in the stamens and pistil are merely secondary effects of the cecidogenetic changes in the perianth. The galls of *Dasyneura sisymbrii* SCHRANK on *Barbarea vulgaris* R. BR., *Contarinia loti* DE GEER on *Lotus corniculatus* LINN., *Contarinia medicaginis* KIEFF. on *Medicago sativa* LINN. and *Contarinia onobrychidis* KIEFF. on *Onobrychis sativa* LAMARCK show this condition[360]. In the gall of *Dasyneura sisymbrii* SCHRANK on *Barbarea vulgaris* R. BR. the calyx and corolla are inflated basally but not fused together. The anthers, which are normally dorsifixed, are, however, basifixed in the gall, rugose, swollen two or three times the normal size and indehiscent. The stamens as a whole are stunted and the pistil is

arrested and sterile. The epidermal cells of the calyx are hypertrophied and the epidermis of the corolla is not differentiated. In
the galls on *Lotus, Medicago* and *Onobrychis* the calyx is only
slightly altered, but the corolla is enormously inflated. The larvae
of the gall-midge *Asphondylia dufouri* (KIEFF.) in the gall on *Verbascum floccosum* (W. & K.) are situated among the bases of the
stamens and in this gall also the petals and sepals are swollen[443].
The stamens are swollen in the inflated-corolla gall of *Gephyraulus*
(= *Dasyneura*) *raphanistri* (KIEFF.) on *Raphanus*. The corolla is
inflated in the galls of *Dasyneura phytoneumatis* F. Löw on *Phytoneuma spicatum* LINN. and *Phytoneuma orbiculare* LINN. from
Europe; the inflation is predominantly only at the base of the
flower, but apically the corolla is elongated into a beak-like process.
The bracts and sepals are unaffected, the stamens are stunted,
curled and densely clothed with hairs, but the pistil is almost normally differentiated. The commonest example of a gall of this type
is caused by diverse species of *Copium* (Heteroptera) on *Teucrium*
spp. from Europe and different Verbenaceae from other parts of
the world. In all these galls the corolla is enormously inflated and
elongated, especially in the upper part, to form a globose, utricular
growth, covered by sticky hairs. The stamens are usually swollen
basally and numerous nymphs of the bugs occur within the coriaceous bag (Plate IX,1).

Turning to the galls in which the cecidogenetic reaction is largely
localized in the androecium and gynoecium, cell proliferation may
be predominantly or exclusively in the stamens or in the ovary. In
the galls of *Asphondylia scrophulariae* TAVARES on the flowers of
Scrophularia sambucifolia LINN. and *Scrophularia canina* LINN. from
Algeria, the filaments are, for example, greatly swollen and fused
together basally, with a large larval cavity. The galls are yellowish-
green, globose growths, about 10 mm in diameter. Ovaries are,
however, normally differentiated, but smaller in size. The abnormalities observed in the androecium are generally associated with
predominance of cell proliferation in the corolla and in most cases
do not affect the pistil.

The gynoecium is the centre of localization of the predominant
cecidogenetic reaction in a larger number of galls, characterized by
the occurrence of the cecidozoa mostly within the ovary. In the
galls of Curculionidae on the flowers of *Veronica, Campanula* and
Phytoneuma the ovary is affected predominantly, but the other
parts are relatively little altered. The ovary is swollen to a spherical
or oval mass, 8 mm long and 6 mm thick, in the gall of *Gymnetron
villosus* GYLL. on the flowers of *Veronica anagallis* LINN. and
Veronica anagaloides GUSS. Initially the ovary has a large cavity
and thin walls and is not very different from a normal ovary. Soon
however, the placenta and septa swell up and the development of

the ovules is inhibited. The calyx is usually inflated, but the corolla and stamens are generally only stunted[492, 493]. The flower galls of the Curculionidae *Miarus campanulae* LINN. on *Campanula rapunculoides* LINN. and of the gall-midge *Dasyneura phytoneumatis* F. Löw are remarkable for the pathological alterations of their ovaries predominating over the abnormalities in other parts. The ovaries are enormously swollen, often asymmetrically unilaterally, with one side more or less normal, but lacking fertile ovules (fig. 91 D). The corolla is stunted, green coloured basally and does not open. The stamens are perhaps the least altered in this gall. The placenta is turned into large cushions of fleshy tissue, almost completely filling the locules of the ovary. The ovules undergo more or less profound alterations, their integument and nucellus being strongly elongated, but with the funiculus greatly aborted. The ovary is prematurely much larger than even a normally developed fruit. The ovary turns into an enormously inflated utricular gall, with thick walls, in the flowers of many common Umbelliferae like *Pimpenella*, *Daucus carota*, *Silaus pratensis*, etc. under the influence of the gall-midge *Kiefferia (= Schizomyia) pimpenellae* (F. Löw)[800]. The flower gall, commonly known under the name "Narrentaschen" in German literature, caused by *Taphrina pruni* (FUCK.) on *Prunus domestica* LINN., is remarkable for the ovary being altered into an elongate, somewhat compressed, pointed, solid, fleshy mass, 6 cm long and 1-2 cm thick, yellow or reddish, with rugose surface. In place of the normal stony kernel we find in the gall only an irregular hollow

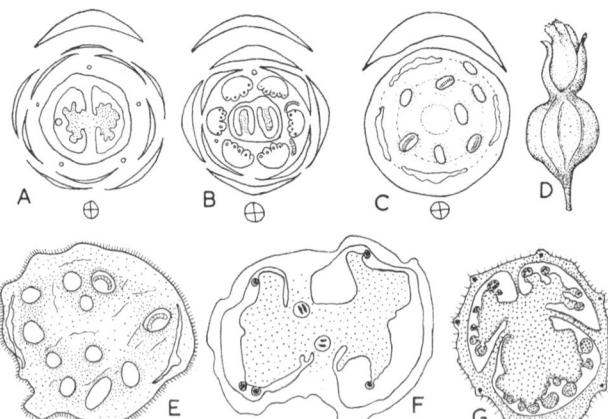

Fig. 91. Structure of flower galls in T.S. A. *Phytoneuma*. B. *Barbarea*. C. *Psophocarpus*. D. Gall of *Miarus campanulae* Linn. on *Campanula rapunculoides* Linn. E. *Psophocarpus*. F. *Phytoneuma* ovary. G. *Miarus* gall on ovary. (Modified from Gambier).

empty space. Similar galls are also found on other species like *Prunus padus* LINN., in which even the calyx tube is more or less altered. *Taphrina rostrupiana* (SADEB.) gives rise to a similar gall on *Prunus speciosa* LINN. The ovary is likewise greatly altered in the gall of *Taphrina johansonii* SADEB. on the flowers of *Populus tremula* LINN. from Europe, with the individual cavities of the ovary becoming greatly swollen and yellow coloured. TROTTER[1177] has described an interesting flower gall of *Eriophyes buceras* TROTTER on *Bucida buceras* LINN. from the Antiles, in which the ovary is enormously elongated into a hollow, corniculate outgrowth, 19 cm long and 24 mm thick, with the cavity lined by fleshy emergences.

The galls of some gall-midges like *Asphondylia tephrosiae* MANI on the flowers of *Tephrosia candida* from India are remarkable for the ovary developing into a kind of fruit-like mass, either utricular or also coriaceous. *Asphondylia pongamiae* FELT gives rise to a most remarkable, spherical, hollow, nut-like, yellowish-green, smooth gall, about 5—8 mm in diameter, developed from the ovary of the flowers of *Pongamia glabra* VENT. from South India (Plate IX, 4). The normal fruit is a flat, elliptical large legume. The sepals, corolla and stamens remain completely arrested. The flower galls on *Papaver* spp. caused by the cynipids *Aylax papaveris* PERRIS and *Aylax minor* HTG. from Europe are characterized respectively by the enormous swelling of the septa of the ovary and of the ovules. The greatly hypertrophied septa bear irregular fleshy emergences, which meet and unite to form a homogenous, parenchymatous tissue, almost completely filling the cavities of the ovaries[802, 731]. In the gall of *Aylax minor* HTG. on *Papaver rhoeas* LINN. the ovule and the funiculus swell up enormously, become fused with the other swollen ovules and fill up the cavity of the ovary with a homogenous parenchyma mass. The gall of *Apion* sp. on the flower of the Argentinian *Portulaca oleracea* LINN. is a spherical swelling of the ovary and calyx, with the corolla and stamens atrophied[533]. The three carpels in the pistillate flowers of *Tragia volubilis* LINN. from South America become swollen into fleshy, globose masses in the gall of *Tragiicola haumani* BRÈTHES. Each of the swollen carpels bears apically a short, curved and horny appendage and the whole cluster measures about 20 mm in diameter[533].

3. Cecidogenetic reaction not localized

The cecidogenetic reaction is not localized to individual parts of flowers in a large number of cases, but embraces the whole of the flower and often also the thalamus and the upper portion of even the pedicel. Cell proliferation is so extensive that all these parts are swollen. Tissue differentiation in every part of the flower is more or less inhibited. The swollen parts fuse together to give rise to a composite fleshy or woody mass, in which there is no trace of the petal or stamen. Some of these develop as elongate hollow galls and others as globose solid or hollow galls. The midge gall on the flower of *Fagonia cretica* LINN. is for example a globose swelling, about 10 mm in diameter, with greatly inflated calyx and incurved sepals, enormously swollen corolla, deformed stamens and abnormally swollen ovaries. Another interesting midge gall is found on *Funtumia (Kickxia) africana* STAPF. from Africa[520]. In this gall the whole flower is turned into a swollen mass, with abnormally enlarged calyx, cylindrical elongated and swollen corolla measuring about 50 mm long and 4—7 mm thick, malformed carpels and with a large gall cavity. The gall of *Dasyneura dielsi* RÜBS. on the flower of *Acacia cyclops* A. CUNN. from Australia is a cylindrical swollen mass, 10—12 mm long, with numerous larval cavities.

The condition is, however, best illustrated in the gall of the midge *Aschistonyx crataevae* (MANI) on the flower of *Crataeva religiosa* FÖRST. from India. The sepals, petals, stamens and the pistil are each enormously and separately swollen to give rise to a solid, fleshy, discoid or globose, yellowish-white mass (Plate IX, 5), with narrow interspaces containing numerous larvae of the midge. When the filaments alone of the stamens are swollen with other parts, the anthers enlarge and contain some pollen, but usually the anthers are also swollen or arrested and sterile. Of the pistil, only the gynophore is sometimes swollen, with the stunted ovary containing normal ovules. The composite fleshy mass measures 10—15 mm in diameter. The swollen parts are, however, completely fused together to form a solid, fleshy, cordate, pale red coloured or yellowish-green swelling in the gall of *Oxasphondylia floricola* MANI on the flowers of *Indigofera gerardiana* WALL. from the Himalaya. The same is the case with the gall on *Amajoua guinensis traxiliensis* K. SCHUM from South America. The entire flower becomes swollen into a spongy mass, about 23—28 mm long and 25 mm thick, with an axial larval cavity 23—26 mm long and 1 mm thick and surrounded by a sclerenchyma zone. The whole flower is swollen into an irregular, spherical, solid fleshy mass, with verrucose surface and measuring about 2 mm in diameter in the gall of *Schizomyia galiorum* KIEFF. on *Rubia peregrina* LINN. from Algeria[520]. The flower galls on a number of Convolvulaceae like *Ipomea cairica* SWEET, *Ipomea sepiaria* KOEN., etc., caused by species of *Schizomyia* (Plate IX, 2), are also remarkable for the complete lack of individuality of the parts and for the swelling and fusion of the tissues of different parts. The gall of *Schizomyia cocculi* MANI on the flower of *Cocculus hirsutus* DIELS. from India is an irregular, solid, subglobose, fleshy, lobulated, tuberculated and finely villous, brown swelling of the entire staminate flower. All the floral envelopes and stamens are enormously swollen and fused together into a complex mass, with narrow tortuous passages between the lobes. Red coloured larvae occur in numerous, irregularly scattered larval cavities. Occasionally the empty and stunted anthers may be found sticking to the surface of the swollen mass. The gall lacks a true epidermis. The small white flowers of *Ehretia laevis* ROXB. are galled by *Eriophyes ehretiae* NAL. from India (Plate II, 1). The entire flower becomes swollen into a large, globose, solid, fleshy, lobulated and finely granulated, finely pubescent and greenish swelling, measuring about 20 mm in diameter.

A number of flower galls are remarkably hard and woody solid swellings. The gall of *Neolasioptera crataevae* MANI on the flower of *Crataeva religiosa* FÖRST. from India (fig. 92 E) is, for example, a large subglobose, solid, hard, woody swelling of the pedicel, thalamus, calyx, corolla, stamens and pistil, fused together to form a composite mass, surmounted by irregular and minute fleshy tubercles, which represent the tips of the petals and rarely also minute vestiges of anthers. The gall measures 20—30 mm in diameter. The

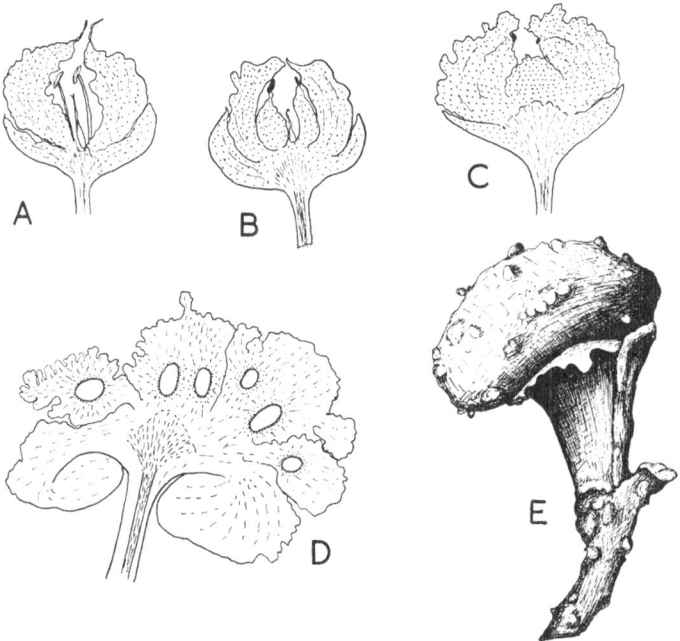

Fig. 92. Structural types of flower galls.   A. swelling localized in calyx and corolla. B. Swelling extending to the perianth whorls and stamens.   C. Swelling extending to all parts, but without fusion of the swollen tissues.   D. Midge gall on *Ipomea*, showing extensive swelling of all parts, with the whole flower turned into a solid, fleshy mass, containing numerous larval cavities.   E. *Neolasioptera crataevae* Mani on *Crataeva religiosa* Forst.

144

galls of *Eriophyes prosopidis* SAKSENA on the flowers of *Prosopis* spp. from India are enormously swollen, irregularly globose or pyriform, lobed, solid, hard masses, that range in size from 5 mm

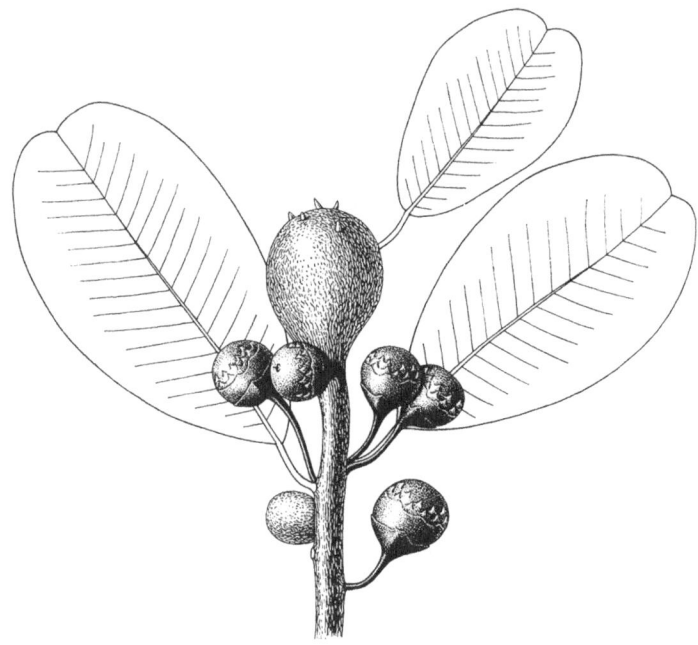

Fig. 93. Gall of *Pruthidiplosis mimusopsicola* Mani on the flower of *Mimusops hexandra* Roxb. Note also the terminal growing point gall and the lateral rindengall caused by another unknown gall-midge.

to over 30 mm in diameter, yellow, brown or reddish-brown and smooth. The irregular hollow spaces found in the young gall become obliterated by irregular fleshy emergences from the wall, so that only narrow labyrinthine interspaces are left and in these spaces the mites occur in enormous numbers. The mites are thus completely enclosed in the full grown gall. None of the parts of the flower may be recognized in the gall and it is indeed difficult even to recognize that it is a flower gall. Similar globose, solid but more or less spongy galls, about 25—30 mm in diameter, are caused by *Eriophyes* sp. on the flowers of *Tamarix articulata* VAHL. from India. In this gall also the mites occur within the gall, in the minute and narrow interspaces that fill the spongy gall tissue. An interesting gall on the flower of *Mimusops hexandra* ROXB., caused by *Pruthidiplosis mimusopsicola* MANI from India, has a most remarkable resemblance to an ordinary stony or woody fruit (fig. 93). This is a globose, smooth, green, solid and dehiscent swelling, with a thick and hard rind of sclerenchymatous cells, enclosing the central core of spongy tissue. All parts of the flower are completely swollen and fused together, with only the tips of the sepals, petals and stamens projecting as minute fleshy or spiny tubercles on the surface. Among the other remarkable flower galls, mention should be made of the echinate, hard, woody gall, about 15—25 mm in diameter, curiously resembling a miniature sea-urchin, caused by an unknown gall-midge on *Hopea wightiana* WALL. from India[745]. Reference must also be made to the flower galls on *Eucalyptus* spp. from Australia, caused by *Brachyscelis* spp. One of these galls is a pyriform swelling, about 50 mm long and 25 mm thick and enclosing a central gall cavity. Others are spherical solid swellings. A depressed spherical gall on flower, with numerous elliptical larval cavities, is caused by an unknown Diptera.

4. Galls on inflorescence

The inflorescence as a whole becomes often swollen into an agglomerate fleshy, solid or hollow mass. The entire inflorescence of *Pistacia mutica* FISCH. & MEY., galled by *Pemphigus* sp. from Persia, turns into an irregular oval, agglomerated mass, 55 mm long and 45 mm thick, with small, branched coralliform excrescences on the surface and partly tinted red. The whole involucre of *Euphorbia* sp. is galled by *Perrisia cornifex* KIEFF. from Algeria and Tunis. The gall is a solid conical, hard red coloured swelling, 10—14 mm long and 2—3 mm thick. Complex abnormal growths of the capitula of several common Compositae like *Crepis biennis* LINN., *Hieracium*, *Cirsium*, etc. are caused by *Eriophyes* spp. The galls of the male catkins of *Salix* spp. caused by *Dorytomus taeniatus* FABR. and *Corylus avellana* LINN. caused by *Contarinia corylina* (F. Löw) from Europe are other examples of galls on whole inflorescences. In the gall of *Rhopalomyia* sp. on the capitulum of *Solidago canadensis* from North America we have a cylindrical or globose, densely pubescent, reddish or green coloured swelling, 6—8 mm long. The inflorescence gall on *Fraxinus* caused by *Eriophyes fraxinivorus* NAL. from Europe involves swelling and agglomeration of the common peduncle and pedicels of the numerous flowers. The most remarkable inflorescence gall is caused by *Asphondylia morindae* MANI on *Morinda tinctoria* ROXB. (Plate IX, 3) from India. This gall is composed of the fused, solid swollen masses of all the flowers and thalamus. The gall measures about 20—30 mm in diameter, fleshy and green coloured. The surface of the gall is marked by characteristic green lines, indicating the limits of the thalamus of the individual flowers. The gall has a curious superficial resemblance to a normal fruit, but the total absence of seeds serves at once to distinguish the gall from the normal fruit of the plant. Other inflorescence galls include those of *Livia juncorum* LATR. on *Juncus* spp., *Eriophyes* on *Bromus erectus* HUDS., *Tylenchus phalaridis* STEINB. on *Phleum phleoides* LINN., *Tylenchus agrostidis* STEINB. on several other grasses like *Agrostis*, *Festuca* and *Poa* and of the fungus *Albugo candida* on Cruciferae like *Sinapis*, *Capsella*, *Raphanus*, *Brassica*, *Sisymbrium*, *Senebiera*, etc.

5. Galls on fruit

In contrast to the gall on the ovary, the gall on the fruit involves the initiation of a cecidogenetic reaction of the plant in the immature fruit, after the pollination of the flower and fertilization of the ovary. At the time of commencement of cecidogenesis, the normal development of the fruit has thus already progressed more or less. In many fruit galls the development of normal seeds is not, therefore, always inhibited as a result of the development of the gall. Depending on the structure and type of fruit—whether a berry, drupe, capsule or legume — and the nature and position of the gall-inducing organism, cell proliferation in a fruit gall may be predominantly localized in the wall of the fruit, septa, placenta or it may extend to the tissues of all these parts. With the cecidozoa situated externally on the surface of the fruit, cell proliferation is restricted mostly to the epidermal and subepidermal layer of cells, resulting in the formation of an erineal or fleshy emergence gall or covering growth gall on the surface of the fruit. The larva of the cecidozoa may also be situated in the septa or in the cavity of the fruit. In drupes and berries the larval cavities are usually situated in the fleshy epicarp or mesocarp and seeds are often normally developed, though fewer in numbers than normal. With the localization of cell proliferation in placenta, the seeds are not developed and their place is taken by a mass of undifferentiated, spongy, parenchymatous nutritive tissue in the gall, often completely filling up the cavities in the fruit. Extensive cell proliferation in all parts of the fruit results in the formation of an undifferentiated mass of gall tissue and total arrest of development of seeds.

We may refer to some outstanding examples of galls on fruits, in which cell proliferation is strictly localized to the epidermis and the subepidermal layer of cells. An undescribed species of *Eriophyes* gives rise, for example, to a fleshy emergence gall on the surface of the winged legume of *Pterocarpus erinaceus* LAMARCK from West Africa[520]. In the subglobose, unilocular midge gall on the fruit of *Zizyphus xylopyra* WILLD. from India (fig. 94 B—C) the covering growth arises from the cell proliferation of the epidermis and subepidermal layer of the epicarp. An ostiole is situated at the tip of a recurved, beak-like prolongation of the gall. The spongy, unilocular, globose gall on the fruit of *Zizyphus orthacantha* DC from Senegal-Nigeria (fig. 94 D) is also the result of such epidermal and subepidermal cell proliferation in the epicarp. Cell proliferation is

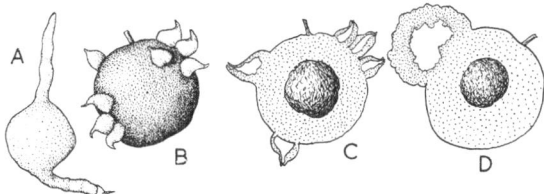

Fig. 94. Galls on fruit.   A. Gall of *Aylax hypecoi* Trotter on *Hypecoum procumbens*. B. Covering gall of an unknown gall-midge on the fruit of *Zizyphus xylopyra* Willd. C. The same in section.   D. Gall of an unknown gall-midge on the fruit of *Zizyphus orthacantha* DC.

localized in the tissue of the cup below the acorn in the acorn-plum gall on *Quercus*, caused by *Amphibolips prunus* WALSH. from North America and the acorn itself is normally developed.

In the galls of *Aylax hypecoi* TROTTER and *Aylax papaveris* PERRIS, respectively on *Hypecoum* spp. and *Papaver rhoeas* LINN. cell proliferation in the placenta leads to the filling of the cavity of the capsule with spongy tissue, in place of seeds.

In a number of fruit galls cell proliferation is generally not localized, so that the galls are characterized by the more or less complete absence of seeds and other characteristic tissues. A fruit gall on *Terminalia catappa* LINN. is a subglobose, solid swelling, about 12 mm in diameter, with irregular, minute and fleshy carinae on the surface representing the vestiges of the wings normally found on the fruit of this plant. Similarly fusiform galls, measuring about 30 mm in diameter, with vestigial wings and a single cylindrical axial larval cavity, are caused by an unknown gall-midge on *Combretum glutinosum* GUILL. & PERR. from Senegal-Nigeria[520]. The gall of *Asphondylia calycotomae* KIEFF. on the legumes of *Calycotome intermedia* DC, *Calycotome villosa* LINN. and *Calycotome spinosa* LINN. from North Africa is a globose, inflated swelling, about 20 mm long. An unknown Chalcid gives rise to a remarkable, irregularly globose, rugose, multilocular, solid gall on the fruit of *Acacia concinna* DC from South India; each gall often attains a diameter of 30—45 mm. A number of cynipids like *Andricus seckendorffi* WACHTL., *Cynips theophrastica* TROTTER, etc. give rise to complex galls on the acorn proper on *Quercus*. The fruit of *Hippocratea myriantha* OLIV. from Africa is swollen into a subglobose gall, about 14 mm in diameter, with three carpels separated distinctly from each other by deep fissures and the remaining carpels aborted. A number of remarkable fruit galls on *Eucalyptus* are caused by *Brachyscelis* spp. from Australia. Some of these are pyriform swellings, 50 mm long and 25 mm thick, with deeply fissured surface and axial central larval cavity 30 mm long and surrounded by a zone of sclerenchyma cells. The legumes of *Acacia leucophloea* WILLD., galled by *Uromycladium* from South India, swell up into enormous, globose, solid, hard, reddish-brown structures, in which none of the parts of the fruit are distinguishable.

PLATE I.
1. Aphid gall on bud of *Vaccinium leschanaulti* Wt. 2. Gall of *Plasmodiophora brassicae*
Wor. on *Brassica* sp. 3. Gall of *Sphaeropsis tumefaciens* Hedw. on stem of *Citrus
medica acida* Linn. 4. Acrocecidia on *Terminalia paniculata* Roth. caused by an
unknown gall-midge. 5. Gall of *Uromycladium notabile* on *Acacia leucophloea* Willd.
6. Acrocecidium on *Tephrosia purpurea* Pers. caused by *Dactylethra candida* Staint.
7. Crown-gall on *Zinnia* sp.

PLATE II.

1. Gall of *Eriophyes ehretiae* Nal. on the flower of *Ehretia laevis* Roxb. 2. Bud gall of *Eriophyes hoheriae* Lamb on *Hoheria populnea* A. Cunn. 3. Gall of *Eriophyes mamkae* Lamb on *Leptospermum scoparium* Forst. 4. Acrocecidia on *Ipomea pes-tigridis* Linn. 5. Cottony gall of an unknown gall-midge on *Artemisia* sp. 6. Aphid gall on *Ulmus wallichiana* Planch. 7. Pouch gall of *Phacopteron lentiginosum* Buckt. on *Garuga pinnata* Roxb.

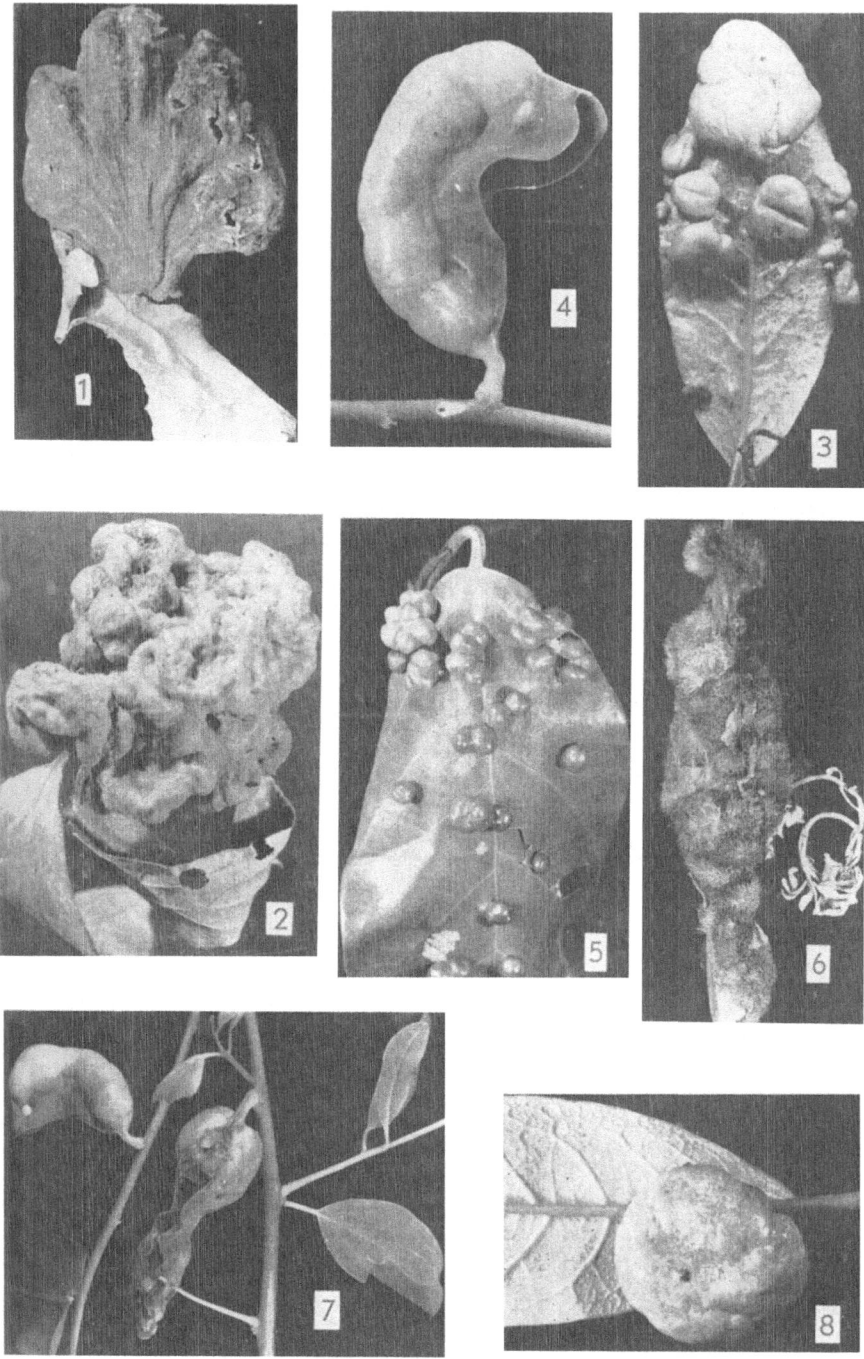

PLATE III.

1. Pouch gall of *Eriosoma taskhiri* Ghulamullah on *Populus alba* Linn. 2. Bud gall of *Austrothrips cochinchinensis* Karny on *Calycopteryx floribunda* Lamarck. 3, 8. Gall of *Exobasidium rhododendri* Cram. on *Rhododendron arboreum* Sm. 4, 7. Fold gall on leaf of *Maerua arenaria* Hook. & Bl. caused by *Schizomyia meruae* Felt. 5. Pouch gall of *Pauropsylla depressa* Crawf. on *Ficus glomerata* Roxb. 6. Midge gall on *Indigofera pulchella* Roxb.

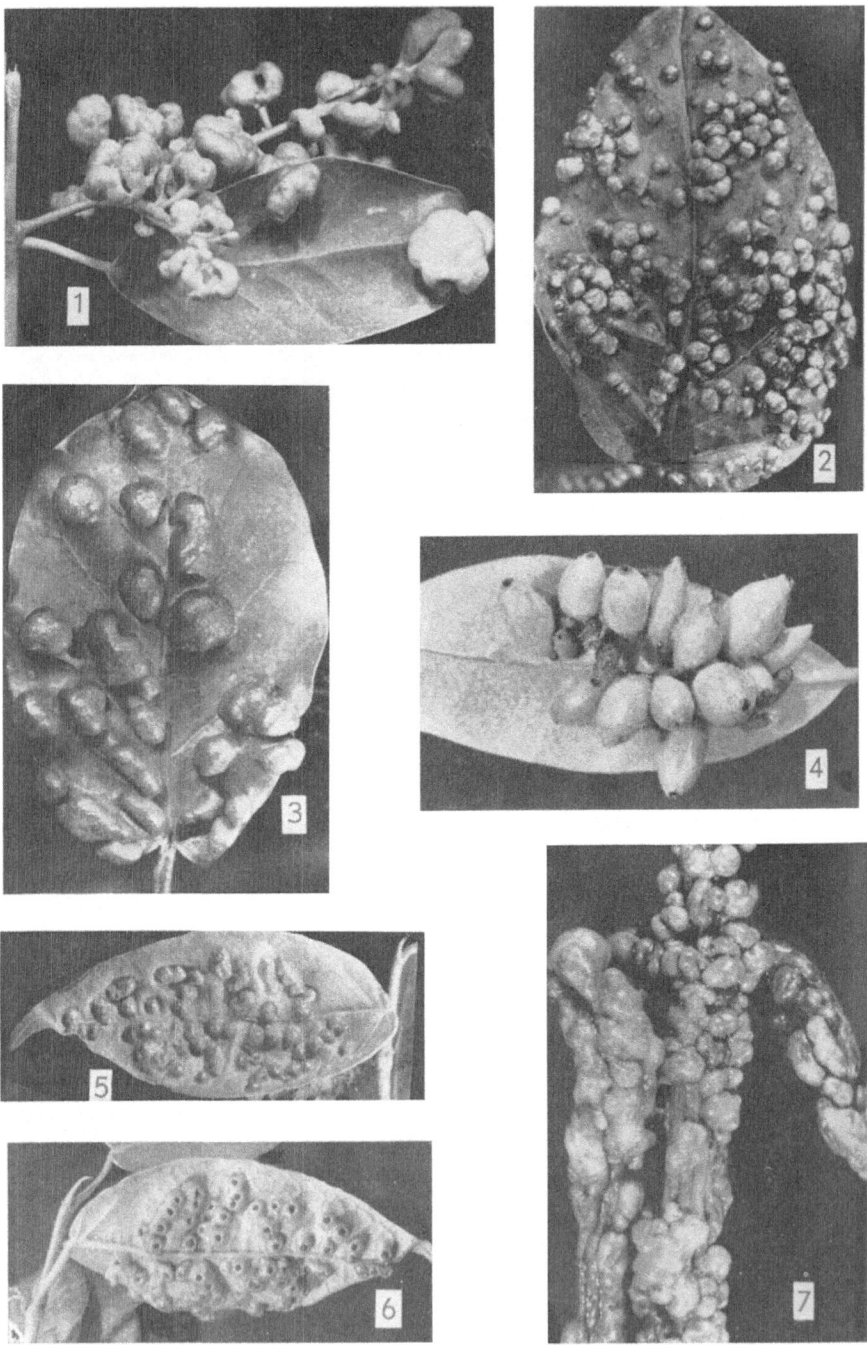

PLATE IV.

1. Gall of *Eriophyes* sp. on *Salvadora persica* Linn. 2. Pouch gall of *Eriophyes gastrotrichus* Nal. on *Ipomea staphylina* Roem. & Sch. 3. Pouch gall of *Eriophyes* sp. on *Holoptelea integrifolia* Planch. 4. Gall of *Neolasioptera* on leaf of *Machilus odoratissima* Nees. 5, 6. Midge gall on leaf of *Ficus foveolata* Wall. 7. Covering gall of *Eriophyes gastrotrichus* Nal. on *Ipomea staphylina* Roem. & Sch.

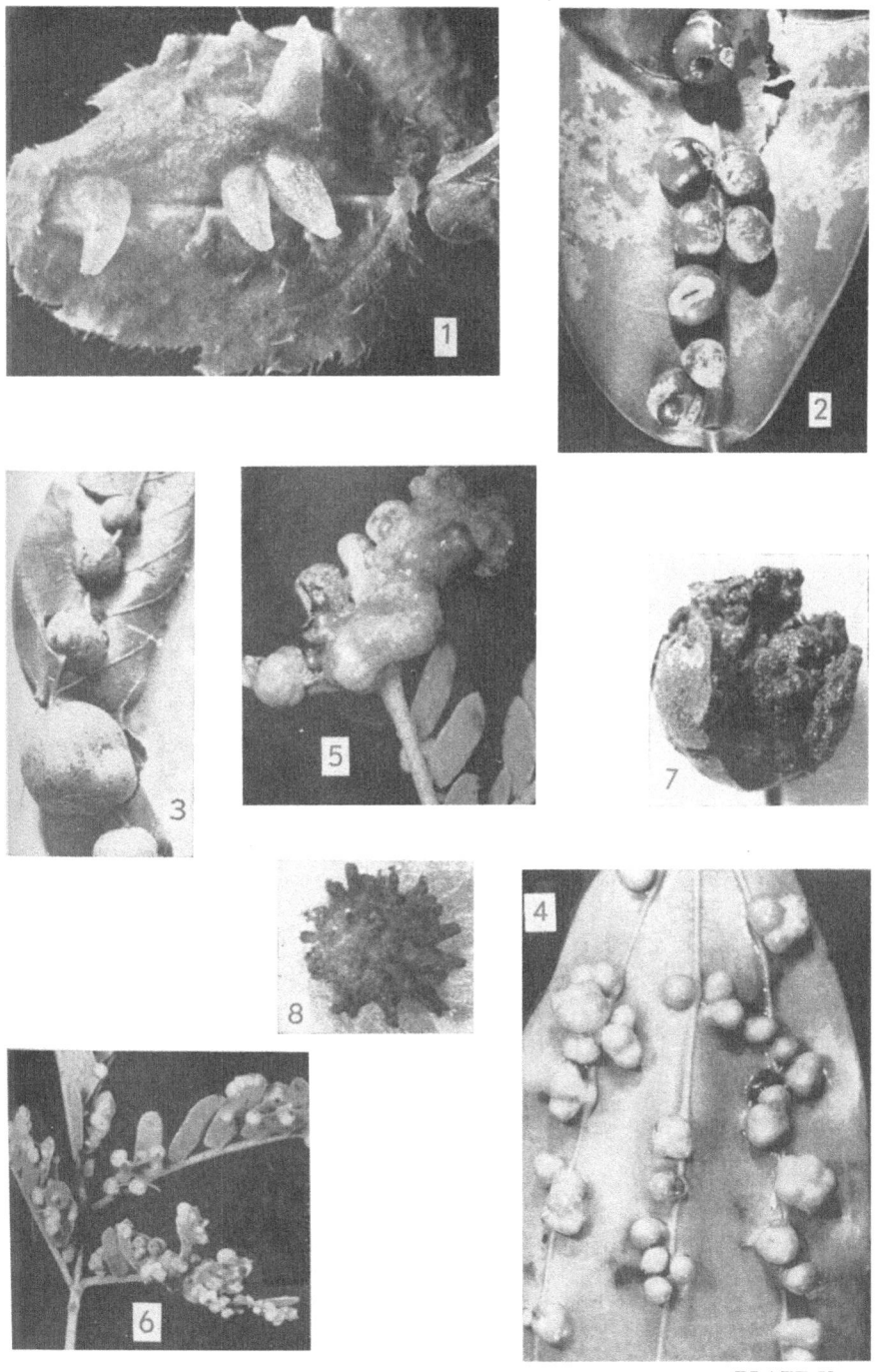

PLATE V.

1. Ceratoneon pouch gall of an unknown *Trioza* on leaf of *Lecanthus wightii* Weed.
2. Midge galls on leaf of *Memecylon edule* Roxb. 3. Midrib gall of *Odinadiplosis odinae* Mani on *Lannea coromandelica*. 4. Psyllid gall on leaf of *Cinnamomum*. 5, 6. Galls of *Eriophyes prosopidis* Saksena on *Prosopis*. 7. Dehiscent gall of *Pruthidiplosis mimusopsicola* Mani on the flower of *Mimusops hexandra* Roxb. 8. Echinate gall on leaf of *Ficus bengalensis* Linn. caused by an unknown Chalcid.

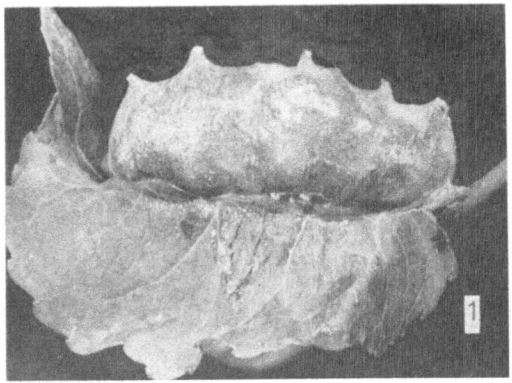

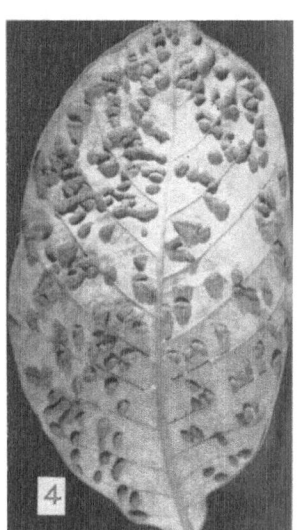

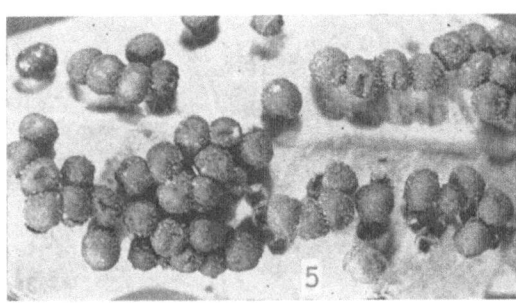

PLATE VI.

1. Gall of *Pemphigus imaicus* Cholod. on leaf of *Populus ciliata* Wall. 3, 4. Chalcid gall on veins of leaf of *Bassia latifolia* Roxb. 2. Cylinder-piston gall of *Lobopteromyia* sp. on leaflets of *Acacia suma* Ham.-Buchn. **5.** Pellet gall by midge on leaf of *Mangifera indica* Linn.

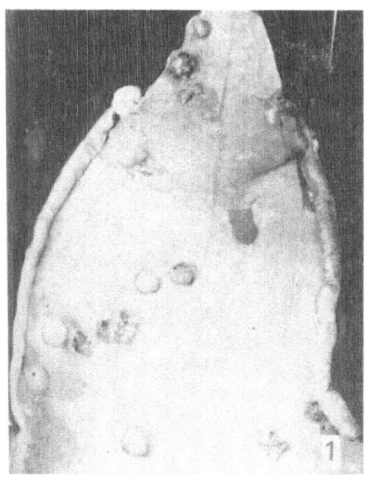
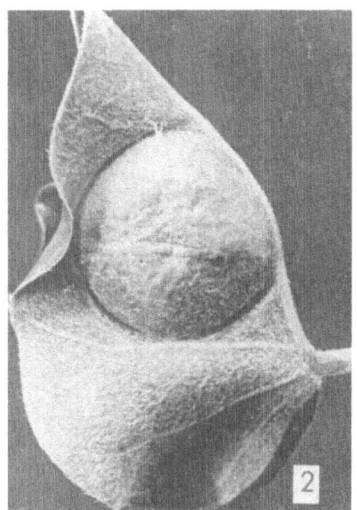

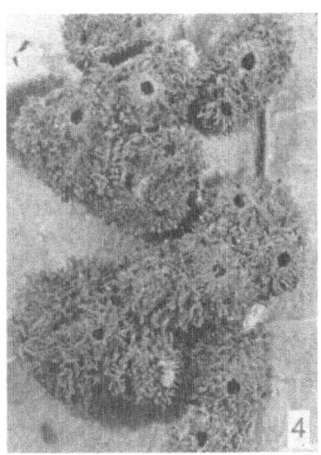

PLATE VII.
1. Leaf margin roll gall of *Gynaikothrips chavicae* Zimmermann and pellet gall by midge on *Piper nigrum* Linn. 2. Leaf fold gall of *Asphondylia riveae* Mani on *Rivea hypocrateriformis* Choisy. 3. Pubescent gall on leaflets of *Acacia leucophloea* Willd. caused by *Schizomyia* sp. 4. Echinate gall of *Amradiplosis echinogalliperda* Mani on leaf of *Mangifera indica* Linn.

1. Gall of *Eriophyes* sp. on leaf of *Millettia*. 2. Bivalve gall of *Lobopteromyia bivalviae* (Rao) on leaflets of *Acacia catechu* Willd. 3. Covering gall of *Eriophyes acaciae* Nal. on leaflets of *Acacia leucophloea* Willd. 4. Barrel gall of *Lobopteromyia ramachandrani* Mani on leaflets of *Acacia suma* Ham.-Buchn.

PLATE IX.

1. Gall of *Paracopium cingalense* (Walk.) on the flowers of *Clerodendron phlomidis* Linn. 2. Flower gall by *Schizomyia* sp. on *Ipomea sepiaria* Koen. 3. Gall of *Asphondylia morindae* Mani on the inflorescence of *Morinda tinctoria* Roxb. 4. Gall of *Asphondylia pongamiae* Felt on the ovary of *Pongamia glabra* Vent. 5. Gall of *Aschistonyx crataevae* (Mani) on flowers of *Crataeva religiosa*.Forst. 6. Gall of *Kiefferia pimpenellae* (F. Löw) on the flowers of *Heracleum canescens* Lindl.

CHAPTER IX

ZOOCECIDIA

Zoocecidia or the galls caused by animals are, as mentioned in the beginning, limited neoplasia, the development and growth of which are strictly dependent upon the continued stimulation of the plant cells by the cecidozoa, so that the gall tissue is not autonomous, like, for example, the crown-gall tissue *(vide* Chapter XV). This limited character of zoocecidia may perhaps be the result of nutritional or structural barriers, rather than an intracellular physiological barrier. Though no genetic changes, in the strict sense of the term, have so far been observed to arise, polarity changes, hypertrophy, hyperplasia, abnormal metamorphoses, etc. in certain zoocecidia, especially the galls caused by Cynipidae, may be caused by an alteration in the genetic products of the plant cells, such as are often induced by viruses or certain nucleoproteins. It has not been possible, at least up to the present time, to raise successfully *in vitro* tissue cultures of fragments of even the most promising zoocecidia. When this can be accomplished, the way will perhaps be opened up for

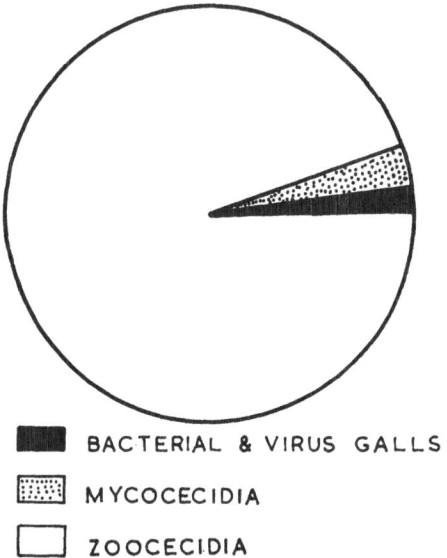

■ BACTERIAL & VIRUS GALLS

▦ MYCOCECIDIA

▢ ZOOCECIDIA

Fig 95. Relative abundance of zoocecidia, mycocecidia and bacterial and virus galls from the world.

investigations, leading to the solution of some of the obscure problems in cecidology. Although, in a sense, pathologically less important than the bacterial crown-gall and perhaps also economically

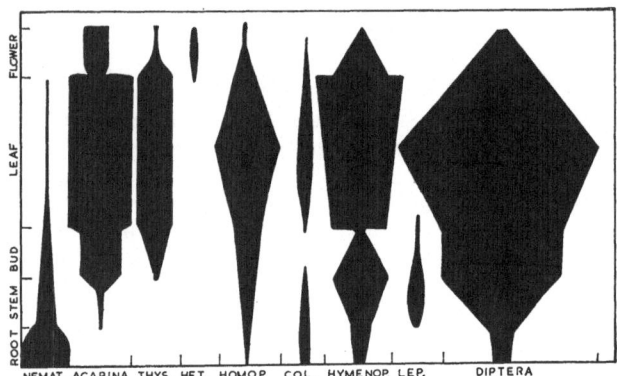

Fig. 96. Relative abundance of galls caused by different groups of cecidozoa on different parts of plants from the world.

less destructive than some of the galls caused by fungi, zoocecidia are, however, of far greater importance than the crown-gall from the point of view of ecology and evolution. Nearly all zoocecidia have a certain fundamental characteristic structure. Some of the outstanding differences between zoocecidia and other galls may be traced to the fact that while other gall-inducing organisms may be intracellular parasites of the plant, this is not so in zoocecidia. The cecidozoa may be situated externally on the epidermis of the plant or may be within the plant tissues among the cells. These differences constitute one of the most important factors that determine the external form and the general histological characters of most zoocecidia. As will be seen later, the morphogenetic centre in zoocecidia lies in the larval cavity of the cecidozoa. There is therefore in nearly all zoocecidia a more or less distinct zonation of cells and tissues (fig. 44). The inner layer of cells, immediately lining the larval cavity, is always abundantly supplied with nutritive material until the cecidozoa are mature and cease to feed. Often a sclerenchyma zone develops outside the nutritive zone. As will be seen further below, this sclerenchyma zone, formerly assumed to serve for the protection of the cecidozoa, arises in the area of neutralization of the morphogenetic fields of the normal plant organ bearing the gall and the new morphogenetic field arising round the larval cavity as a result of cecidogenesis. Compared to bacteriocecidia or mycocecidia, zoocecidia are among the most highly specialized galls. The enormous diversity of their structure, the patterns of development, morphogenesis and their origin, the complex lines of specializations, the highly complicated life-cycles, the mutual adaptations of the plant, cecidozoa and numerous other organisms associated with zoocecidia, the pleomorphic peculiarities, the biochemical and cytological processes underlying their development, the transition from carnivorous to phytophagous, from saprophytic to phytophagous and from phytophagous to carnivorous habits exhibited by cecidozoa in the course of evolution of different zoocecidia, the astonishing parallelism and convergence in the evolution of many zoocecidia and cecidozoa are some of the numerous problems that await further elucidation or need solution. Even the taxonomy of zoocecidia and cecidozoa offers unique opportunities for studies for several decades to come. Zoocecidia are also remarkable for the association of an extremely complex community of species other than the cecidozoa, belonging to very diverse groups, closely inter-linked in a most intricate manner.

1. Galls of Nematoda

Helminthocecidia or the galls caused by Nematoda are among the simplest zoocecidia. The ecologic inter-relations in these galls are never as complex as in some of the galls caused by insects. The Nematode responsible for the gall may not always be found in the gall itself, but may be situated in some other otherwise apparently normal organ, so that the gall is the result of "tele-effect" of the attack by the Nematode. The simplest nematode galls are essentially hypertrophy of the plant organ, with but little anatomical or histological disorganization. The presence of large hypertrophied cells is indeed a characteristic feature of most nematode galls. Nematode galls arise on diverse groups of plants, from mosses to monocots. Although they typically develop on roots, nematode galls on aerial parts are not uncommon.

Nematode galls on roots

Root galls are among the simplest helminthocecidia and are caused by different species of *Heterodera*. The outstanding features of the anatomy, histology and cytology of diverse root galls caused by *Heterodera* have been studied by a number of workers like MOLLIARD[805], TISCHLER[1157], NÈMEC[859], CHRISTIE[196], HOUARD[496], JONES [556], KOSTOFF[613] and others. In the course of the development of the gall the root swells up four or five times the normal thickness and in most cases the cell proliferation is localized in the root cortex. The galls of *Heterodera* on the roots of Dicotyledons are as a rule more or less globose and those on Monocotyledons tend to be fusiform or elongate swellings. The young of *Heterodera* remain within the gall, the size of which thus depends largely on the number of eelworms present rather than on the size of the root. Separate galls arising closely crowded together very often become fused to form large complex masses. The largest nematode gall caused by *Heterodera marioni* (CORNU) GOODY develops on the cauline portion of the stem, just above ground level, on *Thunbergia grandiflora*, *Thunbergia laurifolia*, *Begonia* sp. and *Rheum rhaonticum*[1103]. These giant galls are rugose swellings, in which the progeny of the first generation of worms continue to remain in the gall caused by the parent worm and thus favour its continued enlargement.

CHRISTIE[196] has given an interesting account of a typical root gall of *Heterodera*. The juvenile worm that penetrates the root tissue passes in between the cells of the root cortex, causing very little destruction to the cells at this stage. Later on, however, the worm reaches the plerome. During the first sixty hours of attack, the cells of the central cylinder in the region of the head of the young worm remain still undifferentiated. After about three days, these cells are hypertrophied, with a distinct swelling of their nuclei. Finally the

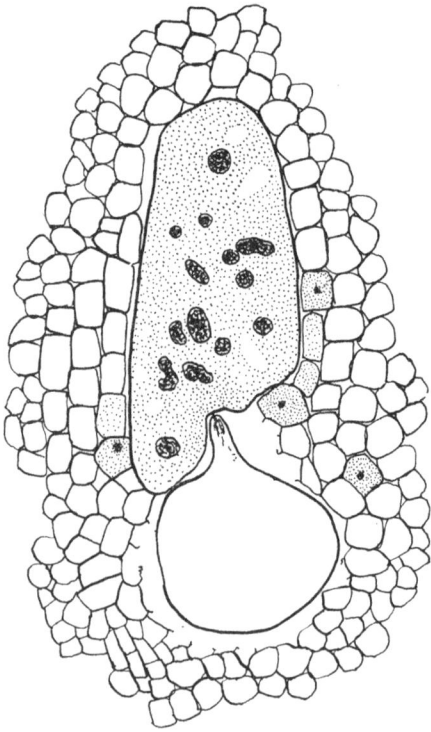

Fig. 97. Part of a section through gall of *Heterodera marioni* (Cornu) Goodey on Dicot root, showing the multinucleate giant cell in front of the head of the nematode.

walls of these cells break, thus giving rise to the multinucleate giant cell, which is a typical character of nematode root galls. COLE & HOWARD[206], who have recently studied the formation of the multinucleate syncitial giant cells in the root galls of *Heterodera rostochiensis* on *Solanum tuberosum*, state that these cells may arise from the cortex, endoderm, pericycle and the parenchyma of the central vascular strand. The first giant cell appears to arise in the cortex and pericycle and in front of the head of the nematode (fig. 97). The formation of giant cells by the parenchyma of the central vascular strand prevents cambium differentiation and secondary xylem is not thus also produced. Very often rows of cells, which would normally develop into a vessel, thus become converted into giant cells. The formation of giant cells soon spreads to adjacent areas. The nuclear membranes rupture, the nuclei coalesce and finally degenerate from the giant cells. Cell proliferation in the pericycle gives rise to a layer of small-celled parenchyma. Phloem-like bundles found near the syncitial giant cells do not actually function as normal phloem[863]. If due to some cause the nematode ceases to feed, the giant cell undergoes certain peculiar changes. The cell wall becomes thickened, the cell contents diminish and the cell finally perishes.

Nematode galls on aerial parts

The nematode galls on aerial parts are of two ecologic groups, viz. the galls in which the nematode is found in the galls themselves and the gall which arise as tele-effect of the presence of the nematode within the root or other underground parts of the plant. The presence of *Heterodera marioni* (CORNU) GOODEY within the roots, gives rise, for example, not only to galls on roots but also curious organoid galls like petalloidy in the flowers[805]. The nematode galls that arise on aerial organs like leaves and flowers are also sometimes associated with singular secondary effects on other remote parts. The gall of *Tylenchus tritici* ROFFR. on the seeds of *Triticum* is thus associated with leaf-spotting caused by the fungus *Diplophospora alopecuri*[33].

The nematode galls on aerial parts comprise rosette galls on moss and liverworts caused by *Tylenchus davani* BAST.[1018, 747]. WARNSTOFF[1232] and DIXON[273] have described such rosette galls on over twenty-five species of liverworts like *Cephalozia* and *Lophozia*. Large numbers of nematodes occur within the gall, particularly in between the stunted and enlarged leaves. Galls on leaves, usually caused by *Tylenchus* spp. are characterized by a very pronounced hypertrophy and cell proliferation of the mesophyll and by the presence of a central cavity, containing the nematodes (fig. 98 A). GOODEY[409] has described, for example, a typical large central hollow space in the gall on the leaf of *Andropogon pertusus* caused by *Tylenchus cecidoplastes* GOODEY from India. The cavity is lined by

154

the so-called nutritive zone of one or two layers of cells with abundant protoplasm, but lacking chlorophyll. Outside of this lie large
simple parenchyma cells. The development and structure of galls
on leaves of *Plantago lanceolata*, caused by *Anguillulina dispasci*
(KÜHN.) GER. V. BEN, are interesting from several respects [401]. The
gall is a simple swelling of the leaf blade, marked by loss of chlorophyll. Many separate galls tend to arise in between veins but soon
spread to the veins also. Stomata are mostly abnormally developed
and there is no differentiation of palisade or spongy tissues in the
mesophyll. The gall cells are on the whole considerably larger than the
cells of the normal mesophyll. There is a conspicuous increase of
collenchyma and intercellular spaces and multinucleate cells are
abundant. The nematodes become free when the galled leaf decays.
The same species gives rise to more or less similar galls on leaves of
nearly 285 different species of plants. In the gall of *Anguillulina
millefolii* (Löw) on the leaf of *Achillea millefolium* LINN., the epidermal cells are much larger than those of the normal leaf. Beneath the
epidermis there is a layer of loose cells with considerable intercellular
space (fig. 98 B). Then follows a compact mass of cells, with some-

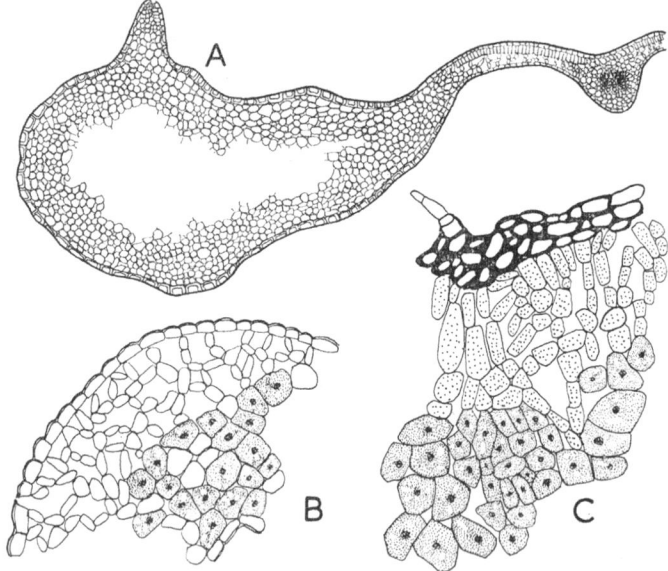

Fig. 98. Structure of Nematode galls on aerial parts. A. Kammergall of *Tylenchus
nivalis* Kühn on leaf of *Leontopodium alpinum* Cass. B. Part of T.S. through the
gall of *Anguillulina millefolii* (Löw) on leaf of *Achillea millefolium* Linn., showing
the plasma-rich hypertrophied cells. C. Part of T.S. through the gall of *Anguillulina balsamophila* (Thorne) on leaf of *Balsamorrhiza sagittata* Nutt., showing
the thick-walled epidermal cells and the hypertrophied cells.

what thick walls, but without evidence of secondary thickening. The cells lining the gall cavity contain granular protoplasm, but many of the cells are also collapsed. Vascular elements are numerous, but irregular, in the gall tissue[808]. It has been shown by GOODEY[402] that in the case of the galls on leaves of *Balsamorrhiza sagittata* NUTT., *Balsamorrhiza macrophylla* NUTT. and *Wyethia amplexicaulis* NUTT., caused by *Anguillulina balsamophila* (THORNE), the young larvae of the worm penetrate the young leaves while the latter are still within the bud and seem quite unable to enter the leaf tissues once the petioles have developed. The point of entry of the worm remains as a discolourized depression on the surface of the gall. Development is rapid and within about three weeks after the entry by the worm, eggs are again deposited. These hatch and the newly hatched larvae develop rapidly until about 1 mm long. By now the galled leaves wither away and the dried galls become buried under the winter snow cover on the ground. The spring thaw is followed by the disintegration of the vegetable tissues and the release of the worm becomes possible. The gall cavity is lined by cells with dense protoplasm, lacking vacuoles, but with giant nuclei. These cells are somewhat smaller than the normal mesophyll cells and are also more closely packed. Many cells are also collapsed. In this gall also we see a pronounced tendency for increase of collenchyma elements. The gall of *Tylenchus* sp. on the leaves of *Miconia* sp. comprises irregular, small bunches of adventitious foliaceous outgrowths, fixed to one of the principal veins. Nematode galls on flowers, especially of Monocotyledons, are characterized by longer glumes, lodicules and suppression of stamens and enormous swelling of the ovaries. The galls of *Tylenchus agrostis* (STEIN.) on the flowers of *Agrostis stolonifera* and *Agrostis tenuis* are common examples. In the gall of *Tylenchus tritici* ROFFR., already referred to, the ovary is swollen and contains often from ten to fifteen thousand dormant juvenile worms. When the gall becomes wetted next season, the worms escape into the soil and seek healthy plants, on which they live as ectoparasites between the leaf sheath and stem for some time. The spores of the fungus, mentioned as causing leaf-spot, stick to the body of these juvenile worms and are thus transmitted to healthy plants. A considerable number of helminthocecidia on aerial parts is remarkable in that the whole of the aerial portion of the plant may become galled. This is, for example, the case with the galls of *Tylenchus dispasci* (KÜHN.) GER. V BEN on grasses. The whole plant is conspicuously stunted, producing prostrate side shoots. The lower part of the shoot is mostly swollen, the leaves curled, crinkled, twisted and stunted and their margins are conspicuously sinuately waved. Ears and spikes, if formed, remain hidden within their sheath. In the flower gall of *Ditylenchus dispasci amsinckiae* on *Amsinckia intermedia*, the ovary swells to sizes larger than even the normal fruit.

156

The formation of vascular elements is greatly retarded and reduced, but the vessels may not be otherwise abnormal. The worms occur in the intercellular spaces. A common example of galling of the whole of the aerial portion is found in *Viola odorata* LINN. by *Aphelenchus olesistus longicollis* SCHWARTZ[395, 399, 695].

2. Galls of Acarina

An outstanding feature of acarocecidia, especially the galls caused by *Eriophyes* spp., is the fact that owing to their minute size the mites are able to attack individual cells of the plant, so that cellular hypertrophy is enormous in most of them (fig. 147). The eriophyid galls are ecologically of exceptional interest on account of the presence of more than one species of *Eriophyes* within the same gall, in addition to numerous acarovorous predators and parasites. Another striking feature is the remarkably close parallelism between eriophyid galls and the galls caused by certain insects like Thysanoptera and Aphids. Most eriophyid galls are initiated by a single gravid female, the progeny of which, however, remain within the

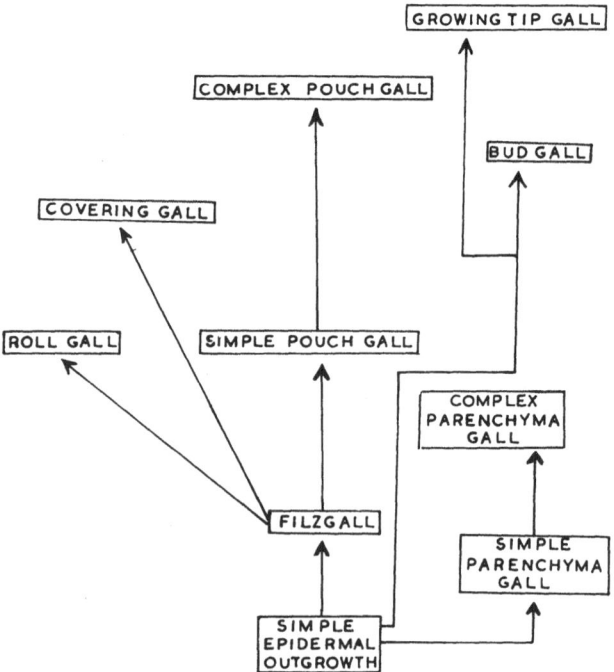

Fig. 99. Structural types of acarocecidia, derived as modifications from primarily simple epidermal outgrowths.

same gall and continue to contribute to its further development and growth. The reader will find further useful information in the contributions of THOMAS[1135-1138], SCHLECHTENDAL[1022-1025], NALEPA[839], CHADWICK[181], HOUARD[507], PERTI[896], TROTTER[1177, 1183], KENDALL[574] and others listed in the bibliography.

An essential character of acarocecidia is the pronounced hypertrophy of epidermal cells, which also undergo very pronounced hyperplasy and give rise to abnormal outgrowths of hairs, fleshy parenchyma emergences, etc. This luxurient hairy outgrowth led the early workers to confuse acarocecidia with fungi. Some of them even went to the extent of describing these galls under generic names *Phyllarium* and *Erineum*, recognizing several species like *Erineum populnium* for the filzgall on the leaf of *Populus tremula* LINN. caused by *Phyllocoptes populi* NAL., *Erineum tiliaceus* for the filzgall on the leaf of *Tilia cordata* MILL. caused by *Eriophyes tiliae* NAL. and *Erineum purpurescens* for the gall of *Eriophyes macrochelus eriobius* NAL. on the leaf of *Acer pseudoplatanus* LINN. Indeed eriophyid galls seem primitively to have been simply hypertrophied epidermal cells, developing later into hairy outgrowths or as parenchyma processes. All other complex galls caused by *Eriophyes* must be considered as secondary effects of the reaction of the epidermal cells to the attack by the mites and have been derived by diverse lines of specialization and modifications (fig. 99). Although unlike in case of insect galls, the position of the mite does not constitute a major factor in the general structural and symmetry pattern, we may recognize the following types of acarocecidia:

I. Galls with *Eriophyes* exposed externally on the epidermis
    A. Galls with *Eriophyes* external
        1. Filzgall
        2. Papillar parenchyma emergence gall
    B. Galls with *Eriophyes* inside the gall cavity
        1. Blister gall
        2. Leaf margin roll gall
        3. Leaf roll gall
        4. Pouch gall
        5. Bud gall
        6. Growing-tip gall
        7. Some flower galls
        8. Galls on the galls of other species
    C. Galls with *Eriophyes* enclosed by covering growth of plant tissue
II. Galls with *Eriophyes* inside the plant tissue
    1. Pockengall
    2. Rindengall
    3. Some flower galls

158

Epidermal outgrowth galls

The epidermal outgrowth galls comprise the filzgalls and parenchyma emergence galls. The account of the development of the filzgalls on *Tilia* by THOMAS[1136] and FRANK[342] is generally typical of other filzgalls also. The gall begins as a slight localized close group of epidermal cells growing into small papillae. These papillae soon elongate into curled hairs. Most filzgalls are hypophyllous and sometimes only epiphyllous and rarely arise on both sides of the blade. The hairy outgrowths are typically thin-walled elongate tubes, with scanty protoplasm and uninucleate. The hairs are variously modified (fig. 100). While the normal trichomes of the plant may comprise simple cylindrical hairs, the filzgall hairs may often be more than 25 cells long[826, 342]. Hairy outgrowths arise not only on leaves, but sometimes extend also to the petiole and main shoot axis. In addition to single epidermal cells developing into hairy outgrowths, multicellular, clavate, fleshy, parenchyma outgrowths arise as emergences from the surface and the outermost layer of cells of these emergences develop into hairy processes (fig. 62, 147). Parenchyma emergences, not bearing hairy outgrowths, constitute the galls induced by several species. This is, for example, the case with the eriophyid gall on *Ficus sycamorus* LINN. We have already mentioned the parenchyma emergence galls on *Commiphora campestris* ENGL. from Africa (fig. 62 A). Such localized irregular fleshy emergences, associated with general swelling of the blade, are known on *Rumex*

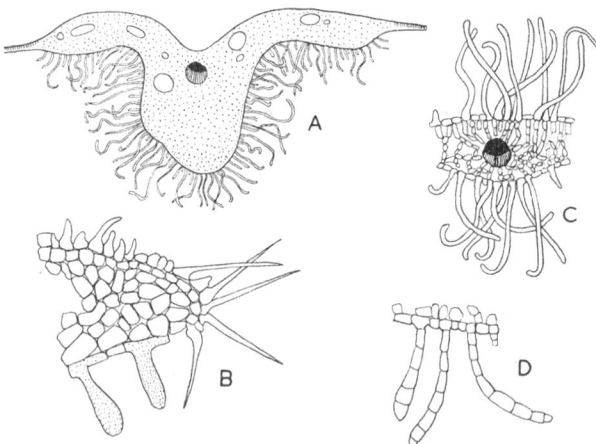

Fig. 100. Erineal galls of Acarina. A. *Cissus kilimandjarica* G. B. Gall of *Eriophyes similis* Nal. on leaf of *Prunus spinosa*, showing the different types of hairs on the outer and inner epidermis. C. Gall of *Eriophyes tiliae* Nal. on *Tilia cordata* Mill. D. Multicellular erineal hairs of the gall of *Eriophyes hippocastani* Fock. on *Aesculus hippocastaneum*.

nervosa usambarensis ENGL. and *Cissus kilimandjarica* GILG. from Africa (fig. 100 A). The development of fleshy emergences is associated with a swelling and cup-like deformation of the leaf of *Hepta-*

Fig. 101. Extensive epidermal emergence gall of *Eriophyes* on *Commiphora caudata* Engl.

pleurum pergamaceum HASSK., described by RÜBSAAMEN[997]. The fleshy emergences become sometimes so extensive that, as in case of the eriophyid gall on *Commiphora caudata* ENGL. mentioned earlier (fig. 101), the whole of the inflorescence axis and even the main shoot axis including most of the leaves are conspicuously swollen, stunted and abnormally shaped.

Leaf roll galls

The development of fleshy emergences and hairy outgrowths from the epidermal and subepidermal cells on leaves is most usually accompanied by secondary reaction of the deeper mesophyll cells, leading to the formation of leaf roll galls. In the tubular leaf roll gall on *Geranium sanguineum* LINN., caused by *Eriophyes dolichosoma* CAN. from Europe, the epidermis of the under side of the blade is very profoundly altered by repeated cell proliferation, at first parallel to the surface of the blade, but later irregular. The interesting leaf roll gall on *Fagus silvatica* LINN. caused by *Eriophyes stenopis typicus* NAL. from Europe consists of the whole length of the leaf margin rolled into a tube, about 1 mm thick, with 1—2 windings on the upper side. In most of the leaf roll galls the epidermal cells proliferate and develop into stout so-called nutritive hairs on the inner surface (fig. 103). The mesophyll of the rolled portion lacks differentiation of palisade and spongy tissues. In some of these roll galls the cells show spiral or annular or reticulate thickenings, giving a superficial resemblance to tracheids[861]. The leaf roll galls of *Eriophyes marginemtorquens typicus* NAL. on *Pirus communis* LINN., and of *Eriophyes galii* NALL. on *Galium mollugo* LINN.[861], are

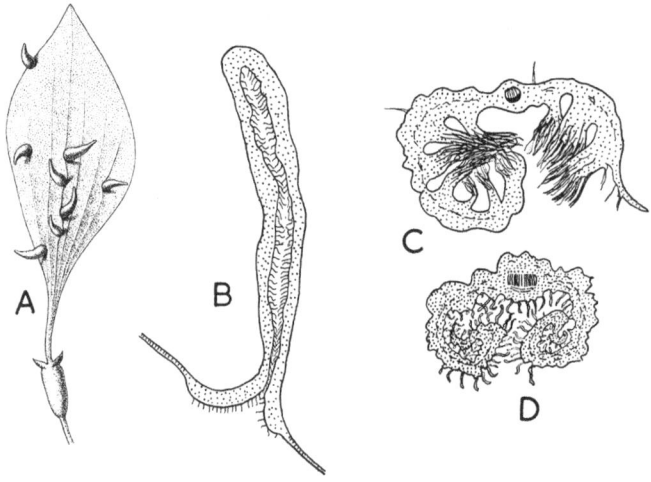

Fig. 102. Some remarkable acarocecidia. A. Ceratoneon gall of *Eriophyes sp.* on the sepal of *Mussaenda*. B. Elongate cephaloneon gall of *Eriophyes* on *Lepidoturus*. C. Eriophyid galls on the fronds of *Nephrolepis exaltata* Schott. D. Leaf roll gall on the frond of *Pteridium aquilinum* Kühn caused by *Eriophyes* sp.

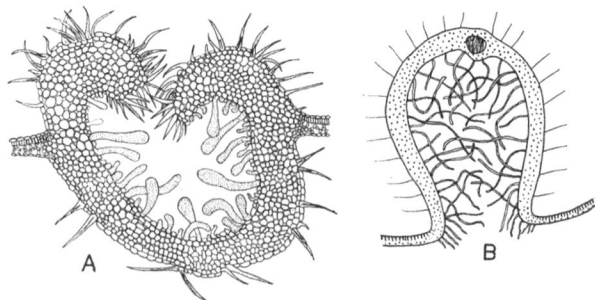

Fig. 103. Structure of typical pouch galls of *Eriophyes*. A. *Eriophyes similis* Nal. on *Prunus spinosa*, showing the gall parenchyma, the nutritive hairs and the typical erineal hairs. B. The spherical pouch gall of *Eriophyes macrorrhynchus cephaloides* Nal. on *Acer campestris*, showing the elongate, unicellular erineal hairs in gall cavity.

characterized by dense erineal outgrowths. The same is also true of
the hypophyllous margin roll gall of *Eriophyes goniothorax typicus*
NAL. on *Crataegus* spp. The hairs in these narrow galls are generally
thin-walled and somewhat clavate. Though in most roll galls the
outer surface is smooth, it is crumpled and partly also folded as in
the margin roll gall on *Lonicera*. The leaf margin roll of *Eriophyes
xylostei* NAL. on *Lonicera xylosteum* is, for example, remarkable for
the upward folding of the margin and the whole rolled portion also
being thrown into irregular folds and curved and pushed outward
by a sharp invagination. For further accounts of leaf roll galls
caused by *Eriophyes* reference may be made to the account of the
anatomy of the leaf margin roll gall of *Eriophyes alpestris* NAL. on
Rhododendron hirsutum LINN. and *Eriophyes tetrastichus* NAL. on
Tilia silvestris DESF. by HOUARD[507].

Pouch galls

Though the greatest majority of pouch galls caused by *Eriophyes*
arise on leaf, some of them appear also on bracts, sepals, etc. We
have, for example, elongate, conical, horn-shaped pouch galls caused
by *Eriophyes* sp. on the single enlarged sepal of the flowers of
Mussaenda tenuiflora BENTH. from Congo[512] and a somewhat similar
gall on *Mussaenda hirsutissima* HUTCH. from India (fig. 102 A).
Though many eriophyid pouch galls arise quite haphazardly on the
leaf blade, a great many of them are, however, typically restricted
to specific parts of the leaf blade, such as, for example, the angle of
the larger veins. The pouch gall of *Eriophyes macrochelus* NAL. on the
leaf of *Acer campestris* from Europe develops in the angle between
the midrib and some of the larger veins. The same is also true of the
gall of *Eriophyes tiliae exilis* NAL. on *Tilia platyphyllops* from Europe.

Most eriophyid pouch galls are conspicuously developed only on the
upper surface of the leaf, but some of them are visible on both sides
and rarely are they nearly equally developed on both sides of the

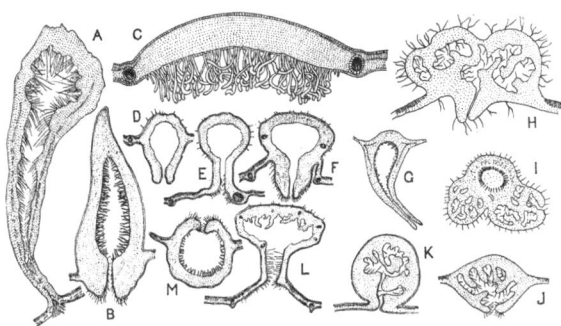

Fig. 104. Some dominant types of pouch galls and covering galls of *Eriophyes.*

blade. The pouch galls range from a simple arching-up of the blade
with erineal growth on the hypophyllous side to complicated horn-
shaped, calvate, spherical, lop-sided and often lobulated and stalked

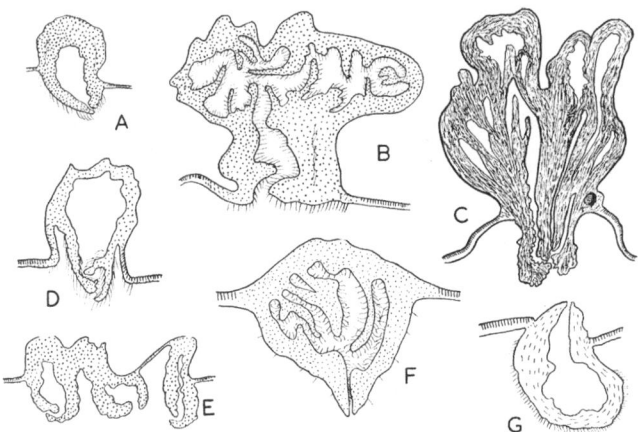

Fig. 105. Some typical acarocecidia. A,B,D,G. Pouch galls on *Salix*. C. Pouch
gall on *Combretum*. E. Covering gall derived from emergence outgrowth on *Salix*.
F. Pouch gall on *Ficus parietalis* Bl.

or sessile outgrowths. The enormous diversity of the external form
of the eriophyid pouch galls is illustrated in fig. 104. The predominantly elongate types of pouch galls are known as ceratoneon galls
and those that are on the whole clavate are termed cephaloneon.
The great majority of eriophyid pouch galls are distinguished by
their asymmetrical and lop-sided development. The ostiole in most
eriophyid pouch galls is very minute and nearly obliterated by
hairy outgrowths and sometimes also totally closed by tissue fusion.
The ostiole is situated on the hypophyllous side of the gall, sometimes within a more or less circular depression or on the summit of a
conspicuous elevated nipple-like process. Only rarely, as in the gall
on *Niconia lateerenata* NAUD described by TAVARES from Brazil[1126],
the ostiole is wide open and surrounded by a broad rim. This gall is
about seven to nine times longer than thick, essentially a hollow
tube, with its tip expanded to form a narrow rim about 2 mm wide
(the gall itself is only 1 mm thick) and the ostiole is situated within
this rim.

The pouch galls are characterized by the small size of the epidermal cells on the outer surface, where the cuticular layer is also often
abnormally thick. The whole mesophyll is characterized by hyperplasy and cell hypertrophy and lack of differentiation of normal
tissues. The peripheral subepidermal layer of cells are generally

closely packed together and contain some chlorophyll. Beneath this lies a layer of thick-walled cells and this is followed by small-sized and closely packed cells. The gall cavity is lined by large cells, with scarcely perceptible cuticle and lacking stomata. The ostiolar region is armed with thick-walled, elongate, straight, acute or rarely clavate hairs. Thin-walled, much longer, curled and twisted and usually clavate hairs inside the cavity constitute the so-called nutritive hairs (fig. 103 A). Vascular bundles, derived from the leaf veins, have distinct xylem and phloem, the latter being inside of the xylem zone. A striking character of eriophyid pouch galls is the presence of more or less irregular, fleshy emergences and septate processes developing from the wall into the gall cavity. Bundles of erineal hairs grow out of these processes. These parenchyma emergences often attain considerable proportions and even nearly obliterate the gall cavity, reducing it to narrow, labyrinthine passages, so that the gall becomes secondarily a solid structure (fig. 104K).

Covering galls

Covering growth of plant tissue to enclose the eriophyid mites is rare. It is seen typically in the covering gall of *Eriophyes gastrotrichus* NAL. on the petioles and tender branches of *Ipomea staphylina* ROEM. & SCH. from the Oriental Region (Plate IV, 2). This mite attacks both leaves and tender branches and on the former gives rise to pouch galls. On the branches and petioles, however, epidermal parenchyma emergences arise, grow over and enclose the mites in a hemispherical gall, with a minute ostiole situated on the summit. We have already mentioned the remarkable covering gall on the leaflets of *Acacia leucophloea* WILLD. caused by *Eriophyes acaciae* NAL. from India.

Other types of galls

Reference has already been made to the pockengall caused by *Eriophyes tristriatus typicus* NAL. on the leaf of *Juglans regia* LINN.[860] Some of the bud galls caused by *Eriophyes* spp. are, as pointed out earlier, typical rosette galls, characterized by fleshy emergences. Mention has also been made of the remarkable solid, fleshy bud gall of *Eriophyes hoheriae* LAMB on *Hoheria populnea* A. CUNN. from New Zealand (Plate II, 2), characterized by its convoluted and lobulated mass, superficially resembling a miniature human brain in external appearance. We have also referred to the enormous, irregular parenchyma gall on the growing point of the shoot axis of *Zizyphus* caused by *Eriophyes cernuus* MASSEE from Africa and India (fig. 38), remarkable for the complete absence of differentiation of an epidermis, so that the gall has the appearance of a mass of cauliflower-like surface. Particular attention may also be drawn to the rindengalls on *Pinus silvestris* LINN. and *Pinus montana* MILL.

caused by *Eriophyes pini* NAL., to which we have already referred in an earlier chapter. Unlike in most eriophyid galls, the mites occur here within the plant tissue. The intercellular spaces are considerable

Fig. 106. The lop-sided cephaloneon pouch gall of *Eriophyes cheriani* Massee on the leaf of *Pongamia glabra* Vent.

and irregular and thus impart a spongy texture to the gall. It is within these intercellular spaces that the mites continue to breed uninterrupted for several years[1190, 807, 485]. We have also dealt with the solid, hard gall of *Eriophyes prosopidis* SAKSENA on the flowers of *Prosopis*, ranging in size from 5 mm to over 30 mm. In this gall also the mites occur wholly within the plant tissues, in between the intercellular spaces and in the labyrinthine passages. *Eriophyes tamaricis* NAL. causes solid, woody lateral swellings on branches of *Tamarix articulata*, often reaching to 20—30 mm in diameter, with deep anfructosities and spongy substances containing irregular narrow spaces. The mites occur within these spaces.

Among the other eriophyid galls mention may be made of the galls of *Eriophyes dispar* NAL. on *Populus tremula* LINN. and of *Eriophyes löwi* NAL. on *Syringa* spp. *Eriophyes hemi* NAL. converts the whole growing axis into a gall on *Atriplex (= Obione) portulacoides* LINN. In the gall of *Eriophyes* sp. on *Ipomea pes-tigridis* LINN.

we have irregularly tuberculated and lobed masses of solid fleshy swellings on stem, petioles, bracts, etc., lacking a true epidermis (Plate II, 4, fig. 107). Some of the eriophyid galls on flowers have a curious bud-like external appearance[903, 397]. Others like the flower gall of *Eriophyes ehretiae* NAL. on *Ehretia laevis* ROXB. are solid, fleshy swellings of the whole flower (Plate II, 1). TROTTER[1177] has given an interesting description of a mite gall on the flower of *Bucida buceras* LINN. caused by *Eriophyes buceras* TROTTER from the Antilles. This gall consists of a greatly elongated slender, hollow, horn-shaped outgrowth, about 19 cm long (instead of the normal fruit measuring only 5—6 mm) and about 2—4 mm thick. The gall cavity contains numerous fleshy irregular emergences. Some of the galls on inflorescence develop into witches-brooms. Other inflorescence galls are solid fleshy swellings. Such a solid, fleshy, irregularly agglomerated mass is caused by *Eriophyes* sp. on *Ipomea scindica* STOCK. from India, in which the whole of the inflorescence, vegetative branch, leaves and buds, is fused together into an irregular solid mass, about 25 mm thick (fig. 51 D).

Not all acarocecidia are caused by eriophyid mites. As mentioned in an earlier chapter, *Tarsonemus phragmitidis* SCHLECHT. causes a curiously cigar-shaped gall on the growing-tip of *Phragmites*

Fig. 107. Rindengall of *Eriophyes* on *Ipomea pes-tigridis* Linn., with the mites external, representing essentially a highly specialized emergence gall.

communis TRIN. The development of the whole shoot is arrested, the leaf sheaths become enlarged and swollen, twisted and folded, thus producing an irregular cabbage-shaped and compressed solid mass. In this gall the development of erineal hairs is curiously restricted to regions of vascular bundles. The hairs arising from adjacent sheaths meet together apically, become flattened and even fused together[419, 425].

3. Galls of Thysanoptera

Galls caused by Thysanoptera may be said to have been first discovered by RÜBSAAMEN during his travels in the Tuchler Heide in 1896–1897. He published interesting accounts of several typical galls and other gall-like abnormalities induced by thrips on leaves of *Galium verum* LINN. and *Stellaria media* CYR. In 1909 GREVILLIUS described thrips galls on *Vicia cracca* LINN. and also discussed the propriety of applying the term gall to some of the growth abnormalities caused by thrips[417, 418]. Some of the salient characters of thrips galls have been discussed by GREVILLIUS *(op. cit.)*, KARNY [560–564], KARNY & DOCTERS VAN LEEUWEN[566], DOCTERS VAN LEEUWEN[301, 303, 308], DOCTERS VAN LEEUWEN & KARNY[288], ZIMMERMANN[1284], HOUARD[523, 524, 527, 528], TAKAHASHI[1121], BAGNALL[49] and others. The bulk of the thrips galls known at present, comes from the tropics, especially Java.

The thrips galls are predominantly leaf galls, but some remarkable bud galls and unique stem and flower galls are also known. Without exception the galls are mantel galls. The thrips occur in large numbers externally on the epidermis. Though they are external, in nearly every case, the thrips come to be enclosed in the gall cavity. This closing-in of the thrips constitutes the principal cecidogenetic growth movement in the thrips galls. Of considerable ecologic importance is the fact that frequently the same species is a gall-inducing agent in one gall and an inquiline in the gall of another species. *Gynaikothrips chavicae* ZIMMERMANN gives rise to a gall on leaf of *Piper retrofractum* VAHL. and *Piper betle* LINN., but is an inquiline in the gall of another thrips on the leaf of *Melastoma malabathricum* LINN. It is frequently very difficult to determine which of the half a dozen or so different species of thrips found in a given gall is the true gall-maker. We find, for example, in the leaf gall on *Ficus retusa* LINN. a number of species like *Gynaikothrips uzeli* ZIMMERMANN, *Gynaikothrips elegans* ZIMMERMANN, *Mesothrips jordoni* ZIMMERMANN and *Leptothrips constrictus* KARNY. *Gynaikothrips uzeli* ZIMMERMANN, *Mesothrips jordoni* ZIMMERMANN, *Leptothrips constrictus* KARNY and *Haplothrips aculeatus* (FABR.) are similarly found in the same gall on the leaf of *Ficus benjamina*. The complex inter-relations of the different species occurring within the same gall are discussed by TAKAHASHI[1121].

In sharp contrast to the acarocecidia, the galls caused by Thysanoptera are predominantly fold and roll galls, in which the cell hypertrophy and cell proliferation are by no means restricted to the epidermis. On the other hand the epidermis may not be greatly altered in some cases and only exceptionally the epidermal cells are enlarged as in the gall on leaf of *Piper retrofractum* VAHL. or they may also grow into minute acutely pointed papillae as in the gall on *Vitis papillosa* BAKER. The differentiation of tissues of the mesophyll is generally inhibited in most thrips galls. Unlike again *Eriophyes* spp. many species of thrips occur on the upper surface of the leaf blade, so that fold and roll galls have an epiphyllous development.

The following are the principal structural types of thrips galls[566]:

I.   Mere curling of the two halves of the leaf blade on either side of the midrib, without the leaf margins coming into contact with each other; the thrips occur exposed on the underside of the leaf; curling hypophyllous. Examples: *Ardisia cymosa* BL., *Eugenia polyantha* WIGHT, *Ficus glomerata* ROXB. var. *elongata* KING.

II.   Folding epiphyllous, with the margins on either side of the midrib meeting together and in contact.
    1. Without very pronounced anatomical abnormalities. Examples: *Ficus benjamina* LINN., *Ficus retusa*, *Ficus retusa nitida*, *Mallotus philippinensis* MUELL. ARG., *Melastoma malabathricum polyanthum*, *Veronica cinerea* Less.
    2. With more or less pronounced swelling of the folded or curled leaf and characterized by conspicuous anatomical abnormalities. Example: *Spatholobus littoralis* BL.
    3. With swelling of the blade in the immediate vicinity of the infected vein, mostly the midrib. Examples: *Piper betle* LINN., *Piper nigrum* LINN., *Smilax* sp. (fig. 64).

III. Folding or rolling of the leaf blade either hypophyllous or epiphyllous and usually extending to the whole blade.
    1. Without pronounced swelling of the blade. Examples: *Cordia suaveolens* BL., *Eurya japonica* THUNB., *Ficus cuspidata* REINW., *Justicia procumbens* LINN.
    2. With more or less conspicuous swelling of the leaf blade. Examples: *Conocephalus suaveolens* BL., *Loranthus pentandrus* LINN., *Memecylon intermedium* BL.

IV. Sac-like epiphyllous or hypophyllous out-pocketing or pouch galls, often with simultaneous rolling and curling of the blade. Examples: *Aporosa microcalyx* HASSK., *Mallotus repandus* MUELL. ARG., *Schoutenia ovata* KORTH, *Vitis papillosa* BACKER.

V.   Ceratoneon pouch gall. Example: *Heptapleurum ellipticum* SEEM.

VI. Emergence galls. Example: *Conocephalus suaveolens* BL.

The leaf fold and roll galls caused by Thysanoptera are readily distinguished from those of Eriophyidae by the insignificant cell hypertrophy and cell proliferation in the epidermis, the relative dominance of cell proliferation and inhibition of tissue differentiation in the mesophyll, absence of epidermal erineal hairs or emergences, etc. Only exceptionally obscure wart-like emergences are found on thrips galls.

The simplest thrips gall, such as that on the leaf of *Bridelia laurina* BAILL. from New Caledonia [524], is a bag-like malformation, due to the failure of opening of the leaf rudiment from the bud, with simultaneous swelling. The halves of the blade on either side of the midrib remain folded together on the upper side, with a rugose outer surface. The gall cavity is thus open to the outside along the leaf margin, apically and basally. In the leaf fold gall of *Gynaikothrips pallipes* KARNY on *Piper sarmentosum* ROXB. from Java[303], the thrips attacks the tender leaves in the bud even before they have opened and gets in between the fold of the leaf. The blade is more heavily infested on one side of the midrib than on the other side, so that the former half rolls upward and the latter half curves over the rolled margin of the former like a sheath. Pronounced rugosities and coriaceous warts, some of which are 1 mm large, appear in great numbers on either side and along the midrib. The leaf roll gall of *Gynaikothrips chavicae* ZIMMERMANN on *Piper betle* LINN. from Java is a narrow, hypophyllous, subcylindrical infolding of the blade along the midrib, so that the major part of the blade remains normal on either side (fig. 64 B). Attention may also be directed to the leaf roll of *Eothrips aswamukha* on *Jasminum pubescens* from India, mentioned in an earlier chapter. In this gall the entire leaf is spirally twisted and rolled to give rise to a verrucose, fusiform swelling, about 75 mm long.

In comparison to those of Eriophyidae, the pouch galls of Thysanoptera are extremely simple. Absence of emergences, hairy processes and the presence of a permanently open and usually wide ostiole distinguishes them at once from the pouch galls of mites. The simplest pouch gall caused by thrips is an utricular bag, with a spacious cavity. An epiphyllous pouch gall, about 2—4 mm high and somewhat recurved apically, is caused by *Cryptothrips conocephali* KARNY on the leaf of *Conocephalus suaveolens* BLUME from Java. *Onychothrips tepperi* UZEL causes a globose, yellow pouch gall, about 10 mm in diameter, on the leaf of *Acacia aneura* MÜLL. from Australia[520]. An epiphyllous or also hypophyllous pouch gall on the leaf of *Aporosa microcalyx* HASSK., caused by *Dolerothrips tryboni* KARNY from Java, develops in great numbers and converts the whole leaf into a globose vesicular mass. The gall of an unknown species of thrips on the leaf of *Memecylon umbellatum* BURM. from India also develops in such large numbers that the entire leaf

becomes a contorted, rugose, multi-chambered coriaceous bag, about 40 mm in diameter. An interesting pouch gall by an unknown thrips arises on the leaf of *Terminalia* sp. from India (Plate I, 4).

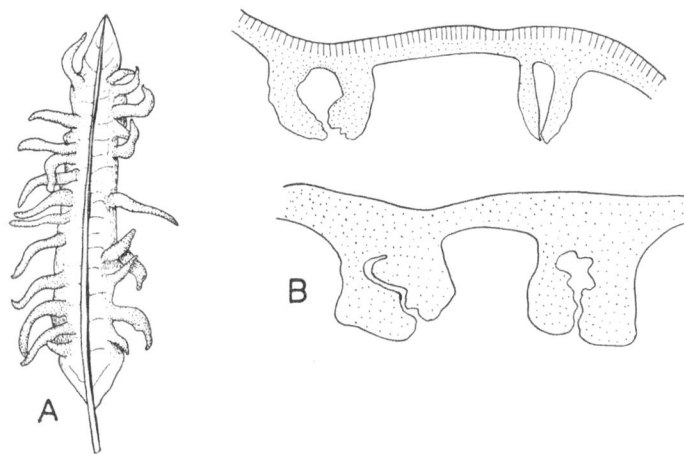

Fig. 108. Galls of Thysanoptera. A. Ceratoneon pouch gall on leaf of *Schefflera hexapetalum*. B. Covering gall on leaf of *Ilex mitis erythraeus* Fios. (After Trotter).

This is a thick-walled, hypophyllous, pyriform or subglobose, cephaloneon gall, about 25 mm long and 10—15 mm thick, developing mostly near the leaf tip. Docters van Leeuwen[305] has described some very interesting pouch galls of Thysanoptera. *Gynaikothrips chavicae heptapleuri* Karny causes an epiphyllous ceratoneon gall, about 18 mm long, on the leaf of *Schefflera odorata* Mer. & Ro. from the Philippines[1198]. The same species causes a contorted cylindrical epiphyllous ceratoneon gall on leaf of *Heptapleurum ellipticum* Seem from Java, Celebes and parts of Australia. Ceratoneon pouch galls are also caused by thrips on *Schefflera divaricata* Bl., *Schefflera hexapetalum* (fig. 108 A) and *Schefflera polybotrya* Harms from Java.

Covering galls caused by thrips are rather rare. We have, for example, an interesting record[1178] of an undescribed species of thrips as giving rise to a hypophyllous, subcephaliform or crateriform covering gall, arising as fleshy outgrowths, with ostiole in the middle of an umbilicus-like depression, on the leaf of *Ilex mitis erythraeus* Fios. from Eritrea (fig. 108 B).

Bud galls are among the largest of thrips galls and typically develop as utricular growths. We have already cited the example of the curiously cabbage-like bud gall of *Thilakothrips babuli* Ramakr. on *Acacia leucophloea* Willd. from India (fig. 51 C). The gall of *Austrothrips cochinchinensis* Karny on the bud of *Calycopteryx floribunda*

LAMARCK from India (Plate III, 2) is an enormous bag-like growth, to which we have earlier drawn attention.

4. Galls of Homoptera

The distinctive characters of development and structure of the galls caused by Homoptera are largely correlated with the habits of feeding of these insects. In sharp contrast to Eriophyidae and Thysanoptera which suck the liquids from a number of plant cells and moving in succession, the Homoptera are permanent sedentary ectoparasites, at least during their immature stages, on the epidermis and continue to suck the plant juices at the same spot and even reproduce fixed to the place. The cecidogenetic action of Homoptera spreads, however, over a much larger area than in the galls of Acarina and Thysanoptera and is also far more intense. Much larger numbers of plant cells react to the cecidogenetic action. These differences are correlated to the larger size of Homoptera. Although the Homoptera are situated externally on the epidermis, the depth to which the piercing stylets penetrate within the plant tissue is considerably greater than in case of Acarina and Thysanoptera. As the nymphs of Homoptera grow, they drive the piercing stylets deeper and deeper and pour increasing quantities of the salivary juice into the plant tissue, thus continually extending the centre of cecidogenetic action. The piercing stylets usually penetrate cambial zones and other meristematic tissues. The galls of Homoptera are therefore relatively complex in structure. As in the case of acarocecidia and thrips galls, the galls of Homoptera are predominantly leaf galls, especially mantel galls. We may recognize three large ecologic groups of Homoptera galls, viz. the galls of Psyllidae, galls of Aphids and the galls of Coccids. The fundamental pattern of Homoptera galls exhibits interesting modifications in each of these groups.

Galls of Psyllidae

The bulk of psyllid galls are pit galls and pouch galls on leaves of Dicotyledons, but some leaf margin roll galls and covering galls are also known. The pit and pouch galls of psyllids differ from those of Eriophyidae in the general absence of erineal and emergence outgrowths, the greater regularity of shape, in the thicker walls, the obliteration and closing of the ostiole, the dehiscence of the mature gall and in the extremely limited number of individuals of the psyllid in each gall—never more than one or at the most two nymphs of the psyllid being found in a single gall. Unlike Eriophyidae, it is the newly hatched immature nymph of the psyllid that is concerned in the initiation of the gall and the progeny do not occur in the gall of the parent. As an example of the simplest pouch gall we may refer to the pit gall of *Trioza* on the leaf of *Ficus religiosa* LINN., mentioned in an earlier chapter. Another simple pouch gall, caused by *Trioza fletcheri* CRAWF. on the leaf of *Trewia nudiflora* LINN. from the Orient, is a coriaceous, epiphyllous, utricular out-pocketing, with the blade only slightly swollen, and with a large wide-open ostiole below. Another pouch gall with permanently open ostiole is the curiously horn-shaped, elongate-cylindrical, slender, delicately membranous, pale yellow or white gall on *Lecanthus wightii* WEED caused by an unknown species of *Trioza* from the Himalaya (Plate V, 1). All other pouch galls caused by psyllids are a typically thick-walled, usually fleshy, epiphyllous, often agglomerate mass, with the ostiole completely closed. Some of them, like for example the gall of *Trioza jambolanae* CRAWF. on the leaf of *Eugenia jambolana* DC from India, are hemispherical growths, marked by a slight depression below. Other pouch galls are generally more or less developed on both sides of the blade, though the major part of the gall is epiphyllous. Some like the gall of *Pauropsylla tuberculata* CRAWF. on leaf of *Alstonia scholaris* R. BR. from the Orient (fig. 109 A) are

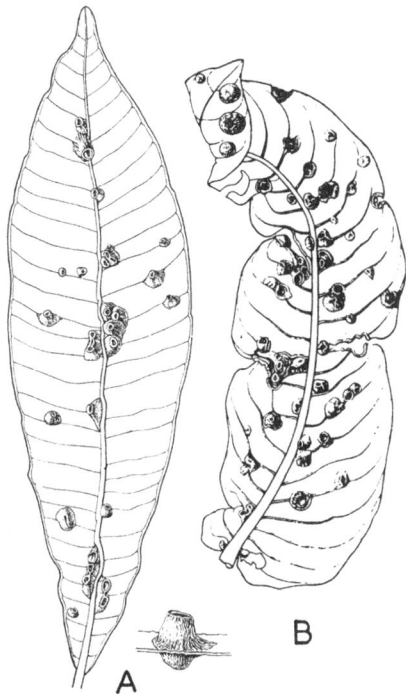

Fig. 109. Galls of Psyllidae. A. Truncated cone-shaped pouch gall of *Pauropsylla tuberculata* Crawf. on leaf of *Alstonia scholaris* R.Br. B. Pouch gall of *Trioza* on leaf of *Terminalia*.

nearly equally developed on both sides. Many of them like the galls of *Pauropsylla depressa* CRWF. on leaf of *Ficus glomerata* ROXB. (Plate III, 2) and of *Trioza* on leaf of *Terminalia* (fig. 109 B) arise very closely crowded together in such large numbers that they become fused together to form large, multi-chambered, fleshy agglomerate masses. The modified pouch galls include the nearly solid, conical or cordate gall of *Neotrioza machili* KIEFF. on the leaf of *Machilus gamblei* KING and the nut-like galls of *Phacopteron lentiginosum* BUCKT. on the leaf of *Garuga pinnata* ROXB. (Plate II, 7) and the gall of an unknown psyllid on leaf of *Rhus semialata* from the Orient. RÜBSAAMEN[997] has described an extremely interesting pouch gall, caused by an undescribed psyllid on the leaf of *Trichilia* sp. This gall is developed on both sides of the blade. The epiphyllous part is a cylindrical truncated growth, topped by a flat umbrella of the epidermis of the leaf, covering the gall like a hood above. Beneath this hood the gall bears short, stout, curved hairs (fig. 77 C). The hypophyllous portion of the gall is larger, but it is deeply sunk inside the upwardly folded leaf, so that a circular furrow surrounds the cylindrical gall on the under side of the leaf. The lower end of the gall is covered by long stretched cells that ultimately become hairs at their tips. The gall evidently arises from localized cell proliferation within the mesophyll and in which the upper epidermis does not take an active part, so that when sufficiently grown the gall breaks through the upper epidermis and carries it up like a flat hood.

Leaf margin roll galls caused by psyllids are generalized structures when compared to those of Eriophyidae and Thysanoptera and are also more localized. The commonest examples of margin roll galls are caused by *Psyllopsis fraxini* LINN. on *Fraxinus* and by *Trichochermes (= Trioza) walkeri* (FÖRST.) on *Rhamnus cathartica* LINN. and *Rhamnus frangula* LINN. (fig. 68 A). The last mentioned gall is remarkable for developing between two major side veins and for its abnormal hairy outgrowth. A singular looking leaf margin gall is caused by an unknown psyllid on *Ficus nervosa* HEYNE from Formosa (fig. 70 B)[1181]. The leaf margin is folded into a conical rounded expansion, projecting about 7—8 mm and constituting a sort of marginal pouch on the under side of the leaf and pointing out at right angle to the general leaf axis. In the leaf margin gall of *Trioza alacris* FLOR. on *Laurus nobilis* LINN. from the Mediterranean region[111], we have a flat coriaceous rolling of the margin (fig. 22). After oviposition along the leaf margin on the under side, the female remains at the site to feed. This feeding activity of the female initiates the development of the margin roll gall, so as to enclose the eggs. The nymphs that hatch continue to feed from within the roll and thus contribute to the further growth of the gall. The mesophyll of the galled part is but little altered and the epidermal cells are generally hypertrophied and lack normal stomata[1141]. Only rarely, as in the case of the gall of *Phylloplecta hirsuta* (CRAWF.) on the leaf of *Terminalia tomentosa* W. & A. from India, is the leaf margin roll gall epiphyllous and extends the whole length of the leaf.

Some interesting rosette galls are also caused by psyllids like *Trioza* on *Cerastium, Valerianella*, etc. Mention must be made of the remarkable rosette gall of *Livia juncorum* LATR. on *Juncus* spp. [150, 397, 1209]. This gall has at first the appearance of a viviparous inflorescence, but is in reality composed of the closely packed rosette of red or brown foliaceous growths. We have also referred earlier to the curious fir-cone-like gall of *Apsylla cistella* (BUCKT.) on *Mangifera indica* LINN. (fig. 50).

Covering galls by psyllids are rather rare. An unknown species of *Trioza* often deposits its eggs on the tender branches or pedicels of the inflorescence or even the young immature fruit of *Terminalia* spp. from India. In all these cases typical covering galls arise, with the plant tissue growing around and upward over the developing nymph, which ultimately becomes enclosed in a hemispherical gall, with a minute ostiole on the summit. *Pauropsylla tuberculata* CRAWF., which ordinarily gives rise to pouch galls on the leaf, sometimes oviposits on the tender fruits of *Alstonia scholaris* R. BR. in certain parts of India. A series of closely crowded covering galls arise on the elongate slightly compressed young fruit, which eventually becomes swollen into an elongate, stout, cylindrical, fleshy, multilocular agglomerate mass.

Galls of Aphidoidea

The peculiarities of galls caused by aphids, that serve to distinguish them from those of psyllids, include the great anatomical and histological complexity and the fact that nearly all aphid galls are initiated by a single female, the progeny of which, however, continue to remain within the gall of the parent. The complex heterogony and polymorphism of aphids are also associated with differences in their galls. Aphid galls may indeed be said to be among the most specialized of insect galls.

Aphid galls are predominantly covering galls, leaf fold and roll galls, pouch galls and krebsgalls.

An outstanding example of a covering gall is the gall of *Pemphigus bursarius* (LINN.) on the petiole of *Populus nigra* LINN., *Populus tremula* LINN. and *Populus pyramidalis* LINN., mentioned in an earlier chapter. The fundatrix of the aphid, hatching from the overwintered egg, settles down on the tender petiole in early spring and at this point the petiole becomes bent more or less strongly. With the continued sucking of the plant juices by the fundatrix, there is an increase in parenchyma cells at the site of the sucking by the aphid. This parenchyma mass eventually grows over and encloses the aphid (fig. 71). The covering gall of *Pemphigus spirothecae* PASS. on the petiole of *Populus pyramidalis* LINN. (fig. 1 C) also starts with the feeding activity of the fundatrix hatching from the over-wintered egg. We have already given a brief account of the structure and development of this gall. The fundatrix moults for the first time after four weeks and after the fourth moult about the middle of June it produces about 20—30 apterous young aphids. The second generation appears usually in the first half of July. The aphids of the second generation moult four times and become alates, which eventually escape from the gall in the first half of August. The alates reach crevices in the bark of *Populus* and deposit 6—8 eggs, from which males and females hatch. These females and males moult three times before mating. The mated female deposits the winter egg[379]. We have also mentioned another remarkable covering gall caused by aphids, viz. the ananas galls of *Chermes (= Adelges) abietis* LINN. and *Cnaphalodes strobilobius* KALT. on *Picea excelsa* (LAMK.) LINK. The sexuales of the former species appear in autumn and after mating, each female lays a single egg, from which the fundatrix hatches. The fundatrix settles down in late autumn on the young shoot of *Picea* near a winter bud and starts sucking the sap from the inner cortex (fig. 52 A)[153, 154]. In April or early May about 100—160 parthenogenetic eggs are deposited by the fundatrix under a protective cover of waxy threads and then the fundatrix perishes. About the same time the buds of *Picea* start unfolding and the gall already initiated by the feeding activities of the fundatrix develops rapidly further. The main axis and the bases of the needles become swollen, but a series of narrow, elongate spaces remain between the shield-like bases of the swollen needles, with narrow slit-like ostioles. The young aphids enter these hollow spaces and start sucking the sap from the tender tissues surrounding the cavity. The shield-like swollen bases of the needles begin to swell up further and soon obliterate the slit-like ostioles. The aphids are now completely enclosed. About the middle of August the flow of sap to the gall tissue diminishes and the gall begins to shrivel up, resulting in the reopening of the slit-like passages. The aphids, which have in the meantime moulted a few times, escape from the gall through the reopened openings and moult once again on the surface of the needles to become alates. Some of the alates continue to remain on *Picea* but others migrate to *Larix*, on which they reproduce. The progeny from *Larix* come back to *Picea* later[105, 882].

The general characters of fold and roll galls of aphids may be observed in the galls of *Aphis atriplicis* on *Atriplex* and *Chenopodium*, *Prociphilus poschingeri* HTG. and *Prociphilus bumeliae* SCHRK. on *Fraxinus excelsior* LINN., *Pachypappa vesicalis* KOCH on *Populus alba* LINN. and of *Aploneura lentisci* PASS. on *Pistacia lentiscus* LINN. (fig. 1B). *Forda formicaria* HEYDEN gives rise to a leaf margin fold gall on *Pistacia terebinthus* LINN. The fold galls of aphids are readily distinguished from those of other Homoptera and Thysanoptera by their pronounced thick-walled tissue and complete inhibition of the differentiation of normal tissues.

Pouch galls caused by aphids are distinguished by their persistent ostiole, which though reduced to a size that does not permit the escape of the aphid, is not completely obliterated. Other striking features of these pouch galls are the characteristic asymmetrically formed bulges, in which the dehiscence of the gall occurs in a lacerated rupture of the wall; the bulge is usually situated in the lower portion on one side. The pouch gall is initiated by the feeding activities of the fundatrix, but its progeny continue to crowd in and feed in the same gall, thus contributing to its full development. Although some hairy outgrowth may be present, the aphid pouch galls differ from the eriophyid pouch galls in the absence of erineal and emergence outgrowths. Another fundamental difference lies also in the fact that the seat of cell proliferation is not predominantly localized to the epidermis in aphid pouch galls. The gall of *Byrsocrypta gallarum* (GMELIN) on the leaf of *Ulmus* is a typical example of an aphid pouch gall[814, 635, 1297]. The fundatrix settles down on the under side of the leaf and thus initiates the development of the gall. At the site of its sucking, the blade becomes slightly swollen and

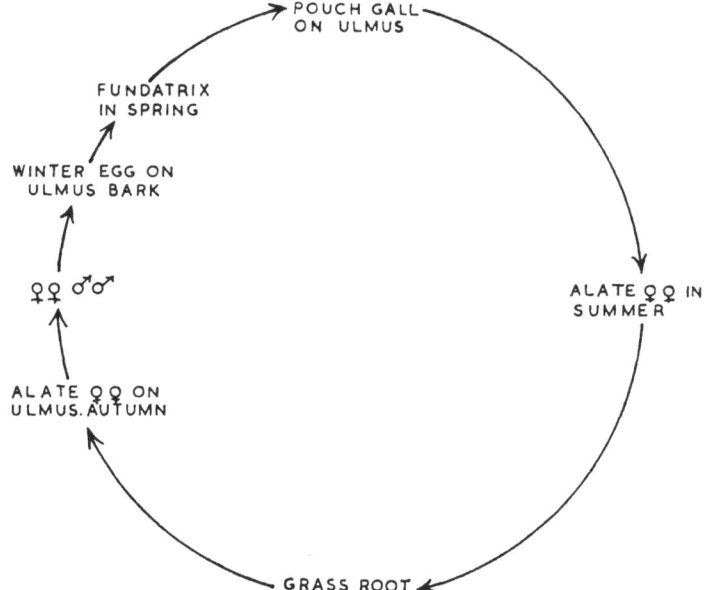

Fig. 110. Life-cycle of *Byrsocrypta gallarum* (Gmelin) from the pouch gall on *Ulmus*.

bulged upward. It is interesting to observe that for the development of the gall it is necessary that the fundatrix should suck near a vascular bundle. Continued infolding from below and bulging upward result in the formation of a clavate, hollow pouch gall, with a short, slender neck-like stalk. The fundatrix, now completely enclosed, moults about four times within about a fortnight and sucks at the upper end of the gall. It reproduces parthenogenetically

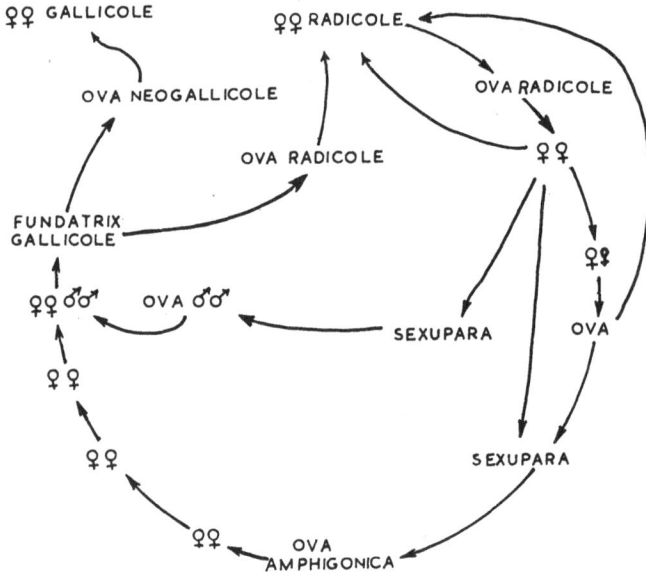

Fig. 111. Life-cycle of *Peritymbii vitifolia* Fitch from its galls on root and leaf of *Vitis vinifera* Linn.

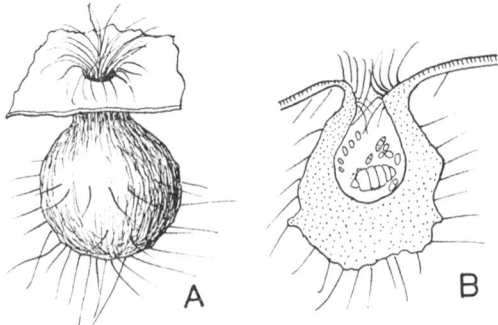

Fig. 112. Pouch gall of *Peritymbia vitifolii* Fitch on leaf of *Vitis vinifera* Linn. A. Single gall. B. L.S. through gall.

to give rise to about 40 viviparous young aphids. The apterous young soon grow and also reproduce parthenogenetically within the gall and produce apterous females. These apterous offspring of the fundatrix crowd in large numbers in the lower part of the gall, where an asymmetrical bulge develops externally. The alates appear by end of June or early July. At this time cracks arise in the bulge and many of the alates escape from the gall and migrate to roots of grasses like *Poa, Festuca, Cynodon, Panicum, Hordeum, Eragrostis*, etc. The return to *Ulmus* occurs again in autumn and after partheno-genetic reproduction, the sexuales arise. After mating, the winter egg is deposited, from which the fundatrix hatches.

The commonest krebsgall caused by aphids is the gall of *Eriosoma (= Schizoneura) lanigerum* HAUSM. on *Pirus malus* LINN. These arise as knotted, irregular swellings of branches, sometimes the crown and rarely the aerial portions of roots. The reddish-brown, apterous aphids secrete copious waxy-wool and live gregariously. The viviparous female reproduces parthenogenetically in Europe, bringing about 30—40 young each. Though about ten generations are produced annually, alates appear only exceptionally in Europe in June-July and in autumn. The cambium is penetrated by the piercing stylets, thus causing spongy growth of elongate, thin-walled cells. Sieve tissue and primary phloem are little affected. The cortex is however generally swollen and fissured. In North America, the alates migrate in autumn to branches of *Ulmus americana* LINN., on which the sexuales appear. The fundatrix next spring gives rise to a leaf roll gall on *Ulmus americana* LINN. but such a gall does not develop in Europe. Another common krebsgall by an aphid is caused by *Peritymbia (= Vitea = Phylloxera) vitifolii* FITCH (formerly *Phylloxera vastatrix* PLANC.). As pointed out earlier, this gall is a nodular swelling at the ends of thin side roots of *Vitis vinifera* LINN. The swelling is the result of cell proliferation in the root cortex. The cells in the immediate site of sucking by the aphid are small, so that the swelling assumes a curious reniform shape. If the aphids attack other roots, the so-called tuberosities arise. The root gall aphids reproduce parthenogenetically, often for six generations under fa-vourable conditions. Each aphid lays about 60 eggs. The same aphid can under certain circumstances give rise to pouch galls on leaves of *Vitis vinifera* LINN. (fig. 112). The sexuales arise in autumn and the fundatrix hatches from the winter egg in spring. The fundatrix re-produces on leaves, giving rise to hemispherical, hypophyllous pouch galls, 3—4 mm large, with its ostiole covered by a corona of hairs. Reproducing parthenogenetically, the aphids produce progeny, which give rise partly to leaf galls and partly to root galls.

Some remarkable bud galls of aphids are also known. We have already referred to the gall on *Vaccinium leschanaulti* WT. from India (plate I, 1). SASAKI[1008] has described a curious bud gall on

Styrax japonicum SIEB. & ZUCC., caused by *Astegopteryx neckoashi* SASAKI. The gall consists of a flat swollen fleshy cap, with a double peripheral row of stunted leaves, about 60 mm in diameter. A remarkable flower gall is caused by *Astegopteryx styracophila* KARSCH on *Styrax benzoin* DRYAND from Java[520]. All the parts of the flower are greatly altered, the peduncle is swollen and the receptacle is enlarged and surmounted by elongate, cylindrical bracts.

Galls of Coccoidea

Mention has already been made of the gall caused by *Asterolecanium* spp. on diverse plants like *Quercus, Templetonia*, etc. A useful list of galls of *Asterolecanium* was published by LINDINGER[696]. The galls caused by the coccid family Brachyscelidae from Australia present a number of peculiarities, including sexual dimorphism of the gall. The galls of the male coccid are typically in the form of tubular swellings, with an apical ostiole. The galls of the female coccid are variable, but generally more oval or elongate-ellipsoidal, sessile and isolated, with an apical ostiole. They arise on different species of *Eucalyptus*[352, 1034, 1127]. The gall of *Brachyscelis floralis* FROGGATT is an ovoidal swelling of the flowers, 30—40 mm long and 18—20 mm thick. *Brachyscelis pomiformis* FROGGATT gives rise to the so-called "blood-wood-apple" gall. An elongate, horn-shaped, subcylindrical gall, measuring nearly 65 mm long, narrowed at both ends and also somewhat curved, is caused by *Brachyscelis sloanei* FROGGATT. A pedunculated, clavately swollen bud gall, about 75 mm long, is caused by *Brachyscelis pedunculata* OLLIFF. (fig. 113 A—C). Dense bunches of short, cylindrical bud galls are formed by

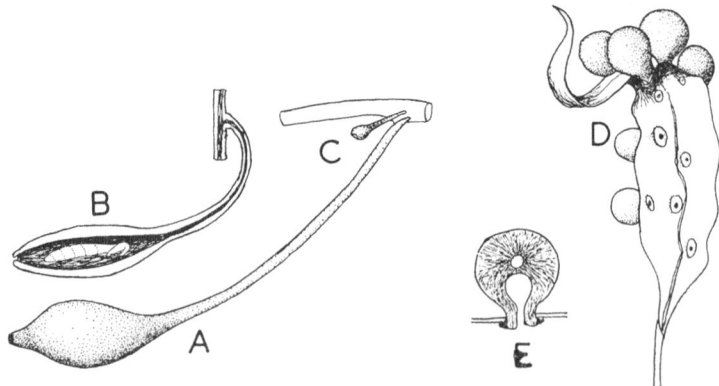

Fig. 113. Galls of Brachyscelidae. A–C. Galls of *Brachyscelis pedunculata* Olliff. A. Gall of the female. B. The same cut open to show the gall cavity and the coccid at its bottom. C. Gall of the male. D. Gall of *Ascelis praemollis* Schr. on leaf of *Eucalyptus*. E. L.S. through the same.

Brachyscelis urnalis TEP. A subglobose gall, about 5 mm thick, is caused by the female of *Sphaerococcus rugosus* MASK. on the branches of the Australian *Leptospermum* sp. The epiphyllous spherical gall of *Ascelis praemollis* SCHR. on *Eucalyptus* (fig. 112 D—E) is about 12 mm in diameter and has a thick fleshy wall. A circular swollen disc indicates the site of the gall on the under side of the leaf. The ostiole is situated at the centre of this disc and leads into a pyriform atrium. Above the atrium lies the minute gall cavity. The galls of the Australian *Cylindrococcus* on *Casuarina* (fig. 53C) present certain interesting features. These galls are mostly cylindrical or subconical hollow swellings of the growing tip of the shoot axis, surrounded by dense growth of expanded scales, with ample cavities containing each a cylindrical nymph of the coccid. *Cylindrococcus casuarinae* MASKELL causes, for example, an elongate cylindrical gall, about 6 mm long at the growing tip and covered with imbricating scales. A single large gall cavity contains the female of the coccid. The gall of *Cylindrococcus spiniferus* MASKELL on *Casuarina quadrivalvis* LABILL. is composed of an isolated group of subpyriform swellings, about 9—25 mm long and 6—13 mm thick, covered with acute rough scales, with an axial elongate gall cavity, in which lies the female of the coccid.

5. Galls of Coleoptera

Relatively few characters are exclusively typical of the galls of Coleoptera. The bulk of these galls are generally diffuse fusiform, solid or generally hollow swellings, which usually persist long after the escape of the adult beetle. The majority of these galls are caused by Curculionidae, but a number of interesting root galls are caused by Buprestidae. We have given the example of the root gall of *Ethon corpulentum* BOH. on *Dillwynia ericifolia* SM. (fig. 31 A) in an earlier chapter. One of the commonest Curculionid galls is caused by *Ceutorrhynchus pleurostigma* MARSHALL on the roots of Cruciferae like *Raphanus* and *Brassica*. The adult weevil generally feeds on flower buds and copulates during spring. The females then drop off to the ground for egg laying. Using the snout, the female scoops out a suitable cavity in the root cortex, into which the egg is then thrust. The injury thus caused is soon closed by growth of tissue, so that the egg comes to be completely within the plant tissues. As the larva starts feeding, the part of the root swells up and the gall matures in about four weeks. The larva now bores its way out to pupate in soil, immediately below the ground level. A second generation arises, so that galls develop both during summer and during autumn. As the egg is situated in the innermost layer of the cortex near the cambium and as the larval cavity arises here, the normal development of the central cylinder is greatly interferred with. The cells resulting from the proliferation of the cambial zone remain thin-walled simple parenchyma elements. The larval cavity later becomes deeper and sometimes sinks even into the medulla. The medullary rays enlarge considerably, breaking up the vascular bundle ring. As the larval cavity is situated in the cambial zone, it may be observed that there is no special nutritive zone around the larval cavity. In the gall of *Nanophyes niger* WETT. on the stem of *Erica* spp., HOUARD[513] has described the development of a zone of nutritive tissue around the larval cavity, new vascular bundles and abundance of secondary tissue. The galls of *Nanophyes* spp. are also interesting for several other reasons. A number of species like *Nanophyes durieuri* LUCAS on *Umbilicus* sp. and *Nanophyes nesae* RÜBS. on *Nesaea sagittifolia glabrescens* give rise to galls[340]. The best known of the galls of *Nanophyes* is caused by *Nanophyes pallidus* OLIV. on the fruits of *Tamarix africanus* and *Tamarix gallica*; these are the well known jumping galls[1124]. Other interesting galls include those of *Gymnetron linariae* on the roots and underground stem of *Linaria vulgaris* and *Linaria striata*[920], *Pachycerus varius* on the root of *Cynoglossum cheirifolium* and *Cynoglossum pictum*, *Anchonoides bonariensis* BRÈTHES on *Sagittaria montevidiensis immaculata* HICK from Argentina[533], *Chromoderus fasciatus* MÜLL. on roots of *Chenopodium album* LINN., *Gymnetron hispidum* BRULLÉ on the stem of *Linaria* sp. and *Apion prosopidis* KIEFF. on the leaf of *Prosopis* spp. from South America. The last mentioned gall is a woody, greenish or reddish swelling, about 60 mm long and 10 mm thick and generally cylindrical or ellipsoidal[533]. Composite fused masses of leaf galls are caused by certain Curculionidae on Cruciferae like *Allyssum maritimum* LAMARCK. Mention may also be made of the leaf gall on *Raphanus raphanistrum* LINN., caused by *Ceutorrhynchus leprieuri* BRIS.[1183]. MOLLIARD[809] has given an interesting account of his "facultative gall" on the male catkin of *Salix*, caused by *Dorytomus taeniatus* FABR. The gall is characterized by dense phyllomes (modified stamens) and a spacious larval cavity. Reference may also be made to the gall of *Miarus campanulae* LINN. on the flower of *Campanula* and *Phyteuma*, described in an earlier chapter.

Some of the hollow galls on the shoot axis caused by Coleoptera are interesting for the complex community of successori, especially diverse species of ants (vide Chapter XI).

6. Galls of Lepidoptera

Although Lepidoptera are taxonomically widely separated from Coleoptera, on account of the considerable structural and ecological similarities of the galls of these two orders of insects, the galls caused by Lepidoptera are best considered at this stage. The eggs are always deposited on the surface of the tender plant part, though often within the bud, but never within the plant tissue. The newly hatched larvae bore their way into the tissue, so that eventually the larva comes to be within the gall. Most of these galls arise on the shoot axis, preferably near or at the growing tip and are often surmounted by curious tufts of stunted rudiments of branches or leaves. An interesting stem gall of *Aegeria uniformis* SNELLEN develops on *Commelina communis* LINN. from Java (cf. fig. 118D). The larva of this Sesid moth burrows inside the tender stem and gives rise to lateral swelling, partly ensheathed by the bracts. The gall often measures 20 mm in length and 10 mm in diameter. A rinden-gall on *Strobilanthes crispus* BLUME is caused by an unknown Lepidoptera from Java[299]. The gall has a swollen pyriform basal portion, which is broadly sessile and 10 mm in diameter, containing the larval cavity. Apically there is an abruptly long, slender beak-like process. The outstanding features of the gall of *Cecidoses eremita* CURTIS on the South American *Schinus dependens* ORTEGA include the peculiar pyxidial dehiscence, described elsewhere. A somewhat similar gall is caused by *Eucecidoses minutanus* BRÈTHES on the same plant in South America. KIEFFER & JÖRGENSEN[593] have described an interesting gall of *Dicranoses capsularis* KIEFF. & JÖRG. on the branches of *Schinus dependens* ORTEGA from South America. This gall is a subcylindrical swelling, about 10 mm long, constricted at the base and dehiscing by an apical cap-like lid. The surface of the gall is sometimes crowded with stunted bunches of leaves. The gall of *Dactylethra candida* STAINT. on the growing tip of the shoot axis of *Tephrosia* (Plate I, 6) has already been referred to. Irregular swelling of the branches of *Populus* spp. is caused by *Sciapteron tabaniformis rhingiaeformis* HÜBN. from Algeria. The gall of *Pamene pharaonana* KOLLAR on the branches of *Tamarix* spp. from North Africa is a spongy, multi-chambered, irregular, green swelling, about 20 mm large. *Amblylapis olivierella* RAGONOT also gives rise to a woody, oval or subcylindrical, hollow, thick-walled swelling of the branches of *Tamarix* spp. in North Africa. This gall often measures 40 mm long and 30 mm thick. *Oecocecis guyonella* GUENEE from North Africa causes hard, hollow, spherical gall on the tender branch or peduncle of *Limoniastrum guyonianum* DAR. The stem gall on *Rubus bogotensis* H.B. & K., caused by *Sesamia cecidogona* KIEFF. from South America[533] is an irregular, globose swelling, superficially resembling the common midge gall on *Rubus* from Europe. Another interesting stem gall caused by *Evertia (= Tortrix) resinella* (LINN.) on the tender shoot axis of *Pinus* is the well known "Harzgall" of the German literature. The larva burrows in the cortex and medulla and the stem becomes thus swollen unilaterally. A conspicuous accumulation of oozing gum on the opposite side is built into a sort of an irregular globose cover by the larva.

Some remarkable leaf galls are also caused by Lepidoptera. We have, for example, the record of a Lepidoptera gall on the leaf of *Sonneratia acida* LINN. from Cochin-China, Java and Siam[520]. An unknown Microlepidoptera is reported as giving rise to a midrib gall on the leaf of *Erioglossum edule* BL. from Java[997]. The outstanding examples of leaf galls caused by Lepidoptera are found, however, on plants of the Natural Order Melastomaceae. HOUARD[520] has, for example, described the curious gall on the leaf of *Henrietta succosa* DC. This gall consists of spherical masses, with radial whorls of stunted bracts, forming a plume on the gall. The gall is inserted in the axil of two leaves, with which it is partly confluent. An unknown Lepidoptera gives rise to a spherical gall, about 12 mm thick, in the axils of leaves of *Rolandra argentia* ROTH. The gall is clothed with a tuft of stunted foliaceous processes, giving it the superficial appearance of a hairy fruit or inflorescence of the plant. Mention must be made of the remarkable spiny gall on the leaf of *Tibouchina* sp. (Melastomaceae), described by TAVARES[1125] as caused by an unknown Lepidoptera from Brazil (fig. 80 C). This gall is a hypophyllous growth, inserted by a short stalk, and about 30 mm in diameter, with a central globose swelling, bearing numerous rays of subconical prolongations of variable size and ending in recurved tips, spongy or also sublignous. The larval cavity is excentric, basal and measures 4—7 mm in diameter.

7. Galls of Hymenoptera

The galls of Hymenoptera are among the most specialized of plant galls and are remarkable for their great complexity of structure and development. The same species often causes fundamentally different types of galls on different organs and in alternate generations. They are also interesting for the extremely complex ecologic inter-relations of species other than the actual cecidozoa occurring in them. Considerable literature exists on the development, structure and ecology of these galls. The Hymenoptera galls are readily distinguished from those of the foregoing groups of cecidozoa by a number of well recognized characters. The bulk of the Hymenoptera galls arise on the higher Natural Orders of Dicotyledons and on many Monocotyledons, including Gramineae. The Hymenoptera galls fall under four well defined ecologic groups, viz. the galls of Tenthredinidae, the galls of Cynipidae without heterogony, the galls of heterogonous Cynipidae and the galls of Chalcidoidea.

Galls of Tenthredinoidea

The outstanding feature of the galls caused by tenthredinids lies in the fact that the first and early phase of development of the galls is initiated even when the eggs have been inserted inside the plant organ. Some of these galls, like that of *Arge enoidis* LINN. on the leaf of *Rosa*, develop as globose or elongate parenchyma pustules, from which the newly hatched larva escapes. These are strictly "procecidia" mentioned in the beginning of this work. The most common tenthredinid galls are, however, caused by *Pontania*. The galls of *Pontania* are typically kammergalls, which are fleshy, thick-walled, hollow growths on leaves, with the larva in the gall cavity (fig. 1 D). Unlike other Hymenoptera, the larvae of *Pontania* eject fecal matter, which accumulates inside the gall cavity. The half-grown larva gnaws a hole in the thick fleshy wall and expels the fecal accumulations through this hole. The full grown larva issues out of the gall through such a hole, hangs down by a fine silken thread to finally drop off and pupate on bark or underground in a silken cocoon. Some of these galls are typically hypophyllous, but others like the gall of *Pontania vesicator* BREMI is epiphyllous. Being strongly constricted basally, the gall is inserted by a relatively small portion. The galls are mostly coloured bright yellow, brown or red. The greatest majority of them are spherical, but some oval or reniform galls, about 20 mm long and 15 mm thick, with relatively thin walls, are also known. The epidermis is composed of small cells, without stomata, but often with lentices. The large reniform galls often contain a subepidermal zone of sclerenchyma cells. The bulk of the gall tissue is composed of simple, large parenchyma cells. The uninjured cells lining the gall cavity grow out as clavate elongate processes. BEIJERINCK[77] has described the development of adventitious roots inside the *Pontania* gall from the fleshy wall. These roots grow into the gall cavity and eventually through the hole made by the larva. All *Pontania* galls develop on diverse species of *Salix*.

8. Galls of Cynipoidea

One of the most striking structural peculiarities of the galls of Cynipoidea is the presence of a series of concentric zones of differentiated cells around the larval cavity. The outermost layer of cells is usually a more or less typical epidermis. Beneath this is a zone of parenchyma surrounding sclerenchyma cells or the so-called "protective zone" (fig. 44). Within this sclerenchyma zone lies the so-called nutritive zone of small-sized cells, rich in reserve food material and amino acids. The cells of the nutritive zone actually line the larval cavity. In the galls of some species on *Quercus*, the innermost zone, surrounding the larval cavity, becomes loosened from the rest of the gall tissue in the course of the growth of the gall, so as to give rise to a so-called "inner gall" (fig. 114). The external form of cynipid galls varies within very wide limits. Some of them are curiously pyriform growths, borne on long, extremely slender stalks (fig. 124C) and others are elongate pedunculated club-shaped structures (fig. 86B) and still others are complicated agglomerated masses.

In many of them there are accessory gall cavities in addition to the larval cavity proper (fig. 85). A number of them are covered on the surface by hairs, fleshy and spiny growths, scaly processes, etc. (fig. 86). The structure and development of the galls of a number of cynipids have been studied by various workers like LACAZE-

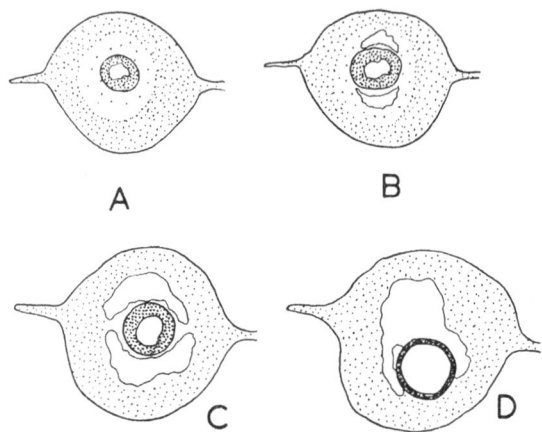

Fig. 114. Development of the "inner gall" of *Andricus curvator curvator* Htg. bisexual generation on leaf of *Quercus*. A. Early stage, with sclerenchyma tissue and spongy parenchyma surrounding it. B. Shrinkage of the softer parenchyma begins. C–D. Stages in the separation of the larval cavity with its sclerenchyma wall from the rest of the gall parenchyma.

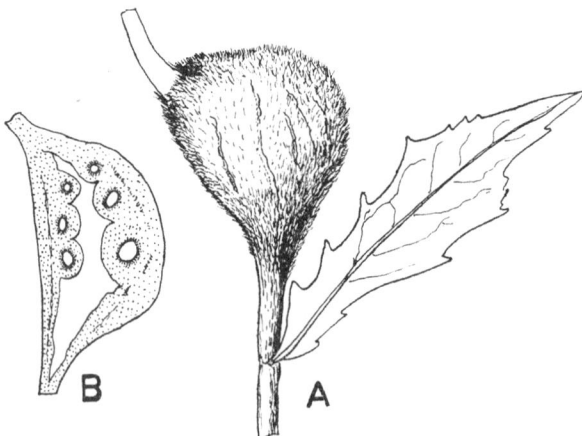

Fig. 115. Gall of *Aylax hieracii* (Bouché) on the stem of *Hieracium*. A. General view of the gall in situ. B. Gall cut open lengthwise showing the larval cavities and the central hollow space.

DUTHIERS[648], BASSET[61], FOCKEU[336], HIERONYMUS[462], WEIDEL[1243], KÜSTER[643], PRILLIEUX[917], MAGNUS[731], KÜSTENMACHER[634], GREVILLIUS & NIESSEN[419], KINSEY[597] and others listed in the bibliography. HOUARD[487, 488] dealt with the anatomy of the gall of *Xestophanes potentillae* and TROTTER[1170] similarly gave an account of the histology of the gall of *Cynips fortii* TROTT. COOK[213] studied the anatomy of the galls of *Amphibolips* and an important contribution on the anatomy and histology of the gall of *Andricus radicis* on *Quercus* is that of DENIZOT[247]. LERA'S[660] studies on the gall of *Cynips kollari* HTG. are an important contribution to our knowledge of the anatomy of cynipid galls in general.

As already mentioned, the great majority of cynipid galls arise on leaves and buds, but many develop also on roots and shoot axis. The cynipid galls show a decided preference to the plants of the Natural Orders Fagaceae and Rosaceae. Most cynipid galls are of considerable interest from the point of the inter-relations of the numerous parasites, predators and inquilines associated with (vide Chapter XI).

Galls of cynipids without heterogony

The gall of *Diplolepis (= Rhodites) rosae* (LINN.)* on *Rosa* spp. is perhaps the commonest example of a gall of this class. This is the well known bedeguar, developing on leaflets, petioles and rarely on branches, pedicels and even fruits (fig. 1 E). The larger bedeguars are as a rule agglomerate, multi-chambered swellings, characterized by the presence of a dense covering of long thread-like and often branching emergences, containing vascular elements connected with those of the gall-bearing organ. The larval cavity is surrounded by a small-celled nutritive zone. The full grown larva over-winters in the gall and the adults emerge in May-June. The males are generally rare and reproduction is mostly parthenogenetic[1, 75, 892, 634, 731]. The globose, often spiny, unilocular gall of *Diplolepis eglanteriae* (HTG.) on *Rosa* also develops without heterogony. Another common cynipid gal without heterogony is caused by *Aylax hieracii* BOUCHÉ on the shoot of *Hieracium*. This unilaterally swollen gall, often several cm long, arises both on the vegetative and on the inflorescence axis and may also involve the flower-heads. A number of larval cavities are scattered in the fleshy mass of the gall, around a central empty space (fig. 115). Around each larval cavity is a nutritive zone, surrounded outside by a sclerenchyma zone. Short tracheids and thin-walled vessels are formed from the gall parenchyma and pass through the sclerenchyma zone to connect with the nutritive cells. When the larval cavity is situated near one of the primary vascular bundles, the vessels become irregularly scattered. Due, however, to

---

* The names of Cynipids adopted here are based on WELD[1242].

the enormous swelling of the shoot, the vascular bundles are always widely separated and the medullary rays are hypertrophied. The pupation of the cynipid takes place in the gall during spring and the adults emerge soon afterwards. The gall of *Aylax papaveris* (PERRIS) on the septa of the fruits of *Papaver* spp. and that of *Aylax minor* HTG. on the ovules of the same plant are other common examples. In all these cases the galls of successive generations are identical.

Galls of heterogonic cynipids

The cyclic heterogony of cynipids is remarkable for the regular alternation of a bisexual and female (usually called unisexual or agamic) generation, the individuals of which differ so very much from each other that they have frequently been described as distinct species and genera. The galls of bisexual and the unisexual generations are strikingly and often fundamentally different. The gall of heterogonic cynipids is known after the type of adult that emerges from it and not after the adult which has given rise to it. Most of these heterogonic galls arise on *Quercus*. The parthenogenetic females of *Diplolepis quercus-folii* (LINN.) emerge from the oak-apple gall (fig. 1 A) on leaf of *Quercus* and this gall is thus known as the agamic gall. The adults of the bisexual generation of the same species emerge from bud galls (fig. 119), which are similarly known as the galls of the bisexual generation. The oak-apple gall is a spherical, hard, solid swelling, about 20 mm in diameter, inserted by an exceedingly short and slender stalk to one of the larger veins on the under side of the leaf blade. The eggs are deposited by the mated females of the bisexual generation inside the vein, among the cambiform cells of the sieve tissue. Arising initially as a hemispherical bulge, the young gall soon breaks through the subepidermal layer of cells and also the epidermis and grows into a spherical mass. A number of obscure humps and stumpy tubercles may be found on the surface of the immature gall, but when full grown, the gall has a smooth surface. The small central larval chamber is surrounded by the nutritive zone of small, thin-walled cells. The gall epidermis is composed of small, thin-walled cells, containing abundant tannin and chlorophyll. A few stomata occur on the stumpy tubercles on the immature gall, but with its maturing, these soon lose their identity. The general mass of the gall is composed of thin-walled parenchyma cells, which are closely packed in the gall periphery but tend to be radially elongated in the deeper layer, with increasing intercellular spaces. Outside of the nutritive zone and within the general parenchyma there is often a layer of rounded cells, thick-walled at the sides and on the inner side. The thickening of these walls is due to cellulose at first, but later lignin is also deposited. Pupation occurs in the gall in autumn and the adults emerge from galls in December to February. Though the adult emerges from the pupa, it does not escape from the gall at once, but after gnawing a cylindrical passage to nearly the surface, waits till the external climatic conditions become favourable. The adults, which are all only females, escape by simply pushing off the thin epidermis to the outside. These females reproduce parthenogenetically and oviposit in the dormant buds. Both males and females emerge from the bud gall in June. Both males and females of *Biorrhiza pallida* (OLIV.) emerge from the bud gall, but only females from the root gall on *Quercus*. The bud gall of the bisexual generation is a globose or irregularly spherical mass, about 4 cm in diameter, with numerous spherical lobes, multi-chambered and with exit holes on the surface when old (fig. 55). Though fleshy at first, the gall is spongy when about to mature. According to the observations of BEIJERINCK[75], MAGNUS[731], FRÜHAUF[354] and others, the long ovipositor of the parthenogenetic female pierces the base of the bud, boring through the bud scales and other covering and through parts of the bud and penetrates right to the growing point (fig. 116). The upper part of the bud that would have normally developed into leaves thus becomes completely isolated by the injury caused in oviposition and may be observed as a minute cap on the summit of the young gall. The lower part or the stump of the bud, surrounded by the bud scales, develops into the gall. The larva hatching from the egg becomes surrounded by a layer of active meristematic cells,

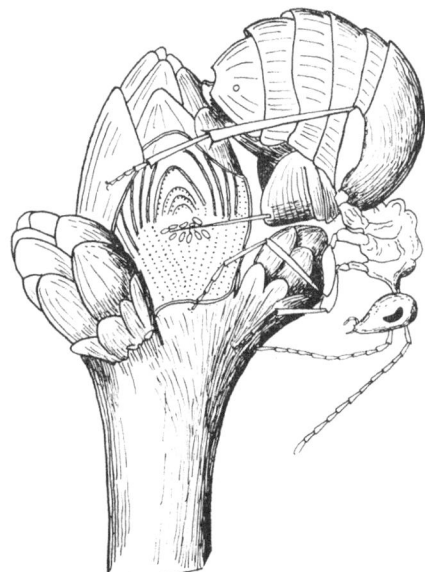

Fig. 116. Oviposition by *Biorrhiza pallida* (Ol.) inside the bud of *Quercus*. (After Magnus)

which now give rise to the larval cavity. The gall is at first composed of nearly simple parenchyma cells, arranged radially outward from the base of the gall. The larval cavity is surrounded by a nutritive zone of primary and secondary nutritive layers. The growth of the gall is at first due to cell proliferation but as development proceeds, cell hypertrophy also becomes evident. Much branched vascular elements traverse the gall parenchyma and extend to the tissue of the larval cavity. The gall is remarkable for lacking a true epidermis, absence of stomata or hairs. On the peripheral limits of the nutritive zone, the walls of several layers of cells become gradually thickened, often only unilaterally. At a later stage, the thickening of the cell walls disappears and there is free passage of substances into the secondary nutritive zone. When the gall is nearly mature, sclerotization and lignification may be observed here and there in the parenchyma, particularly in the vicinity of the larval cavity. This thickening develops nearly uniformly on all sides and as the larval cavities are usually closely crowded together, the sclerenchyma tissue becomes fused together to form a continuous mass. Typically alate males and apterous or brachypterous females emerge from the bud gall in July (fig. 14). The gall does not dry even after the escape of the adults. After copulation, the female seeks the soil to deposit eggs on roots. The spherical gall, about 5 mm in diameter, often becomes agglomerated into grape-bunch-like masses on roots (fig. 33 A). The root gall develops relatively slowly and begins to increase in size rapidly only during the second year. Only apterous females emerge from the root gall (fig. 14) in the winter of the second year, ascend to buds and give rise to bud galls (fig. 121).

The lenticular gall of *Neuroterus quercus-baccarum* (LINN.),* to which we have already referred, also belongs to this class. Only the parthenogenetic females emerge from the lenticular gall (fig. 84 A) during spring. Oviposition by these females in the bud gives rise to kammergalls, 5—8 mm in diameter. Unlike *Biorrhiza pallida* (OLIV.), this species deposits its eggs in any part of the bud, with the result that the kammergalls often arise on leaf, tender branch, spindles of male catkin, etc. Another interesting example is provided by *Andricus fecundator* (HTG.). Both females and males emerge from the gall on the staminate flower of *Quercus*, but only parthenogenetic females from the oak-rose gall, which is a brown swelling of the bud, about 2—3 cm in diameter, densely clothed with scaly outgrowths. The gall of some heterogonic cynipids develops on various other plants also. We have, for example, the gall of *Pediaspis aceris* (GMELIN) *(P. aceris* FÖRST. of authors) on *Acer pseudo-*

---

* **vide** also an interesting account of the ecology of button-shaped and spangle galls of *Neuroterus* by ASKEW, R.R. 1962. The distribution of galls of *Neuroterus* (Hym. Cynipidae) on oak. *J. anim. Ecol.*, 31(3): *439-455*, fig. 7.

platanus LINN. (fig. 33, D—E). Both males and females emerge from the hypophyllous kammergall on the leaf (sometimes also on the petiole, tender shoot axis and even fruit), often agglomerated in large masses. After mating, the females seek the roots for oviposition and give rise to a globose kammergall, 7 mm thick, on them.

9. Galls of Chalcidoidea

The salient features of the galls caused by Chalcidoidea are illustrated by a number of remarkable galls. We have a number of different galls on the buds and growing point of the shoot axis of *Acacia* spp., characterized by their solid, hard, woody swelling, with numerous, usually superficially situated larval cavities in the cortex. The larval cavities are surrounded by typical nutritive zones. We have already mentioned the remarkable chalcid gall on the growing tip of the shoot axis of *Acacia leucophloea* WILLD. from India (fig. 36). Another notable example is the gall of *Trichilogaster acaciae-longifoliae* FROGGATT on *Acacia* spp. from Australia. Attention may also be drawn to the gall on the stem of *Prosopis* from India, mentioned in an earlier chapter. In this hard, woody and solid, globose gall, as many as 200—500 oval or spherical larval cavities, surrounded by a continuous sclerenchyma tissue, are embedded in the cortex, beneath the thick epidermis. An unknown Chalcid gives rise to a curious reddish echinate gall on the under side of the leaf of *Ficus bengalensis* LINN. from India. The echinate processes are

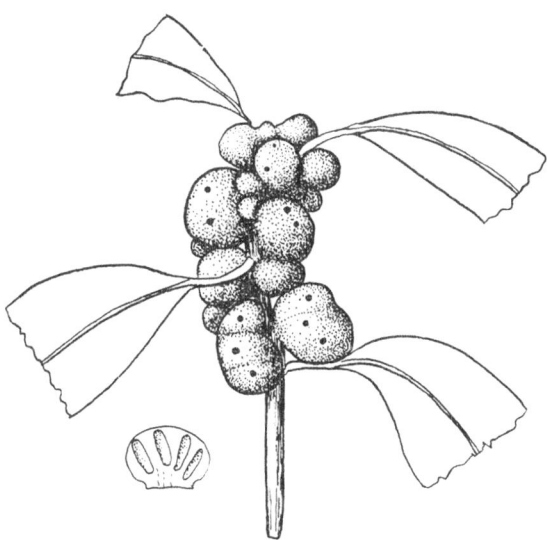

Fig. 117. Chalcid gall on stem of *Eucalyptus sieberiana* F. v. M.

composed of parenchyma emergences of bluntly conical growth. The paired larval cavities are surrounded by nutritive and sclerenchyma zones. We also mentioned earlier the fusiform gall on the veins of the leaf of *Bassia latifolia* ROXB. from India (Plate VI, 3,4). Among other remarkable chalcid galls on Dicotyledons, the gall on *Eucalyptus sieberiana* F.v.M. from Australia (fig. 117) is remarkable for agglomeration of oval, sessile swellings, with numerous elongate larval cavities. The gall of *Eurytoma felis* GIR. on the branches of *Citrus* from Australia is another interesting example. Chalcid galls are perhaps exceptional among the galls caused by Hymenoptera for occurring on Monocotyledons. *Isthmosoma* spp. are, for example, associated with remarkable galls on different grasses. *Isthmosoma hyalipenne* (WALK.) (formerly known as *Isosoma graminicola* GIRD.) gives rise to a tubular gall on the stem of a number of grasses like *Agropyrum repens* PB. The eggs are deposited in May-June inside the growing point. Although the gall is tubular, its wall is nearly three times thicker than an ordinary internode of a normal stem. The gall is composed of thin-walled parenchyma cells; there is also conspicuous increase of sieve tubes, which are also enlarged[291, 488]. Some species of *Isthmosoma* give rise to galls on the inflorescence of the couch grass *Triticum repens* in parts of Europe[547]. An interesting gall of *Isthmosoma* on the stem of *Paspalum scrobiculatum commersonii* STAPF., described by TROTTER[1187] from Ethiopia (fig. 118) is a hollow, fusiform, unilateral swelling, about 15 mm long and 5—6 mm thick. Further records of chalcid galls may be found in NOBLE[875, 876] and HEDICKE[451, 452].

10. Galls of Diptera

The dipterocecidia embrace three major ecologic groups viz. 1. the galls of Itonididae (= Cecidomyiidae or gall midges), 2. Agromyzidae and 3. Trypetidae, recognized by distinctive structure and characteristic communities. The galls caused by Itonididae are perhaps the most important, dominant and also most highly specialized types. They exhibit enormous diversity of form and structure and arise on an extremely wide range of plants, including Monocotyledons, particularly Gramineae. As pointed out elsewhere, the greatest majority of midge galls are leaf galls; then follow the galls on buds, stem, flower and root. They range from the simplest types to the most complex, with differentiation of gall tissues, reminiscent of the cynipid galls. A number of midge galls appear to have evolved on remarkably parallel lines with certain types of acarocecidia or cynipid galls, so that it is often difficult to distinguish between them in the absence of the cecidozoa. They include every developmental and structural type like fold galls, roll galls, pouch galls, covering galls, rindengalls and other complex types. Although the galls of successive generations of a species are identical, alternation of different types is also known in many species. The spring generation of *Rhabdophaga heterobia* H. Löw, for example, causes galls on the male catkin of *Salix triandra* LINN. The adult midges, which emerge from this gall in summer, do not, however, find any more young catkins on the plant. They oviposit, therefore, on the actively growing vegetative buds and thus give rise to a typical rosette gall in summer. *Asphondylia sarothamni* H. Löw, emerging from the gall on the ovary of *Sarothamnus scoparius* WIMM. do not naturally find any more flowers for oviposition and therefore now deposit their eggs in vegetative buds, thus giving rise to bud galls in late summer (the midge emerging from the bud gall was formerly called *Asphondylia mayri* LIEB.). As many species of gall midges have an extremely complex life-history, associated with cyclic parthenogenesis and often also paedogenesis, the gall types and their life-histories are likewise greatly complicated. Nearly all midge galls are remarkable for the highly complex community of species associated with them, both during the active phase of development and after the escape of the gall midge. Many midge galls are wholly fleshy and decay rapidly, but others are remarkably resistant, hard and woody and persist for long periods and serve for hibernation, aestivation and pupation of the gall midge. We have thus two principal ecologic classes of midge galls, viz. midge galls in which the gall midge pupates and midge galls in which pupation does not occur. The structure, development, communities and other characters of these two classes are marked by considerable differences. To the former belong most leaf fold galls, pouch galls, most bud galls, stem galls, most flower galls, fruit galls, root galls, etc. The latter class include leaf roll galls, some simple leaf fold galls, some covering galls, some bud galls and some flower galls. The literature on midge galls is very extensive, but the more important works include the monographs of RÜB-SAAMEN & HEDICKE[999].

Midge galls on Monocotyledons

As in the bulk of the midge galls on Monocotyledons the growing point is affected, they are typically acrocecidia. These acrocecidia constitute, however, a distinctive class by themselves and differ fundamentally from typical acrocecidia on Dicotyledons. The outstanding features of acrocecidia caused by midges on Monocotyledons are illustrated by the galls of *Orseoliella javanica* KIEFF. & DOCTERS VAN LEEUWEN on *Imperata cylindrica* BEAUV., *Oligotrophus ischaemi* KIEFF. on *Ischaemum pilosum* TRIM., *Courteia graminis* KIEFF. & DOCTERS VAN LEEUWEN on *Panicum nodosum* KUNTH, *Pachydiplosis oryzae* MANI on *Oryza sativa* LINN., etc. Some of these galls are fistular elongate growths, surrounded by bunch of stunted leaves and others are naked elongate tubes and still others are ovoid bags. The gall of *Poamyia poae* Bosc. on *Poa nemoralis* Linn. from Europe is remarkable for the curious growth of adventitious roots, parted neatly lengthwise in two groups, curiously like the hair-parting on the human head[76]. In all these cases pupation takes place within the gall.

Root galls

The root galls caused by gall midges differ from those of Nematoda and Coleoptera in the presence of a definitive larval cavity surrounded by nutritive zone and often also some sclerenchyma elements.

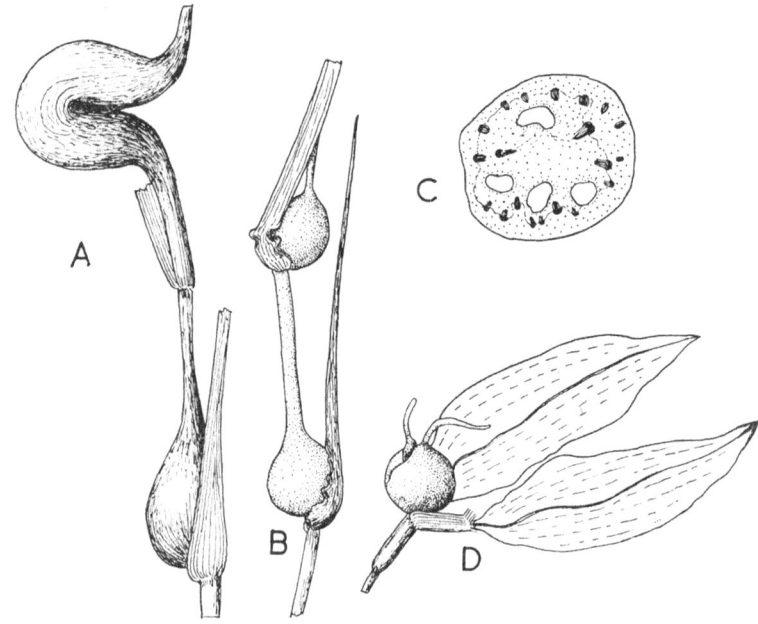

Fig. 118. Zoocecidia on stem of Monocots. A. Gall of *Isthmosoma* on *Paspalum scrobiculatum commersonii* Stapf. B. Diptera gall on stem of *Aristida stipoides* Lamarck. C. Section through the same. D. Gall on the growing tip of *Commelina capitata* Benth. caused by an unknown Lepidoptera (After Houard).

Mostly only the superficial and smaller roots are galled and the galls are predominantly also rindengalls. We have given a number of examples of root galls caused by midges in an earlier chapter. Some of them like the gall of *Asphondylia strobilanthi* Felt on the root of *Strobilanthes cernuus* Bl. (fig. 32, B—C) have a hairy cover. This gall is a curious pyriform swelling of the root cortex, with a central larval cavity surrounded by a distinct nutritive zone. The salient characters of midge galls on roots are also illustrated by the gall of *Pseudohormomyia subterranea* KIEFF. & TROTTER on *Carex divulsa* GOOD from Italy.

Shoot axis galls

A number of examples of shoot axis galls caused by gall midges have been mentioned in earlier chapters. The general structure and development of these galls depend on the position of the larval cavity with reference to the cambial zone: larval cavity superficially in the cortex, larval cavity in deep cortex, larval cavity in medullary rays and in the parenchyma of vascular bundles and larval cavity in medulla. A number of these galls are rindengalls; this is for example the case with the gall of *Acroëctasis campanulata* MANI on *Sabia campanulata* WALL. (fig. 40 A). Others develop initially wholly within the cortex, without the epidermis and the subepidermal layer of cells being affected, so that eventually the gall breaks through and erupts on the surface. The larval cavity is surrounded by nutritive and sclerenchyma zones. Pupation takes place within the gall. Most of these galls are spherical, oval or fusiform solid, fleshy or woody swellings of the branches, with a smooth surface or also clothed with various processes like hairs, scales, stunted leaves and branches. In the majority of them the larval cavity is an elongate tunnel that terminates immediately under the epidermis, which is, however, intact. The shoot axis pleurocecidia of gall midges thus differ in this respect from those caused by cynipids.

Bud galls

Bud galls of midges are characterized by the very profound alteration and swelling of the parts. Some of them are remarkably similar to those caused by *Eriophyes*, but are readily distinguished by the absence of erineal outgrowths. Others are similar to the bud galls caused by Thysanoptera, but differ in the very limited number of larvae of the gall midge. The bud galls in which the larvae are situated externally in between the bud rudiments are on the whole simple and pupation takes place outside the gall. Others in which the larvae are within the plant tissue are far more complex than the bud galls of mites and thrips. They range from typical rosette galls like those on *Salix*, *Crataegus*, *Euphorbia*, *Veronica*, *Medicago*, etc. to cabbage-like growths caused by *Contarinia helianthemi* HARD on *Helianthemum salicifolium* PERS. from Algeria[520] with a single central larval cavity. The bulk of midge galls on buds are fleshy solid swellings, with often outgrowths of stunted leaves, etc. This is, for example, the case with the gall of *Allodiplosis crassa* KIEFF. & JÖRG. on *Gourliea decorticans* GILL. (Fig. 54, A-B). We have already mentioned the bud gall of *Rhopalomyia millefolii* H. Löw on *Achillea millefolium* LINN.

Roll galls

In sharp contrast to the roll galls of other groups of cecidozoa, those of gall midges are strongly developed, fleshy or coriaceous swollen masses, which completely enclose the larvae during their trophic phase. In most roll galls, the larvae leave the gall when full grown, drop off to the ground for pupation. Although erineal hairs, typical of roll galls of mites, are absent, some like the roll gall of *Dasyneura marginemtorquens* WINN. on *Salix viminalis* LINN. have a hairy covering. In the roll gall of *Dasyneura tiliamvolvens* RÜBS. on *Tilia silvestris* DESF. the hairy covering is dense. Some typical roll galls include those of *Wachtiella persicariae* (LINN.) on *Polygonum*, *Macrodiplosis dryobia* F. Löw on *Quercus* and *Macrodiplosis volvens* KIEFF. on *Quercus*. A striking feature of some of these roll galls is the presence of sclerenchyma elements (fig. 67B).

Fold galls

The leaf fold galls of gall midges differ conspicuously from those of Acarina or Thysanoptera in the enormous swelling of the leaf blade and in the tissue fusion of the folded parts. The general structure of these galls is illustrated by reference to the galls of *Schizomyia meruae* FELT on *Maerua arenaria* HOOK. & BL. and of *Asphondylia riveae* MANI on *Rivea hypocrateriformis* CHOISY (fig. 65) mentioned under leaf galls (Chapter VII). These are solid spongy or fleshy swellings of the leaf blade, with numerous larval cavities imbedded in the middle of the flesh. There is, however, no distinct nutritive zone or sclerenchyma zone round the larval cavities (Plate VII, 2).

Petiole and vein galls

The midge galls on petioles and leaf veins are solid, fleshy or woody swellings, generally spherical, but also fusiform and extensive. Some of the common examples include the galls of *Zygobia carpini* (F. Löw) on *Carpinus betulus* LINN., *Dasyneura fraxini* KIEFF. on *Fraxinus excelsior* LINN., *Pipaldiplosis pipaldiplosis* MANI on *Ficus religiosa* LINN., *Odinadiplosis odinae* MANI on *Lannea coromandelica* and *Arnoldia nervicola* KIEFF. on *Quercus*, mentioned earlier. Pupation takes place within the gall in all these cases. All petiole and vein galls are generally characterized by a profound alteration of the normal anatomical and histological structures of these organs.

Pouch galls

The pouch galls of midges differ basically from those of mites, psyllids and aphids in a number of respects, particularly in the complete obliteration of the ostiole and the greater diversity and complexity of anatomical and histological structures. Most of them are globose, pyriform, fusiform or subcylindrical solid or nearly solid structures, smooth or also clothed with pubescence or other processes. Some of them like the gall of *Mikiola fagi* (HTG.) on *Fagus silvatica* LINN. (fig. 73), mentioned in an earlier chapter, have a special mechanism, facilitating the breaking loose of the mature gall. Many of the pouch galls are hard and nut-like and fall off with the larva inside. A number of pouch galls are greatly depressed structures, similar to lenticular galls of *Neuroterus* on *Quercus*. We have, for example, a curiously flat, disc-shaped gall, about 5 mm in diameter and 1 mm high, on the leaf of *Millettia sericea* W. & A. from Java. The gall is inserted by a slender short stalk to one of the veins on the underside of the leaf blade. Other lenticular galls of midges occur on *Ardisia attenuata* WALL. and *Luchea speciosa* WILLD.

Other remarkable midge galls

We have already described the unique midge galls on *Acacia*, comprising fusion of two or more leaflets. Reference may be made to the galls of *Asphondylia trichocecidarum* MANI, *Schizomyia acaciae* MANI, the bivalve gall of *Lobopteromyia bivalviae* (RAO), the cylinder-piston gall of *Lobopteromyia* sp. and the barrel gall of *Lobopteromyia ramachandrani* MANI on *Acacia* spp. The outstanding flower galls, characterized by total swelling, include the galls of *Kiefferia* on Umbelliferae, *Schizomyia cocculi* MANI on *Cocculus hirsutus* DIELS, *Pruthidiplosis mimusopsicola* MANI on *Mimusops hexandra* ROXB., *Schizomyia cheriani* MANI on *Ipomea*, *Neolasioptera crataevae* MANI on *Crataeva religiosa* Forst., *Asphondylia morindae* MANI on *Morinda tinctoria* LINN. and others mentioned in Chapter VIII.

Galls of other Diptera

We have already referred to the galls of Trypetidae like *Euribia cardui* LINN. on the shoot tip of *Cirsium arvense* SCOP. (fig. 49, B—C) and to the gall of *Lipara lucens* MEIG. on *Phragmites communis* TRIN. Mention also may be made of galls of Agromyzidae like *Agromyza marelli* BRÈTHES on *Alternanthera philoxeroides* from South America[533]. HERING* has recently described some interesting galls of the Agromyzids *Phytobia (Paraspedomyia) pittosporocaulis* HER. on leaf and *P. (P.) pittosporophylli* HER. on the shoot axis of *Pittosporum undulatum* ANDR.

---

*HERING, E. M. 1962. Galls of Agromyzidae (Dipt.) on *Pittosporum undulatum* Andr. *Proc. Linn. Soc. N S. Wales*, 87 (378) (1): 84-91, fig. 4, pl. II

CHAPTER X

ETHOLOGY OF ZOOCECIDIA

Although it is not possible at the present state of our knowledge to discuss in detail the diverse problems of ethology of zoocecidia, we may nevertheless draw attention to some of the more outstanding facts, in so far as they relate to the general ecology of galls. The zoocecidia represent the product of an unique interspecific relation between a plant and another organism, constituting a closely integrated ecologic complex. The zoocecidia, the gall-bearing plant, the cecidozoa and the complex communities of other organisms associated with them make an ecosystem. The general course of the life-history of the gall is thus intimately bound up with the peculiarities of distribution, habits, development, mutual adjustments and other features of the ethology of both the gall-bearing plant and the gall-inducing organism.

1. Phenology

With few exceptions, zoocecidia are characteristically periodic. Some galls are, however, irregular and appear in a given locality in great abundance for some years in succession and then abruptly disappear or become extremely scarce for a long interval. NALEPA[845] found, for example, the gall of *Eriophyes ulmicola* NAL. on the leaf of *Ulmus campestris* LINN. to be abundant in some years, becoming totally absent for several years afterwards. Earlier workers have similarly observed other acarocecidia behaving in a similar manner. NALEPA attributes these irregular fluctuations in abundance of galls to climatic factors, unfavourable to the mite at the time of the unfolding of the buds. According to him, the stage of development of the bud at the time of the wandering of the mites from the winter buds is of decisive importance. The irregularities and fluctuations in the abundance of zoocecidia are hardly explained on the basis of the local decimation of the cecidozoa by an abnormal increase of parasites, peculiarities in the life-cycle involving changes in sex-ratio and climatic factors.

Many galls such as, for example, that of *Trioza alacris* FLOR. on the leaf of *Laurus nobilis* LINN., appear as long as new leaves are put out on the plant. The galls of *Perrisia veronicae* VALLOT on *Veronica chamaedrys* and of *Wachtiella persicariae* (LINN.) on *Polygonum* also continue to appear for several months.* The greatest majority of zoocecidia develop strictly at a definite time of the vegetative season. Some of them arise in spring, others in early summer, still others in mid summer and several others

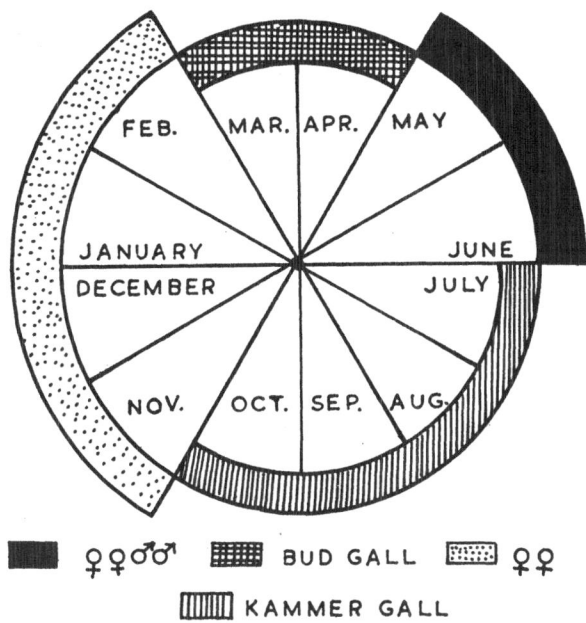

Fig. 119. Phenology and life-cycle of *Diplolepis quercus-folii* (Linn.) and its galls on *Quercus*.

in late summer or even autumn. The galls of *Byrsocrypta gallarum* (GMELIN) on the leaf of *Ulmus*, *Mikiola fagi* (HTG.) on the leaf of *Fagus silvatica* LINN., *Dasyneura marginemtorquens* WINN. on the leaf of *Salix viminalis* LINN., *Pemphigus spirothecae* PASS. on *Populus nigra* LINN., the bisexual generation galls of *Trigonaspis megaptera* PANZ., *Biorrhiza pallida* (OL.), *Diplolepis quercus-folii* (LINN.), *Neuroterus quercus-baccarum* (LINN.), *Andricus curvator curvator* HTG., *Andricus inflator* HTG., *Diplolepis longiventris* (HTG.), *Amphibolips confluentus* (HARRIS), *Neuroterus albipes albipes* (A. SCH.), *Neuroterus numismalis* (FOURC.) and *Neuroterus aprilinus* (GIRAUD) appear in spring. The gall of the bisexual generation of *Biorrhiza pallida* (OL.) on the bud of *Quercus* matures in early summer. Other typical summer galls include the galls of the unisexual generation of *Diplolepis quercus-folii* (LINN.)

---

* Some bacterial and fungal galls may also be said to be phenologically indifferent and may thus be found developing at any time of the year, at least in the tropics. The gall of the fungus *Uromycladium* on *Acacia* develops, for example, almost throughout the year.

and *Ceutorrhynchus pleurostigma* (MARSH.) on roots of Cruciferaceae. Late summer and autumn galls include those of *Andricus inflator* HTG. (unisexual generation), *Andricus curvator curvator* HTG. (unisexual generation), *Andricus quercus-radicis* (FABR.) (bisexual generation), *Biorrhiza pallida* (OL.) (unisexual generation) on the root, *Diplolepis quercus-folii* (LINN.) (unisexual generation) on the leaf and the galls of the unisexual generation of several other species of cynipids.

The time of appearance of most zoocecidia synchronizes with the time of active growth of the part on which the galls develop. It also marks the time of emergence of the adult cecidozoa, oviposition or hatching of the young larva. We have, for example, already explained that the eggs of the spruce gall aphid hatch at the time when the young leaves begin to appear and the newly hatched aphids establish themselves on the actively developing tender leaves, which have already become somewhat swollen basally under the influence of the fundatrix. The leaf gall caused by *Phylloxera* on the North American hickory appears in spring while the leaves are still young, but on the closely related pecan, the gall arises throughout

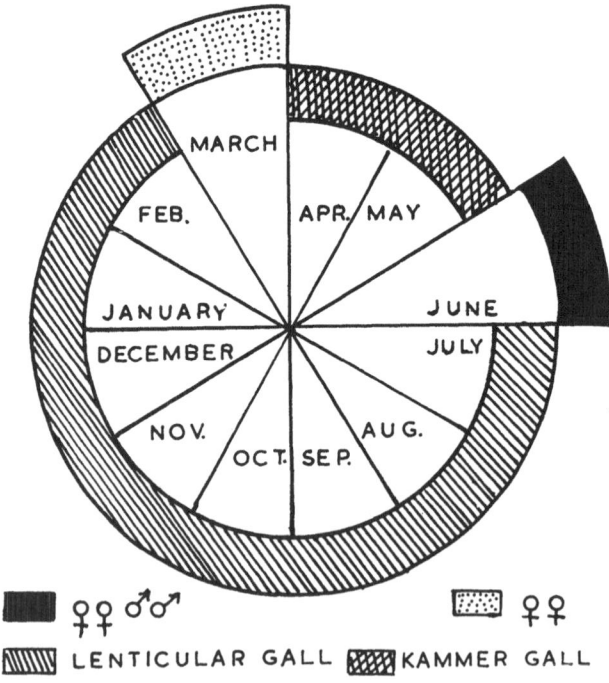

Fig. 120. Phenology and life-cycle of *Neuroterus quercus-baccarum* (Linn.) and its galls on *Quercus*.

the summer. This difference is related to the fact that pecan produces leaves throughout summer. The formation of the pear midge gall is similarly related to the time of appearance of pear blossoms, with which the emergence of the adult midges from under ground also synchronizes.

The phenological peculiarities of zoocecidia have been interpreted as indicating the capacity of cecidozoa to stimulate the formation of galls only when the plant tissue is actively growing. This view has, for example, been repeatedly emphasized by THOMAS and others, inspite of numerous observations to the contrary. The general view that only meristematic tissues are able to react to the cecidozoa is now known to be erroneous. In the light of recent advances, it is suggested that only when a plant tissue is in active state of growth the cecidozoa will find in it optimal conditions for development. The intense protein synthesis in the actively growing plant tissue seems to constitute an important factor in the nutrition of the cecidozoa. Only in a meristematic tissue the action of the saliva of cecidozoa brings about the formation of amino acids, which are ultimately concerned in the induction of cell division (*vide* Chapter XIV). The cecidozoa can therefore feed only on actively growing tissue, but this has no direct relation to gall induction. As the cecidozoa are incapable of utilizing the nutrition except from meristematic tissue, they are in effect seasonally greatly isolated.

2. Life-history

The life-history of zoocecidia comprises an early phase, a trophic phase and post-trophic phase. In some zoocecidia the life-history ends with the termination of the trophic phase, which also marks the maturation of the gall and the escape of the cecidozoa. In most others, however, the termination of the trophic phase does not mark the end of the life-history, but is followed by complex series of post-trophic changes.

Early phase

The starting point of the complex series of developmental changes and growth reactions involved in the formation of galls is usually the oviposition by the cecidozoa. Differences in the habits of oviposition introduce significant changes in the early phase of life-history of the gall. The majority of cecidozoa oviposit on or within the organ destined to bear the gall. The oviposition may be associated with more or less profound mechanical injury to the plant tissues or the eggs may be deposited inside leaf folds, buds, etc. without injury. In the case of some zoocecidia oviposition does not, however, mark the beginning of the early phase, but the attack by the young larva as in Nematode galls or by the fundatrix of the aphid or by the mite. There is usually a more or less short latent period between oviposition or the attack by cecidozoa and the appearance of the first external symptoms of gall formation. This period embraces the

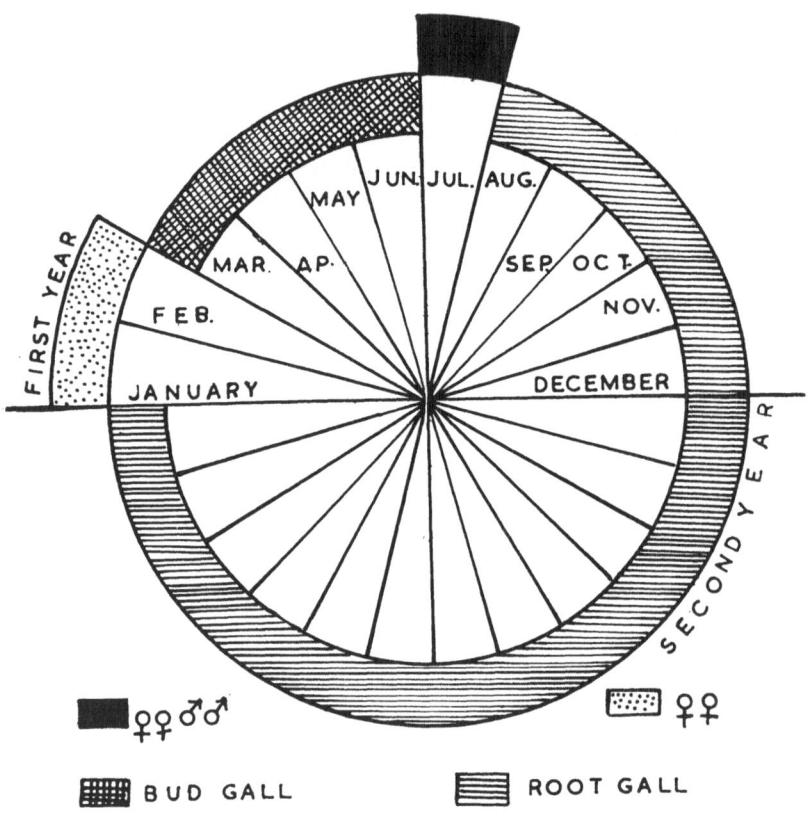

Fig. 121. Phenology and life-cycle of *Biorrhiza pallida* (Ol.) and its galls on *Quercus*.

embryonic development of the cecidozoa, the feeding by the newly hatched larva and the onset of reaction by the plant cell. This period may vary from a few hours to several days. ZWEIGELT[1296] has shown, for example, that in the pouch gall of *Byrsocrypta gallarum* (GMELIN) on the leaf of *Ulmus campestris* LINN. the first external indication of galling, viz. the evaginated bulging of the leaf blade, appears three days after the fundatrix has commenced sucking the sap. In most zoocecidia the early phase in the life-history marks a relatively insignificant development, but in some there is very pronounced growth during this phase. The early phase represents indeed an important part in the life-histories of the galls of *Pontania*. As mentioned already, considerable cell proliferation and growth of gall take place even before the larva of *Pontania* has hatched and starts feeding.

Trophic phase

Irrespective of the differences in oviposition and the concomitant differences in the development of the gall, its continued development and growth depend on the feeding activities of the cecidozoa. The active phase in the life-history of zoocecidia synchronizes with the active feeding period of the cecidozoa and hence this phase may be called the trophic phase in the life-history of the cecidozoa-gall complex. The trophic phase coincides with the larval period of the cecidozoa like Coleoptera, Lepidoptera, Psyllidae, Hymenoptera and Diptera, but in some cases extends even into the adult stage. During this phase of development of some zoocecidia the aphids, thrips and mites, which induce gall formation, often reproduce within the gall and progeny also continue to feed. Depending on the mouth-parts, cecidozoa either feed on solids or suck liquids from the plant tissues. In sucking liquids, the piercing stylets either penetrate the cell (as in *Eriophyes*) or pass in between the cells (as in aphids). At the time of feeding, the cecidozoa pour their salivary fluid on the plant cells, either to dissolve the cell wall or to liquify the cell contents. Active cell proliferation, cell enlargement, differentiation of gall tissues and growth of the gall take place. The gall also attains its characteristic form and mean size. The trophic phase may thus be short or also rather prolonged, sometimes extending to more than two or three years. In some galls it is only a few days, but in the gall of the unisexual generation of *Biorrhiza pallida* (OL.) it extends to about 16 months. The maturing of the gall marks the end of the trophic phase. The maturation of the gall involves more or less profound physical, chemical, cytological, structural and functional changes in the gall tissue, indicating an abrupt transition from highly optimal to pessimal conditions for the cecidozoa. With approaching maturity of the gall, cell proliferation slows down rapidly and finally ceases altogether and hypertrophy of most of the cells has also reached its maximum. The growth of the gall has attained the limits of the cecidogenetic field (*vide* Chapter XIV) and there is thus a weakening and disappearance of the cecidogenetic gradient. The flow of sap to the gall stops, the intensity of protein synthesis reaches its zero level, cell walls often become thickened and the cell contents also disappear or undergo other changes. Shrinkage of tissue, changes in the pH and enzyme compositions, reduction of nutritive material, reduction of free permeability of the cell wall and lowered turgidity are other changes. The accumulation of metabolic products of intense cell proliferation has apparently reached such a proportion that they stimulate the synthesis of hormones in the cecidozoa favouring quiescence and dormancy. The degenerative changes in the gall tissue truly represent part of the life-history of the gall and these very degenerative changes act as factors that determine the future course of biology of the cecidozoa. As is well known, degenerative changes take place more readily in pathological tissues than in normal tissues. Hypertrophied and polyploid cells, for example, undergo degenerative changes more quickly than ordinary cells. While in the normal plant organ, the end of an actively growing period is marked merely by a slowing down of metabolism and eventual dormancy, in the gall the conditions approach more ageing than dormancy. The cecidozoa henceforth fail to secure the essential optimal conditions of nutrition and enzyme complex in the gall tissue and are thus literally stimulated and compelled to cease feeding, become dormant or leave the gall, which has so far provided them with both food and shelter. As the gall matures at different stages in the development of the cecidozoa, it is conceivable that in many cases the life-history of the cecidozoa-gall complex ends at gall maturity. The maturity changes not only compel but also facilitate the escape of the cecidozoa from the gall. A number of larvae escape thus when full grown from the gall and pupate outside on the plant or underground. The larvae of the midges *Atrichosema aceris* KIEFF. from the fusiform unilateral gall on the branch of *Acer campestris*, *Contarinia aceplicans* KIEFF. from the leaf roll gall on *Acer*, *Macrolabis hieracii* RÜBS. from the galls on *Hieracium*, *Jaapiella medicaginis* RÜBS. from the gall on *Medicago sativa*, *Contarinia quercina* RÜBS. from the gall on *Quercus* and *Ceutorrhynchus pleurostigma* (MARSH.) are examples of cecidozoa leaving their galls at maturity. The life-histories of their galls terminate thus with the escape of the larvae. We have already mentioned the full grown larva of *Pontania* as leaving the mature kammergall on *Salix*. It has generally been believed that maturing of the gall is a function of the development of the cecidozoa, so that the escape of the larva or its ceasing to feed brings about the degenerative changes and arrests the growth of the gall. As pointed out above, it is, however, now recognized that the maturity of the gall is not a function of the development of the cecidozoa, but on the other hand, the development of the cecidozoa is a function of the gall. The gall decays not because the cecidozoan escapes from it, but the latter escapes from the gall just because the gall is decaying and it does not find any more optimal conditions in it. The entire life-history of the cecidozoa is a function of the nutritional and developmental conditions in the course of the life-history of the gall. Whether a gall matures at the end of the larval stage of the cecidozoa, as in the above mentioned examples, or the gall maturity synchronizes with other stages as in aphid galls, thrips galls and mite galls, it is important to emphasize that it marks the end of the active feeding phase in the life-history of the cecidozoa.

Post-maturity phase

All galls do not necessarily decay and disintegrate on reaching maturity and may not even fall off from the gall-bearing organ. Although the cecidozoa cease feeding, they do not also escape as larvae from the gall, but remain within the gall and hibernate or aestivate or also pupate and complete metamorphosis. This is the case with the galls of many midges, all cynipids and chalcids and some Lepidoptera. In the case of the galls of psyllids, the last non-feeding semi-quiescent nymph remains within the gall. The hibernation of *Aylax hieracii* (Bouché) takes place, for example, within its gall. Over-wintering in *Neuroterus* spp. is also in the galls. The larvae of *Mikiola fagi* (Htg.) not only hibernates but also pupates inside the gall. The emergence of the adults takes place within the gall. In some cases, as for example, *Diplolepis quercus-folii* (Linn.) the adult remains within the gall for a long time; it emerges from pupa in October, bores a tunnel right to the epidermis, but does not immediately escape. It spends the whole of autumn within the gall and escapes only in the middle of December. Galls with a well defined post-trophic phase of life-history, either persist on the gall-bearing organ or break loose from it and fall to the ground. In either case the gall is hard, with a resistant outer surface and serves as effective shelter for the cecidozoa during winter. Galls on the shoot axis, acrocecidia on shoot, fruit galls, and many leaf galls, like the galls of *Mikiola fagi* (Htg.) and the lenticular galls of *Neuroterus*, fall off to the ground. When, however, the gall with pupa remains on the plant, the gall tissues remain alive, but no growth or cell division take place. The post-maturity phase embraces also the fate of the gall after the escape of the cecidozoa. The galls which do not fall off the plant before the escape of the cecidozoa may then either decay rapidly under the action of fungi and bacteria or may also become dry and persist for a long time. The root galls, most succulent and fleshy galls on leaves, buds, flowers, etc. decay rapidly. The hard stem galls, bud galls and many leaf galls remain dry for long periods. White[1263] has recently reported the interesting case of large tumor galls on *Picea glauca* persisting for six or seven years and in some cases 75—100 years on the tree in some localities in Canada and Alaska. In some cases the plant reorientates the growth and activity of the gall tissue, which soon becomes merged indistinguishably with the other normal parts.

Total duration of the life-cycle

The duration of the complete life-cycle of zoocecidia varies within wide limits, depending on the species of both plant and cecidozoa. The most decisive factor that determines the life of a gall is the physiological death of the gall-bearing organ. Some galls, like the gall of *Pemphigus spirothecae* PASS., fall off with the gall-bearing organ, but others decay long before the natural death of the gall-bearing organ. The galls of *Byrsocrypta gallarum* (GMELIN) on the leaf of *Ulmus, Neuroterus quercus-baccarum* LINN. and others on *Quercus* decay before the death of the leaf. The galls of *Neuroterus numismalis* FOURC. and *Diplolepis (Dryophanta) longiventris* (HTG.) on *Quercus* remain alive for much longer periods. The kammergall of *Pontania* on *Salix* has been artificially kept alive and green in the laboratory long after the escape of the larva from the gall.* Depending on the duration of the life-cycle, we may recognize the following gall types: 1. Galls with an annual life-cycle and 2. galls with a 2—3-year life-cycle. The former class of galls include three sub-types: i. Galls with a short life-cycle, perishing in the same summer in which they arise; galls in which the cecidozoa do not pupate; galls in which the cecidozoa pupate, but do not hibernate. ii. Galls which perish in winter, cecidozoa not hibernating in the gall. iii. Galls which perish in winter, cecidozoa hibernating and pupating within the gall.

Galls with annual and short life-cycles arise predominantly on herbaceous plants, but many of them develop on perennial shrubs and trees. The total life-cycle is in many cases hardly ten days and does not exceed on an average 10—15 weeks. Most of these galls start in spring or early summer and before the summer is ended the entire life-cycle is completed. There are often more than two or three overlapping generations within summer, at least in the tropics. The best known examples of this class are found among the galls on buds, leaves, flowers, fruits, etc. Most succulent galls, bacterial, fungal, nematode galls on roots are other examples. A number of galls caused by Eriophyidae, Thysanoptera, Psyllidae, Aphidoidea, Coleoptera, Lepidoptera, many Chalcids, some cynipids and Itonididae also complete the whole life-cycle within the summer. Although the cecidozoa may pupate in the gall in a great many of these cases, hibernation does not naturally take place in the gall. The galls of many Eriophyidae, Thysanoptera, Aphids and Itonididae arise in spring or summer and perish in the winter of the same year. The total life-cycle is thus completed within about 20—25 weeks. This class includes some bud galls, many leaf galls and most shoot axis galls, not only on shrubs but also on herbaceous plants. Hibernation may take place within the gall, as in the case of *Mikiola fagi* (HTG.) on the leaf of *Fagus silvatica* LINN.

The life-cycle of galls of heterogonic cecidozoa is complicated by the alternation of gall types, such as bud galls and root galls, bud

---

* Crown-gall tissue lives and grows long after the complete elimination of the causative bacteria.

galls and leaf galls, etc. and extend from spring-summer to the spring-summer of the next year. The bud gall of the sexual generation of *Biorrhiza pallida* (OL.), initiated in April-May, matures in June, the adults emerge in July and the root gall that alternates with the bud gall matures in October-November and the adults of the unisexual generation emerge from December to February of the next year. The life-cycle of the gall complex of *Diplolepis quercus-folii* (LINN.) extends from the appearance of the bud gall of the bisexual generation in May to the escape of the adults of the unisexual generation from leaf galls in December. The same is the case with the gall complex of *Diplolepis longiventris* (HTG.). The life-cycle of the gall complex of the North American *Amphibolips confluentus* (HARR.) extends from May, with the leaf gall of the bisexual generation, emergence of adults in June, the beginning of the leaf gall of the unisexual generation in late summer and emergence of the adults from these galls in October of the same year to early spring next year. The life-cycle of the leaf margin gall and lenticular gall complex of *Neuroterus albipes albipes* (A. SCHENK) commences in May with the leaf margin gall of the bisexual generation and ends with the emergence of the adults of the unisexual generation from the lenticular galls in March next year. *Neuroterus quercus-baccarum* (LINN.) and *Neuroterus numismalis* (FOURC.) have also similar life-cycles. The life-cycles of other heterogonic galls are of 2—3 year

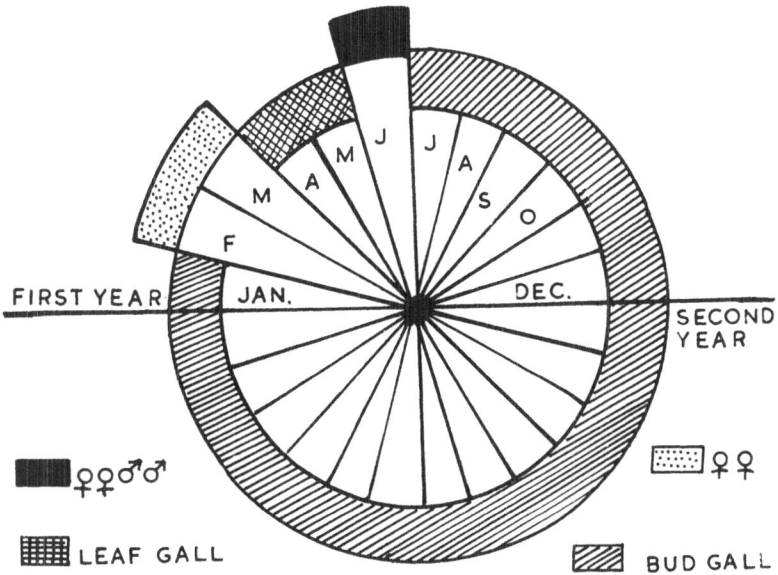

Fig. 122. Phenology and life-cycle of *Andricus curvator* Htg. and its galls on *Quercus*.

206

durations. In the case of *Neuroterus aprilinus* (GIRAUD) the bud gall of the bisexual generation begins in March, matures in mid April, adults emerge in May, the acorn galls of the unisexual generation fall down in May-June and the adults emerge in July-August of the first or also of the second year. The life-cycle of the gall complex of *Andricus testaceipes testaceipes* HTG. lasts for 2—3 years. The gall of the bisexual generation matures in August-September, but the gall of the unisexual generation matures only in April of the third year. In the case of the gall complex of *Andricus quercus-radicis quercus-radicis* (FABR.) the life-cycle extends from early summer, when the petiole and midrib galls begin, to April of the third year, when the adults of the unisexual generation escape from root galls. The adults of the bisexual generation escape from the petiole-midrib gall in August-September of the first year and also the second year. The root gall of the unisexual generation matures in autumn of the second year but the adults escape only in April of the third year. The life-cycle of *Andricus curvator curvator* HTG. ends also in February-March of the third year. In the case of the gall complex of *Andricus inflator* HTG. the life-cycle is completed only in the third or fourth year. The shoot tip gall of the bisexual generation matures in the middle of May and the adults emerge in July. The bud gall of the unisexual generation appears in September, matures in October, falls down to the ground and the adults escape in April of the second, third and fourth year. The galls of some cynipids survive thus for 14—16 months and of others like that of the bisexual generation of *Cynipis quercus-operator* O.S. on the shoot axis of *Quercus ilicifolia* remain alive for twenty-one months[61]. Many witches-broom galls also usually require several years to mature.

3. Escape of cecidozoa

As mentioned above, the very changes in the gall tissue, which induce the cecidozoa to leave the gall, facilitate also the process of escape of the cecidozoa. There is thus a most remarkable coördination and synchronization of events in the life-cycles of the gall and of the cecidozoa. The escape of the cecidozoa from their galls differs in different species and depends primarily on the structure and stage of development of the gall. We may distinguish the following conditions: 1. Indehiscent galls, with the cecidozoa incompletely enclosed, so that the gall cavity is open to the outside and thus permits the ready escape of the cecidozoa when the conditions in the gall become unfavourable to them. 2. Dehiscent galls, with the cecidozoa completely enclosed within the gall tissue or where at least the ostiole is too narrow, so that the escape of the cecidozoa is possible only by dehiscence of the gall or by the active breaking through of the gall tissue by the cecidozoa if the gall is indehiscent.

3. A complex mechanism involving the dehiscence of gall tissue overlying the larval cavity and subsequent tunnelling of a passage by the cecidozoa. There are therefore two major ecologic groups of galls, viz. the indehiscent and dehiscent galls, which determine the escape of cecidozoa from their galls.

Escape of cecidozoa from non-dehiscent galls

From the point of view of the escape of cecidozoa, the non-dehiscent galls represent a homogenous ecologic class. Here belong the bulk of the fold and roll galls, many pouch galls and some covering galls, a number of bud and flowers galls caused by Thysanoptera, Acarina, some Psyllidae and Aphidoidea and a few species of Itonididae. In a great majority of the non-dehiscent galls, the gall cavity is not completely cut off from the outside, but is open by more or less wide and persistent openings. The cecidozoa merely crawl out through the existing passages like the wide open ostioles of pouch galls or in between the margins of fold and roll galls on leaves. The escape of the cecidozoa may be in the larval stage, just before pupation or as in many cases pupation may be completed in the gall and the adults escape through these natural openings. The escape is often facilitated by more or less partial shrinking of the gall tissue. In the leaf fold galls of *Gynaikothrips* spp. on *Piper*, mentioned in an earlier chapter (fig. 64) and in the gall of *Cryptothrips intorquens* KARNY on the leaf of *Smilax* sp., the gall cavities are incompletely closed and the thrips simply crawl out when the conditions within the gall become unfavourable for them. Rosette galls and often also other bud galls caused by mites, thrips, some aphids, etc. do not completely enclose the cecidozoa. The pit galls of some mites, aleurodids and psyllids also do not enclose the cecidozoa. The filzgalls of mites and other galls, in which the cecidozoa are situated externally, exposed on the epidermis, offer no obstruction to the escape of the cecidozoa. In all these cases the cecidozoa are theoretically in a position to move out of the gall at any time, but in reality they do so only when the gall has matured, and the gall tissues do not any more provide the necessary optimal conditions. The Eriophyid mites, thrips and other species escape through the permanently open ostioles of pouch galls. In a number of galls, the ostiolar and other openings become more or less narrowed during the growth of the gall or the openings may be obstructed by inwardly directed hairs. In some leaf fold and roll galls the gall cavity is tightly closed. On maturation of the gall, the ostiole becomes wider by partial shrinkage of surrounding tissue or by the wilting of the obstructing hairs, partial uncoiling and unrolling of the parts, etc. The gall of *Pemphigus spirothecae* PASS. undergoes, for example, a gradual loosening of the spirally twisted ribbon-like petiole on maturation in autumn and thus permits the escape of the aphids. Similar condi-

tions are met with in the galls of *Forda formicaria* HEYDEN (=
Pemphigus semilunaris PASS.) on *Pistacia terebinthus* LINN., *Schizoneura ulmi* (LINN.) on *Ulmus, Perrisia fraxini* (KIEFF.) on *Fraxinus, Chermes (= Adelges) abietis* (LINN.), etc. In most of these galls,
with maturation of the gall and the cessation of the flow of sap,
the larger turgid and succulent cells and tissues shrink more than
thick-walled and smaller cells, so that the resulting unequal short-
ening and deformation lead to reopening of the passages to the
outside.

Other examples of escape of incompletely enclosed cecidozoa may
be mentioned. The bivalve gall of *Lobopteromyia bivalviae* (RAO) on
the leaflets of *Acacia catechu* WILLD. (fig. 88 C), facilitates the escape
of the midge by the two tight-fitting valves moving apart and exposing the gall cavity with the pupa. This is also true of the bivalve gall
on *Acacia venosa* HOCHST (fig. 88 A), described by TROTTER[1178] from
Eritrea. There is in most galls of *Brachyscelis* a distinct ostiole,
through which escape of the cecidozoa is possible at the right moment.
In some cases the ostiole becomes narrowed during growth of the gall,
but after maturity, there is once again a widening of the ostiole,
mainly due to the shrinking of the surrounding tissue.

In the non-dehiscent galls which we have so far considered, the
cecidozoa merely utilize existing openings in the gall for their escape.
There is, however, a large and interesting subdivision of the non-
dehiscent galls, which either enclose the cecidozoa completely or in
which the ostiole is too narrow to permit the exit of cecidozoa.
The cecidozoa escape from these galls by biting their exit holes,
usually at some predetermined spot in the wall of the gall. Most
Coleoptera, Lepidoptera, Hymenoptera and Diptera escape from
their galls in this manner. The exit hole is situated, for example, on
the summit of the galls of *Andricus inflator* HTG. but laterally in the
gall of *Andricus testaceipes* HTG. It is the adult that bites its way
out in the case of Coleoptera and many Hymenoptera. In a number
of examples it is, however, the larva that makes the exit hole for
the future adult, which is incapable of biting out the hole in the wall
of the gall. In all Lepidoptera galls, the larva bites its way to the
epidermis of the gall, but does not pierce the epidermis. Pupation
then takes place just beneath the thin epidermis. When the
adult emerges, the thin epidermal operculum-like cover is also
dry and readily pushed off by slight pressure from inside. In
the gall of *Dactylethra candida* STAINT. on *Tephrosia* (Plate I, 6)
the full grown larva gnaws away the thick wall, all except the thin
epidermis, which is left intact to form a circular, bulging lid on the
summit of the gall. As the moth emerges, the epidermis is dried and
loosened from the surrounding tissue and readily pushed off. In
most galls of Itonididae also, the larva bores its way nearly to the
epidermis and pupates underneath. At the time of emergence of the

adult midges, the pupa wriggles around, rising up to the epidermal covering, pierces the now dry epidermal lid by the antennal horns, sticks out on the surface of the gall and then the pupal skin bursts open for the emergence of the adult midge. This is for example the case with *Oligotrophus betulae* WINN. on the fruit gall of *Betula* and *Didymomyia reaumurianus* (F. Löw). The pupa of *Braueriella phillyreae* F. Löw. also sticks out on the surface of the gall on *Phillyrea media* LINN. from the exit hole prepared by the larva. The pupa of the midge from the singular looking gall on the leaf of *Vitex* from East Africa (fig. 77 A)[997] breaks through the exceptionally thin floor of the flattened larval cavity just before the emergence of the adult. The floor is very thin, delicate and fragile in the mature gall. The pupa pushes itself among the elongate, slender hairs, which have so far formed a tight plug for the evaginated pouch gall, but have now partially shrunk. We have already mentioned the habit of the larva of *Pontania* of cutting through the fleshy wall of its kammergall on *Salix* and dropping off to the ground, suspended by a silken thread. Even in the case of some species which have well developed biting and chewing mouth-parts, the adults escape by pushing off thin and cap-like lids of the larval cavities. The gall of *Trichilogaster* on *Acacia leucophloea* WILLD. (fig. 36) from South India has the numerous larval cavities beneath the epidermis in the flat, cup-like top of the gall. Each larval cavity is plugged by small, hemispherical friable tissue. The adults simply push off these lids and escape from the gall.

Escape from dehiscent galls

In the dehiscent galls, the cecidozoa are completely enclosed within the plant tissue and the ostiole present in the beginning becomes completely obliterated by the growth of surrounding tissue. The cecidozoa are therefore unable to escape and also cannot bite their out through the gall tissue. Their escape becomes possible only after the gall tissues crack and expose the gall cavity. The dehiscence of the gall takes place in different ways, but the mechanism involved is not different from other dehiscent structures like fruits. The dehiscence may be effected by the irregular peeling off of the bark from the gall, thus exposing the cecidozoa and permitting their escape. Other methods of gall dehiscence include irregularly lacerated rupture of the gall, opening at predetermined spot, breaking off nearly the whole or major portion of the gall from the base and falling off of cap-like or plug-like lids or pyxidial dehiscence.

The solid chalcid gall on the stem of *Prosopis* from India mentioned in an earlier chapter (fig. 37) has numerous larval cavities, imbedded peripherally in the central hard core, immediately beneath the rather thick bark. The metamorphosis is completed and the beautiful metallic green adult chalcids emerge while the gall is still live and fresh, but are completely covered by the bark. They remain within their narrow larval cavities. Several weeks later, the natural process of drying of the gall results in the cracking and irregular peeling of the bark in pieces, which fall off and thus expose the open larval cavities, with the adults ready to escape. The gall of *Pruthidiplosis mimusopsicola* MANI on the flower of *Mimusops hexandra* ROXB. (Plate V, 7, fig. 93) is a solid, nut-like spherical mass, with a hard thick rind and central spongy core. When mature, the outer rind cracks open irregularly and exposes the numerous pupae, sticking out imbedded in the spongy core. On emergence, the adult midges are thus free to fly off. The galls which develop wholly sub-epidermally and later erupt through the epidermis also belong to this group. The galls on *Indigofera dosua* HAM. and *Indigofera pulchella* ROXB. are remarkable for the irregular bursting of the bark. As mentioned earlier, these galls arise wholly at first from within the cortex, the epidermis and sub-epidermal cells not taking an active part in the development. When near maturity, the bark peels off from the swollen stem and exposes the numerous ovoid and sclerenchymatous larval cavities. The larvae tunnel an exit hole on the apices of the larval cavity before pupating, so that the adult midges escape without difficulty.

The galls of many psyllids like *Trioza jambolana* CRAWF. dehisce by breaking open irregularly. The full grown nymph crawls out of the open gall on the leaf, moults once again before emerging as adult. In the gall of another psyllid *Phacopteron lentigenosum* BUCKT. from India (Plate II, 7) the adult emerges within the gall, but remains imprisoned in it for some time, often for more than a week. The gall now cracks open irregularly and exposes the gall cavity with the adult. Dehiscence by lacerated rupturing, usually at some predetermined place is met with in the galls of *Byrsocrypta gallarum* (GMELIN), *Pemphigus vesicarius* PASS. and *Baizongia pistaciae* (LINN.) (= *Pemphigus cornicularius* PASS.).

The breaking loose of the gall from the gall-bearing organ does not always have anything to do with the escape of cecidozoa, but if the gall breaks to several pieces at the same time and leaves behind only a small basal portion on the plant, this would undoubtedly result in exposing the gall cavity and also permit the escape of cecidozoa. The piece that breaks loose in this way may be small or large and the residue left behind on the plant can also be minute and insignificant. Some of these galls break off neatly at such a point that a basal part of the gall is left behind with the cecidozoa. The pouch gall of *Rondaniella* (= *Oligotrophus*) *bursarius* (BR.) on the leaf of *Glechoma hederacea* falls off from the plant as a result of the formation of an annular zone of macerated cells at the point of the insertion of the gall on the leaf blade. The cells become rounded off and loosen themselves. The residue left on the leaf is a cylindrical ring. Increase of atmospheric humidity seems to favour the dehiscence of this gall. Another interesting example of the breaking loose of the gall and thus indirectly facilitating the future escape of the adult cecidozoa is the gall of *Mikiola fagi* (HTG.) on the leaf of *Fagus silvatica* LINN. As the gall starts to dry in summer, the midge larva is already full grown. The presence of small, thin-walled parenchyma cells in the constriction between the basal annular part and the upper portion in which lies the larva, facilitates the breaking of the gall loose from the leaf. There is either a separation of the middle lamella of the cell walls of the dehiscent tissue or a rupture of the drying membranes. The annular basal portion has a characteristic sclerenchyma zone. At the constriction and below it there is a strip of thin-walled small cells or the so-called "Trennungsschicht", representing the residue of the gall meristem[905]. The loosened upper part, with the mature larva inside, falls to the ground. The sclerotic

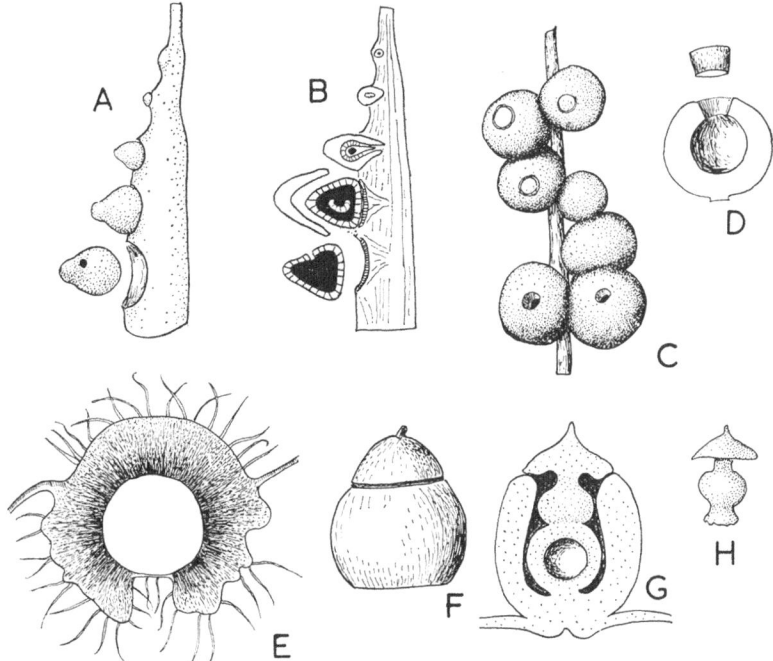

Fig. 123. Dehiscent galls. A–B. The dehiscent gall of *Andricus testaceipes testaceipes* (Htg.) unisexual generation on *Quercus*. C. Pyxidiocecidium of *Cecidoses eremita* Curt. on *Schinus*. D. The same cut open. E. Midge gall on leaf of *Ficus*, with a preformed operculum, thin enough to be readily pushed off by the midge (After Rübsaamen). F–H. Pyxidiocecidium on leaf of *Newtonia insignis* Baull., caused by an unknown Psyllid. F. General view of the gall. G. Gall cut open. H. The peculiar stopper-shaped lid of the gall (After Houard).

walls and a waxy coating on the surface of the gall make it very resistant to the winter conditions. The circular aperture in the gall is covered by webbing by the larva before pupation. The so-called "jumping galls" are unique among the galls which break loose from the plant. These galls are set in motion by the jerking movements of the larva inside. The common examples of jumping galls are those of *Nanophyes pallidus* OL. and *Neuroterus saltans*. Useful information on the dehiscence of these galls may be found in MÜHLDORF[825], PFEIFFER[905], TAVARES[1124], GIBBS[388] and LEACH[653]. In the gall on the leaf of *Parinarium obtusifolium*, the presence of mechanical tissue favours the breaking off of a cap-like lid. The upper part of the gall consists of a very hard sclerenchyma lid, lined only by the epidermis and resting on a lower petri-dish shaped flat sclerenchyma floor. At the time of dehiscence of the gall, the upper part separates cleanly off from the lower part and thus exposes the larval cavity (cf. fig. 26).

Pyxidiocecidia

The pyxidiocecidia are characterized by the presence of a cap-like or plug-like covering of the gall cavity. The larval cavities in the gall of *Callirhytis elliptica* WELD on the shoot axis of *Quercus alba* LINN. from North America open out, when small, oval, circular, button-like lids fall off when the gall is mature. In the midge gall on branches of *Combretum* sp. (fig. 42 D—E), described by HOUARD from West Africa[505], the axial cylindrical larval cavity is closed apically by a conspicuous, neatly fitting, peg-like lid, projecting on the surface like a stumpy process. At maturity of the gall, the tissue of the peg-like stopper shrinks more rapidly than the body of the gall and becomes loosened. The best known pyxidial gall is caused by *Cecidoses eremita* CURT. on *Schinus dependens* ORTEGA from South America[567]. This gall dehisces by a circular, stopper-like lid, about 4—5 mm in diameter, falling off and leaving a neat circular aperture for the escape of the adult moth (fig. 123, C—D). HOUARD[512] has described a curious pyxidial gall on *Newtonia insignis* BAILL., caused by an unknown psyllid from Congo (fig. 123 F—H). A conical process, fitted to a ball-like stopper, forms an accurately shaped lid in an axial cylindrical terminal passage leading into the gall cavity. During dehiscence, the whole of the stopper-like lid falls off.

CHAPTER XI

ECOLOGIC INTER-RELATIONS IN ZOOCECIDIA

1. The gall community

We have in zoocecidia a highly specialized and complex community, in which the character species is the cecidozoa. With its localized concentration of highly nutritive substances and its marked succulence, a gall provides extremely favourable breeding conditions to a great variety of organisms. Nearly every gall is associated with diverse organisms, animals, fungi and bacteria, other than the true cecidogenetic species. The gall of *Biorrhiza pallida* (OL.) has, for example, over seventy-five different species of insects, associated with it. In 1883 BEAUVISAGE applied the term *locatari* for the cecidocole and cecidophile Arthropods, other than the true cecidozoa. STEFANI-PEREZ[1094] and STEGAGNO[1097] devoted considerable attention to these cecidocole and cecidophile species, with special reference to the gall of cynipids on *Quercus*. The locatari include 1. accidentals, 2. harmless visitors, 3. parasites on cecidozoa, 4. hyperparasites on cecidozoa, 5. inquilines, 6. parasites on inquilines, 7. hyperparasites on inquilines, 8. predators in cecidozoa, 9. predators on inquilines, 10. predators on harmless visitors, 11. parasites on harmless visitors, 12. hyperparasites on harmless visitors, 13. parasites on predators, 14. hyperparasites on predators, 15. symbiotes, 16. predators on symbiotes, 17. parasites on symbiotes, 18. hyperparasites on symbiotes, 19. cecidophags, 20. predators on cecidophags, 21. parasites on cecidophags, 22. hyperparasites on cecidophags, 23. parasites on predators of cecidophags, 24. hyperparasites on predators of cecidophags, 25. succesori, 26. predators on successori, 27. parasites on successori, 28. hyperparasites on successori, 29. parasites on predators of successori, 30. hyperparasites on predators of successori, 31. myrmecophile of ant successori, 32. parasites on myrmecophiles and 33. species of uncertain position in the community.

The inter-relations of these diverse cecidocole and cecidophile species are extremely complex and it is not always easy to assign a given species precisely to any of the above mentioned groups. It is common experience that a species is an inquiline in the gall of one species and a true cecidogenetic species in another gall. The size, character, composition and integration of the communities vary in different galls. Strictly speaking, the cecidocole community is not a single unit, but actually represents two fundamentally different, but nevertheless closely integrated communities. The primary community is associated with the gall during its active phase of devel-

opment and growth, before the escape of the cecidozoa. This primary community is regularly succeeded by an ecologic complex of
species, viz. the *successori*, which inhabit the gall after the escape
of the cecidozoa. The member species of these communities are
ecologically closely dependent on the cecidozoa and the gall. The
inter-relations in zoocecidia are thus extremely complex and include the inter-relations of the cecidozoa and the gall-bearing plant,
the inter-relations of cecidozoa and other cecidocole organisms and
the inter-relations of the gall and the cecidocole organisms.

2. Inter-relations of cecidozoa and
the gall-bearing plant

Host specificity

Galls represent the mutual adaptations of the cecidozoa and the
plant. It is conceivable that all plants are not equally capable of reacting to the attack of the cecidozoa by producing galls. Cecidozoa are
therefore predominantly bound to a definite group of plant and in
many cases to a given family, genus or even species. Although
Heterodera marioni (CORNU) GOODEY gives rise to galls on a large
number of species of plants of diverse Natural Orders, there is a
definite race phenomenon associated with this apparent polyphagy.
No other group of cecidozoa is comparable to the root-knot Nematode in its capacity to induce the development of galls. It is relatively rare that a species gives rise to galls on more than one species
of plant. The greatest majority of species are characteristically
restricted to definite groups of plants. The galls of *Chermes* arise, for

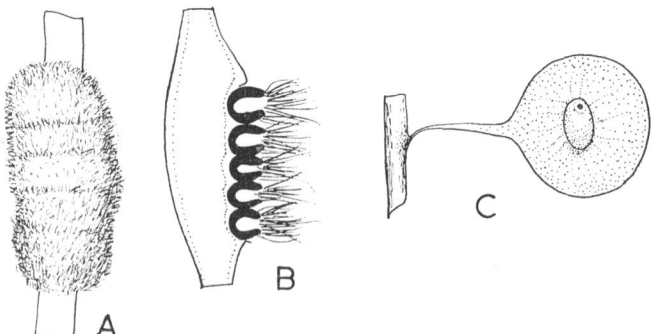

Fig. 124. Two remarkable cynipid galls. A. Hairy erupting rindengall on *Quercus*.
B. Mature galls with hairy outgrowths, surrounding the exit hole. The hairs separate
in the mature gall to permit the escape of the adult that bites its way out of the
sclerenchyma wall of the larval cavity. C. Stalked gall on stem of *Quercus* with
a central sclerenchyma larval cavity.

example only on Coniferae. The galls of *Clinorrhyncha* Löw, *Cysti-phora* KIEFF., *Rhopalomyia* RÜBS., *Tephritis* LATR., etc. are simi-larly confined to Compositae. Cynipid galls are predominantly found

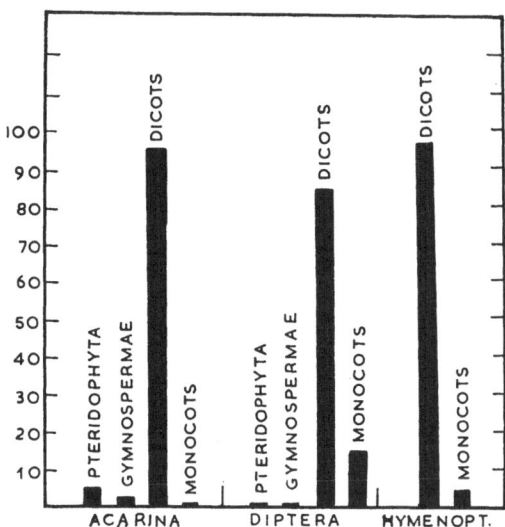

Fig. 125. Percentage frequencies of zoocecidia on different groups of higher plants in the Oriental flora.

on *Acer, Centaurus, Glechoma, Hieracium, Hypochoeris, Papaver, Potentilla, Quercus, Rubus, Rosa, Sonchus,* etc.

In the vegetable kingdom a certain definite phylogenetic anti-quity seems to constitute an optimum for production of galls. Older groups of plants do not seem to react to the action of cecidozoa as readily as the younger ones. Progressive adaption can only lead to specialization on the part of the cecidozoa: from polyphagy to oligophagy and from oligophagy to extreme monophagy.

It is well known that an organism which is in some respect adap-ted to another species must be phylogenetically younger and also more plastic than the other. Species of animals which feed on all kinds of plants in general are on the whole phylogenetically older than those that are adapted to definite species for nutrition. Among the monophagous forms, species which depend for their food on phylogenetically older groups of plants, such as for example Pteri-dophyta or the Gymnospermae, are older than those that are nutritionally adapted to phylogenetically younger groups of plants, like the Dicotyledonae and the Monocotyledonae. This concept is best illustrated by reference to KARNY[565]. KARNY investigated the frequency of galls on diverse groups of plants in the Javanese flora.

He measured the frequency of gall in terms of the gall index. The gall index represents the number of galls, caused by different groups of cecidozoa, per 100 species of plants in each Natural Order. The gall index is of importance in larger Natural Orders, but not very reliable in case of the smaller ones. KARNY has shown that the oldest groups of plants like Pteridophyta bear on the whole fewest galls. The small numbers of galls known on them are also caused by cecidozoa, more than half of which belong to Acarina. Extremely few Thysanoptera galls are met with and these are also caused by Terebrantia; Tubulifera, so frequently associated with higher plants, do not give rise to a single gall on the Pteridophytes. High holometabolic insects like Coleoptera, Hymenoptera, Lepidoptera, etc. are not so far known to give rise to galls on Pteridophyta. The Natural Order Polypodiaceae, comparatively young and rich in species, bears the bulk of the known galls on the Pteridophyta. Gymnospermae, which are also relatively old groups of plants, exhibit nearly similar characters. Taxaceae and Gnetaceae alone among them bear any galls of Acarina, Thysanoptera and Diptera. Some Tubulifera also give rise to relatively simple types of galls on these plants.

The legion Dicotyledonae bears the greatest bulk of all known galls, caused by every class of cecidozoa (fig. 125). The Monochlamydae, generally considered as rather primitive and most generalized among the Dicotyledonae (though some workers consider them as derived from the Ranales), provide interesting data in regard to the host-parasite mutual adaptation and specialization (fig. 129). KARNY gives the gall index of the whole group of Centrospermae (Monochlamydae) as 62.7 for the Javanese flora. The gall index for the individual Natural Orders of Monochlamydae are Polygonaceae 6.6, Aizioaceae 11.1, Nyctaginaceae 18.1, Amaranthaceae 18.7, Myrsinaceae 22.5, Olacaceae 25.0, Fagaceae 32.1,

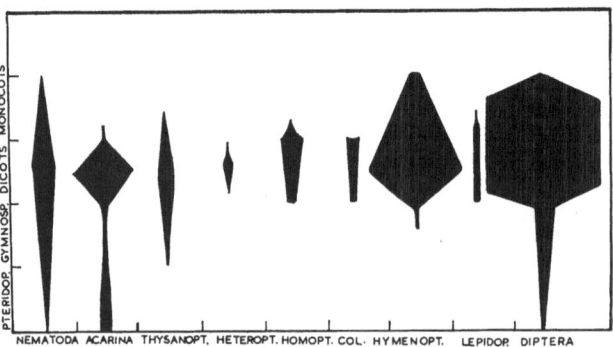

Fig. 126. Relative abundance of galls of different groups of cecidozoa on higher plants from the world.

Chloranthaceae 22.5, Piperaceae 36.3, Ulmaceae 50.0, Urticaceae 53.5, Loranthaceae 60.0, Juglandaceae 100.0, Moraceae 110.0, Casuarinaceae 133.3, Proteaceae 140.0 (fig. 129). It may be observed

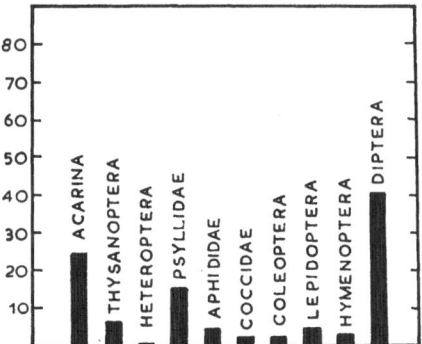

Fig. 127. Percentage frequency of galls of different cecidozoa on Dicots from the Oriental flora.

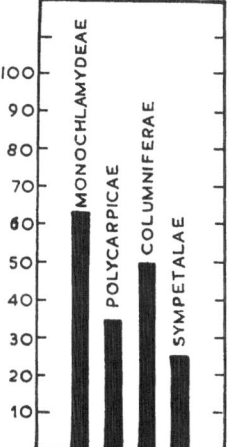

Fig. 128. Gall indices in Dicots.

that the gall index is highest in the younger Natural Orders and lower at the bottom of the geneological tree of Monochlamydae. We thus find that the abundance of galls increases proportionately to the stage of evolutionary development of the plant. In accordance with the view that Casuarinaceae are derived types, it is interesting that a number of galls by Hymenoptera are known; Hymenoptera are phylogenetically a young group of insects. The gall index of the whole group of Ranales (Polycarpicae) is 35.2. The gall

218

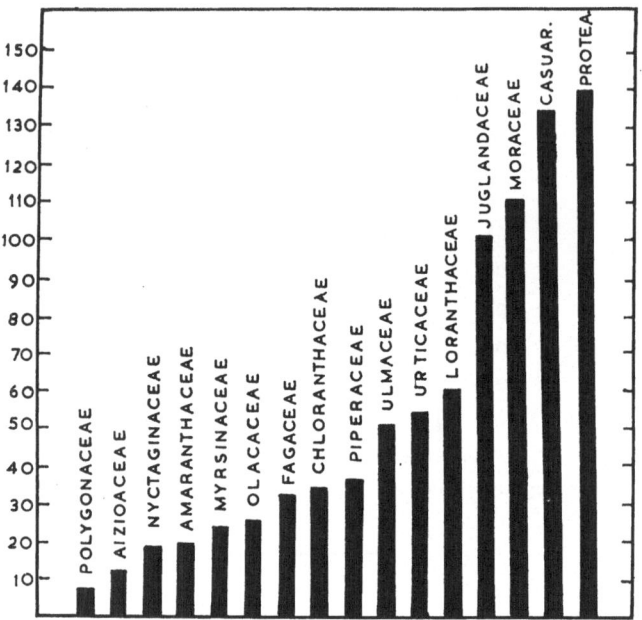

Fig. 129. Gall indices in Monochlamydae.

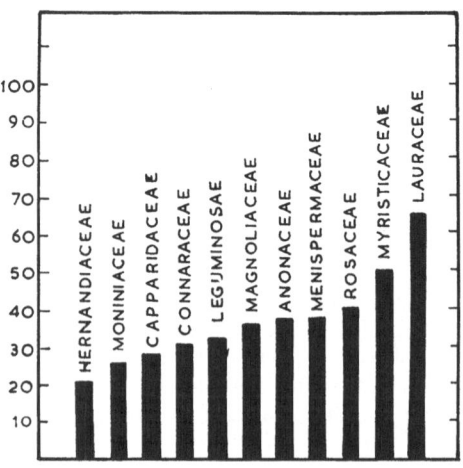

Fig. 130. Gall indices in Polycarpicae.

index of the individual Natural Orders of Ranales is Hernandiaceae
20.0, Moniniaceae 25.0, Capparidaceae 27.7, Connaraceae 30.0, Leguminosae 32.8, Magnoliaceae 35.7, Anonaceae 36.2, Menispermaceae

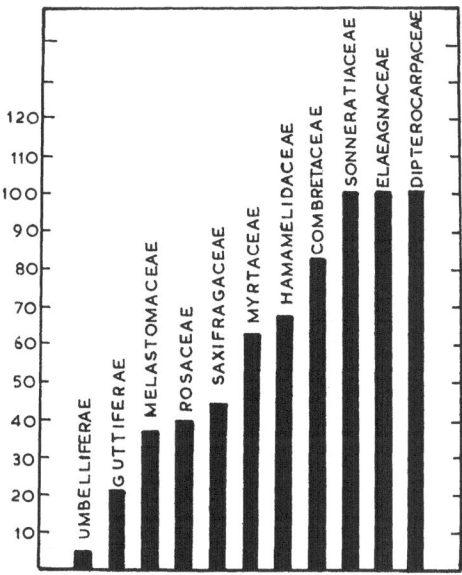

Fig. 131. Gall indices in Umbelliferae-Dipterocarpaceae.

36.8, Rosaceae 40.0, Myristicaceae 50.0, Lauraceae 63.1 (fig. 130).
Here also the older Natural Orders like Magnoliaceae, Moniniaceae
and Ranunculaceae have relatively low gall index. In the higher
Natural Orders like Lauraceae, Myristicaceae, Rosaceae and Menispermaceae the gall index is relatively high. The low gall index of
some of the higher Natural Orders like Leguminosae and Capparidaceae from Java is explained by KARNY as indicating that these
groups are as yet too young for many cecidozoa to become adapted.
The low gall index is in these cases the result of too high a specialization. Inspite of this fact, it is remarkable that every group of
cecidozoa gives rise to galls on Leguminosae, while the more generalized and phylogenetically older Natural Orders, which also have
low gall indices, are characterized by the fact that none but Acarina
and some Diptera give rise to galls on them. The gall indices of
Dipterocarpaceae 100.0, Elaeagnaceae 100.0, Sonneratiaceae 100.0,
Combretaceae 83.3, Hamamelidaceae 66.6, Myrtaceae 63.0, Saxi-
fragaceae 44.4. Rosaceae 40.0, Melastomaceae 37.1, Guttiferae 20.8
and Umbelliferae 4.5 are again of considerable interest (fig. 131).
It may be pointed that the gall index of the ancestral Rosaceae

220

corresponds with that of the entire group and the gall indices of the derivatives increase at first, but become distinctly smaller in the very recent descendents like Leguminosae, Cornaceae

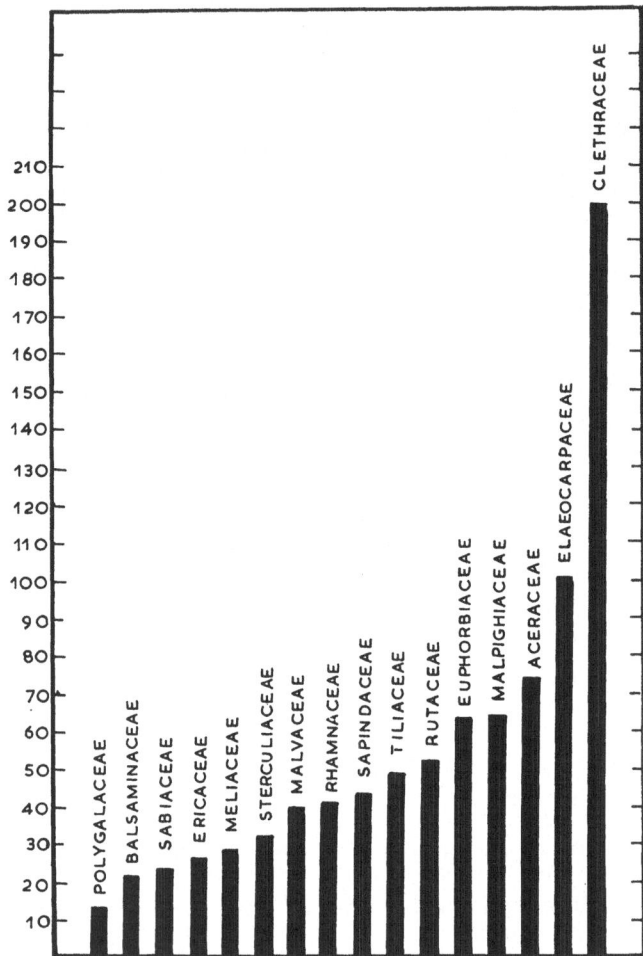

Fig. 132. Gall indices in Columniferae.

27.2, Thymelaeaceae 25.0 and strikingly low in the final member of the entire branch, viz. Umbelliferae. In the Columniferae (fig. 132) branch the total gall index is 47.9 and for the individual Natural Orders we find Clethraceae 200.0, Elaeocarpaceae 110.5, Aceraceae 72.5, Malpighiaceae 66.6, Euphorbiaceae 63.0, Rutaceae 52.6, Tiliaceae 48.8, Sapindaceae 42.8, Rhamnaceae

41.1, Malvaceae 40.0, Sterculiaceae 32.4, Meliaceae 28.1, Ericaceae 26.0, Sabiaceae 22.2, Balsaminaceae 21.4 and Polygalaceae 12.7. The more primitive groups like Tiliaceae, Malvaceae and Sterculiaceae have a relatively low gall index, approaching that of the branch itself and various groups of cecidozoa are also associated with these, but the Hymenoptera are on the whole only sparsely associated as cecidozoa with these Natural Orders. The gall index rises in Euphorbiaceae and is nearly as high as in Malpighiaceae. The branch from Simarubiaceae to Aceraceae shows irregular changes in their gall indices. There is then a rise in the gall index with noticeable regularity from Celastraceae to Vitaceae and Aquilifoliaceae (fig. 132). KARNY attributes the low gall index of Ericaceae to the high specialization, as is evidenced by the fact that Lepidoptera are associated with them as cecidozoa and Heterometabola and Acarina are not thus associated. The general gall index of Sympetalae is 24.4 (fig. 133) and the gall indices of the individual Natural Orders are as follows: Scrophulariaceae 4.2, Asclepiadaceae 5.4, Bignoniaceae 7.2, Labiatae 10.9, Oleaceae 13.1, Loganinaceae 15.0, Convolvulaceae 21.9, Boraginaceae 24.0, Compositae 24.2, Rubiaceae 32.5, Apocynaceae 36.5, Cucurbitaceae 38.2, Acanthaceae 42.3, Symplocaceae 46.1, Pedaliaceae 50.0, Verbenaceae 69.6, Sapotaceae 93.7, Caprifoliaceae 100.0, Styracaceae 750.0. The general gall index is on the whole less than in other Dicotyledonae and this peculiarity is attributed to the higher organization, especially if we consider the galls of all groups of cecidozoa. The Diptera are the most abundant and the cecidozoa associated with these Natural Orders include both the relatively generalized Dipterous family Itonididae, and the specialized Trypetidae. The Monocotyledonae, which are highly specialized forms derived monophyletically by some authors and polyphyletically by others from the Dicotyledonae, show a general gall index of only 7.8 (fig. 134). The highest gall index among the Monocotyledonae is found in Pandanaceae and the lowest in Orchidaceae, with the Gramineae midway (7.5 gall index in the Javanese flora). It is also remarkable that acarocecidia are practically absent on Monocotyledonae, but galls by the higher Diptera-Brachycera are common.

Summarizing his results, KARNY concludes that a certain definitive phylogenetic antiquity constitutes an optimal condition for the production of galls. The older classes of plants have remarkably few galls; these are also found on the younger members of the old groups, which alone seem to be sufficiently plastic enough to undergo the mutual series of adaptations with cecidozoa necessary for the development of galls. Acarocecidia are therefore very much in dominance among these groups of plants. Among the Angiospermae the Ranales have not yet attained the optimal phylogenetic stage in regard to gall formation and the wealth of gall increases towards the end of their geneological tree. The Monochlamydae exhibit the optimum for gall adaptations, having a gall index of 62.7. In regard to the increase of the gall indices towards the end series of the Monochlamydae, the group behaves as if it were a primitive one, but it must be remembered that Hymenoptera galls are already becoming more dominant among them, but particularly among Casuarinaceae and Moraceae. Among the higher groups, the younger branches are still too young to have become mutually adapted with cecidozoa for gall formation and accordingly the gall index begins to decrease once again. Acarina are undoubtedly the oldest group of cecidozoa and it is therefore interesting to remember that more than half the known galls on Pteridophyta and Taxaceae are actually acarocecidia (fig. 126). Acarocecidia are also quite abundantly represented among the Ranales and Monochlamydae, but begin to decline with the Columniferae and Rosales. They are conspicuously insignificant on Monocotyledonae. Diptera are well represented as cecidozoa on nearly all groups of plants and the younger lines of this order seem to have retained their plasticity and adaptability right down to the present and they have thus become adapted as cecidozoa to the youngest Natural Orders. We have also already mentioned the fact that the youngest families of Diptera like Trypetidae are associated with younger plant groups like Compositae and Monocotyledonae. Psyllidae are common as cecidozoa on relatively primitive Natural Orders like Lauraceae, Monochlamydae (especially

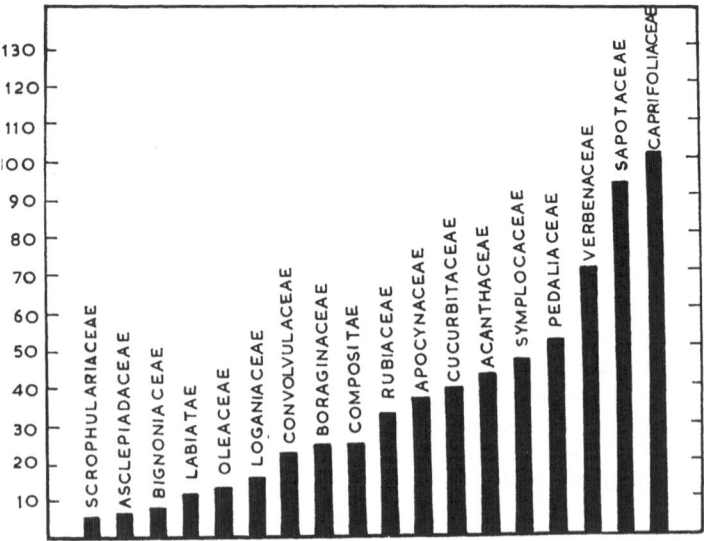

Fig. 133. Gall indices in Sympetalae.

Moraceae) and also partly Columniferae (Myrtaceae in particular). They are not also known to be associated with Pteridophyta and Gymnospermae and they again recede to the background on the young-

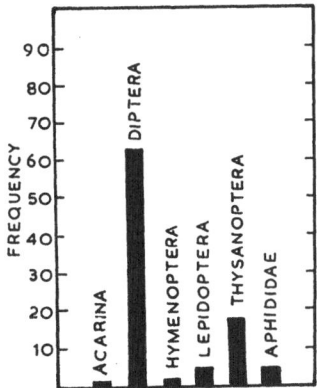

Fig. 134. Percentage frequency of gall of different cecidozoa on Monocots.

er Natural Orders. Aphid galls are most abundant on Styracaceae in the Javanese flora. Many coccid galls are also known on Gymnospermae in Australia and also on Columniferae. Among Thysanoptera, the suborder Terebrantia is more generalized than Tubulifera. Thysanoptera galls on the Pteridophyta are caused by Terebrantia and Tubulifera appear first as cecidozoa only on Gymnospermae. They give rise to galls on all higher plants, except perhaps the Monocotyledonae. Coleoptera galls are absent on Pteridophyta and Gymnospermae, but some are found among the Ranales and Monochlamydae. There are also no Lepidoptera galls on Pteridophyta and Gymnospermae, but they occur on Ranales and most abundantly on Euphorbiaceae. Only on the Monocotyledonae (fig. 134) Commelinaceae do we find any gall by Lepidoptera. Hymenoptera galls are also not known on the Pteridophyta and Gymnospermae and Ranales and the bulk of the Hymenoptera galls occurs on Monochlamydae. It is thus seen that the oldest groups of cecidozoa are associated with the oldest groups of plants and those which are phylogenetically younger are similarly associated with the younger groups of plants.

Inter-relations of gall and plant

The inter-relation of the gall and the gall-bearing plant has been comparatively neglected. A consideration of this relation covers the outstanding effects of the formation of the gall on the plant. The effect of gall formation on the plant includes injury to cells and tissues, abnormal nutritional conditions and osmotic changes. Abnormal over-production of organs, new formations of organs, especially the formation of abnormal adventitious roots and leaves, oolysis or the modification of ovules into stunted growths are other striking effects of formation of galls on plants. Remote organs are also frequently more or less affected. The effects on remote organs usually take the form of disturbances in the chemico-physiological functional capacities of these organs. Deviations in the normal direction of growth, rupture of vascular elements, deflections of sap flow, blocking up of the conducting tissues and slowing down or total stoppage of the sap flow, material and functional exhaustion of the gall-bearing organ and other abnormalities have also been described. Although the development of the gall may not directly involve the growing tip of the shoot, the growth of the shoot as a whole is usually more or less abnormal or its growth may also be arrested. KÜSTER[639] has shown, for example, that in the shoot axis gall of *Contarinia tiliarum* KIEFF. on *Tilia*, the part of the shoot axis above the gall is usually very stunted and conspicuously attenuated and often also falls off prematurely. Similar conditions are also reported in the case of the gall of *Rhabdophaga salicis* SCHR. on *Salix*. Even if the conducting elements are affected, alternate channels arise, so that often the part beyond the gall is not apparently abnormally under-sized. The deficiency of nutrition of the part of leaf beyond the gall is readily observed in the gall of *Mikiola fagi* (HTG.) on *Fagus silvatica* LINN. As already mentioned, the gall is inserted on the midrib or some of the larger side veins and this affects the vascular elements in such a way that the part of the leaf blade beyond the gall does not dry up, but loses its normal green colour and becomes conspicuously pale yellowish-green. A rhombus-shaped part of the leaf blade apically of the gall becomes yellow coloured. The nutritional deficiency of this part of the blade becomes even more pronounced in autumn, when the last residue of chlorophyll disappears much earlier than in the rest of the leaf blade. Nutritional deficiency and chemico-physiological exhaustion probably underlie the death of the part of the blade where the lenticular gall of *Neuroterus quercus-baccarum* (LINN.) is inserted, even when the rest of the blade is still fresh. Drought causes the wilting of gall-bearing leaves much earlier than normal leaves. According to BÜSGEN[163], three galls of *Mikiola fagi* (HTG.) drain off, on an average, the total assimilate synthesized by a single leaf blade. As each leaf of *Fagus silvatica* bears usually from six to eight galls, the drain of material must naturally affect a number of other leaves also.

Although the development of a gall involves more or less pronounced localized enrichment of nutritive material, to an abnormal degree, it is not always that the extra flow of sap to the gall-bearing organ results in its becoming more vascular than normal. Some well known instances of such vascularization have been reported. The petiole of the leaf of *Rosa* bearing the gall of *Diplolepis rosae* (LINN.) was observed by KÜSTER to be characterized by considerable strengthening of the conducting tissues. One of the effects of gall formation by *Neuroterus quercus-baccarum* (LINN.) on the inflorescence axis of *Quercus* is prolongation ot its life and development of secondary tissues. Similar conditions may also be observed in the case of the gall of *Rhabdophaga heterobia* H. Löw on the male catkins of *Salix*. The stipules of leaves of *Quercus* bearing cynipid galls are described by BEIJERINCK[75] to be longer lived and become more foliaceous than normal ones. According to KESSLER[581], the leaf blade of *Populus dilatata* AIT. is heavier than normal when the gall of *Pemphigus spirothecae* has developed on its petiole.

MOLLIARD[800] and GIARD[386] have described interesting examples of indirect castration as an effect of gall formation.* In the gall of *Perrisia lotharingiae* (KIEFF.) on *Cerastium vulgatum* the pollen grains become uni- or binucleate or also hypertrophied uninucleate. Fasciation of the part of the shoot axis above the gall of *Aylax hieracii* (BCHÉ) on *Hieracium* has also been observed.

Significaenc of gall formation

Rigid adherence to the ideas of natural selection has so far prevented a correct understanding of the significance of gall formation. The material which the plant accumulates in the tissues of a gall undoubtedly serves to supply the essential nutritive requirements of cecidozoa. With exceedingly few exceptions, cecidozoa are without doubt parasites in every sense of the term on the plant on which they induce the formation of galls. The benefit is evidently wholly unilateral to the cecidozoa and the plant suffers considerable injury as a result of the association with cecidozoa. It is only exceptionally that some loose symbiotic relationship between the cecidozoa and the gall-bearing plant has been observed. MATTEI[763] has, for example, observed that in the gall of *Cynips mayri* KIEFF. on *Quercus leptobalanos* GUESS. from Italy a peculiar exudation, having the characteristic odour of *Hyacinthus orientalis*, accumulates on the surface. This exudation attracts large numbers of Microlepidoptera, Diptera and Coleoptera, many of which may be found engulfed and dead in the viscid mass. MATTEI believes that these dead victims are partly digested and the dissolved material is absorbed by the plant and thus forms a source of nitrogen to the plant tissue. This ia a "carnivorous gall". The formation of the carnivorous gall establishes a sort of indirect symbiotic relation between cecidozoa and the gall-bearing plant. We have already referred to another symbiotic relation in the case of the galls of Agaontidae *(Blastopha-*

---

* Castration and sex reversal are known in many mycocecidia (Vide Chapter XII).

226

ga) on *Ficus*. These insects give rise to galls on the pistillate flowers and the gall insects at the same time bring about the pollination of the flowers.* It must, however, be emphasized that even in these cases of so-called symbiotic galls, the primary relation between the cecidozoa and the gall-bearing plant is one of true parasitism. The bacterial nodular galls on Leguminosae and the *Blastophaga* galls must in the final analysis be considered as pathological growths and the organisms evoking these growths are true parasites.

The pathological nature of a gall and the parasitic relation of the gall inducing organism have indeed been recognized from the beginning. Most of the earlier workers seem to have been greatly impressed by this extra-ordinary unilateral action. The predominant view with regard to galls is aptly summarized by ROMANES[968] as follows: "the one and only case in the whole range of organic nature where it can be truly said that we have unequivocal evidence of a structure occurring in one species for the exclusive benefit of another." These ideas have also been the fundamental basis of the definition of gall.

Although as early as 1890 COCKERELL[204] felt that phylogenetically considered the formation of a gall must be equally important to the cecidozoa and the plant, it is only comparatively recently that it has come to be recognized that the benefit may not after all be so one-sided. Recent advances have shown that the formation of the gall is the result of a specific reaction on the part of the plant to the attack of the cecidozoa, to counteract the harmful effects of cecidozoa. It has been known that there is a more or less pronounced tendency on the part of the plant to digest the parasite and to utilize the material, an action somewhat similar to the phagocytic behaviour in animal tissues. HILTNER[473] found that in the case of the root nodule gall on Leguminosae there is a close struggle between the bacterium and the plant tissue, evident even from the very early phase of the development of the gall. Depending on the virulence of the bacterium, the result of this struggle differs within wide limits, so that not every infection results in a gall. Weak strains of bacteria are soon resorbed by the cell nucleus of the plant tissue and thus only a very insignificant swelling of the root appears. ADLER[1] found that out of thirteen plants of *Rosa* on which *Diplolepis rosae* (LINN.) was able to oviposit, only five developed galls. These differences are believed by him to be probably the result of diverse conditions of growth of the different plants. KÜSTER does not, however, believe that the formation of a gall is an attempt on the part of the plant to encapsulate the parasite. ZWEIGELT[1295] has put forward the view that

---

* Another example of symbiotic relation is provided by the fixation of atmospheric nitrogen in bacterial nodule galls of Leguminosae and other nodule galls of *Alnus, Eleagnus,* etc.

gall formation is in most cases the result of specialization and race formation by the cecidozoa and hence the immunity from a gall is a passive one. The immunity of *Vitis vinifera* LINN. to phylloxera leaf gall is thus explained on the basis that its leaves could not acquire active immunity to phylloxera, which did not exist in Europe before its introduction by man. The fundatrix that attacks the leaves is extremely specialized to the American vine, but the virgo attacking the roots are less specialized and produce galls on roots to nearly a hundred percent, while the leaves of the same plant are free.

Available evidence shows that the plant produces the gall primarily not to advance the interest of the cecidozoa as has been emphasized by KÜSTER and other earlier workers. The primary reaction of the plant to the attack by cecidozoa is not even indirectly in the interest of the cecidozoa; it is one of self interest for the plant from the very first. The primary reaction on the part of the plant is a defensive one, involving a struggle of resistance between the plant tissue and the cecidozoa, the ultimate object of which is the neutralization of the toxic effects of the cecidozoa on the plant cells. In so far as the reaction of the plant promotes this defensive effect, the primary significance of the gall is for the plant and not for the cecidozoa. The adherents to the doctrine of natural selection have tried to find in every structural and developmental peculiarity of a gall some teleological consideration for the cecidozoa. The empty atrial cavities surrounding the larval cavity in many galls, to which we have already drawn attention in earlier chapters, have, for example, been explained as an adaptation against the attack by parasites on the larva of the cecidozoa. It has, however, been repeatedly observed that the presence of atria and numerous other structural peculiarities of the gall does not have any selective value at all and does not indeed confer any immunity from attack by parasites and inquilines. The presence of these structures depends on the morphogenetic character of the gall and the organ bearing the gall. The sclerenchyma zone surrounding the larval cavity in midge and cynipid galls has similarly been described, though quite without justification, as "protective tissue" by all workers and it has been taken for granted that this zone serves to protect the larva of the cecidozoa within. The sclerenchyma zone can be explained entirely on mechanical and chemical conditions prevailing during cecidogenesis and the zone is at any rate not related to the advantage of the cecidozoa. The teleological significance of the sclerenchyma zone surrounding the larval cavity is for the plant. The dehiscence of galls is not for the benefit of the imprisoned cecidozoa, so as to set it free at the right moment in its life-history, but it is purely a physiological process of the plant tissue. The cecidozoa has on the other hand been in a sense forced to take advantage of the dehiscence of the gall and to readjust its life-history, so as to synchronize its escape with dehiscence. The physico-chemical changes preceding the dehiscence of the gall serve as decisive factors, which stimulate and regulate the development of the

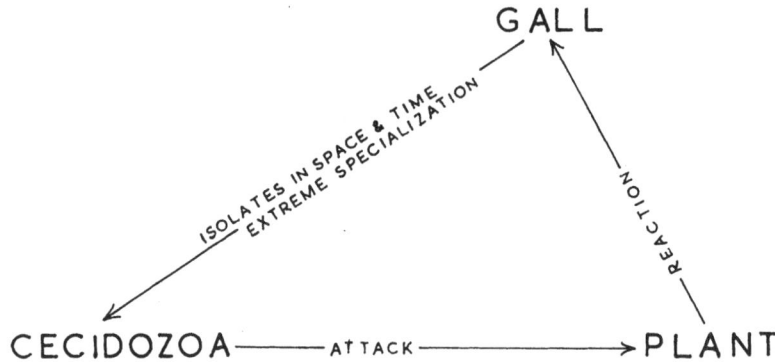

Fig. 135. Inter-relations of cecidozoa, gall and the plant.

cecidozoa. By producing the gall, in response to the attack by cecidozoa, the plant has literally compelled the cecidozoa to extreme specialization. The complex ecologic relations of cecidozoa, the gall and the gall-bearing plant may be diagrammatically illustrated as in fig. 135. The plant reacts to the attack of cecidozoa by producing a gall. By being compelled to breed in the gall, the cecidozoa is driven to extreme specialization. The gall can only develop on a certain specific part or organ and only at definite times. Against this background, it is obvious that the significance of the gall is the fact that the plant has successfully restricted the parasite to be a specialized parasite, localized in space and time. It is the cecidozoa which has in reality undergone more changes than the plant in the course of the evolution of the cecidozoa-plant complex. The extreme specialization has the effect of isolation of cecidozoa in space and time. Isolation in space embraces the extreme monophagy and inability to breed on normal plant tissue and restriction to the gall for nutrition. Emergence, escape from the gall, oviposition limited by the structure and development of gall, gall-bearing organs, etc. are part of the severe phenologic isolation. Other effects of being bound to the gall include the cyclic parthenogenesis, elimination of males, heterogony, obligatory and permanent parthenogenesis and elimination of males, geographic isolation, limits to dispersal, etc. The nutrition of cecidozoa is bound up with substances found only in meristem-like tissues, so that optimal conditions for development exist only in the gall tissue.

3. Inter-relations of cecidozoa and other cecidocole organisms

Predators and parasites

The number of secondary cecidocole organisms in galls is often enormous and the bulk of them are usually parasites. It is thus common experience in breeding experiments with galls that while the percentage of adults of the cecidozoa emerging from any random collection of galls is very small, the parasites are in great abundance. The predator-parasite complex of galls is often considerably larger than the inquilines. While many galls indeed lack inquilines, exceedingly few, if any indeed, are free from parasites. Out of the about 177 locatari recorded by STEGAGNO[1097] from cynipid galls on *Quercus* from Italy, over 139 are parasites and predators. The predators and parasites attack not only the larvae of the cecidozoa, out also those of the inquilines and other locatari and many of them are indeed hyperparasites. The predator-parasite complex of galls belongs to nearly every group like Acarina, Nematoda, Thysanoptera, Heteroptera, Coleoptera, Lepidoptera, Diptera and Hymenoptera.

In most acarocecidia we find the predatory midge *Arthrocnodax* and the parasite *Tetrastichus*. *Arthrocnodax coryligallarum* TARG.-Toz. is, for example, found in the gall of *Eriophyes avellanae* NAL. on *Corylus avellana*. *Tetrastischus eriophyes* is a common chalcid parasite in acarocecidia. *Bremia aphidivora* ROND., a zoophagous gall midge, is found in many aphid galls. The dominant parasites in aphid galls are Diptera, particularly Syrphidae. *Pipiza* spp. occur

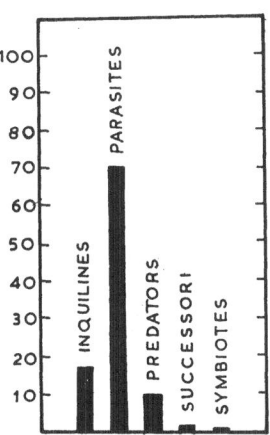

Fig. 136. Mean percentage frequencies of different categories of locatari in cynipid galls on *Quercus*.

in galls of *Schizoneura lanuginosa* (HTG.) on *Ulmus campestris* LINN. Other gallicole parasitic Diptera belong to the genus *Leucopis*. *Leucopis bursaria* ROND. is for example an important parasite in the galls of *Pemphigus bursarius* PASS. on *Populus nigra* LINN, *Byrsocrypta gallarum* (GMELIN) on *Ulmus, Baizongia pistaciae* (LINN.) on *Pistacia terebinthus* LINN., etc. The common predators and parasites in most aphid galls are chalcids, syrphids, Chaememyidae, Coccinellidae like *Coccinella, Scymnus*, mites, chelifers, etc. RONDANI has given an account of the predators and parasites found in aphid gall on *Pemphigus*. The caterpillars of *Cecidiptera exocoerariae* live in the gall of adelges on *Exocoeraria biglandulosa*. *Lestodiplosis liviae* RÜBS. is a common predator in the gall of *Livia juncorum* LATR. The predator-parasite complex in the gall of *Schizoneura lanuginosa* (HTG.) includes *Leucopis puncticornis* MAGN., *Pipiza aphidiphaga* COSTA, *Pipiza vitripennis* MAGN., *Anthocoris nemoralis* FABR., etc. *Baizongia pistaciae* (LINN.) is associated with the predator-parasite complex of *Leucopis bursaria* ROND., *Leucopis ballesterii* ROND., *Leucopis palumbii* ROND., *Pipiza vitripennis* MAGN., *Pipizella heringii* ZIFF., *Anthocoris pistacinus* ROND., *Pempelia gallicola* STGR. (Lepidoptera), *Pempelia palumbiella* ROND., *Stathmopoda guerini* STT., etc. A number of Microlepidoptera are known to be predatory forms in galls of aphids and mites. *Pempelia gallicola* STGR. is, for example, predaceous on *Baizongia pistaciae* (LINN.), *Tortrix* sp. is predaceous on *Eriophyes stefanii* NAL. in its gall on *Pistacia*. The dominant parasites in insect and mites galls are, however, Hymenoptera. Many different families including some interesting parasitic Cynipidae are among the parasites in this class. The cynipid *Allotria aphidicida* (ROND.) D. TORRE is for example a common parasite in the gall of *Aphis persicae*. A number of Ichneumonidae like *Pristomerus* CURT. and *Meniscus* SCH. and Braconidae like *Apanteles* FÖRST. and *Bracon*; chalcids like *Tetrastichus* HALIDAY, *Holcopelte* FÖRST., *Entedon* DALM., *Eupelmus* DALM., *Olinx* FÖRST., *Eulophus* GEOFF., *Beatomus* FÖRST., *Pteromalus* SWED., *Semiotellus* WESTW., *Copidosoma* RATZB., *Ormyrus* WESTW., *Megastigmus* DALM., *Torymus* DALM., *Decatoma* SPIN., *Eurytoma* ILL.-ROSSI; the family Platygasteridae, several Scelionidae, etc. are among the other important gallicole parasitic Hymenoptera. The parasitic Hymenoptera found in the gall of *Asphondylia conglomerata* STEF. include, for example, *Beatomus rufomaculata* (WALK.), *Copidosoma boucheanum* RATZB., *Eupelmus bedeguaris* RATZB., *Eurytoma contraria* WALK., *Heteroxys stenogaster* WALK., *Proctotrupes ater* (NEES), *Pteromalus puparum* LINN., *Rhopalotus cothurnatus* (NEES), *Sactogaster curvicauda* FÖRST., *Semiotellus tarsalis* (WALK.), *Synopeas prospectus* FÖRST., *Torymus auratus* MAYR., etc. As may be expected, some of these species are primary parasites of the gall midge, but others are hyperparasites. The common parasites in the gall of *Andricus panteli* KIEFF. on *Quercus* are *Chrysoideus chrysidiformis* DE ST., *Decatoma biguttata* (SWED.) CURT., *Decatoma strigifrons* THOMS., *Eupel-*

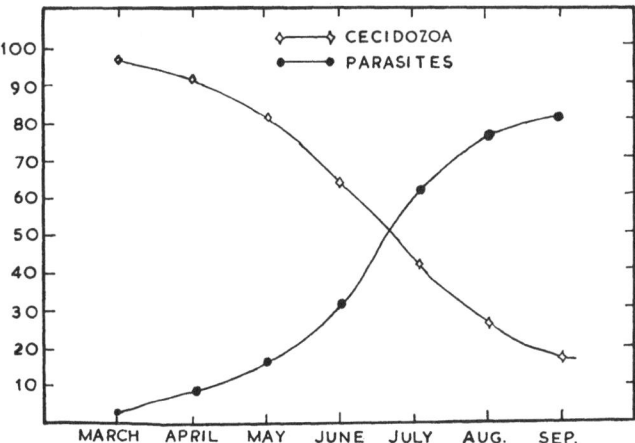

Fig. 137. Mean fluctuations in the relative percentage abundance of cecidozoa and parasites-complex in cynipid galls on *Quercus* with the advance of the vegetative season.

mus kiefferi DE ST., *Eurytoma aterrima* (SCHRK.) LATR., *Megastigmus dorsalis* (FABR.), *Megastigmus stigmationus* (FABR.), *Olinx scianeurus* (RATZB.), *Ormyrus sericeus* NEES and *Pteromalus bimaculatus* NEES. In the gall of *Biorrhiza pallida* (OL.) the parasites are *Bracon immutator*, *Callimome costata* ROND., *Decatoma biguttata obscura* WALK., and *Torymus auratus* MAYR. Some of these are parasites on the psenid cynipid and others on the inquiline cynipid and still others are in all probability hyperparasites. Some Nematodes have also been observed to be parasitic in zoocecidia. MORGAN[822] has for example described the association of a nematode and the Diptera *Fergusonia* in the gall of the latter. The larva of the gall midge *Asynapta citrina* KIEFF. is said to be parasitic in the anguillulid gall on *Carpinus*. The abundance of parasites in a gall usually increases with the advance of the vegetative season, so that the galls appearing in early spring have relatively few parasites, but the galls appearing in late summer and autumn are remarkable for the high percentage of parasitism (fig. 137).

Symbiotes

A number of reports of possible symbiotic relations between cecidogenetic aphids, Itonididae and other organisms like bacteria, ants, Lycaenidae, etc. exist in literature[7, 21, 52, 655, 720]. The association of symbiotic organisms with cecidozoa and zoocecidia has been known for a long time and it has also been suggested that such symbiotic organisms are ultimately involved in the etiology of a number of insect galls[38, 720, 918]. A bacterium has been constantly found to be associated with the aphid *Hamamelistes spinosus* SCHIMER in the bud gall on hazel from North America and symbiotic relation has been attributed to the bacterium. It has been suggested that the symbiotic bacteria and probably also certain virus give rise to the gall now attributed to insects and the insects which are vectors of the bacteria and virus feed on the gall tissue.

An outstanding case of symbiotic relation in zoocecidia is the so-called "ambrosia gall". In the ambrosia gall we find an association of certain fungi and Itonididae. NEGER[853–856] suspects such symbiotic relation between fungi and gall-midges. DOCTERS VAN LEEUWEN[282] has given an account of the gall of *Asphondylia bursaria* FELT on *Symplocos fasciculata* ZOLL. with a symbiotic fungus. The gall matures in December, the midge larva is completely surrounded by the fungal hyphae, black on the surface and white deeper below. The deeper white portion of the fungal mycelia is consumed by the midge larva before maturity. The eggs of the midge are naturally contaminated with the spores of the fungus, which begins to grow rapidly as soon as the gall tissue becomes differentiated. GOIDANICH[400] has also ascribed a symbiotic relation between the midge *Ischnonyx pruniperda* (ROND.) *(Asphondylia prunorum* WACHTL.) and a deuteromycete fungus *Sphaeropsis* sp. in the bud gall on *Prunus* and records that the midge larva feeds on the fungus mycelia. The best known of these "ambrosia galls" is the gall of *Asphondylia sarothamni* H. LÖW on the buds of *Sarothamnus scoparius* WIMM. This is an acutely pointed, oval gall, about 12 mm long, with a short, curved beak-like process at the tip. Inside the spacious gall cavity there is a single larva of the gall midge. This larva lies in the lower portion of the gall cavity in the early stages of development of the gall. It derives its nourishment from the vascular elements supplying the bud. In the upper portion of the gall cavity a coating of fine mycelial threads is visible even to the naked eye. The mycelia are at first confined to the lowest part of the larval chamber and are also white coloured, but as the development and growth of the gall progress, the mycelia cover the whole of the inner surface and also turn grey. Eventually the entire surface of the gall cavity is covered completely by a layer of multicellular palisade-like hyphae, the tips of which become constricted into spherical or clavate conidial bodies. The fungal layer lies closely applied to the surface of the gall tissue but does not at this stage send haustoria or hyphae into the intercellular space of the gall tissue. The fungus is at this stage strictly saprophytic [980, 981]. It must be observed that the gall midge larva is thus not in direct contact with the surface of the gall tissue, but is separated by the mycelial layer of the fungus. However, it seems to derive its nourishment from the plant tissue through the mycelial layer and also to provide the necessary stimulus for cell proliferation, leading to the continued growth of the gall tissue. According to Ross[980], the midge larva develops normally even if the mycelia of the fungus occasionally fail to appear inside the gall cavity or if the layer is only poorly developed. This would seem to show that the presence of the fungus is not absolutely necessary either for the development of the gall or even for the normal development of the midge larva. The fungus does not also behave as an enemy of the midge larva and the pupation of the midge takes place normally among the dense and luxurient fungal mycelial threads. The fungus does not seem to gain an upper hand in the gall as long as the midge larva is active and feeding. The larva presumably secretes something that inhibits the dominance of the fungus. When, however, the larva ceases to feed before pupation, the situation changes rapidly and the fungus soon attains a most luxurient growth and nearly fills the entire gall cavity. The pupa is still not in any way adversely affected by the fungal mycelia. The adult midges emerge and escape from the gall by the end of May or early June. With the gradual shrinking of the gall tissue, the fungal mycelia penetrate into the intercellular spaces of the gall tissue and soon become even intracellular. From now onwards the fungus becomes parasitic on the gall tissue in the ordinary sense. It is not yet clearly understood whether the fungus occurs in the gall of the same midge in the flowers of *Sarothamnus*. The midge gives rise to a gall on the fruit of *Sarothamnus* and the inner surface of the fruit gall is also clothed with a layer of the fungal mycelia, exactly as in the bud gall. The behaviour of the fungus in the fruit gall is also similar to that in the bud gall. In course of time the fungus produces pycnidia, which discharge the conidia to the outside[853-856]. Ross[980] has demonstrated the presence of fungi in the galls of *Asphondylia* spp. on the bud in one generation and fruits in another generation, on many other Leguminosae like *Calycotome, Coronilla, Cytisus, Dorycnium, Genista, Ononis*, etc. He has also given an account of another *Asphondylia* gall on the flower of *Echium vulgare* LINN. with mycelial coating on the inner surface. The gall of *Contarinia echii* KIEFF. on the flower of the same plant does not have a mycelial cover. Fungal mycelial cover has been found inside the galls of *Lasioptera carophila* F. Löw on the inflorescence axis of several Umbelliferae. NEGER[853], who was the first to study the occurrence of the association between gall-midge and fungi in the gall of the former, attributed a symbiotic relationship between the fungus and the gall-midge larva and, on the analogy of the Bostrychid-fungus association, gave the name "ambrosia gall" to this type of gall. He believed that the fungus is adapted to the gall midge and the transmission of the fungus to the new gall occurs through the egg of the gall-midge. Ross[981] failed, however, to demonstrate such a symbiotic relation. He does not therefore accept the name ambrosia gall, but prefers the use of "verpilzte Mückengallen". According to him, the fungus is a true plant inquiline, comparable to other insect inquilines so commonly found in cynipid and other galls. The fungus is not in point of fact adapted to the specific gall-midge, but only to the plant species on which the gall appears. The spores of the fungus may doubtless be passively carried by the gall-midge larva, when it bores into the plant tissue soon after hatching from the eggs deposited on the surface of the plant. The fact that occasionally normal galls develop and the midge larva also grows normally and pupates in total absence of fungal mycelia in the gall is proof against the idea of a symbiotic relationship. Ross does not also agree with the view that the gall-midge actively serves as the transmitting agent of the spores of the fungus. Neger made pure cultures of the fungus from the fruit gall on *Coronilla emerus* Linn. and identified the fungus as belonging to a new species of *Macrophoma*, but Ross believes the differences from *Phoma* to be too trivial to warrant erection of a new genus. Although it has not been possible so far to make pure cultures of the fungi on any nutritive media and get the spores, the general appearance of the whole fungus growth seems to indicate that some species of *Cladosporium* are probably thus associated with midge galls. The fact that there is more than one species is again an indication that we are not dealing with a case of true symbiosis. Ross has given a list of midge galls, in which he found a similar fungal association with gall-midge. The majority of these galls are caused by different species of *Asphondylia*, but many are also caused by *Lasioptera* spp. The list includes the galls of *Asphondylia adenocarpi* Tav. on *Adenocarpus intermedius* DC, *Asphondylia hieronymi* F. Löw on *Baccharis salicifolia* Pers., *Asphondylia sarothamni* H. Löw on *Calycotome spinosa* Lamarck, flower gall of *Asphondylia capparidis* Rübs. on *Capparis spinosa* Linn., shoot gall of *Asphondylia coronillae* Vall. on *Coronilla* spp., fruit gall of *Asphondylia jaapi* Rübs. on *Coronilla* spp., bud gall of *Asphondylia cytisi* Frauenh. on *Cytus* spp., flower gall of *Schizomyia galiorum* Kieff. on *Galium* spp., shoot gall of *Lasioptera carophila* F. Löw on *Petroselinum sativum* Hoffm., etc. Schmidt[1028] has discussed the association of fungi in the galls of *Blastophaga* on *Ficus*, *Melampsora* in the gall of *Rhabdophaga heterobia* H. Löw, *Cystopus candidus* in aphid gall on Cruciferae, etc. In a recent contribution, Meyer[786] has shown that in the so-called ambrosia galls of *Lasioptera* spp., especially *Lasioptera rubi*, there is no separate nutritive cell zone, since the midge larvae derive their nourishment from the fungal mycelia.

4. Inter-relations of zoocecidia and cecidocole organisms

Inquilines

Inquilines are cecidocole mites and insects, which are incapable of giving rise to galls themselves but their lives are closely bound up with galls of other species. The transition from a true cecidogenetic species to an inquiline is often not sharply defined and it is known that inquilines frequently introduce far-reaching structural and developmental modifications in the galls they invade. The presence of inquilines is known to prevent the normal dehiscence of a gall. Also galls invaded by inquilines do not mature and decay in the usual manner. Furthermore the transition from an inquiline and a psenid is not really very sharply defined. It has been repeatedly observed, for example, that the same species may sometimes be a true cecidogenetic agent in one gall, but an inquiline in the gall of another species[83, 867]. A number of such cases are known among *Eriophyes* and cecidocole Thysanoptera. Although inquilines occur in nearly all galls, those found in the galls of Acarina, Thysanoptera, Hymenoptera and Diptera are ecologically of great importance.

The inquilines in acarocecidia include both mites and insects. The inquiline mites include *Phyllocoptes gallatus* NAL. and *Phyllocoptes heteroproctus* NAL. found in the gall of *Eriophyes brevipunctatus* NAL., *Phyllocoptes obtusus* NAL. in the gall of *Eriophyes salviae* NAL., etc. Several species of inquiline mites have been recorded in the galls of *Eriophyes avellanae* NAL. on *Corylus*, *Eriophyes macrorrhynchus* NAL. on *Acer pseudoplatanus* LINN., *Eriophyes thomasi* NAL. on *Thymus* and *Eriophyes macrostichus* NAL. on *Carpinus*. *Eriophyes vermiformis* NAL. gives rise to a characteristic leaf crinkle gall on *Corylus avellana* and may also be found as an inquiline in the gall of *Eriophyes avellanae* NAL. on the same plant. *Eriophyes tetanothorax* NAL. gives rise to a pouch gall on the leaf of *Salix* sp. and often also occurs as an inquiline in witches-broom gall on the same plant and in galls on other plants. The habit of being a cecidogenetic and an inquiline species seems to be rather widespread among mites. In addition to *Eriophyes* spp., Gamasid mites have also been found as inquilines in some acarocecidia. According to STEGAGNO[1097], the mites *Tyroglyphus minutus* TARG.-TOZ., *Glycyphagus domesticus* DE GEER, *Gamasus vepallidus* KOCH and *Galigonus virescens* TARG.-TOZ. are habitually met with as inquilines in the galls of *Eriophyes avellanae* NAL. on the leaf of *Corylus avellana*. The possibility of some of these mites being true predators must not, however, be overlooked. The leaf margin roll gall of *Eriophyes stenaspis plicator* NAL. often contains aphid inquilines. Some inquiline Thysanoptera have also been found in the same gall.

Inquilines in midge galls belong also to a wide group. The midge

236

Perrisia iteophila H. Löw is an inquiline in the galls of *Rhabdophaga rosaria* (LINN.) and *Rhabdophaga strobilina* on *Salix. Clinodiplosis botulariae* (WINN.) KIEFF. is possibly an inquiline in the gall of *Perrisia fraxini* KIEFF. on *Fraxinus excelsior* LINN. Other inquiline gall-midges include *Clinodiplosis liebeli* KIEFF. in the galls of *Macrodiplosis dryobia* F. Löw and *Macrodiplosis volvens* KIEFF. on *Quercus sessiliflora, Quercus pedunculata* and *Quercus pubescens.*

The number and complexity of inquilines found in cynipid galls is truly astounding. These inquilines belong not only to cynipids, but also to nearly every other group of cecidozoa. *Clinodiplosis galliperda* (F. Löw) is, for example, gall-midge inquiline in the galls of *Neuroterus quercus-baccarum* LINN. on *Quercus. Clinodiplosis biorrhizae* KIEFF. is an inquiline in the gall of *Biorrhiza pallida* (OL.) and *Clinodiplosis gallica* KIEFF. and *Arnoldia gemmae* (RÜBS.) are similarly met with in the gall of *Andricus foecundatrix* (HTG.) (bisexual generation). *Clinodiplosis galliperda* (F. Löw) occurs in the gall of the unisexual generation of *Neuroterus tricolor* HTG. and *Neuroterus albipes* (A. SCHENK). The gall of *Biorrhiza pallida* (OL.) contains as many as twenty different species of inquilines and the gall of *Diplolepis rosae* (LINN.) on *Rosa* contains twenty-five species of inquilines The bulk of the inquilines in cynipid galls are cynipids themselves. The inquiline cynipids belong to the genera *Synergus* HTG., *Saphonecrus* DALLA TORRE & KIEFF., *Ceroptres* HTG., *Sapholitus* FÖRST. and *Periclistus* FÖRST. Over sixty species of *Synergus* HTG. are known from the world. The same species of *Synergus* HTG. is inquiline in the galls of several species of cynipids. *Synergus umbraculus umbraculus* (OL.) is for example an inquiline in the galls of about twenty different species like *Andricus clementinae* (GIRAUD), *Andricus foecundatrix* (HTG.) bud gall of unisexual generation, *Andricus lucidus lucidus* (HTG.) unisexual generation, *Andricus mayri* (WACHTL.) catkin gall of unisexual generation, *Trigonaspis mendesi* TAV. unisexual generation gall on leaf, *Cynips amblycera* GIRAUD unisexual generation bud gall, *Cynips quercus-calicis* BURGSDF. unisexual generation on acorn, *Cynips caliciformis* GIRAUD unisexual generation bud gall, *Cynips caput-medusae* HTG. unisexual generation bedeguar-like gall, *Cynips conglomerata* GIRAUD unisexual generation, *Cynips coriaria coriaria* HTG. unisexual generaation bud gall, *Cynips coronata* GIRAUD unisexual generation bud gall, *Cynips glutinosa glutinosa* GIRAUD unisexual generation bud gall, *Cynips kollari kollari* HTG. unisexual generation bud gall, *Cynips lignicola* HTG. unisexual generation bud gall, *Cynips mayri* KIEFF. unisexual generation, *Cynips polycera polycera* GIRAUD unisexual generation axillary bud gall, *Cynips gallae-tinctoriae* (OL.) unisexual generation bud gall and *Cynips quercus-tozae* BOSC. unisexual generation bud gall. The adults emerge in April-May, rarely in June-July of the second year. *Synergus reinhardi* MAYR is inquiline in the galls of *Cynips caliciformis* GIRAUD, *Cynips quercus-calicis* BURGSDF., *Cynips caput-medusae* HTG., *Cynips glutinosa glutinosa* GIRAUD, *Cynips kollari kollari* HTG., *Cynips mitrata* MAYR bud gall of unisexual generation, *Cynips gallae-tinctoriae* (OL.), *Cynips quercus-tozae* BOSC, etc. The adults emerge in May-June of the second year. *Synergus gallae-pomiformis gallae-pomiformis* (FONSC.) occurs as inquiline in the galls of *Andricus albopunctatus* (SCHLECHT.) bud gall of unisexual generation, *Andricus quercus-radicis quercus-radicis* (FABR.) branch gall of the bisexual generation, *Andricus curvator curvator* HTG. galls of both unisexual and bisexual generations, *Andricus glandulae* (HTG.) unisexual generation, *Andricus foecundatrix* (HTG.) unisexual generation, *Andricus nudus* AL. catkin gall unisexual generation, *Andricus quercus-ramuli quercus-ramuli* (LINN.) catkin gall of unisexual generation and two other species of *Andricus*, *Diplolepis megaptera* (PANZ.) bisexual generation, *Diplolepis quercus-folii* (LINN.) bisexual generation, *Biorrhiza pallida* (OL.) bisexual generation, *Dryocosmus gallae-ramulorum* (FONSC.), *Cynips glutinosa glutinosa* GIRAUD, *Cynips coriaria coriaria* HTG., *Neuroterus quercus-baccarum quercus-baccarum* (LINN.) bisexual generation, *Neuroterus tricolor* (HTG.) unisexual generation. *Synergus vulgaris* HTG. is an inquiline in the galls of no less than twenty-two different species of cynipids belonging to *Andricus* HTG., *Callirhytis* FÖRST., *Cynips* LINN. *Neuroterus* HTG. and *Diplolepis* GEOFF.

MAYR[765] devoted considerable attention to the inquilines occurring in cynipid galls on *Quercus* and recognized the following groups on the basis of their biological peculiarities:
1. Inquilines that live in the larval cavity of the cecidozoa and subdivide the larval cavity by a slimy membrane into as many smaller cells as there are individuals of the inquiline and the cecidozoa is killed. Examples: *Synergus incrassatus* HTG. in the galls of *Andricus quercus-radicis* (FABR.) and *Synergus melanopus* HTG. in the gall of *Cynips caput-medusae* HTG. 2. Inquilines which partly destroy the larval cavity of the cecidozoa and also part of the surrounding gall tissue; the larva of the inquiline is enclosed in a membranous wall. Examples: *Synergus melanopus* HTG. and *Synergus vulgaris* HTG. in the galls of *Cynips ligniperda* and *Cynips gallae-tinctoriae* (OL.) and *Synergus reinhardi* MAYR in the gall of *Cynips caliciformis* GIRAUD. 3. Inquilines which occupy the larval cavity of the cecidozoa after enlarging it more or less and which do not in any way interfere with the otherwise normal development of the larva of the cecidozoa. Example: *Synergus melanopus* HTG. in the gall of *Cynips polycera* GIRAUD. In the gall of *Cynips cerricola* there is an outer atrium containing numerous inquilines, separated by a membrane from the larval cavity proper. MAYR found in this gall 19 individuals of the inquilines *Synergus thaumacera* DALM., two of *Synergus variabilis*

MAYR and three of *Eurytoma* sp., which last mentioned are probably parasites. 4. Inquilines with their cavities situated within the parenchyma of the gall and the larva of cecidozoa may be normally alive or also dead prematurely. Examples: *Synergus reinhardi* MAYR in the gall of *Cynips kollari* HTG. and *Sapholytus undulatus* and *Synergus variabilis* MAYR in the gall of *Cynips cerricola*.

ADLER[2] believes that the habits of *Andricus curvator* HTG. seem to be highly suggestive of the possible transition from a true cecidogenetic species to an inquiline. This species is known to show decided preference to having its eggs deposited on buds, in which another species viz. *Andricus pilosus* ADL. has already oviposited. In consequence of this peculiar habit, we often find small galls of *Andricus curvator* HTG. developing on those of *Andricus pilosus* ADL. ADLER believes that a further step in the direction has in probability led to the origin of inquilines. It would thus appear that inquilines were primarily cecidogenetic species, which have secondarily become specialized to life in the gall of some other cecidogenetic species. Some inquilines seem to have also become still further specialized to parasitism on the primary cecidogenetic species. This view is supported by the observation that many species of *Eriophyes* and cecidocole Thysanoptera are habitually both cecidogenetic and inquilines.

Cecidophags

The cecidophagous species are an ecologically highly specialized group that feed either preferentially or obligatorily on galls of various other species. They differ fundamentally from inquilines in the fact that the cecidophagous habit is primary and not an acquired habit in a cecidogenetic ancestor. Starting from species that cut open a gall or burrow into it in order merely to reach the cecidozoa or the inquiline, we have an astonishing array of transitional forms leading eventually to species which are exclusively dependent on galls for their food. Not only insects but a great variety of other organisms including even birds feed on the rich nutritive and succulent gall tissues. According to BEIJERINCK[75] the galls of *Diplolepis (Rhodites)* on *Rosa*, *Cynips quercus-calicis* BURGSDF. and *Neuroterus quercus-baccarum* (LINN.) on *Quercus* are regularly eaten by finks, domestic hen and various other birds. The gall of *Biorrhiza pallida* (OL.) is tunnelled by the ant *Camponotus ligniperda*. In a number of cases the cecidophagous insects bore into the tissues of galls, eating the entire flesh and leaving only the outer skin in the form of an empty bag, inside which they even pupate. The larva of an unidentified moth behaves in this manner in the gall of *Odinadiplosis odinae* MANI (Plate V, 3). The larva of another moth similarly eats all the spongy mass of the gall of *Asphondylia riveae* MANI on the leaf of *Rivea hypocrateriformis* CHOISY (Plate VII, 2). The larvae of the Microlepidoptera *Alophia combustella* H.S. and *Stathmopoda guerini* STAUD. feed on the tissues of the galls of the aphids *Geoica utricularia* (PASS.) and *Forda* on *Pistacia*. MANSON[746] has also recorded observing the caterpillars of an undescribed Noctuid as feeding on the tissues of aphid galls. PING[911] similarly found the beetle *Mordelistina unicolor* LEC. breeding in the gall of *Eurosta solidaginis* FITCH on *Solidago canadensis*. In the same gall are also found the cecidophag larvae of *Gelechia* and Tortricid moth. Many fruitflies regularly breed in the galls of Itonididae, especially the fleshy galls on stems of herbaceous plants. Other well known cecidophags include *Phthoroblastis amygdalana* DUP., *Phthoroblastis costipunctata* HW., *Phthoroblastis motacillana* HÜB., *Phthoroblastis argyrana* HÜB., *Ephestia interpunctella* HÜB., *Ephestia ficulella* BARR., *Lithosia complana* LINN., *Steganoptycha corticana* HÜB., *Penthia profundana* SV. and *Sesia melliniformis* LASP. and *Balanius villosus* in the galls of *Biorrhiza pallida* (OL.).

The gall tissues seem also to be an excellent medium for the development of a number of saprophytic and parasitic fungi. Either perhaps the localized abundance of highly nutritive material or also the succulence of the gall tissues or some other as yet unsuspected metabolic condition of the gall tissue seems to particularly favour such growth. The fact that fungi, which normally grow only rather sparsely on the healthy part of the same plant, are capable of growing abnormally luxuriantly in the gall tissue is significant. Diverse galls on *Salix* spp. are astonishingy heavily infected with *Melampsora*, even it this fungus does not occur on other normal parts of the plant in the locality. This is, for example, the case with the kammergalls of *Pontania proxima* LEPEL and *Pontania vesicator* BREMI and the gall of *Rhabdophaga heterobia* H. LÖW on *Salix* spp. The rhodites gall on *Rosa* is generally attacked by the fungus *Phragmidium subcorticum* far more heavily than the healthy parts of the plant. A number of fungi are remarkable for being found exclusively in the galls of cynipids on *Quercus*. *Exoascus cecidomophilus* ATKINSON occurs, for example, in the fruit gall on *Prunus virginiana*. *Sphaerotheca phytoptophila* KELL. Ew. is known to develop only in the gall of *Eriophyes* on *Celtis occidentalis*. A number of saprophytic fungi arise on the surface of galls and some of them often also invade the larval cavity, atrial cavity, etc. Abundant growth of *Pennicilium glaucum*, *Cladosporium herbarum* and other fungi is found in the gall of *Pemphigus spirothecae* PASS. Even mycocecidia are attacked by other non-gall-forming fungi. The Protobasidiomycete fungus *Urobasidium rostratum* occurs exclusively in the gall of the fungus *Taphrina cornu cervi* on *Aspidium rostratum*. *Peronospora parasitica* seems also to prefer the gall tissues of *Albugo candida* on *Sinapis arvensis*, *Erysium cheiranthoides*, *Raphanus raphanistrum* and *Capsella bursa-pastoris*.

Successori

The successori represent the inhabitants of the empty gall after the escape of the cecidozoa and inquilines. In a number of cases the gall does not decay after the escape of the cecidozoa, but the dry gall provides shelter for a variety of Arthropods and breeding ground for fungi[911], [1230]. The successori constitute the Nachfolger of the German literature.

The successori embrace diverse plant-nesting ants, myrmecophilous insects like aphids, coccids, etc., parasites of myrmecophilous insects, spiders and their parasites, free living and parasitic mites, chelifers, beetles, thrips, hibernating caterpillars, bugs, plant-nesting wasps (solitary) and bees, Corrodentia, etc. Some of these inhabitants of old galls enter the gall through the exit hole of the cecidozoa, but others bite their way in. A few species slightly enlarge the exit hole and also close the hole with webbing, mud, slime, gum and other suitable material. The successori do not therefore in any way interfere with the normal development of the cecidozoa, the gall or of the inquiline, but only move into the gall after the cecidozoa and inquilines have departed. They find in the gall more or less suitable ready-made nests, hiding places, lures for trapping prey, etc. Many species utilize the empty galls as brood nests and have indeed become wholly specialized for exclusive gall-nesting. Some species of the solitary bees of the genus *Osmia* are, for example, habitual gall-nesters. Other gall-nesting Hymenoptera include the solitary wasps like *Pison*, *Trypoxylon* and solitary bee *Halictus*. *Halictus provancheri* occurs, for example, in the gall of *Eurosta solidaginis* FITCH on the stem of *Solidago canadensis*; in the same gall is also found *Trypoxylon politum* SAY. *Trypoxylon figulus* nests in the gall of *Cynips quercus-tozae* BOSC. on *Quercus robur*.

Ants are perhaps the most important and dominant inhabitants of old galls. Nearly every old hollow gall shelters one or more species of ants[895], [1250]. The more common gall-nesting ants are *Cataulacus intrudens* SM., *Cremastogaster solinopsides flavida* MAYR, *Cremastogaster chiarini* EMERY, *Cremastogaster admota* MAYR, *Cremastogaster gallicola* SJÖST., *Cremastogaster sjöstedti* MAYR, *Cremastogaster mimosae* SANT., *Leptothorax fortinodis*, *Leptothorax obdurator*, *Olopsis etiolata*, *Cremastogaster clara*, *Camponotus decipiens*, *Camponotus rasilis*, *Leptothorax curvispina*, *Lasius umbratus* NYL., etc. Some of these ants occur inside the hollow stipular thorns of *Acacia* spp. and others in various other hollow stem galls.

Honeydew galls

The galls of a number of Eriophyids, Homoptera, Hymenoptera and some Diptera have gummy or sugary exudations on their surfaces, attracting large numbers of flies, ants and other insects.

The ant *Myrmecocystus melliger* is attracted, for example by the honeydew-like exudations on the gall of *Cynips quercus-mellariae* on *Quercus undulata*. The ribbed bud gall of *Callirhytis gemmaria* ASHM. on *Quercus* from North America secretes honeydew from a gland situated apically. The wool sower galls of *Callirhytis seminator* HARR. on *Quercus alba, Quercus montana* and *Quercus prinus* from North America have very sticky exudations on their surface when young. Other galls known to produce such exudations include those of *Callirhytis carmelensis* WELD, *Callirhytis balanaspis* WELD, *Callirhytis balanopsis* WELD, *Callirhytis balanosa* WELD, *Callirhytis balanoides* WELD, *Callirhytis congregata* ASHM., etc. The bud galls of *Andricus atractans* KINS., *Disholcaspis monticola, Disholcaspis chrysolepidis* BEUTENM. and *Neuroterus vernus* GILL. also secrete honeydew. The gall of *Disholcaspis eldoradensis* BEUTENM. on *Quercus lobata* from North America secretes copious quantities of honeydew, which attracts honeybees. The bees often gather as much as 36 kilograms of the sweet exudation from this gall and store it in their hives.

5. Ecologic succession in zoocecidia

From the foregoing account of the cecidocole organisms, it is apparent that there is a characteristic ecologic succession of species. The climax of this succession is always marked by the dominance of parasites and hyperparasites, both in the communities before the escape of the cecidozoa and in the communities of the successori. The dominance of the individuals of the cecidozoa marks the earliest phase in this ecologic succession. The appearance and development of each member species of the gall communities serve to create the ecologic optima for the next species complex in the succession.

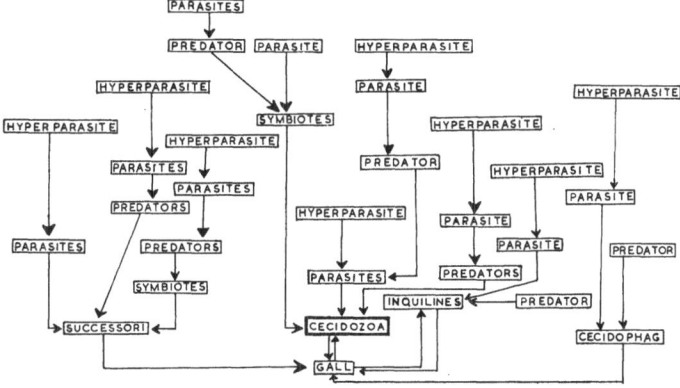

Fig. 138. The complex composition and ecologic succession in cecidocole community.

When a gall appears first in a locality, there is total predominance of cecidozoa and the inquilines and parasites are practically absent. After the passage of some time, there is a conspicuous diminishing in the abundance of the cecidozoa and a corresponding increase in the abundance of inquilines and parasites. Still later, there is a much larger proportion of foreign organisms than the cecidozoa. The general pattern of ecologic succession in galls is diagrammatically represented in fig. 138. The cecidozoa prepare and provide for the optimal conditions for inquilines, which in turn ensure the optima for their parasites. The activities of these groups serve to create the necessary conditions for the appearance of cecidophags. The activities of the communities of the gall during its active phase of life-history tend to create the conditions that favour the life of the successori. All these events are in turn governed and controlled by the general developmental growth peculiarities of the gall, such as its dehiscence, presence of atria, etc. The factors that control the ecologic succession and the community regulating mechanisms are thus found in the gall itself and are indeed extremely complex.

CHAPTER XII

MYCOCECIDIA

It will be wholly outside the scope of the present volume to deal with even the more important problems of mycocecidia. We shall therefore confine ourselves to some of the outstanding features, in so far as they are directly related to the discussion on the ecology of galls.

Mycocecidia are perhaps only second in importance to zoocecidia. They are characterized by their abundance, wide distribution, the great diversity of plants on which they occur and the complexity of their structure. Mycocecidia arise on every group of plants, ranging from Algae to Monocotyledonae, but are perhaps most frequent on Dicotyledonae. Although mycocecidia develop practically on all parts, leaves would appear to be the preferred seat. Nearly 60% of the known mycocecidia on Angiosperms from the world are, for example, leaf galls, about 20% arise on flowers, especially the ovary and the rest develop on various other parts of the plant (fig. 139). Mycocecidia are remarkable for their truly enormous size,

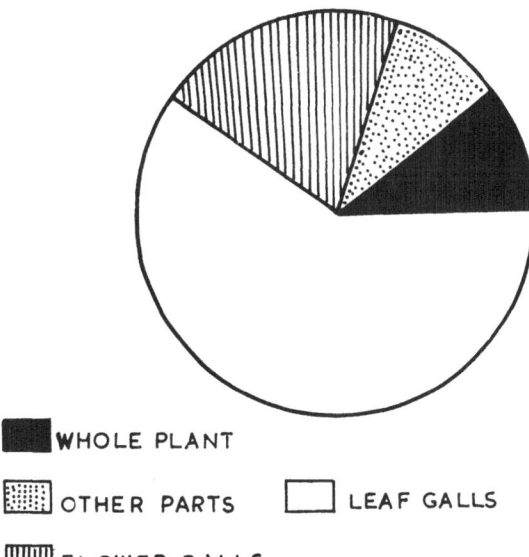

WHOLE PLANT

OTHER PARTS   LEAF GALLS

FLOWER GALLS

Fig. 139. Relative abundance of mycocecidia on different parts of higher vascular plants from the world.

ranging from simple enlarged cells to swollen masses of 50 cm in diameter. The largest plant galls are also mycocecidia. The common cecidogenetic fungi are found among *Plasmodiophora, Sorosphaera, Olpidium, Urophlyctis, Synchytrium, Taphrina, Gymnosporangium, Puccinia, Uromycladium* and *Exobasidium*. With very few exceptions, the cecidogenetic fungi are intercellular parasites in the gall tissue; only some species of *Synchytrium* are strictly intracellular parasites. The histological and anatomical structure ranges from simple cellular hypertrophy of single cells to complex hyperplasia, inhibition of tissue differentiation, cytolysis, etc. Most mycocecidia are remarkable for their large simple parenchyma cells and some of them consist wholly of such cells. A number of mycocecidia have relatively unlimited growth: while in most zoocecidia the cell proliferation is intense and localized in the middle of the gall, around the larval cavity, in mycocecidia the most intense cell proliferation is peripheral, immediately below the surface of the gall. Unlike most zoocecidia, the same species of fungus often gives rise to nearly the same type of gall on more than one organ of the plant or may gall the whole of the aerial portion of a plant in toto. While the majority of them are pale coloured, some mycocecidia are coloured remarkably vivid yellow, orange, red or violet, especially when exposed to the direct sun rays. Chlorophyll is absent in most mycocecidia, but BARTLETT[57] has reported that the nodular galls on the roots of Cruciferae caused by *Olpidium radicicolum* DE WILDEMAN turn green when exposed to the sun above ground and to give rise to leafy branches. Mycocecidia are generally smooth when immature, but mostly dusted with a powdery bloom when mature. The best known of the powdered surface galls arise on Neotropical Lauraceae like *Nectandra, Ocotea, Cryptocarya*, etc., caused by Basidiomycetes[86].

The literature on mycocecidia and the causative fungi, their taxonomy, structure and cytology is very large. A useful review of fungal galls, especially those caused by Uredinae may be found in FISCHER[334]. TOBLER[1159] has published a monographic account of *Synchytrium* and its gall. The life-history and cytology of the same genus are discussed by CURTIS[234] and RYTZ[1002, 1003]. The histological anatomy of a number of mycocecidia is described by TROTTER [1175]. Other important papers dealing with mycocecidia are listed in the bibliography[3–5, 39, 89, 158, 178, 202, 324, 376, 482, 633, 658, 659, 897, 898, 1017, 1079, 1106, 1213, 1275, 1276, 1277].

1. Simple mycocecidia

Some of the simplest mycocecidia are only more or less pronounced enlarged cells attacked by the fungus. Such hypertrophy of single cells is caused by *Synchytrium*, usually on leaves but sometimes also on the stem. The fungus is intracellular and is most often confined to a single epidermal cell, which becomes enormously swollen (fig. 140 A). The cells in the immediate neighbourhood, but

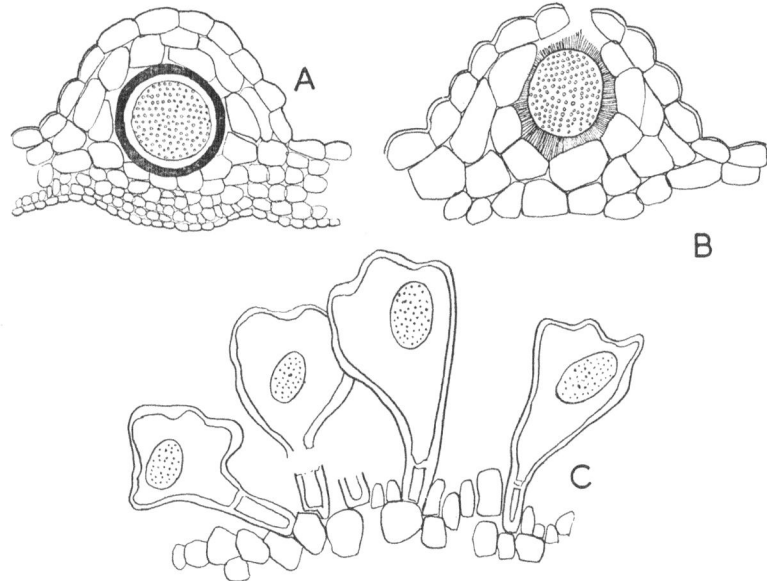

Fig. 140. Galls of *Synchytrium*. A–B. Stages in the development of the gall and spore formation. C. Epidermal cells gall of *Synchytrium papillatum* Farl. on leaf of *Erodium cicutarium* (After Magnus).

particularly those that are in close contact with the attacked cell, also react soon by becoming hypertrophied and thus come to surround the originally attacked cell like a sheath. This results in the formation of a conspicuous pustule-like bulge on the leaf surface (fig. 140 B). The gall therefore consists of a localized malformation of the leaf blade, confined wholly to one side. *Synchytrium mercurialis* LIEB. is responsible for such a simple gall on the leaf epidermis of *Mercurialis perennis* LINN. Other common examples include the galls of *Synchytrium succisae* DE BARY & WOR. on *Succisa pratensis* MOEN and *Synchytrium taraxaci* DE BARY & WOR. on *Taraxacum* spp. The gall of *Synchytrium papillatum* FARL. on the leaf of *Erodium cicutarium* (fig. 140 C) consists of clavately swollen and abnormally elongated outgrowths from the epidermal cells.

2. Mycocecidia on roots

Mycocecidia on roots seem to be curiously confined to certain groups of plants, but at the same time widely distributed. Compared to the zoocecidia on roots, the mycocecidia involve a rather very extensive portion of the root system and usually comprise solid irregular swellings of the underground portion of the shoot and crown also.

We have already given several examples of mycocecidia on roots. Nodular galls, greatly resembling the bacterial root nodules of Leguminosae, are caused by fungi on the roots of a number of plants. TROTTER'S[1175] account of the fungal root nodule galls on *Alnus cordata* may be taken as typical of other nodular mycocecidia on roots. Reference may also be made to the account of the root nodule mycocecidium on *Myrica gale* by HARMS[437]. SCHAEDE[1013] has given an account of the mycocecidia on the roots of Podocarpaceae, particularly *Podocarpus chinensis* and *Podocarpus nubigena*. *Olpidium radicicolum* gives rise to globose nodular galls on the roots of Cruciferae[57]. Root mycocecidia are generally caused by *Urophlyctis, Physoderma, Schinzia (= Entorrhiza)*, etc. *Plasmodiophora brassicae* WOR. gives rise to the most remarkable root gall on Cruciferae. As mentioned in an earlier part, this gall develops particularly on the side roots and only rarely on the main root (fig. 28 A). The fungus is an intracellular parasite and occurs within the vacuoles of the gall cells. The spores develop from naked plasmodia and are liberated in the soil by the decay of the gall[216, 482].

We may also consider here mycocecidia on the subterranean shoot.

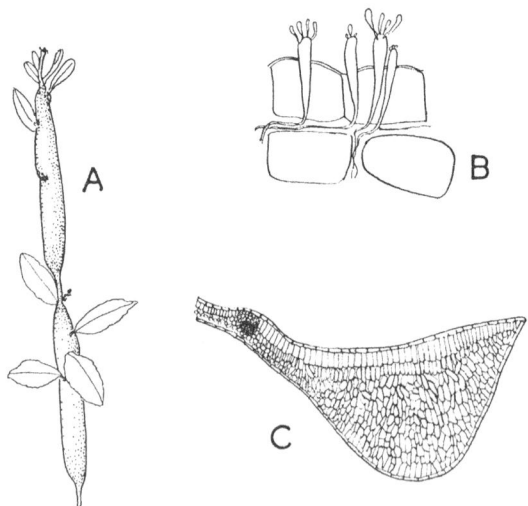

Fig. 141. Mycocecidia. A. Gall of *Calyptospora goeppertiana* Kühn on the stem of *Vaccinium vitis idaea*. B. Part of T.S. of the gall of *Exobasidium vaccinii* on leaf of *Vaccinium vitis idaea*, showing the basidia bearing spores. C. Gall of *Exobasidium vaccinii* on leaf of *Vaccinium vitis idaea* (After Woronin).

Synchytrium endobioticum (SCHIL.) PERC. is responsible for the well known potato cancer gall. The underground parts, especially the tubers, rarely, however, also the aerial portion, develop typically verrucose, cauliflower-like swellings of variable size. The surface of the gall is at first warty and bright brown, but later black. The hibernating winter sporangia of the fungus from the soil revive in spring and produce numerous motile swarming spores, which enter the epidermal cells of the young underground parts. Mycelia are not, however, formed. The host cells in the neighbourhood of the attacked cells undergo rapid mitoses. Spherical summer sporangia arise in the gall cells and these again produce swarming spores, some of which attack other healthy cells. The summer spores enter the soil and also infect new tubers. Winter sporangia arise in the galls in autumn in the form of thick-walled, golden-yellow, small bodies of spherical shape. These over-winter in the tubers or in soil. Similar galls are caused on *Solanum nigrum, Solanum dulcamare* and *Solanum lycopersicum*.

3. Mycocecidia on shoot axis

The galls of *Uromyces pisi* (PERS.) SCHRÖT. on the stem of *Euphorbia cyparissias* and *Euphorbia esula* represent perhaps the simplest of mycocecidia on the shoot axis. The mycelia remain in the rootstock and from here penetrate the young shoot in its very early stages of development. Penetration from the rootstock into the shoot does not, however, seem possible at later stages. The mycelia are intercellular and do not extend to the outer layer of cells, but remain within a few layers beneath the surface. TISCHLER[1158] showed that the mycelia do not attach their haustoria to the very young cells, filled with plasma completely, but only to the cells in which vacuoles have already formed. The development of the gall is accompanied by an abnormal elongation of internodes, suppression of normal branching and flowering. The ascidial cups (spermatogonia) develop on the underside of the abnormally shortened but broadened and swollen leaves. Although the fungal mycelia are abundantly present in the region of the vascular bundles, remarkably enough, no haustoria arise here. Very little anatomical abnormality is caused in the course of the development of the gall. The leaves are, however, much altered, especially the mesophyll, and there is a curious increase in the number of stomata. The palisade tissue, normally only 3—4 cells deep, becomes twice this number and each cell is also greatly elongated to at least three times their normal length. The cells of the spongy parenchyma are greatly hypertrophied and the intercellular space is considerably increased. The whole mass is interpenetrated by mycelia. Rarely if flowers at all arise on the galled stem, they show a remarkable tendency for virescence, absence of stamens and pistil.

The whole shoot of *Cirsium arvense* SCOP. develops into an enormous gall when attacked by *Puccinia suaveolens* PERS. There is a more or less pronounced elongation of the shoot axis, but the leaves

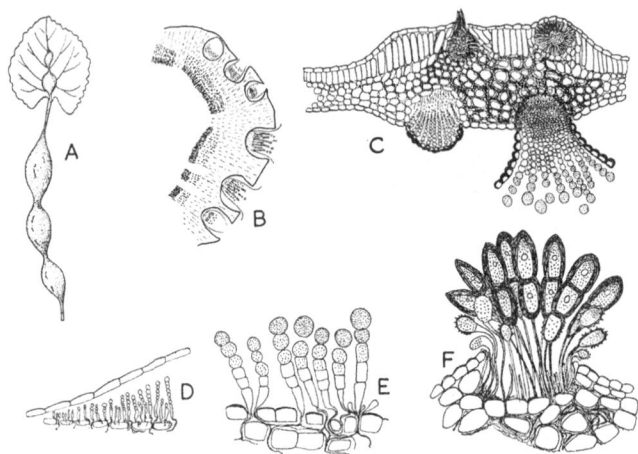

Fig. 142. Mycocecidia. A. Gall of *Urophlyctis violae* on stem of *Viola odorata*. B. T.S. through part of the gall of *Puccinia* on stem. C. *Puccinia* gall on leaf. D. T.S. through gall of *Albugo candida* Pers. E. Ripe conidia on gall. F. *Puccinia graminis* Pers. with uredospores and teleutospores.

are simpler, smaller and less spiny than normal. Witches-broom galls are caused by *Taphrina, Puccinia* and *Melampsorella. Taphrina betulina* ROST. causes the witches-broom gall on *Betula pubescens* EHR. and *Taphrina turgida* (ROST.) on *Betula verrucosa* EHR. The gall of the teleutospore generation of *Gymnosporangium* on *Juniperus* arises at first as a diffuse unilateral swelling of the cortex of the young branch and soon extends to the stele. The growth progresses slowly, so that the gall usually takes several years to mature, the swelling becoming more and more pronounced each year. The annual rings of vascular bundles tend to be irregular, the medullary rays are strongly formed and there is usually a pronounced abundance of simple parenchyma cells. *Gymnosporangium sabinae* DICKS gives rise to one of the commonest galls on *Juniperus.* The aecidia generation of this fungus occurs on *Pirus communis.*

The gall of the teleutospore generation of *Calyptosora goeppertiana* KÜHN on the shoot of *Vaccinium vitis idaea* is a greatly swollen, elongate mass, often reddish coloured, conspicuous above the normal shoot. The leaves on the galled part usually remain small-sized and often also fall off prematurely The teleutospores arise in the epidermal cells of the gall and aecidia in the seedlings of *Abies alba.* The mycelia are intercellular in the cortex of the gall on *Vacci-*

nium and haustoria penetrate the cells. It is only in the next spring that the fungus extends to the neighbouring tissues. Uredospores and spermatogonia have not so far been found in the gall.

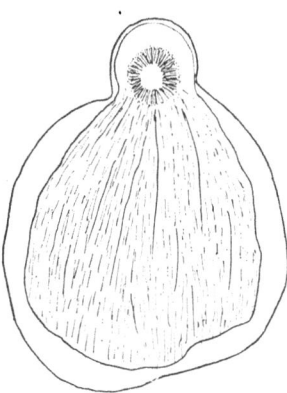

Fig. 143. Rindengall of *Puccinia* on stem of *Berberis lyceum* Royle.

In the spherical gall of *Melampsorella caryophyllacearum* SCHRÖT. on the branches of *Abies alba*, spores are not formed. The mycelia penetrate into a bud that may happen to arise in the galled part of the branch, causing the needles formed from the bud to be swollen, stunted, soft and pale coloured. The aecidia develop on these in June and July. The spermatogonia often also appear as orange spots on the surface of the gall. The uredospores and teleutospores develop in Caryophyllaceae like *Stellaria* and *Cerastium*, *Moeheringia trinerva*, etc. Reference has already been made to the spherical gall on branches of *Citrus* caused by *Sphaeropsis tumefaciens*. One of the most remarkable mycocecidia on branches is caused by *Uromycladium* on *Acacia*. This is a conspicuously red coloured large irregular solid, hard swelling, appearing often in hundreds and persisting for several years on trees. An excellent account of the anatomy of the gall and cytology of the fungus *Uromycladium tepperianum* from the gall on *Acacia stricta* WILLD. may be found in BURGERS[158].

4. Leaf galls

The outstanding features of mycocecidia on leaves are best illustrated by referring to two distinctive types, caused respectively by *Puccinia* and *Exobasidium*.

The galls of *Puccinia* on leaves are commonly found on Malvaceae, Boraginaceae, *Jasminum*, *Berberis*, *Urtica*, *Rosa*, etc. In the simplest case, the gall consists of a localized, cup-shaped swelling of the blade, brilliantly yellow, orange or carmine-red coloured. The description of anatomy, histology and cytology of the gall of *Puccinia caricis* SCHUM. on *Urtica dioica* LINN. given by CAVADAS[178] is typical of these galls. The aecidia generation of *Puccinia graminis* PERS. gives rise to pustule galls on the leaf of *Berberis vulgaris* and *Berberis lyceum*. The spermatogonia develop on the upper side, without causing serious anatomical deformation of the leaf. *Puccinia fusca* RELH. galls the whole leaf blade on *Anemone nemorosa*. The petiole becomes elongated, the blade remains small but greatly swollen and coriaceous. The epidermis is very greatly modified on both sides of the blade. The palisade cells are somewhat hypertrophied, but the spongy parenchyma cells are very greatly altered and it is in this tissue that the mycelial threads occur. Dark reddish spermatogonia arise on either side of the leaf and brown teleutospores in masses on the under side of the gall.

The galls of *Exobasidium* are remarkable for the truly enormous size to which the leaves become swollen, the abundance of simple large parenchyma cells and considerable intercellular space. Nearly every gall develops on the under side of the leaf blade and the site of the gall is hardly indicated on the upper side. Most of these galls are white, but often also red or violet on exposed portions and when mature powdered with spores. The mycelia are intercellular and are also largely confined to the peripheral layer of the gall. Immediately beneath the gall epidermis hymenia are formed, from which the basidia arise. These penetrate in between the epidermal cells and finally burst through the cuticle. The basidial tip projects outside above the general surface of the gall. On this projection arise four short sterigmae, which constrict off the spores. The histological changes in the course of the development of this gall comprise essentially enormous hypertrophy and hyperplasy of the spongy parenchyma of the leaf, disappearance of intercellular spaces and inhibition of the development of chlorophyll. Normal gland cells are replaced by numerous oxalate crystal cells. The gall cells do not lignify, so that both sclerenchyma and collenchyma are absent. All the cells of the gall are typically thin-walled. Even the formation of sieve vessels in the veins is exceedingly poor. Although the mesophyll is thus profoundly altered, the epidermal cells are, curiously enough, almost normal, except that they are generally smaller and longer. The layer of cells below the epidermis is however very strongly altered.

Some of the common *Exobasidium* galls include the following: *Exobasidium vaccinii* (FUCK.) on leaves (sometimes also on branch and flowers) of *Vaccinium vitis idaea* (fig. 141 B—C), *Exobasidium hesperidum* MAIRE on *Rubus oxycantha* from Marocco[741] and *Exobasidium camelliae* on *Camellia japonica* from Japan. The best known *Exobasidium* gall is, however, found on the leaf of *Rhododendron*. This is caused by *Exobasidium rhododendri* CRAM. (Plate III,

3, 8, fig. 144). Other galls include those of *Exobasidium discoideum* on *Rhododendron indicum* from South America[926], and *Exobasidium vaccini* on *Rhododendron hirsutum* from different parts of the world [110, 1282]. The galls of *Exobasidium* on *Rhododendron* are mostly irregular, globose, soft, succulent, solid outgrowths, almost sessile but also sometimes with a very short and narrow stalk. Although vascular bundles are found in the highly parenchyma tissue of the gall, there is no trace of arrangement characteristic of the normal leaf. Mycelia are specially concentrated in the gall periphery and are at first formed between the epidermis and the subepidermal layer of cells, but later penetrate the epidermis to break through the outer cuticle to the surface of the gall. When the spores are formed, the surface of the gall is typically powdery. The detailed anatomy of these galls is described by GUTTENBERG[424]. Reference may also be made to the account of BOEDIJN[102] on the mycocecidium on the leaf of *Hypolytum*.

The gall of *Ochrospora sorbi* OUD. involves swelling of the entire leaf blade of *Anemone nemorosa*; aecidia generation is associated with the gall. White spermatogonia develop on the upper surface of the gall. Uredospores and teleutospores develop, however, on the under side of the leaf of *Sorbus*. Mycocecidia appear to be relatively rare

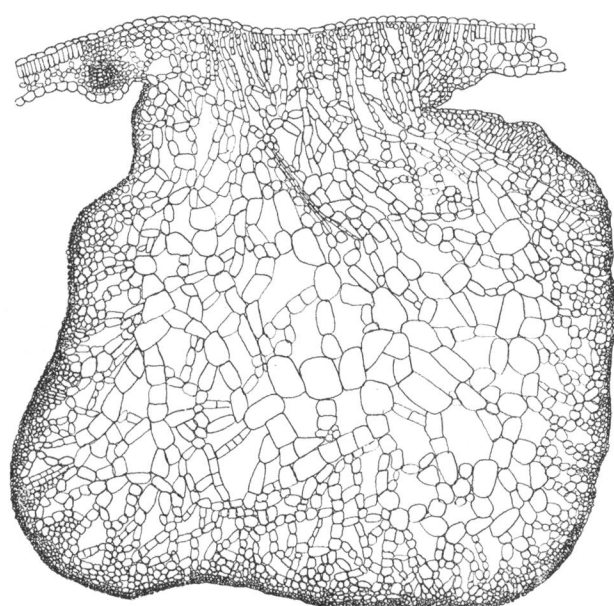

Fig. 144. T.S. through the immature gall of *Exobasidium rhododendri* Cram. on leaf of *Rhododendron arboreum* Sm.

on petioles. *Urocystis violae* Sow. gives rise to a conspicuous swelling
of the petioles of *Viola*[888]. *Ustilago* gives rise to globose pustule galls
on leaves of *Rumex*, *Polygonum* and other plants. In the early
stages of their development, these galls have a curious similarity to
certain mite galls.

5. Flower galls

Mycocecidia on flowers are often remarkable for the sex reversal
in the flower. In the gall of *Ustilago violacea* PERS. on the flower of
Caryophyllaceae, the fungal spores arise in the anthers. On dioe-
cious plants like *Melandrium album* GARCKE the formation of the
flower gall is in the staminate flowers. In a pistillate flower, the gall
formation results in the appearance of stamens. In case of *Cintractia*
spores develop in the pistil of *Carex* and if staminate flower is infected, the pistil appears as a consequence of gall formation. This
structural hermaphroditism seems to be common in many of these
mycocecidia. In the gall of *Ustilago antherarum* on the flower
of *Melandrium album* and *Melandrium rubrum*, there is an
abnormal development of stamens[385]. In both these species the
flowers are normally either staminate or pistillate and both types
of flowers are galled by the fungus, but the formation of a gall is
conspicuous only in the pistillate flower. The wholly insignificant
rudiments of the stamens of the pistillate flowers develop into more
or less normal-sized stamens, within the anthers of which the spores
of the gall fungus mature. A short stretch of the floral axis between
the calyx and corolla becomes visible in the gall of the pistillate
flower, but this is not found in the normal flower. ROZE[990] found that
in the gall of *Ustilago caricis* FUCK on the staminate flower of *Carex
praecox*, the pistil arises. According to the observations of KELLER-
MAN & SWINGLE[571], there is an abnormal appearance of ovary in the
gall of *Tilletia buchloëana* on the staminate flower of *Buchloë dacty-
loides*. MAGNIN[725] reports that in the gall of *Ustilago vaillantii* on the
inflorescence of *Muscari comosum*, there is also an abnormal development of stamens, which are normally only vestigial and the anthers become filled in the gall with the spores of the fungus.

Mycocecidia on flowers are characterized by the general tendency
to involve the ovary more than any other floral organ. This is best
seen in the gall of *Taphrina pruni* (FUCK.) TUL. on *Prunus domestica*.
This gall is a typically yellow or red coloured, flattened, short, somewhat curved bag-like growth, about 6 cm long and 1—2 cm thick,
with a warty or rugose surface, often also powdery during summer.
In place of the normal stony seeds, the gall contains merely an
empty cavity. The mycelia are concentrated mostly among the
peripheral cells and in due course form hymenia between the epidermis and the cuticle. Each hymenium is typically a clavate tube,

about 30—40 microns long and 8—15 microns thick, imparting a superficial hairy appearance to the gall. Similar galls are also found on *Prunus padus*, caused by *Taphrina pruni*. In most galls of *Taphrina* on *Populus* the pistil becomes swollen into an oval or subglobose mass, with enlargement of scales into red coloured fleshy ribbon-like structures. The gall has a curious compressed and corrugated appearance and is nearly fifteen times larger than the normal organ. Mycocecidia on the flowers of Cruciferae like *Capsella, Raphanus, Sinapis, Brassica, Sisymbrium, Senebiera*, etc., caused by *Albugo candida* have been studied by EBERHARDT[313]. The calyx is mostly greatly swollen and also bears numerous flask-shaped or cylindrical emergences. The corolla is likewise swollen, but the stamens are mostly very profoundly altered, with the filaments enormously swollen, stunted and often also flattened leaf-like. The anthers are either reduced or enlarged, but do not contain pollen grains. The pistil becomes leafy and swollen. The mycelia ramify profusely in between the cells and attach short or globose or bladder-shaped haustoria to the gall cells. The conidiospores arise from under the gall epidermis, push the epidermis upward and finally break through it. The parasite fungus *Peronospora parasitica* is also often met in the gall of *Albugo candida*. The ovaries of grasses are frequently galled by *Claviceps purpurea* FRIES and by *Fusarium* spp.[659]. *Tilletia* spp. and *Ustilago* spp. also cause galls on fruits of various other plants.

CHAPTER XIII

VIRUS AND BACTERIAL GALLS

1. Virus and virus gall

Viruses give rise to diverse growth abnormalities in plants, such as leaf curling, leaf crinkling, fasciation, bunching, etiolation, swelling, etc. in addition to various other structural and functional disturbances. Whether any of the growth abnormalities should be included under galls depends partly on the limitations of the customary definitions of galls and partly on the vagueness of our concept of virus. It does not seem to have been satisfactorily decided whether a virus is an organism in the ordinary sense or only a giant proteid molecule.

Since its discovery nearly seventy-five years ago, virus has continued to be the centre of considerable controversy and inspite of the intensive research in recent years virus is but still imperfectly understood. Briefly stated, a virus is essentially something of a giant proteid molecule, exhibiting an astonishing variation in size from a little smaller than some of the smallest bacteria down to the molecular grades of magnitude. Some of them are rounded bodies about 20—30 millimicrons in diameter and others are elongate needle-shaped rods, reaching to 125—1500 millimicrons in length. Some of the larger viruses have a complex composition and contain fat, carbohydrates and even copper, as well as nucleoproteins; desoxyribose nucleic acid is normally present in such viruses. Some of them have been crystallized in the chemical laboratory. The most remarkable feature about a virus is, however, its infective character — it multiplies inside an organism, either a plant or an animal, and gives rise to more or less far-reaching disturbances in the metabolism of the host. Specific nucleoproteins isolated from plants infected with virus are found to be highly infective even at great dilutions, indicating clearly that these nucleoproteins are identical with the virus.

A virus differs fundamentally from even the simplest definitive microörganism in its great relative simplicity of composition. The simplest bacterium contains, for example, a wide variety of complex components, including naturally nucleoproteins and its composition is also continually changing. A virus on the other hand is just nucleoproteins, the nucleic acid and proteins of which are united to give a fixed regular structure. In their chemical simplicity and internal structure, a virus resembles very closely some of the complex substances synthesized by an organism, rather than a whole organism. The purified virus crystal shows nothing in the laboratory to distinguish it from preparations of other nucleoproteins. It is only its ability to multiply inside a susceptible host cell – in other words, its infectivity – that marks it off but this infectivity is often readily lost by certain treatments, such as irradiation by ultra-violet. "Mutation" has also been attributed to virus. Unlike a bacterium, a virus cannot, however, be cultured in artificial culture media, at least in the present state of our knowledge.

The general effects of virus infection on a host, especially plants, have been described as disturbance of the metabolic activity of the host cell, by the virus acting perhaps as a phosphatase or other type of enzyme, which pushes the host cell processes in an abnormal direction. In doing this, the virus multiplies itself and thereby greatly increases the protein content of the host cell. It piles up its byproducts of inert proteins, while in normal and healthy cell metabolism, there is minimum of such an accumulation. The virus in a way starves out or at least deforms the normal protein synthetic processes in the cell and replaces them by unusual ones. Such a situation has also been observed to greatly benefit bacterial infection of the plant cell. In green plants, the plastids and pigments suffer most as a result of virus infection. With its specialized activity, a virus usually favours special tissues and cells of the host plant. The relation of various viruses one to another presents certain remarkable peculiarities. Many viruses have no effect on each other within the same host and are thus neutral or independent. In some cases, as in the rosette disease of *Nicotiana tabacum*, a coöperation of viruses has been observed[1068]. Sometimes antagonism is also known to be exhibited, especially between strongly related or mutated strains of the same virus. BAWDEN[66] found, for example, that severe etch replaces potato-virus Y and *Hyoscyamus*-virus 3.

The same virus may often exist in several different strains. No less than fifty such strains have, for example, been found in the case of the *Nicotiana* mosaic virus. Viruses have also been observed to become modified by rapid passage through highly susceptible hosts[480]. There is yet another remarkable feature about a virus, which is illustrated by the *Abutilon*-viruses. When the variegated *Abutilon striatum* and the green *Abutilon indicum* are grafted onto different parts of a third species *Abutilon arboreum*, the virus passes from *Abutilon striatum* through *Abutilon arboreum* without affecting it, enters *Abutilon indicum* and causes in it variegation. The virus thus multiplies in *Abutilon arboreum*, without injuring it and thus behaving as if it were a normal protein of *arboreum*.

Although a great deal is thus known about viruses and their effects on organisms, we are still surprisingly ignorant about the true nature of the virus. Virus has been considered as a transitional stage between molecules and organisms — as an "upward" stage by some and a degenerate "downward" stage by others. They are believed to represent organisms that have lost their capacity for synthesizing complex material as a consequence of being constantly supplied with ready-made complex organic compounds by their host organisms, thus losing the functions no longer needed and naturally also the associated structures. Viruses have also been considered as merely invisible stages in the life-history of a larger and more complex organism. Transformation of a virus into a bacterium has also been reported to have actually been observed in the laboratory. LIESKE[693] believes, for example, that filterable viruses, gram-negative monotrichous rods (like *Bacterium tumefaciens* or the crown-gall bacterium), variable gram-positive rods, *Streptococcus*, etc. are merely cyclic expressions of one and the same organism. A virus is thus an obligatory parasite, whatever else it may be. Convincing evidence has also been adduced by DUBUY and his collaborators[310], showing that at least the phytopathogenic viruses are derived from mitochondria of normal cells. Loss of part or all of the chlorophyll is considered to bring about the formation of light green or yellow plastids and further loss of the carotinoid pigments leads to colour-less plastids or mitochondria and finally of smaller submitochon-drial units, which become viruses. It is indeed well known that both in plants and in animals, a number of diseases arise not by infection, but become infective afterwards. A distortion of a normal protein metabolism in an individual results in the genesis of a virus; a new protein forms instead of the normal proteins. One step further from these so-called *intrinsic viruses* are those, which are normal proteins in one individual, that is the individual of their origin, but unfriendly in an alien cell. The virus, producing the ascending myelitis in man bitten by a healthy monkey, is a case in illustration. The protein molecules of the monkey have become virus in man by *transplantation*. Even in plants, grafting of different varieties and species have been definitely known to cause the origin of new viruses. It has been pointed out by DARLINGTON[242] that a protein produced in the cell of one organism can propagate itself injuriously in the cell of others or even in the same organism. A virus is thus in a sense a foreign protein molecule. It is a protein molecule that when introduced into one individual disappears, in another individual multiplies to produce neutral equilibrium, into yet another, multiplies without limit and produces detrimental results. The virus can therefore be considered as a protein molecule taken out of an organism and introduced into another to which it has not been adapted: it is a protein out of place [1007]. Since such a giant protein molecule exhibits the property of self synthesis by autocatalysis, we are compelled to consider the virus as a living organism.

The bulk of the viruses, which cause diseases, including galls, are usually transmitted by diverse insects, which act as vectors. Sucking insects, especially Thysanoptera, Psyllidae, Jassidae, Aleurodidae, Aphididae, Coccidae, etc. are perhaps the most important vectors of viruses. Eriophyid mites have also been reported to act as vectors of viruses. The cushion-gall virus of cocoa is for example transmitted by the nymphs and adult females of *Pseudococcus jalensis* and *Ferrisiana virgata*. BLACK[90, 91] has stressed the rôle of wounds and other mechnical injury in the transmission of tumor-inducing viruses in the root of sweet clover and other plants. He also demonstrated the possibility of transmitting the virus by experimental tumor grafting. He grew the sorel tumor in sterile tissue culture for nearly two years and the culture contained the virus.

BLACK *(op. cit.)* has described the tumor gall induced by the virus *Aureogenus magnivena* on the roots of a number of plants. Some of these galls are globose, hard and woody and often measure up to 10 mm in diameter. They develop either singly or in more or less crowded clusters. Sometimes several of them become fused together to form large composite masses. He was also responsible for the description of the virus gall on the stem of *Melilotus alba* and other species. In most cases the virus gall consists of irregular swellings of leaf veins and hypophyllous overgrowths. ALTENBURG[11] has described the relation between tumor formation and the origin of virus and suggests that the formation of tumor is a possible mechanism by which viruses become established. Proceeding from certain well known observations, KENNEDY[575] has recently put forward the suggestion that at least in some cases the insects give rise to galls on plants through the agency of a virus, which they transmit and then feed on the plant material predigested or conditioned by virus activity. *Piesma quadrata* (Heteroptera) is in reality the vector of a virus that causes the leaf curl gall on *Beta vulgaris*. The adult bug sucks the sap and thus inoculates the leaf, discolourizing the puncture spot. Associated with this galling is found a pronounced etiolation of the foliage.

2. Bacterial galls

Though not as abundant as zoocecidia and mycocecidia, the bacteriocecidia are widely distributed and arise principally on Dicotyledonae. Some of the outstanding problems of bacterial galls are best illustrated in the crown-gall, caused by *Phytomonas (= Pseudomonas) tumefaciens* (SM. & TOWNS.). This gall has been studied intensively in recent years and an enormous amount of literature has accumulated. The crown-gall arises at ground level on stem or root (crown) on a great variety of plants. The striking peculiarities of the crown-gall are summarized in Chapter XV.

Besides the crown-gall, a number of other interesting bacterial galls are known. The best known of these bacterial galls are the root nodules of Leguminosae, caused by *Bacterium radicicola* BEIJ. and other species. The bacteria develop in the epidermal and cortical cells. These nodule galls have been extensively studied and the more important references are listed in the bibliography. The structure, development and evolution of the bacterial nodule galls are considered by DANGEARD[241] and ZIPFEL[1286]. The life-cycle of the bacterium, with special reference to infection of the root, is described by THORNTON & GANGUELLE[1151]. The formation of pyruvic acid[20], the production of antibodies[176] and the relation between bacteriophage and the root nodule bacteria[651] are some of the more important recent findings. It has also been shown recently that the legume root nodule bacteria are capable of becoming adapted to give rise to galls on various non-leguminous plants[98]. The symbiotic utilization of nitrogen is believed to be occurring in the root nodule galls on non-leguminous plants also[1215].

Other interesting bacterial galls include the gall on the stem of the hygrophile leguminous plant *Aeschynomene aspera*, developing both on the submerged and aerial portion[429], the bacterial gall of *Beta vulgaris* caused by *Bacterium beticola*[311], bacterial tumor on the thallus of *Saccorrhiza bulbosa* and transmitted by phytophagous mollusca[175], *Gracilaria confervoides*, *Cystoclonium* and *Chondrus crispus*[186]. RANFIELD[925] has described an interesting bacterial gall on branches of *Rubus occidentalis*, in which the bacteria are intercellular and occur in pockets and strands. The cells in contact with the bacterial mass undergo cytolysis. In the early phase of development of the gall, the remote cells undergo division and then the bacteria migrate there; then more distant cells divide and the process goes on till the gall is fully formed.

An important bacterial gall occurs on branches of the olive tree in Mediterranean lands and parts of the United States of America [1269]. It is caused by *Phytomonas savastoni*, introduced by the oviposition by the olivefly *Dacus oleae*. Mention may also be made of the gall of *Bacterium pseudotsugae* on Douglas fir and transmitted by *Chermes cooleyi* and developing as rough, globose swellings of the branches[435]. Bacterial nodule galls on leaves of about 370 species of plants are also supposed to have nitrogen fixation significance[537]. BREMEKAMP[135] has described bacterial nodule galls on leaves of forty-two species of *Psychotria* (Rubiaceae) from Africa. STEC[1091] has given an account of the interesting gall on the anthers of *Solanum tuberosum*, characterized by intumescent growths on the outer walls. According to some workers[52, 720], some of the galls now generally attributed to aphids may eventually turn out to be bacterial galls, the aphids acting merely as vectors of the causative bacteria.

CHAPTER XIV

DEVELOPMENT AND GROWTH OF THE GALL

1. General characters of cecidogenesis

The fundamental difference between development and growth is now widely recognized. While growth embraces changes in size and weight, development is a sequence of internal adjustments undergone in ontogenesis of the organ or tissue. The rates of growth and of development are governed by environmental conditions, independent of each other. The conditions favourable for growth of gall may be unfavourable for its development and vice versa and growth and development are usually accelerated entirely differently. These differences are illustrated by recent investigations on the photoperiodic induction of development and photoperiodic inhibition of growth in plants. The rate of development is not only independent of the rate of growth, but may also not be concurrent with or subsequent to growth. Development may be initiated and progress normally in a rather slowly growing organ. Long before the perceptible growth of the gall, diverse internal readjustments occur within the plant tissue and cell. Whatever their nature, these changes are presumably connected with the physiological state of the cells and tissues at which the capacity to grow is required. Although the rate of increase in size of the gall is independent of the rate of development, at least in the early phase development depends on growth to some extent and yet growth must be considered as one of the functions of development. In the final analysis, a gall is the result of a specialized reaction of the plant cell to the changes in its internal environment caused by the gall-inducing organisms. Cecidogenesis embraces thus a complex series of interactions of the plant cell and the gall-inducing organism. The development and growth of the gall are therefore essentially ecologic problems.

Cecidogenesis does not involve processes unknown in normal ontogenesis. The cells and tissues of galls are abnormal only in so far as they appear in abnormal abundance in unusual places and at unusual time or in combinations other than those met with under normal conditions of development of the gall-bearing organ. The abnormal tissues of galls, but with the cells similar to those of normal organs (homeoplasy) and galls with the cells differing from those of the normal organ (heteroplasy) arise essentially by the same histogenetic processes that give rise to the normal organ itself. The various histogenetic and morphogenetic processes, however, take place at intensities different from normal or the individual processes combine in some abnormal manner. The histogenetic processes follow each other in a regular and a definite sequence in the development of the normal organ but, in cecidogenesis exhibit more or less pronounced independence of each other. The formation of new cell wall membrane, for example, normally keeps pace with increase in the volume of the growing cell, but in cecidogenesis the growth of the cell may be initiated and inhibited to a greater or lesser extent, though the production of the cellulose wall may continue unabated. Cell division and nuclear division may progress independently of each other and nuclear division may be inhibited even if the growth of the cell is not affected to an appreciable degree. Although the individual histogenetic processes in cecidogenesis are identical with those in normal ontogenesis, some of the processes arise prematurely early or lag behind abnormally, so that there is a localized inbalance.

GÖBEL[397] recognized these characters when he expressed the view that galls contain neither something "new" nor anything that does not ordinarily occur in the tissue components of the plant. What we usually call "new" in gall formation is in fact only a new combination of the properties, capable of being exhibited by the plant under other conditions. The normal course of development and pattern of growth are in some way or other disturbed, even including the development of the apparently new growth of erineal hairs in acarocecidoa. The erineal hairs of acarocecidia may differ from the normal hairs and yet are not strictly "new". Even in complex galls, with considerable tissue differentiation, the anatomical elements are the same as may be found in the plant under otherwise normal conditions[263a]. Some workers like BEIJERINCK[75], however, believed that new structures actually arise in cecidogenesis. It does not seem to have been decided as to how much the gall cell should be abnormal in order to be described as something "new" for the plant. The so-called "new" may be expressed in different directions, such as abnormality in size relation, shape or internal composition. We have already pointed out that most gall cells are not related in size to those of the normal organ. KÜSTER[641] believes that gall cells also do not stand in relation to normal cells in their form. The biramous stellate hairs on the lenticular gall of *Neuroterus* (fig. 84 A) are never found on *Quercus* under normal conditions. It has been repeatedly observed that, under certain conditions, cells of diverse kinds and origin are capable of developing abnormally, with little relation to their original structure and function. The development of chlorophyll in cells normally devoid of it has been repeatedly described in the course of development of a number of galls. Similarly thin-walled cells are modified into thick-walled cells. Closer examination, however, reveals the fact that chlorophyll is not formed in the gall cells of plants that do not normally have this pigment and thick-walled cells do not also arise in the galls of plants lacking such elements. The capacity of the protoplasm to produce chlorophyll or of forming around itself a thick cellulose wall is thus a necessary condition. Considered from this point, the formation of abnormal structures is not "new". The gall only achieves what the normal cells of the plant are capable of doing under another set of conditions, though the cells of the tissue on which the gall arises do not ordinarily develop chlorophyll. The latent faculties of the cells seem to be thus simply evoked in wrong context and in wrong combinations. Although the gall cells seem only to realize the same inner potentialities, which are realized even in normal ontogenesis, all sorts of possible cells arise in cecidogenesis. The differentiation processes also comprise those normally realized in some other cells or some other organ under normal conditions. In the gall we have thus abnormal combinations of normal components and in this interpretation we have in mind not the gross structure but the individual part processes. If we compare, however, the cells as a whole with one another, we then find that in the course of development of galls new structures do arise. This new structure is really the outcome of normal elements. Localization, abnormal intensity and combinations of normal components and part processes result, however, in diverse conditions. It is interesting in this connection to refer to KÜSTER: "Aus jedem Gewebe kann alles werden" each tissue is capable of developing into every other tissue under abnormal conditions. In the gall of *Hartigiola annulipes* (HTG.) on *Fagus silvatica* LINN., for example, the uppermost layer of palisade cells of the leaf can develop into typical hairs (fig. 76). In endogenous cynipid galls on *Quercus* a typical stomata-bearing epidermis is formed. The phellogen derived from the parenchyma can produce typical epidermis; typical parenchyma develops into vascular bundles. In leaf galls, all types of thick-walled and thin-walled cells are formed from typical mesophyll cells. It has also been found that cells and tissues normally not characteristic of the species on which the gall arises, but nevertheless occurring in some other closely related species, sometimes develop in the course of cecidogenesis. Such structures are sometimes found only in unrelated plants. Half-side thickened stone cells, normally occurring in Lauraceae have, for example, been described in many cynipid galls on *Quercus*. Such apparently atavistic reappearance of structures has been described in several other galls also. According to GERTZ[380], in the gall of *Isthmosoma graminicola* on *Triticum junceum* stomata revert to phylogenetically earlier types. We have described earlier the ridge-like outgrowths at the bases of the needles in the gall of *Chermes abietis* beneath the site of attack. This portion of the needles exhibits a wholly different course of development in the gall from the other part of the needles. It has been shown that in the needle of the Japanese *Sciadopitys verticillata* such lateral outgrowths of the leaf parenchyma normally produce bilobed processes, similar to those found in the gall formation by *Chermes* on the European species. In the simplest case, the histogenetic processes may almost be completely normal, but development as a whole is abnormal in so far as an organ is formed in a false position. Such heterotopy is well illustrated in the gall of *Dasyneura (= Perrisia) crataegi* (WINN.) on *Crataegus*, in which gland organs develop on the surface of the leaf blade, although normally no such glands are found on the surface of the blade but only along the leaf margin. MAGNUS[731] has described an interesting example of heterotopy in the gall of *Pontania* on *Salix*, consisting of abnormal development of stomata in the second epidermal layer of cells, instead of the outermost layer.

2. Histogenesis in the gall

The more important histogenetic processes in cecidogenesis include hypoplasy, redifferentiation and retrograde differentiation, form anomaly, hypertrophy, infiltration, anomalies of cell division, hyperplasy, tissue stretching and tissue rupture, tissue fusion, cell fusion, tissue degeneration and necrosis, cytolysis and differentiation.

Hypoplasy is inhibition of normal development, characterized by incomplete development of cells, incomplete differentiation of tissues or the inhibition of typical tissue results in the formation of homogenous parenchyma. Incomplete formation or thickening of cell walls, incomplete development of chlorophyll and crystals are also often met with. Redifferentiation or metaplasy, rare among plants, may be associated with retrogressive differentiation and such changes like formation of chlorophyll in cells, which are normally chlorophyll-free. Metaplasic changes include also abnormal accumulations of starch and proteids. The metaplasic transformation of thin-walled cells to tracheids has been described in certain galls. Galls inhabited by inquilines exhibit a peculiar metaplasic modification, consisting of the thick-walled cells becoming turned into thin-walled cells. Without growth or cell division, the thin-walled cells become modified into stone cells in the gall of *Diplolepis quercus-folii* (LINN.). Growth anomalies arise from localized acceleration or inhibition of growth and results in abnormal shape of cells, as in the gall of *Notommata werneckii* on *Vaucheria*.

Hypertrophy is cell enlargement without division and is an important event in cecidogenesis, including even the galls that are formed predominantly by repeated cell divisions. Cells of nearly all tissues undergo hypertrophy, such as, for example, epidermis, parenchyma and vascular bundles. In the development of some galls (fig. 153) hypertrophy is the exclusive process or at least the predominant histogenetic process, but in most other cases hypertrophy marks a definitive phase in cecidogenesis. Exclusively hypertrophy galls are usually unicellular galls on Algae and some galls of the fungus *Synchytrium* on Phanerogams. In the gall of *Synchytrium papillatum* FARL. on the epidermal cells of *Erodium cicutarium* single epidermal cells are enormously enlarged into clavate and irregularly lobed galls (fig. 140). Most erineal galls of *Eriophyes* spp. are multicellular, but single cells also become enormously hypertrophied, without division. Hypertropy of the parenchyma cells is common in many galls of Diptera and Hymenoptera. Even in galls which develop by pronounced cell division, there are zones of cells of low cecidogenetic gradient *(vide infra)* and remote from the gall-inducing organism, where the cecidogenetic influence is not apparently strong enough to induce cell division, but adequate to bring about a simple increase in size of the cells. Hypertrophy also represents the last phase of development of the gall and sometimes continues after cell divisions have practically ceased. According to WEIDEL[1234], the increase in size of the lenticular gall of *Neuroterus numismalis* FOURC. on the leaf of *Quercus*, after the gall breaks off from the gall-bearing organ, is the result of such hypertrophic changes in the gall parenchyma. Cells that undergo hypertrophy occasionally become also lobed, with the lobes lengthening and penetrating in between the other neighbouring cells. Cellular infiltration of this type was observed by MAGNUS[728] in the gall of *Urophlyctis leproides* on *Beta vulgaris*. JENSEN[548] was one of the first to describe this type of metastatic growth in *Beta vulgaris*. Enormous tissue masses of tumor arise in various parts and affect the normal growth of the tuber and often even exceed in size the tuber itself. These tumors are comparable with metastatic malignant neoplasia on grafting. Whether large infiltrating tissue masses and cells result in malignant neoplasia in plants, comparable to what we find in human malignant tumors, has been the subject of controversy. As we shall show elsewhere, the malignant nature of the tumor tissue of *Beta* has been demonstrated by grafting of differently coloured tumor tissues and healthy tissues.

Anomalies of cell division in cecidogenesis are wide spread and embrace very diverse conditions like abnormal direction of cell division, unequal cell division, division of nuclei without simultaneous division of the cell thus resulting in multinucleate giant cells, etc. The term hyperplasy was applied by VIRCHOW to cell proliferation. Hyperplasy is perhaps the most important histogenetic process in cecidogenesis. The number of cells that initially begin to divide repeatedly depends on a great variety of factors and may range from a few to over two thousand[75]. Although theoretically all cells are capable of undergoing hyperplasy, they are not all equally active in this respect. Most intense cell proliferation is usually found in parenchyma, the cambial part of the vascular bundles and medullary rays, but only weakly in the epidermis. Exceptionally, even epidermal cells proliferate very profusely. Cell proliferation may be observed in tissues that are normally unsuited for cell division. The cysto-liths of *Ficus vogeli*, for example, undergo repeated division in a gall caused by Diptera[494].

Enormous localized growth of deeper tissue usually results in the abnormal stretching and ultimate rupture of the subepidermal and epidermal tissues in many galls. In galls in which the peripheral cell layers undergo division and growth at a greater intensity than that of the deeper layer, a folding of the peripheral part may be observed. The formation of the socalled "inner galls", mentioned earlier (fig. 114) is essentially the result of such tissue stretching and tissue rupture.

In cell fusion, the cell walls dissolve more or less completely, thus permitting the cytoplasm to flow from one cell into the other and lead to large masses of symplasia. Cell fusion has been observed under different pathological conditions. In galls of some *Synchytrium* sp. the cell wall of the host dissolves to form syncitial, multinucleate masses[52a, 646a]. We have already described this condition in the galls of *Heterodera*. Tissue fusion is common in galls.

3. Cecidogenetic factors

The mechanism that makes the plant cells to divide without re-ference to the general morphogenetic influence of the organ — the problem of what initiates the abnormal course of histogenesis, parti-cularly hyperplasy, and keeps it going in the course of the develop-ment of the gall — still remains obscure. We do not even know with any degree of certainty whether the hyperplasy in cecidogenesis in-volves mitosis or amitosis. The problem is also complicated by the fact that our knowledge of the mechanism and factors governing normal mitosis, especially the rôle of cytoplasmic changes in the initiation of nuclear divisions, origin of polyploidy and heteroploidy, the basic peculiarities of morphogenesis in plants and the conditions of normal growth and differentiation and of the underlying factors is still largely incomplete. Normal growth and development are largely phasic processes, in which even slight abnormality at any phase is likely to introduce a chain of abnormalities. There is, how-ever, very little doubt that normal mitosis is governed by a com-plex chain of factors, controlled and coördinated by the activities of neighbouring cells. A cell may escape from this coördinating and controlling influence either due to inhibition of this influence or due to some alteration within the cell itself, so that it is no longer subject to the morphogenetic effects of the plant body. This alteration in the cell may either be temporary or also permanent as in unlimited neoplastic galls and may embrace changes in the nucleus, cytoplasm or both. Viscosity changes in the cytoplasm are, for example, known to largely govern nuclear changes leading to normal mitosis.

Conditions like polyploidy, polynuclear cells, endomitosis, polytene states, etc. are increasingly being recognized to be merely secondary to an increased "stiffness" of cytoplasm and are no more than signs of such a change[823]. The nuclear changes on which so much emphasis has all along been laid are now widely recognized to be initiated and controlled by cytoplasmic changes. The principal cytoplasmic changes that initiate nuclear changes are perhaps physico-chemical in their nature, presumably correlated with altered metabolism and probably also involve enzyme reactions. A precise knowledge of the cellular interactions during normal differentiation should prove useful in a correct understanding of the problems underlying the abnormal hyperplasy and inhibition of normal differentiation during cecidogenesis. The factors underlying cecidogenetic hyperplasy are unquestionably complex.

While theoretically speaking, any cell may become abnormally hyperplastic or in other words neoplastic, the development of a gall represents a close mutual interaction of the plant cell and the gall-inducing agent. It is common observation that not all nematodes, mites, insects and fungi cause galls on plants, so that certain specific factors or combinations of factors alone seem to be capable of inducing the plant cell to undergo abnormal division. It is also highly significant that different cells react differently to the same stimulus. Even if we exclude the correlating and integrating influence of the neighbouring cells and tissues, the behaviour of cells in this respect is remarkably different. The capacity for reaction, division, growth and differentiation are very profoundly modified by the age of the cell and all normal cells are also characterized by a definitive polarity. The cell polarity emphasizes the fact that under identical conditions the poles of cells react differently to the same stimulus, either unequally rapidly or also qualitatively unequally[792a]. The reaction capacity of different cells of diverse tissues to the gall-inducing stimulus is both qualitatively and quantitatively different. Modifications of this capacity continually arise with the progress of cell reaction and environmental conditions.

In attempting to elucidate the complex factors governing cecidogenesis, earlier workers seem to have been greatly impressed by the evidence of comparative pathological anatomy. Galls with a structure resembling typical wound callus were, for example, supposed to have been caused by the mechanical injury inflicted by cecidozoa. Galls were thus grouped etiologically as traumatomorphose, chemomorphose, osmomorphose, etc.[639]. It must not, however, be ignored that even in the formation of a typical wound callus tissue, the mechanical injury is only an insignificant element in the chain of factors. The injury brings about the release of cells and tissues from turgidity conditions, abnormal osmotic changes are caused and there is also escape of cell contents. Furthermore, we also know that in many mycocecidia abnormal hyperplasia sets in only as a result of the fungus penetrating the plant tissue after wounding. In this case neither the previous wounding nor the chain effects of wounding are evidently the determining factors. The concept of wound hormones as determining factor in cecidogenesis is also not satisfactory[426]. Recent studies have also shown that in some cases the point of reversal of electric potentials of cells becomes the site of a distinct tumor growth[719, 752a, 1229].

Cecidogenetic hyperplasy is conceivably determined by a complex of physico-chemical changes in the cell induced by the cecidogenetic agent. The source of the cecidogenetic agent is apparently different in different galls. As gall development is in the final analysis based on a complex chain of chemical changes, it is conceivable that cecidogenesis is likewise determined by chemical factors. The physico-chemical nature of the mitosis factor is indeed supported by recent observations on chemically induced galls.

4. Chemically induced galls

Galls and gall-like outgrowths have been experimentally produced on diverse plants by the action of various chemicals, ranging from a simple dilute solution of ammonia to complex carcinogens. By spraying a dilute solution of cupric sulphate or cupric chloride, SCHRENK [1036] was able, for example, to produce intumescence of cells in *Brassica oleracea*. According to SILBERBERG[1047a], the action of N/12 and N/14 solutions of zinc sulphate produces the formation of callus cells and wound cork on tubers of *Solanum tuberosum*. A large callus is reported to arise on tubers of *Daucus carota* treated with 8% magnesium chloride and, 12% manganese sulphate or 7% magnesium chloride and 14% manganese sulphate for 10—50 minutes and washed in running water afterwards[913a]. BLUMENTHAL & MEYER[97] report similar results after treatment with 1% lactic acid on tubers of *Daucus carota*. The growth exceeds the spontaneous callus and resembles rather the tumor caused by the crown-gall bacterium *Phytomonas tumefaciens* (SM. & TOWNS.). PERTI[898a] induced the formation of swellings in *Vitis vinifera* by injecting the plant with a 0.2% solution of sodium glycolate. LEVINE[682] has published a useful review of the effects of diverse carcinogens on plants. Out of about seven hundred different substances tested on animals for their carcinogenic action, nearly one hundred and seventy have been found to be capable of inducing more or less pronounced galls on plants. The carcinogens are applied on plants either with or without previous wounding. The effects of coal-tar, polycyclic and azo compounds, lactic acid, ammonium compounds, tannic acid, sodium chloride, lipids, phosphatides, polysaccharides found in the crown-gall bacterium, organic sulph-hydril compounds, gonadotrophic hormones, etc. on plants are compared by Levine with the typical characters of development of the crown-gall. He stresses the important fact that the reaction of plants to these very diverse chemicals is specific to the plant and not to the chemical. While the carcinogens typically give rise to malignant tumors in animals, a constant stimulus by these chemicals is, however, required in the case of plants and the principal effect of carcinogens on plants is necrosis rather than the formation of tumors. Lanolin pastes containing different fractions of coal-tar, 1,2-benzopyrene, etc., placed on plants in a layer of about 1 mm thickness and about 4 sq. cm area, without causing injury to the tissue, cause the formation of abundant adventitious roots on *Nicotiana suaveolens* and *Lycopersicum esculentum*[598]. Deep wounds arose on *Sambucus nigra*, but healed up rapidly under the same treatment. Tumors were not, however, induced in any of these plants. Immersion of the seedlings of *Vicia faba* and *Pisum sativum* in coal-tar solution for brief periods causes vacuolization of the cytoplasm and nuclei and the development of giant galls follows[610]. Cell proliferation and irregular small masses result by painting with solutions of scharlach-red, coal-tar, 1,2,5,6-benzopyrene, etc. The tumors induced by these chemicals are, however, smaller than those arising under the influence of *Phytomonas tumefaciens* (Sm. & Towns.). Anomalies in cell division of the root meristem and tumor induction within twenty-four hours in the submeristem zone are described by Gavaudan & Gavaudan[371] in *Triticum vulgare* treated by apiol. Diploids, tetraploids and giant nuclei, in which it is difficult to establish the degree of polyploidy, are also described in this tumor. The effective constituent of apiol is believed to be allyl-1-dimethoxy-2-5-methylenedioxy-3-4-benzene. A tumor has also been produced on *Phaseolus* by the application of a 2% mixture of L-tryptophan with lanolin. Concentrations of 0.02% induce marked histological changes and produce distant tumors and concentrations of 0.002% cause but slight histological change[832]. Differences of reaction to L-tryptophan and indole-3-acetic acid include vascularization of the cortex, endomitosis and certain peculiarities in the medulla. Levine[681] found that a similar application of 3,4-benzopyrene and methylchloranthrene induces necrosis. Under the action of diverse concentrations of the cecidogens 1,2,5,6-dibenzanthracene, 3,4-benzopyrene and 20-methylchloranthrene, meristem tissues *in vitro* culture produce tumors, which, though not malignant, are not very much unlike those induced by the crown-gall bacterium [684-687]. Hamner & Kraus[434] have described the development of large galls, both on the stem and pods of *Phaseolus* by applying mixtures of indole-acetic acid and lanolin and Gautheret[367] found similar tumors on *Lycopersicum esculentum*. These "chemical galls" are characterized by their limited hyperplasy. Hough[536] has recently suggested that while growth hormones like indole-acetic acid can themselves produce overgrowths, they normally, however, play a secondary rôle in gall induction. Desoxyribose nucleic acid in combination with the auxins appear to be responsible for tumor induction. Although these experiments serve to show that the cecidogenetic factors are largely if not exclusively chemical, the conditions in cecidogensis are without doubt far more complex. As none of the chemicals tested so far give rise to hyperplasia comparable to the condition met with in any gall, the question naturally arises as to whether there is a specific cecidotoxin.

5. Is there a cecidotoxin?

The concept of a specific cecidotoxin may be traced back to MALPIGHI, who was also perhaps the first to recognize the chemical nature of the gall-inducing factor and trace it to the secretions of cecidozoa. He believed that a poisonous fluid, injected into the plant tissue by the cecidozoa at the time of egg laying, initiates the formation of galls. We have seen that the action of a fluid injected by the female of *Pontania* with its egg into the plant tissue is responsible for the initiation of cell proliferation in the development of the kammergall on *Salix* (fig. 2). Even if the eggs are destroyed, the initiation of cell proliferation does not stop, so that there seems to be little doubt about the cecidogenetic properties of the liquid secreted by *Pontania*[77, 730]. In most other cases, if any liquid is injected by cecidozoa at the time of oviposition, it does not seem to have any significant effect on gall induction. In the majority of zoocecidia, the immediate stimulus for cell proliferation proceeds from only the larva or at least the active feeding stage in the life-history of the cecidozoa. Though RÖSSIG[983] believed that the malpighian tubules are the source of the gall-inducing secretion, it is now known that cecidogenesis is associated with the salivary secretions of cecidozoa. The active cecidogenetic substance is soluble in water and has the capacity to penetrate the membrane of plant cells. In some cases previous wounding of the plant tissue seems to be necessary, but even uninjured and remote cells also react to it.

The salivary secretions are primarily digestive liquids. The saliva is either simply poured on the plant cells or is injected into them at the time of feeding by cecidozoa. The mixture of saliva and the cell contents is either sucked up or absorbed by the cecidozoa. A great deal is known about the composition, digestive action, mode of secretion and injection, the mechanism of suction and the reaction of the plant cells to the chemicals contained in the salivary fluid of diverse insects. The contributions of MAGNUS[730], BÖRNER[106], DEWITZ[266], NIERENSTEIN[868, 869], SCHÄLLER[1014], KLOFT[603, 604], NOURTEVA[878-881], ZWEIGELT[1292], ANDERS[12-19], NYSTERAKIS[883-886], HOPP[483], GODAN[392, 393], WEIDNER[1235] and others listed in the bibliography cover various aspects of the problem of salivary secretions of mites and insects, with special reference to cecidogenesis.

In accordance with its primary function, the saliva of cecidozoa contains several well known digestive ferments like diastase, amidase, protease, etc. KOSTOFF & KENDALL[619] have reported, for example, the presence of abundant diastase, protease, amidase, cytase, etc. in the salivary secretions of the larvae of *Neuroterus* from galls on *Quercus*. According to COSENS[222], the saliva of the larva of the cynipid *Amphibolips confluens* (HARRIS), also causing galls on *Quercus* in North America, contains an enzyme capable of converting starch of the plant cell into soluble sugar. COSENS even supposed that, as a result of the action of the ferments in the saliva, the nutritive zone of the developing gall becomes stored with unusually large amounts of available nourishment, both for the larva and the plant cells. Cell proliferation was thus explained by him on the basis of the fact that the protoplasm of the plant tissue became unusually active, since it receives an abnormally large quantity of nutritive material within a limited and localized area. NIERENSTEIN has also made the extremely interesting observation that, in addition to primarily digestive enzymes, there are other enzymes in the saliva of cecidozoa. The saliva of cynipid larvae contains invertase and diastase, which act on the cell contents, and the products of this reaction are sucked up by the larva. The tannin produced by the plant cells precipitates these digestive and cell dissolving enzymes present in the saliva and thus makes them inactive. The tannin thereby acts as sort of a barrier to the cell dissolving activity of the cynipid larva. The tannin of the plant cell is hydrolysed to gallic acid by tannase, another enzyme present in the cynipid saliva. The gallic acid changes to pyrogallol, which is oxidized to purpurogallin by other oxidizing enzymes, so that the tannin eventually disappears more or less completely in the cell. The disappearance of tannin facilitates the action of invertase and diastase. The saliva of *Pontania proxima* LEPEL giving rise to the kammergall on *Salix caprea* contains also appreciable quantities of tannase. Nearly 60% of the tannin of the plant cell combines with diastase and invertase, thus inactivating them. The production of tannase, however, protects the digestive enzymes. KOSTOFF & KENDALL *(op. cit.)* have also demonstrated the presence of several enzymes in the gall. They believe that the diastase found in certain galls is probably a product of some unicellular organism associated with cynipid larva. The growth of the gall ceases when the antibodies of the plant neutralize the effects of the cynipid secretions. The neutralizing zone is marked by sclerenchyma cells, which are thus apparently the reaction products.

Although the salivary secretions of diverse cecidozoa like Tenthredinidae, Cynipidae, Itonididae, etc. have been intensively studied, the most outstanding results, in so far as cecidogenesis is concerned, are associated with work on aphid galls, especially the grape-vine phylloxera and the elm leaf gall aphid. The path of the piercing stylets of the aphids inside the plant tissue is in most cases intercellular. The plant cells are merely pushed aside, without the cells being directly injured. Thin-walled cells, especially the sieve tissues of vascular bundles, are punctured by the piercing stylets and the contents are thus directly sucked up. Such cells usually undergo profound changes or they may also perish. The saliva flows down through the rostrum and then along the piercing stylets into the wound. When freshly discharged into the plant tissue, the saliva is thin, but soon becomes denser and is readily stained bright red by alcoholic or aqueous safranin or stained bluish-red by carbol fuchsin. It is possible by this means to trace the path of the piercing stylets within the plant tissue. The canaliculous passage pierced by the stylets passes through 8—10 layers of parenchyma cells in the case of *Byrsocrypta gallarum* (GMELIN) on *Ulmus*. The aphid saliva is alkaline and contains, in addition to others, enzymes capable of dissolving starch. The starch granules of the uninjured cells within reach of the intercellular passage pierced by the stylets thus become converted into soluble sugar, which is then sucked up the rostrum. The cells around the canaliculous passage of the piercing stylets also become plasmolysed under the action of the salivary liquid and this seems to have the effect of facilitating the sucking of the cell contents, which are otherwise normally under high pressure. Both cytoplasm and nuclei of the collapsed cells become greatly altered and flow into the passage. The saliva also seems to soften the cell walls, so that, without being mechanically injured, the cells along the path of the piercing stylets become readily displaced to make way for them. Cells differing widely in their specialized contents such as oil, tannin, crystals, etc. are as readily affected by piercing stylets and salivary secretions as ordinary parenchyma cells. The tips of the piercing stylets seem also to be capable of perception of the physical, mechanical and chemical conditions in their path within the plant tissues. The enzymes contained in the saliva appear at first to favour an increased flow of sap to the cells and only later either accelerate or inhibit growth. Recent evidence brought together by ANDERS [12-19] seems to show the presence of proteolytic enzymes like protease and peptidase in aphid saliva. Although some workers have also earlier found proteolytic enzymes in the saliva of aphids, the rôle of these enzymes in cecidogenetic changes had not been suspected before ANDERS.

The demonstration of digestive enzymes in the saliva does not, however, account for the induction of cell proliferation by salivary action, especially when it is recollected that similar enzymes are present in the saliva of even the non-gall-forming insects. It was therefore assumed that the saliva of cecidozoa contains some specific gall-inducing principle or cecidotoxin. With perhaps the exception of ANDERS, all other workers have not only assumed the existence of such a cecidotoxin but have also taken it for granted that it is a secretion of the salivary glands of cecidozoa. BÖRNER[106], SCHÄLLER [1014], DEWITZ[265, 266] and others have suggested an "aphidolysin" present in aphid saliva as altering the growth of young plant cells pathologically. As mentioned above, similar ideas had been put forward by BEIJERINCK[75], who believed that the toxic substances produced by the cynipid *Andricus circulans* MAYR neutralized the plant cell contents. ZWEIGELT[1299] is also of the view that the development of a gall represents the successful expression of "toxin paralysis" or toxin neutralization, enabling the plant cell to survive a sort of phagocytic action within the imperilled zone induced by specific irritation. The suggestion has been made in recent years that the cecidotoxin is either a heterauxin (growth promoting hormone) or at least something in the nature of such a hormone, contained in the salivary secretions of cecidozoa. The action of this hormone has been assumed to induce cell division in cecidogenesis. Though some of these auxins have been found in the secretions of cecidozoa, we have no proof that they are actually produced by cecidozoa and are also concerned in cecidogenesis. Most early workers were unable to obtain adequate quantities of uncontaminated aphid saliva, but experimented with extracts of saliva-honeydew mixtures and found in them certain auxins. From these early investigations, NYSTERAKIS[883–886] and HOPP[483] concluded that indole-3-acetic acid is present in these extracts and that this auxin is the active cecidogenetic principle of aphid saliva. This auxin has been considered to be responsible for polyploidization and club formation in root[483]. Heterauxins are not strictly speaking nutrients but are rather sublethal toxins. It is now generally recognized that growth is controlled by certain substances produced by buds, leaves, tips of coleoptiles, etc. at periods af active growth. These substances are transported to remote parts and growth in these parts depends on the amount of the transported auxins. The chemical nature of these substances is known. Auxin a ($C_{18}H_{32}O_5$ or auxentriolic acid) and auxin b ($C_{18}H_{30}O_4$ or auxenolonic acid) have, for example, been isolated from higher plants and indole-3-acetic acid ($C_{10}H_9O_2N$) from fungi. Other substances like vitamin B_1 are also known to behave like auxins. The auxins move in polar direction from tip to base in stem, leaf, petiole, etc. The auxins promote growth by cell enlargement, induce cell division in the cambium and have also growth inhibitory action. Auxins inhibit, for example, lateral bud development and root elongation. Inhibition occurs mainly at concentrations higher than those at which growth promotion takes place. The crown-gall bacterium is also known to produce abundant heterauxins and some workers have even assumed that the induction of the crown-gall is due to the action of these auxins. The gross visible effect of auxins is a modifition of the extensibility of the cell walls[461] preceded by accelerated protoplasmic streaming. The protoplasmic streaming is dependent on an adequate supply of carbohydrates[1132] and is sensitive to respiratory poisons[109a, 1131a]. Indole-acetic acid has been suspected to have the capacity of gall induction like perhaps colchicin, acenaphtha and other polyploidizing materials. There are several fundamental differences between galls and the tumors induced by the action of indole-acetic acid. To mention only a few, the nucleus is nearly always central in gall cells and is also rounded, but in tumors induced by indole-acetic acid it is peripheral and more or less rectangular[64]. Polyploidization of meristem cells is also not as complete in auxin tumor as in galls. Exhaustive tests have indeed failed to demonstrate the presence of indole-3-acetic acid in the saliva of the grape-vine phylloxera.

That the induction of galls under the action of aphid saliva is not due to indole-acetic acid is evident from several facts. The optimal concentration of indole-acetic acid, applied all around the root tip, is 10^{-6} to 10^{-8} mg/cc for the formation of a clavate swelling. As the aphid injects extremely minute quantities locally and as the saliva is also greatly diluted by the plant juices, the initial concentration of indole-acetic acid must be very much higher if we are to consider it as playing a significant rôle in cecidognenesis. The recent experiments of ANDERS[18] have shown that only in dilutions of 1000 to 10 000 times does the aphid saliva induce root swelling. Undiluted saliva causes, however, necrosis if applied all around the root tip in a sufficient amount. If really such high concentrations of indole-acetic acid are present, it should be possible to detect its presence by suitable chemical methods and yet none of the chemical tests has so far shown the presence of even traces of this auxin in the uncontaminated aphid saliva.

Recent investigations on the gall-inducing principle of the grape-vine phylloxera saliva undertaken by ANDERS *(op. cit.)* have shown that the cecidogenetic principle is thermostable and can be maintained indefinitely at 4 °C, dialysable and resistant to the action of ultra-violet rays. Pure and undiluted aphid saliva fluoresces on excitation by ultra-violet weakly at first, but after about 15 minutes dark blue-violet. The pH is unexpectedly high, viz. 9 (it is interesting in this connection to note that leaf sap has a pH of 3—4, but the gall tissue has a somewhat higher pH than the healthy tissue). The saliva is also highly hygroscopic. ANDERS demonstrated the presence in the saliva of three orthophenols characterized by a blue, blue-green and brownish fluorescence. These are insoluble or only slightly soluble in chloroform, ether or benzene, slightly soluble in alcohol and readily soluble in water and methenol. Treatment with potassium hydroxide solution abolishes the fluorescence and the colour changes to brown. Fresh aphid saliva fluoresces blue and therefore contains pure orthophenols.

ANDERS[19] has given an interesting account of the paper chromatographic analysis of the cecidogenetic principle of aphid saliva and reports the finding of ninhydrin positive bodies like lysin, histidin and tryptophan. In addition to these three dominant substances, he found also two other ninhydrin positive substances, viz. glutaminic acid and valin in smaller amounts. He believes that the saliva contains high concentrations of these amino acids, totalling to nearly 7%. In addition to these amino acids, some sugars have also been detected, but these latter have no cecidogenetic significance. The paper chromatographic investigations seem to indicate that the most important cecidogenetic substances are perhaps strong basic lysin and the two heterocyclic compounds histidin and tryptophan. The high pH, viz. 9, referred to above, would also appear to be readily explained by the presence of these substances.

SCHÄLLER[1014] has recently shown that the amino acids present in the salivary liquid play an important rôle in the trophic relation between sap-sucking insects and the plants on which they feed. He developed a method of investigating the amino acids present in the uncontaminated saliva of *Aphis pomi* by paper chromatography. The amino acids found by him include alanine, glutaminic acid, aspartic acid, valin, serine and two ninhydrin positive compounds not definitely identified. Alanine and glutaminic acid occur in larger amounts than the others. The honeydew of the same aphid contains about twenty amino acids and the amides asparagin and glutamin in high concentrations. Threonine, serine and glutaminic acid occur in high concentrations, but in the saliva asparagin, glutamin and threonine are absent.

ANDERS investigated the cecidogenetic action of the amino acids contained in the aphid saliva and found that each of the five amino acids tested by him is capable of inducing clavate swellings of the root tip in suitable concentrations. The growth abnormalities that thus arise at the root tip are generally specific to the substance. Stout root swellings are indeed, however, caused by only certain amino acids. Most of these substances, however, inhibit the growth, so that root tips become merely slightly swollen. The specific action of lysin, histidin, tryptophan, glutaminic acid and valin are of particular importance in connection with cecidogenesis. The remaining fifteen amino acids found by him are not considered to be of importance to the problem of cecidogenesis. Since glycol occurs in a relatively high proportion in phylloxera saliva, some attention was also paid to it. Tryptophan induces the most pronounced swelling of the root tip. It is significant to note the relatively weak concentration of 0.00006 to the relatively high concentrations of 0.25%, in which it can be applied without any noticeable injury to the plant tissue and give rise to clavate swellings. The observations on experimental tumor production by the application of L-tryptophan on *Phaseolus* are of considerable interest in this connection[624]. It may also be mentioned that the crown-gall bacterium synthesizes indole-acetic acid from tryptophan. Histidin in favourable concentrations also induces a clavate root swelling, but the swelling is never so pronounced as in the case of tryptophan. The seedlings show evidence of toxic effects of histidin in all concentrations. Even the dosage range is narrower than in the case of tryptophan. Histidin occupies the third place after tryptophan, the second being taken by glycin, with regard to the potency of production of clavate swelling of the root tip. Lysin does not on one hand induce any typical clavate swelling like that induced by tryptophan and histidin and on the other hand lysin does not also cause total inhibition of growth. Lysin appears to bring about a strengthening of the normal longitudinal growth, the end effect of which is that the elongated root swelling arises, which can only be described as vaguely clavate swelling. Lysin is able to bring about this condition in concentrations of 0.006–0.2%. Valin and glutaminic acids are rather poor inducing agents and the latter appears merely to retard growth, so that the swelling is generally minimal.

Although experiments with isolated amino acids have shown that more or less pronounced swellings arise, none of these swelling are comparable in size with those induced by total aphid saliva. Mixtures of amino acids appear therefore to be involved in the development of swellings induced by total aphid saliva. ANDERS studied the tumor induction with mixtures of i. tryptophan-histidin-lysin-glutaminic acid-valin, ii. tryptophan-histidin-glutaminic acid, and iii. tryptophan-lysin-glutaminic acid-valin. The results of these experiments show that combinations of tryptophan, histidin and lysin are the principal cecidogenetic agents in the saliva of the grapevine phylloxera. There may also perhaps be other as yet unknown components in the saliva, which are important in cecidogenesis. Cytological and histological investigations have also established the identities of the amino acids in the aphid saliva as the chief cecidogenetic agents. The principal difference between the clavate swellings experimentally induced by amino acid action and the gall caused by the aphid is in the mode of application of the cecidogenetic agent. In experimental galls the amino acids are applied externally on the actively growing root tip, but in natural gall formation the insect injects them into the plant tissue. Amino acids have also been injected with the help of a suitable micromanipulator into plant tissues in concentrations ranging from 0.000 0003 to 10.0%. It has been found that concentrations lower than 0.5% are completely ineffective. The ferment system and the amino acids represent essentially the cecidogenetic agents.

Enzymes have been found not only in aphid saliva but also in the saliva of other cecidozoa and also in the case of cecidogenetic fungi enzymes have been detected. *Ustilago zeae* is known, for example, to produce enzymes capable of dissolving the cell contents of its host. Recent investigations on the interactions of parasitic and saprophytic fungi in different host plants have shown that the metabolic products of the fungi have far-reaching influence not only on each other but also on the host cell. The gall of *Puccinia triticina* develops, for example, in the immediate vicinity of patches of the mildew *Erysiphe graminis* on leaves of *Triticum*, normally highly resistant to the *Puccinia* strain. The mildew fungus splits up some of the complex compounds into a simpler form for *Puccinia*. The mildew fungus develops haustoria in epidermal cells and *Puccinia* in mesophyll, so that it seems that the mildew fungus is able to affect adjacent cells. The demonstration of acquired physiological immunity through secretion of antibodies in plants is also proof of the action of toxic products of parasitic fungi. The inhibitory action of Hyphomycetes, Ascomycetes, Basidiomycetes and Pyrenomycetes in culture media seen in the fact that one fungus dissolves the hyphae of the other at a distance is also strong evidence of enzyme action. Fungi produce both growth promoting and growth inhibitory substances. *Ophiobolus* produces, for example, substances passing through a Chamberland filter and promoting the growth of *Aspergillus niger* and other substances not thus passing through the filter and with growth inhibitory action.

6. Cecidogenesis a specific reaction of the plant cell

Assuming that certain amino acids in suitable combinations are the determining factors in cecidogenesis, the question now arises: Are the amino acids secreted by the salivary glands? SCHÄLLER *(op. cit.)* advanced the view that the amino acids are the breakdown products of proteins in the food and that they are circulated in the haemolymph to reach the saliva and are then injected into the plant. It has, however, been shown by ANDERS that pure uncontaminated aphid saliva does not contain amino acids, but only the protease enzymes. The young aphid thrusts its piercing stylets into the plant tissue in order to suck the nutrient liquids. The aphid does not pierce at random, but selects certain regions, the tissues of which are packed with actively dividing and growing cells, such as young leaves, root tips, etc. and avoid tissues which are not meristematic or at least not sufficiently meristematic. The fact that feeding is effected from only such tissues naturally ensures a high protein content in the food and it is only in such tissues also that secondary protein synthesis is actively going on in the plants, so that a relatively high amount of plant proteins is available to the aphid. Some free amino acids may be present in such tissues, but these seem to occur in such insignificant amounts that we need not consider them as important in cecidogenesis. Since the saliva of cecidozoa contains only proteolytic enzymes, which come into contact with the liquid food before being absorbed, it is evident that the plant proteins are converted by the aphid saliva enzymes into amino acids by hydrolysis and the whole is then absorbed by the aphid. The mixture of plant juice and saliva, with the proteins of the plant converted into amino acids by the proteolytic enzymes of the saliva, seems to serve not only as aphid nutrients but also as cecidogenetic material. The amino acids may be excretions of the aphid or may also represent regurgitated material. ANDERS inclines to the view that the amino acids are simply regurgitated by the aphid into the plant tissues. It

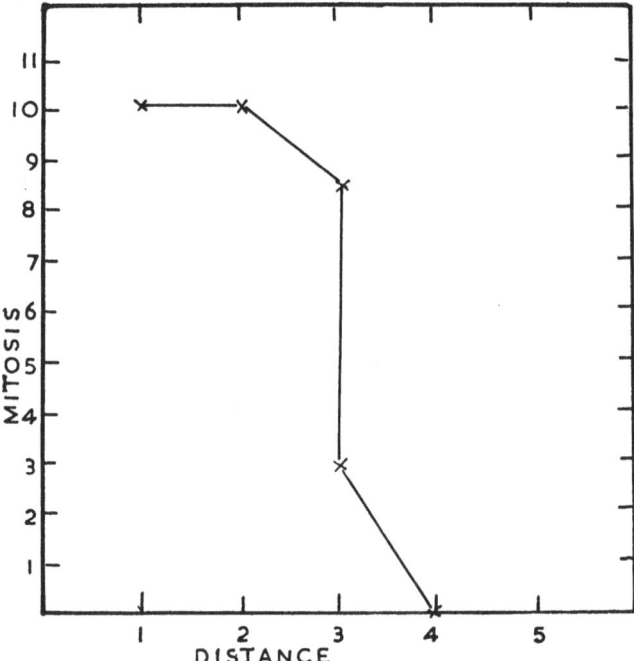

Fig. 145. Gradient of mitosis from the larval cavity in the gall of *Acroëctasis campanulata* Mani on *Sabia campanulata* Wall. Distances are measured in terms of the radius of the larval cavity, along an imaginary line joining the centre of the larval cavity and the centre of the cambial ring in the stem, which represents the axis of symmetry of the stem-gall complex. Mitosis is measured in term of the mean numbers of actively dividing cells along the axis of symmetry in each distance zone.

is important to note that in no case the amino acids represent products of the salivary gland activity: they are not produced by the salivary gland cells of the aphid. The amino acids thus arise not in

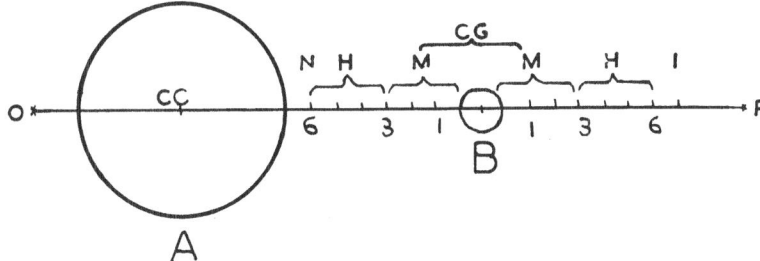

Fig. 146. Diagrammatic representation of the cecidogenetic and the normal morphogenetic fields, with the axis of symmetry of the stem-gall complex OP. A. cambial ring. B. larval cavity. CC. centre of the cambial ring in the stem, which may be considered as the centre of the normal morphogenetic field. CG. centre of the cecidogenetic field, coinciding with the centre of the larval cavity. H. zone of intense cell hypertrophy. I. indifferent zone. M. zone of intense cell division. N. zone of neutralization of the cecidogenetic gradient. The numbers 1-6 are the distances in terms of the radius of the larval cavity measured along the axis of symmetry.

the cecidozoa but in the plant cells under the action of the proteolytic enzymes present in the saliva. The aphid starts regurgitating some time after it has sucked the plant juice. The sucking by the aphid is a discontinuous action and is repeated as long as the aphid continues to feed. The cecidogenetic action of cecidozoa therefore consists only in causing the production of certain amino acids from the proteins present in the plant cells for its own nutrition and then releasing a part of the free amino acids into the plant tissue.

The effect of free amino acids on the plant cells is polyploidization. The polyplodization is preceded by cytoplasmic streaming and certain other cytoplasmic changes. Since the polyploidizing influence persists over prolonged periods in natural gall development, the polyploidization of individual cells is repeated many times, so that the end effect is a high degree of polyploid cells, with an abnormally large size. The local accumulation of polyploid cells is conceivably the first phase in cecidogenesis. The mechanism by which polyploidization arises is only imperfectly understood. The polyploidization induced by high concentrations of amino acids is similar to that observed in c-mitosis (inhibition of spindle formation) and that induced by low concentrations of amino acids resembles true endomitosis (without spindle formation). The endomitosis induced by amino acids gives the impression not of a general inhibitory course of normal mitosis but of accelerated auto-reproduction of chromatic material, with which the spindle formation does not apparently

278

keep pace. It would thus appear that the amino acids injected by the cecidozoa do not in reality play the rôle of a nuclear toxin, but rather act as stimulants, especially in their optimal concentrations and possibly as bricks in the synthesis of nucleoproteins and proteins in gall cells. To the intermediate products, viz. the amino acids, the plant reacts by accelerated and intensified nucleotide and protein synthesis. Spontaneous cell subdivisions are also associated with polyploidization, especially in the cells originally within the range of the influence of the substances from cecidozoa. We thus find the extraordinary situation that there is strictly speaking no specific cecidotoxin at all, either secreted or elaborated by cecidozoa. On the other hand, the factors which induce cecido-genesis arise rather within the plant cell itself and are specific to the

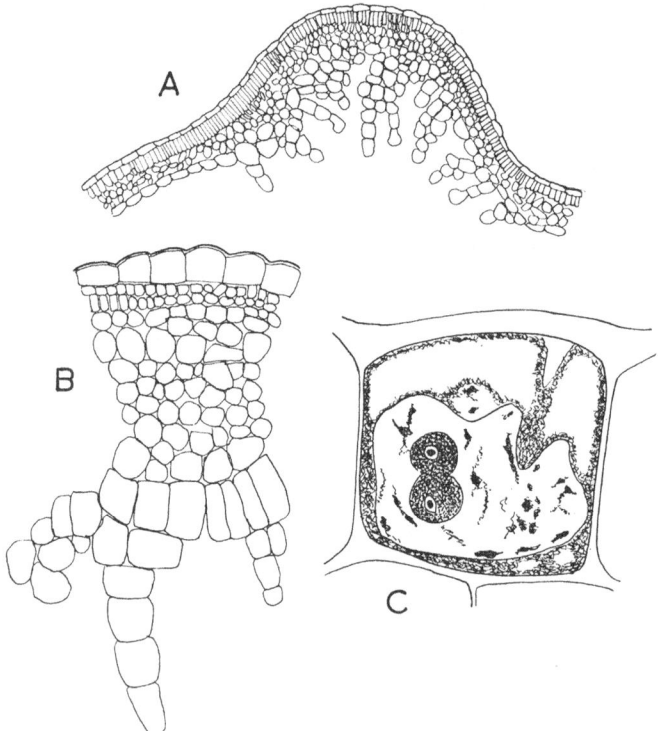

Fig. 147. Cells of acarocecidia. A–B. Regenerative callus-like outgrowths of the epidermal cells in the gall of *Eriophyes* sp. on leaf of *Strobilanthes dalhousianus* Clarke. The cecidogenetic reaction is almost wholly confined to the epidermal cells at first, as only single cells are attacked by the mites. The cecidogenetic reaction spreads later on to only the cells of the spongy parenchyma to a small extent.
    C. Single epidermal cell pierced by *Eriophyes galii* Karp (After Nèmec).

279

plant, its organ and its cells. Cecidogenesis is a highly specific and specialized type of plant reaction to the secretion of the cecidozoa and not every part or every plant exhibits this reaction always. Every phytophagous insect presumably pours its saliva into the plant tissue and in all cases diverse digestive enzymes are doubtless present in the saliva, but every insect cannot bring out the specialized response from the plant. The physico-chemical changes that induce hyperplasy are not also produced by the cecidozoa but by the plant cell itself.

7. General principles of morphogenesis of gall

The problem of morphogenesis of even the principal types of galls would be outside the scope of this work. We may, however, give here a brief outline of the fundamental principles of morphogenesis of galls in general.

Irrespective of the underlying factors and the mechanism of cecidogenesis, the general morphogenesis of the gall is governed by the same fundamental principles which we observe even in normal morphogenesis. Normal morphogenesis is governed by the morphogenetic field, characterized by a graded decrease of develop-

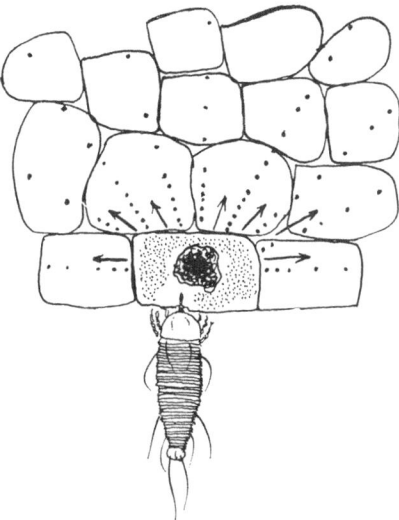

Fig. 148. The cecidogenetic gradient, spreading with diminishing intensity from the cell pierced by *Eriophyes*. The effect of the reaction of this cell spreads to the other cells immediately in contact and only to a lesser extent to the remoter cells. Note the nuclear gigantism of the cell attacked by the mite. There does not appear to be any significant diffusion of the saliva of the mite; but the changes in the cytoplasm of the pierced cell induce the cecidogenetic gradient.

mental activites in all directions from a region of maximum intensity, viz. the morphogenetic centre, passing eventually into an indifferent zone. The spatial developmental pattern is primarily

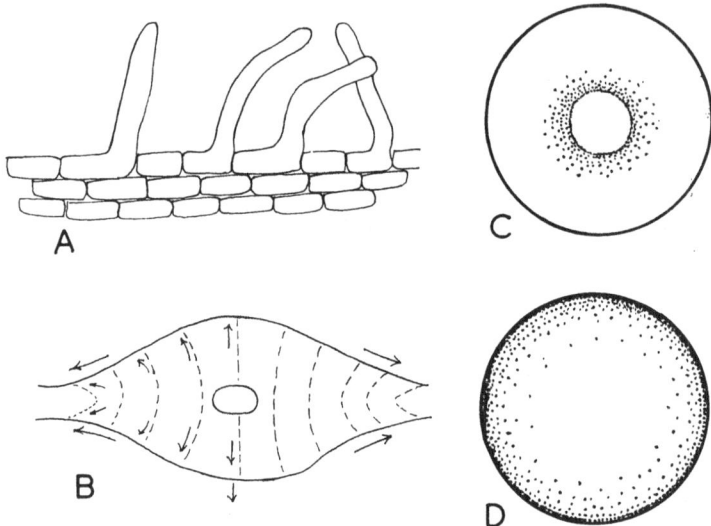

Fig. 149. The direction and localization of cell growth and cell division in cecidogenesis. A. Normal polarity of the epidermal cells persists in cecidogenetic response in erineal growth in the gall of *Eriophyes* on a branch; note that all the cells grow out into hairs from the same pole.   B. Leaf gall with cell division in pericline direction and growth in thickness. With the weakening of the cecidogenetic gradient at the two ends, the dominant growth becomes more and more surface growth.   C. Cell division central, with internal cecidozoa, localized around the larval cavity. The zone of hypertrophy is peripheral and the size and shape of the resulting gall are characteristic.   D. Zone of cell proliferation peripheral, with hypertrophy zone central and the growth and shape irregular.

quantitative and concerns differences in the rates of metabolism. Graded differences exist, for example, in the rates of cell division, degree of differentiation, intake of nutritive substances, respiration, synthesis and breakdown of material, concentrations of enzymes and hormones, etc. The highest region in the gradient — the most intensively active region — is functionally dominant and controls the activity of other neighbouring regions. The dominance of an actively developing mass of cells over another similar mass of cells is expressed as inhibition. In the plant, for example, the actively developing terminal bud inhibits the development of axillary and other buds below it to a certain distance. In his researches on *Phaseolus* Mc CALLUM[768a] showed that the active terminal growing point inhibits other growing points and that this inhibition is not really competition of material and wounding does not also constitute a necessary stimulus to regeneration. Buds inhibit buds and roots inhibit roots but not other

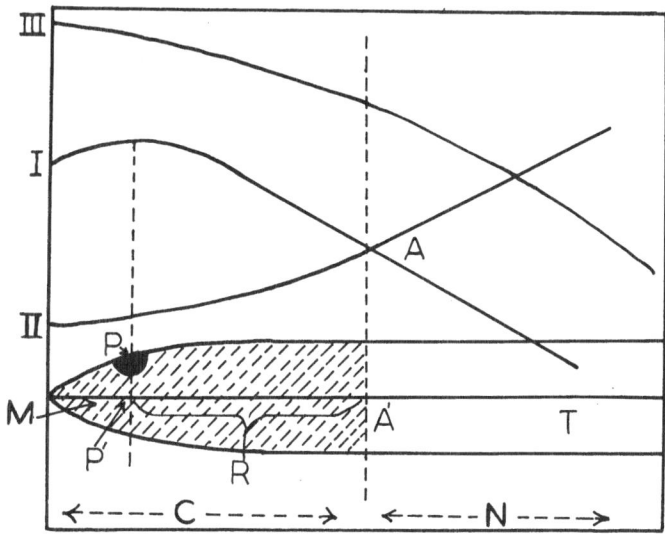

Fig. 150. Diagram of the gradients operating in cecidogenesis according to Prat. I. Chemical gradient. II. Gradient of resistance of the plant tissue to the cecidogenetic influence. III. Gradient of sensitivity of the plant tissue. C. zone of cecidogenesis (shaded). M. apical meristem. N. indifferent zone. P. seat of attack. R. PA cecidogenetic gradient. T. stem.

organs. The cells capable of developing and growing into new organs are held in check by the organ already actively developing; the more actively this develops the greater its check on other similar organs. WENT[1247] demonstrated the so-called apical hormone, now known as auxin, to be probably indole-acetic acid. This is known to inhibit buds, but to stimulate stem and root. (cf. SINNOT [1048a]).

Changes in activities in the course of development give rise to new centres of activity and new gradient systems, often with metabolic processes of a wholly different type. The gradients arise and undergo changes in a definite order and relation one to another and gradients also precede sharply defined and bounded morphogenetic differentiations. Increase of these differences affects the activities in other parts and the functional inter-relations and integrating factors become increasingly effective and varied in character. A gradient arises by the localization of an activated region, from which the activation spreads, with decreasing intensity as distance increases and with different intensities in different directions, depending upon the presence or absence of other gradients. The active region corresponds to a region of excitation. The definite order and sequence of development in any given direction are based on the gradient in that direction. Two or more gradient systems in two or more different directions, operating at the same time, constitute a coördinate system of developmental pattern of symmetry. The resultant of two or more equally intense and same kind of gradients is the direction of the axis of developmental pattern. Several such centres may often have indefinite and over-lapping boundaries. The morphogenetic field, in its simplest form, is thus nothing more than a gradient system, in which the different gradients serve as vectors of the field, determine its extent and orderly relations within it. In the course of development, the potencies of the morphogenetic field are actually localized to the high regions of the gradient system, viz. to the regions which are the most intensively active. The cells along the course of a gradient of any developmental activity represent different levels of activity and development. The polarity of cells has also its roots in these differences. A cell is morphologically polarized if we can recognize formal differences at opposite ends. Of far greater significance in developmental mechanism is the physiological polarity, recognized by differences in the reaction to identical stimulus at the two poles of the cell. The pole of the cell near the gradient centre reacts wholly differently from the distant pole. Cell polarity was first clearly recognized by MIEHE[792a] and the later researches of VÖCHTING[1219] have demonstrated the existence of not only the root-shoot polarity but also medulla-cortex polarity in the cells of shoot axis. According

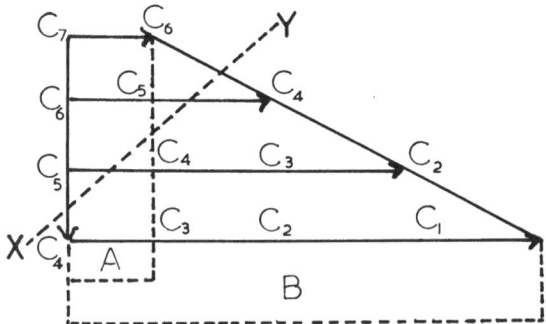

Fig. 151. Graphic representation of the cecidogenetic gradient of chemical action according to Garrigues. On the abscissa A size of the leaf attacked by the cecidozoa. B. the size of the mature gall. Between A and B are intermediate stages of development of the gall. On the ordinate the concentrations of mitoses induced by the cecidozoa represented by C_7, C_6, C_5, etc. The numbers affixed after C represent the relative value of concentrations of mitosis. C_7-C_5 concentration of hyperplasia on the line x-y. C_4-C_1 concentration of hypertrophy along the line x-y.

to VÖCHTING, a number of abnormal structures may be satisfactorily explained in the light of the polarity concept. The cells are not only polar differentiated longitudinally, but it has been observed

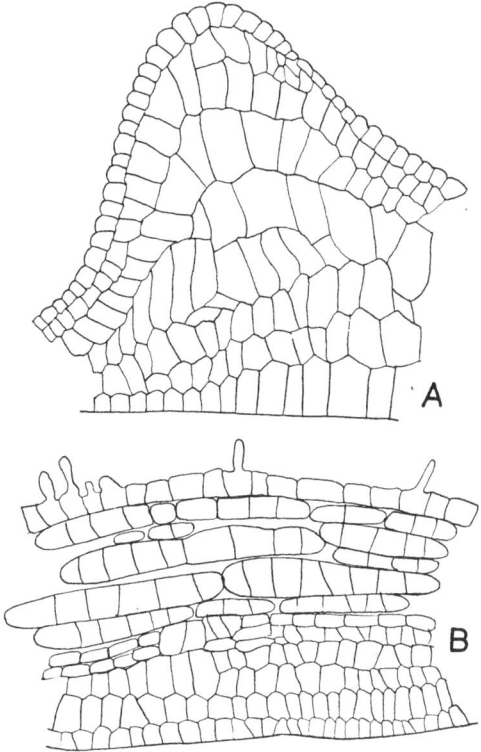

Fig. 152. Direction of cell division and cell enlargement in leaf gall. A. Growth in thickness and surface growth in young gall of *Phemphigus bursarius* (Linn.) on *Populus*. B. Anticline cell division and surface growth in the gall of *Pemphigus filaginis* Fonsc. on *Populus*. (After Küster).

that like poles of cells repel each other, when they grow opposite to each other. Recent advances in cellular interactions in plants and animals have shown that polarity is an integrating factor, in addition to direct cellular interaction, in organization[972a]. The cells of any part are determined and continue to function at that part only so long as they represent a certain relative gradient level and are in functional relation with other parts. When isolated or freed by some means from that relative position, any cell may become the centre of a dominant region. Recent researches have shown that cells are not irreversibly differentiated and the ability to differentiate is by no means lost as a result of differentiation. Alterations, obliterations and determinations of dominance and gradients result in new axiate patters of development. Reorganization of other gradients is associated with the high end of the gradient system. Environmental differential or gradients of different kinds determine new gradient and create new dominance. New patterns are thus determined either by activation from other regions or by external differentials such as light, gravity, electric currents and hydrogen ion concentrations, oxygen tension, concentrations of enzymes, hormones, amino acids, etc. Polyploidy, nuclear gigantism, multinucleate condition, excess of free amino acids, presence of proteolytic enzymes and other situations described earlier as resulting from the cecidogenetic action of the saliva of cecidozoa or of the substances elaborated by cecidogenetic fungi, are indicative of profound chemical, functional and polarity changes in the cell. The cell influences other neighbouring cells within a certain distance all around, with diminishing intensity as the distance increases. Any factor which changes the internal environment of the plant cells eventually results in the localization of a new activated region and the appearence of a new gradient field.

We have seen that the secretions of gall-inducing organisms give rise to the formation of certain substances in the cytoplasm of the plant cell, favouring polyploidy and other conditions that eventually result in repeated cell division. The concentration of these substances, which may for the sake of brevity be called cecidogen, is obviously maximum in the cells immediately surrounding the gall-inducing

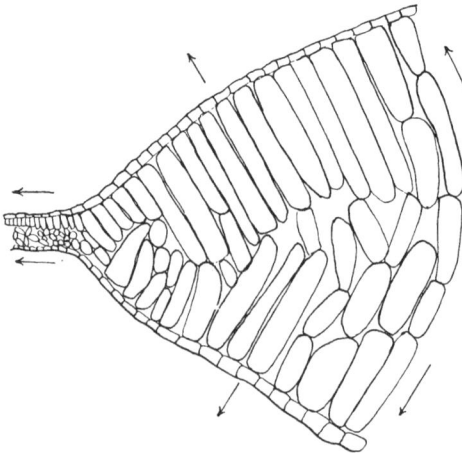

Fig. 153. Cecidogenesis by predominant cell enlargement, showing the direction of elongation of the cells, representing an exaggeration of the normal morphogenetic tendency of the cells.

organism. The affected cells now constitute the peak of a new metabolic gradient and thus become the centre of excitation and acquire dominance over other similar cells, in so far as protein synthesis, amino acid formation and cell division are concerned. This localization of cell division creates a new gradient field, viz. the cecidogenetic field of cecidogenetic gradient or gradient of active cell divisions. The cecidogenetic gradient corresponds to HOUARD's rayon d'activité cécidogénétique[485]. The cecidogenetic gradient determines the sphere of the cecidogenetic influence, the centre of which coincides with the centre of the larval cavity in zoocecidia. Whether a specific cecidotoxin is secreted by the gall-inducing organism or the cecidogen is merely a reaction product arising within the plant cell itself under the action of the substances produced by the gall-inducing organism, its concentration must naturally be maximum at its source, viz. around the gall-inducing organism. We thus find

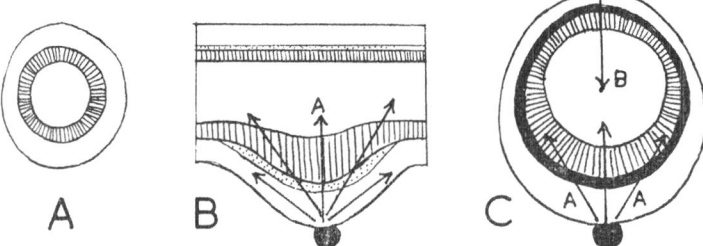

Fig. 154. Cecidogenetic field in stem gall, with the cecidozoa situated externally on the epidermis. A. T.S. normal stem for comparison. B. L.S. of the galled part of the stem. A–A. lines of cecidogenetic action. C. T.S. through gall-stem, B normal morphogenetic field resistance forming the axis of symmetry of the stem-gall complex with cecidozoa external. (After Houard).

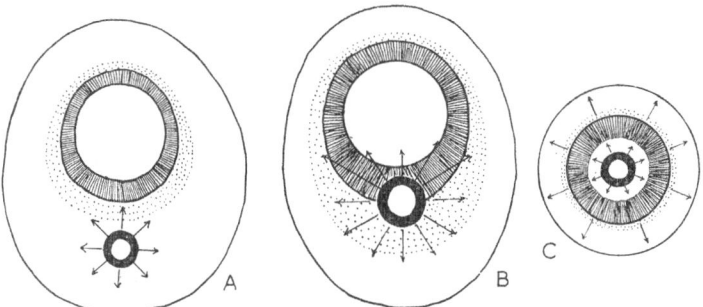

Fig. 155. Cecidogenetic field in stem gall. A. Cecidozoa and the larval cavity positioned within the cortex. B. Cecidozoa and the larval cavity positioned in the zone of the vascular bundles. C. Cecidozoa and larval cavity positioned in medulla. Arrows indicate the lines of cecidogenetic action (After Houard).

that in zoocecidia the most actively dividing cells are found immediately around the gall organism and as the distance increases there
is a corresponding fall in the intensity of cell division. In many
simple parenchyma galls there is a distinct gradient of cell divisions.
Beginning from the zone of actively dividing small-sized cells immediately around the cecidozoa, we find a gradual fall in the number of
dividing cells, with a corresponding increase in the number of larger-
sized cells as the distance increases. The frequency of cell division
falls distinctly with the increase of distance from the cecidozoa. Measuring the distance in terms of the radius of the larval cavity, the
maximum frequency of cell division is within one radius of the larval
cavity from the cecidozoa. Between one and two radii the frequency
falls appreciably and beyond three radial distance there is practically no cell division (fig. 146, 147).

The size of the cecidogenetic field, which determines the size of the
future gall, depends to a large extent on the length of the cecido-

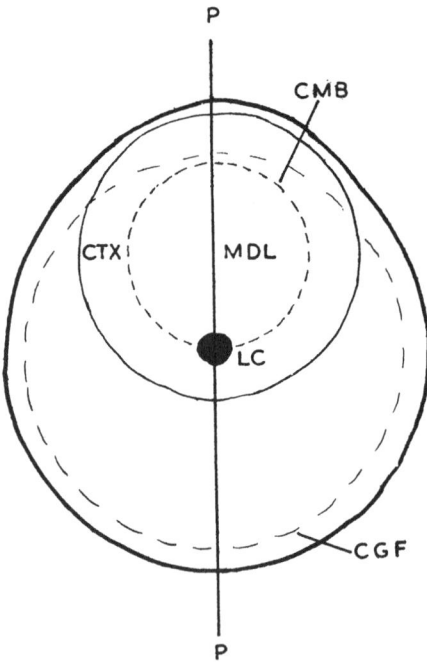

Fig. 156. Diagrammatic representation of the interaction of the cecidogenetic and
the normal morphogenetic field, with the larval cavity positioned in the cambial
zone of the stem. CGF outline of the maximum cecidogenetic field. CMB cambial
ring. CTX cortex of the stem-gall complex. MDL medulla. PP axis of symmetry of
the stem-gall complex, passing through the centre of the cambial ring and LC the
larval cavity (After Houard).

genetic gradient, but a variety of other factors also influences the size and the shape of the cecidogenetic field. The length of the cecidogenetic gradient is determined by the intensity of cell division

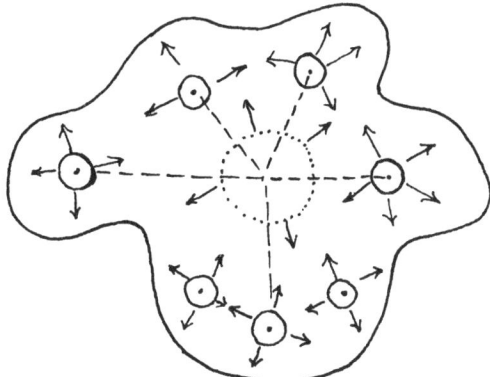

Fig. 157. Cecidogenetic field in multilocular rindengall on stem.

in the cecidogenetic centre and by the rate of possible diffusion of cecidogen. The rate and direction of diffusion of the cecidogen are governed not only by its concentration but also by the structure and metabolic condition of the plant cells, by the size and position of the cecidozoa with reference to the normal morphogenetic field in the gall-bearing organ, etc. The length of the cecidogenetic field is proportional to the nature and size of the gall-inducing organism, its age and stage in life-history, its feeding activity and nutritional requirements, its metabolic rate and the concentrations of enzymes and other material it elaborates. In the case of zoocecidia, the cecidogenetic gradient becomes continually longer with the growth of cecidozoa and as it feeds more intensively and pours in larger amounts of saliva. The cecidogenetic gradient is also proportional to the number of individuals of the cecidozoa present within the plant tissue. The effect of a number of individuals is readily observed in multilocular galls, which are as a rule larger than unilocular galls. Though the eriophyid mites rarely exceed a length of 300 micron, acarocecidia attain relatively enormous proportions because of the large numbers of mites which attack the plant cells. A number of workers like ANDERS, GARRIGUES and others *(op.cit.)* believe that diffusion of the saliva of cecidozoa extends the cecidogenetic field. As indicated earlier, there is, however, ample evidence that the primary reaction of the plant cell is to neutralize the digestive enzymes present in the cecidozoa saliva. There is therefore little or no effective diffusion of the saliva of cecidozoa over any considerable distance, perhaps beyond the cells directly involved. Furthermore,

288

it is certain that at least in the case of mites, the saliva does not diffuse beyond the cell immediately sucked by the mite (fig. 148). The localization of cell divisions and the intense drain of proteins and other nutritive material from the plant cells for the nutrition of cecidozoa have on the other hand the result of protein depletion in the region of the peak of the cecidogenetic gradient. Instead of an outward diffusion of cecidogen from the affected cells we find an inflow of material from other neighbouring cells to the centre of the cecidogenetic field. The cells in the path of the flow of sap become elongated in the direction of the flow and acquire new polarity, sometimes even involving reversal of old polarity. These cells develop eventually into the faisceaux d'irrigation described by HOUARD[485], as ending in the nutritive zone surrounding the larval cavity in zoocecidia. The vessels nearest the cells thus drained become also enlarged and connected with the elongating cells in the path of the flow of sap. The cecidogenetic gradient does not therefore involve either a large-scale outward diffusion of the cecidogen or also of the cecidozoa saliva. We find therefore that the actively dividing mass of cells around the cecidoza is the actual centre of morphogenetic excitation for hypertrophy and spread of hyperplasy into the initially inert zones. It is the localization of the actively dividing cells that acts as the morphogenetic influence in cecidogenesis.

The morphogenesis of the gall is determined not entirely by the properties of the cecidogenetic field, which are more or less profoundly influenced by the normal morphogenetic field that is not wholly obliterated by the cecidogenetic field. The eventual morphogenesis of a gall is thus a resultant of the interaction of the cecidogenetic and

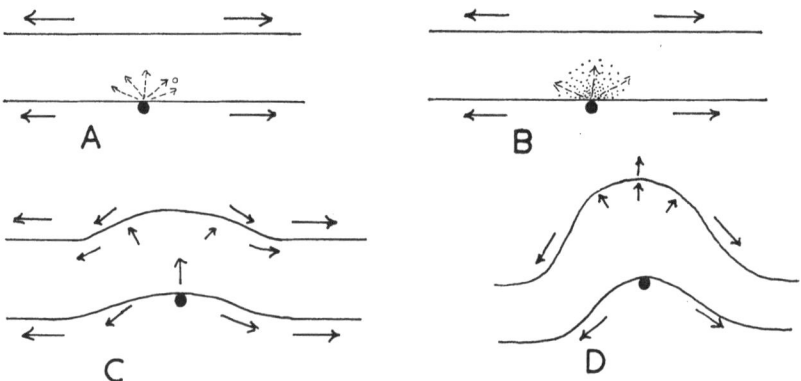

Fig. 158. Cecidogenetic movements in leaf gall, with the cecidozoa positioned on the under side. A–B. Initiation of the cecidogenetic gradient. C. Normal morphogenetic movements become exaggerated locally, giving rise to upward bulge. D. Closing movement.

the normal morphogenetic fields. The length and height of the cecidogenetic gradient and consequently the size and the intensity of the cecidogenetic field are thus very profoundly modified and controlled by the cell polarity and intensity and length of the normal morphogenetic gradients in the organ. For the sake of simplicity, the normal morphogenetic gradient may be considered to counteract and resist the cecidogenetic gradient. The gradient of resistance of the plant tissue to the cecidogenetic action is often defined as the minimum of cecidogenetic agent necessary to just initiate visible physiological or structural modification in the plant cells. As the cecidogenetic gradient falls with increasing distance, the normal morphogenetic gradient also increases until the former is completely neutralized. The zone of neutralization of the cecidogenetic gradient is determined by the lengths of the normal morphogenetic and cecidogenetic gradients. The neutralization is marked in many cases by the presence of thick-walled sclerenchyma cells of the so-called protective zone. The inhibition of cell division by the mitosis hormones from the cecidogenetic centre may also apparently favour the thickening of the cell walls outside the zone of proliferating cells. The cecidogenetic centre acts both as a centre of excitation and also as inhibitory centre on certain organs. MEYER[791] has shown for example that in the gall of *Perrisia urticae* PERRIS on *Urtica dioica* LINN. there is a distant effect of inhibition of growth of certain organs, including inhibition of cytologic differentiation of sex. Some workers have assumed the presence of a gradient of sensitiveness (fig. 150) of the plant tissue to the cecidogenetic action, but this would seem to be unnecessary. The gradient of sensitiveness is in effect reciprocal of the gradient of resistance and normal morphogenesis.

The modifying influence of the normal morphogenetic field on the cecidogenetic field may be readily observed, for example, in the direction in which the cells grow and divide in the course of cecidogenesis. This direction is determined to a great extent by the cell polarities in the region, as these polarities are not always totally abolished. In filzgalls, for example, the hairy outgrowths of the epidermal cells arise in all cells from the same pole (fig. 149). In organs normally characterized by predominance of surface growth the cecidogenetic cell divisions take place horizontally (fig. 152B), thus merely exaggerating the normal destiny of the cells. Where the direction of cell division is inclined to the plane of normal morphogenetic growth, we find that cell division is localized in the palisade parenchyma and galls in this case the direction of cell divisions may also be wholly vertical to the plane of the normal growth (fig. 17M, 77A). In most galls there is, however, a complex combination of surface and thickness growths, involving in part an exaggeration of normal potentialities and in part superimposing these by the cecidogenetic direction. The intensity of cell divisions differs very considerably, depending on the presence and distance from normal morphogenetic fields. These differences are decisive factors in the development of fold gall, roll

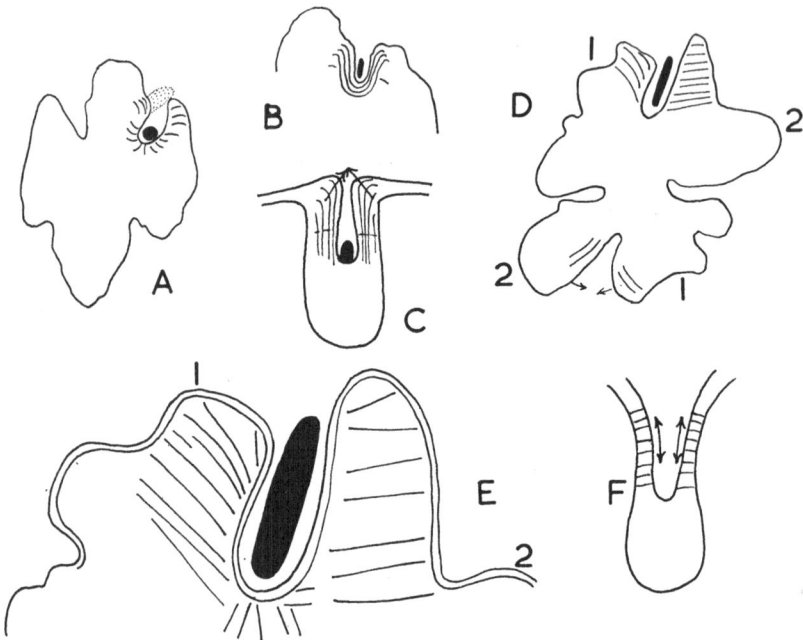

Fig. 159. Cecidogenetic movements. A,B,D,E. Covering growth enclosing the cecidozoa positioned initially in natural depression; a continuation and exaggeration of the normal morphogenetic movement. C,F. Morphogenetic movements on leaf gall, with the cecidozoa in fold above on the midrib; the normal movements result in the closing-in of the cecidozoa, 1-2 covering growths. (After Meyer).

gall, pouch gall, etc. Even in case of galls that arise by pronounced growth in thickness the rate of cell division is not uniform everywhere. We thus find that in many galls the epidermal cells do not divide as rapidly as the parenchyma cells below.

Since the cecidozoa occupies the centre of the cecidogenetic field in zoocecidia, the position of the cecidozoa with reference to the normal morphogenetic field, especially with reference to normal meristematic tissue is an important factor in the morphogenesis of the gall. This relation of cecidozoa to normal meristem covers not only the distance but also its direction. The importance of the position of cecidozoa with reference to the cambial ring has been stressed by HOUARD[485]. In describing the salient features of galls on the shoot axis we have already dealt with the effect of the position of cecidozoa on the general shape and symmetry of galls.

From the stand-point of morphogenesis of galls the position of cecidozoa may either be external on the surface of an organ or within the plant tissue. Where the cecidozoa are situated externally, the early phase of morphogenetic movement in the gall is covering growth — to enclose the cecidozoa within the plant tissue, more or less completely or at least to form around it a mass of cells so that the cecidozoa comes to lie in a depression. In all galls the initial morphogenetic movement is one of differential growth, the effect of which is to surround and enclose the cecidozoa in a mass of actively dividing cells. On a thin organ such as the leaf, the primary gall growth thus brings about the formation of the gall cavity, which also results in the development of the nutritive tissue of the gall. This covering is not indeed directed in relation to the cecidozoa at all, but to the morphology of the gall-bearing organ, resembling more the normal morphogenetic destiny and realizing in the formation of the nutritive tissue. In stout organs such as the stem, which do not take

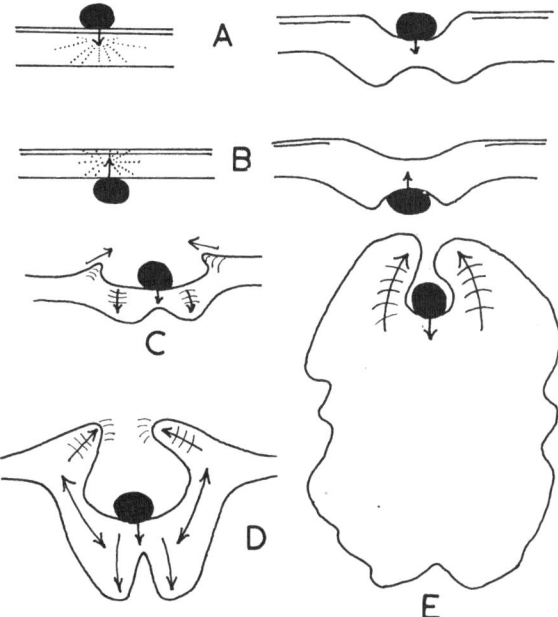

Fig. 160. Cecidogenetic movements. A–D. Enclosing of the cecidozoa positioned on the upper surface of the leaf and on the lower side. E. Covering growth to enclose the cecidozoa positioned externally on thick substratum. Arrows indicate the direction of growth movement. AB. Similarity of centres of different cecidogenetic movements on different sides of blade; A. effective with the cecidozoa epiphyllous and B. ineffective with the cecidozoa hypophyllous. C–D. Direction of movement of growth of gall on leaf. E. Direction of growth movement on stem, covering growth owing to the inertia of the stem. (After Meyer).

part in the morphogenetic movement readily owing to their greater inertia, the deformation proceeds about a fixed point. The cecidogenetic gradient does not extend to the side of the stem opposite to the position of cecidozoa. The gall tissues appear in effect to be pushed away from a point situated within the normal morphogenetic field and developed uniformly in a straight line that determines the plane of symmetry of deformation (fig. 154C). In such cases the presence of a preformed concavity or depression in which the cecidozoa lies or at least an infra-lateral substratum plays an important rôle. If such a preformed cavity exists such as in ribbed petioles and stems, the differential growth is retarded and lobed enveloping growth takes place. The tissue below the cecidozoa either swells up and grows upward, over and around the cecidozoa and the tissue at the sides of the cecidozoa simply exaggerates the normal morphogenetic growth, enlarges and grows over to enclose the cecidozoa. Where the cecidozoa is already within the plant tissue, the same processes of localized cell divisions spread around it and result in the formation of nutritive tissue. In any case, we observe that the primary morphogenetic process is to enclose and isolate the cecidozoa. After this is effected, the line joining the centre of the normal morphogenetic field and the centre of the cecidogenetic field becomes the plane of symmetry of the gall. The intensity of growth of the gall is equal on either side of this axis of symmetry (fig. 160). The form of the gall in transverse section depends upon the form of the transverse section of the normal stem, the cecidogenetic gradient and the position of the cecidozoa. The circumference of the stem and the circumference of the cecidogenetic field overlap each other and combine to form the outline of the transverse section of the gall (fig. 155). The general outline of the transverse section of the gall is thus a curve encircling the circumference of the stem and of the cecidogenetic field (fig. 156).

A careful consideration of the whole course of the development of zoocecidia gives the impression that the normal morphogenetic field moulds and effectively limits and localizes the cecidogenetic field. The departure from the normal morphogenetic control is thus never complete and although the proliferating cells around the cecidozoa may perhaps exhibit maximum independence from the normal morphogenetic control, with increasing distance this independence steadily diminishes, until we find nothing but the normal morphogenetic field. The cecidogenetic and normal morphogenetic gradients are opposite in direction and where the former disappears the latters seems to take over. It is remarkable that even well within the cecidogenetic field, traces of the old morphogenetic field seem to persist. The cell polarities are not, for example, abolished totally, the direction of growth of cells, the direction of cell divisions etc. in the gall often continue to be the same as in the normal organ. The palisade cells of the leaf elongate, for example, and divide vertically to the surface of the blade even in cecidogenetic growth (fig. 63A, 82). Cecidogenesis seems often merely to abnormally accentuate some part of the normal morphogenesis. It is perhaps this tendency that results in the primary morphogenetic movement of enclosing the cecidozoa or surrounding it or at least pushing it to one side. After the cecidozoa is enclosed in a mass of dividing cells, this mass inhibits further cell division outside. Once again we note that the development of gall isolates the cecidozoa.

CHAPTER XV

PLANT GALLS AND CANCEROUS GROWTHS IN ANIMALS

This chapter is in a sense a continuation of the last chapter and deals likewise with the factors, mechanisms and coördination of development and growth but in a somewhat more fundamental aspect.

Being hyperplasia characterized by more or less complete escape from the normal restraining morphogenetic influence, galls are, as indicated in the first chapter, essentially neoplastic growths. As in the case of neoplasia in animals, the departure from the normal to the uncontrolled growth may be total and irreversible or only partial and reversible. In the galls which we have so far considered the escape from the coördinating influence of the normal morphogenetic field seems to be incomplete, so that the cecidogenetic field is eventually neutralized and counter-acted by the normal morphogenetic field, bringing about an effective localization and isolation of the gall-inducing organism. These limited neoplastic galls rather resemble the cysticercus tumors of animals, in which some helminth or other parasitic organism is physiologically and structurally isolated by the tumor tissue and thus rendered comparatively harmless. In a sense, the shaping and the ultimate fate of the gall tissue are determined by the normal morphogenetic field, though perhaps somewhat indirectly. There are, however, certain galls, in which the escape from the coördinating influence of the normal morphogenetic field is complete, so that the gall cells are permanently altered in their nature and the galls are true unlimited neoplasia. The fundamental character of neoplastic growths, whether they are plant galls or animal cancers, is the proliferation of cells, the division of which is normally controlled, coördinated or inhibited by the activity of other neighbouring cells. The problems underlying such a fundamental change in the growth pattern may eventually be found to be the same both in plants and animals. Much of the critical work on some of these basic problems of neoplastic growths in plants has been carried out on the crown-gall. Before dealing with the problem of fundamental similarity of plant galls to animal cancers, we shall therefore give a brief outline of the outstanding results of recent researches on the crown-gall.

1. What is crown-gall?

A crown-gall usually develops on the "crown" or the part of the stem at or just below the ground level on a wide group of plants and is caused by the bacterium *Phytomonas tumefaciens* (Sm. & Towns.)

(= *Bacterium tumefaciens, Agrobacterium tumefaciens, Pseudomonas tumefaciens*). It varies in size from that of an ordinary pea seed to large solid lumps, several centimetres in diameter. Plants inoculated with this bacterium produce, in addition to the primary tumor gall at the site of inoculation, secondary tumors at some distance away and also low ridge-like outgrowths of parenchyma called tumor strands, stretching from the primary tumor to within a short distance of the secondary tumor. Bacteria may be demonstrated by suitable means in the primary tumor and also often in the tumor strands, but the secondary tumor galls are bacteria-free. The crown-gall is characterized by small-celled tissue, endowed with an enormous proliferating capacity, which is not under the general physiological and morphogenetic control of the plant and the continued development of which is detrimental to the plant, both locally and constitutionally. The tumors, even when deep-seated, are incompletely vascularized and quite fleshy. Like mouse cancer, when grafted on to healthy plants of the same species, the crown-gall tissue grows into a tumor.

Since its discovery in 1853, the crown-gall tumor has been extensively studied and an enormous amount of literature has accumulated, particularly in recent years. The gall is of peculiar interest in view of the repeated assertion by SMITH[1059, 1060] and others that the crown-gall is comparable to neoplasia on animals and man. Useful summaries of the results of the numerous investigations on different aspects of the crown-gall problem may be found in SCHILBERSZKY[1020], KOSTOFF[616], BRAUN[122], WHITE[1261,1262] and DE ROPP[261]. Despite nearly half a century of attempt to bring an analogy between crown-gall tumors of plants and cancerous tumors of animals, it has been possible only recently to understand the existance of a close analogy between these widely different pathological growths. This has been made possible by the application of methods of tissue culture to the gall problem. The extension of the tissue culture technique to studies of the physiology of pathological structures has brought many interesting and unexpected results. Following earlier fundamental studies, WHITE[1253, 1254], GAUTHERET[634] and NOBÉCOURT[871] succeeded in establishing true tissue cultures of plant organs. These cultures comprise mostly undifferentiated callus, growing indefinitely on synthetic media [1254]. The literature on the subject has recently been beautifully reviewed by GAUTHERET[364, 364a] and MOREL[820]. Tissue cultures of bacteria-free crown-gall tissues, obtained by a number of isolations from the primary and secondary tumor galls[1264], are capable of unlimited growth *in vitro* on synthetic media.

2. The causal organism

The bacterial origin of crown-gall was established by SMITH &

TOWNSEND in 1907. They described the bacterium under the name
Bacterium tumefaciens[1067]; it is now known as *Phytomonas tumefaciens* (SM. & TOWNS.).

The crown-gall bacteria were described by SMITH & TOWNSEND
as motile, non-gas-forming, short rods, with rounded ends, often
occurring in pairs, with distinct constriction, each about 1 micron
or less in diameter and 2—3 microns long, with 1—3 polar flagella.
They are aerobic with sugars like dextrose, saccharose, lactose,
maltose, mannate and glycerine and produce white colonies on
standard nutrient agar, growing best on top in tubes of bouillon as
stringy rim. The surface colonies on agar are small, round, smooth
and dense. Gelatine is not liquified. The bacteria are capable of
using inorganic forms of nitrogen to produce nitrite from nitrate.
The organism is a gram-negative and capsulated structure that
forms zoogleal masses in old cultures, in which we see faintly staining cocci. Though LEVINE[679] interpreted certain bodies in the zoogleal mass as spores, there is at present no further evidence that the
crown-gall bacterium is spore-forming. In the course of its life-cycle,
Phytomonas tumefaciens forms peculiar star-shaped cell aggregates,
especially in media containing dilute carrot juice and 0.01% each
of ferrous sulphate and manganese sulphate. STAPP & BORTELS[1086]
have given an account of the growing together of the motile bacteria
into the bacterial stars under certain cultural conditions. During
this stage the bacterial cells appear bound together by protoplasmic
connections and the whole star is embedded in a tough slime. The
formation of the star is favoured by the addition of iron and manganese to the medium, by increasing the carbohydrates in relation
to nitrogen, together with decreasing the total concentration of
nutrients and by increasing the acidity. STAPP[1082, 1083] and BRAUN
et al.[128] studied the stellate aggregates under the electron microscope and by Feulgen-positives at the centre of the aggregation.
BRAUN concludes from these observations that there is some suggestion of a simple form of sex in the crown-gall bacterium. The
possibility of sexual fusion of the bacteria during the formation of
stars must, however, remain uncertain until a definitive nuclear
fusion can be demonstrated. The bacterial stars ultimately disintegrate, the slime becomes less tough, the bacteria separate and the
young forms begin to swim. These workers failed, however, to find
evidence of the assertion by ROSEN[873] that at a certain stage in the
life-cycle, the crown-gall bacteria are capable of passing through a
Berkefeld filter. In the presence of ammonia, *Phytomonas tumefaciens*
utilizes ammonium nitrate as the sole source of nitrogen. CONNER
and his collaborators[210, 211] showed that polypeptide and amino
nitrogen of peptone are also readily utilized by the bacteria. He
found further that glucose is fermented, with liberation of carbon
dioxide; addition of manganese accelerates this fermentation. Iron

and zinc are also of importance in the nutrition of the crown-gall organism. The bacterium synthesizes large quantities of biotin, riboflavin, some amount of thiamine and pantothenic acid[773] and probably also other growth factors required for its growth. The conflicting views of SMITH and BLUMENTHAL on the motility of *Phytomonas tumefaciens* are reconciled by KAUFMANN[569], who reports the appearance of motile, flagellate forms in cultures, which have been for long periods completely non-motile. Another variation of this organism is found in alteration of its colonies from those with a characteristic smooth glistening surface to rough-surfaced ones. By filtration and enrichment, KAUFMANN *(op. cit.)* obtained a bacteriophagic lysin, which is believed by him to be responsible for these mutations. Like D'HERELLE and PEYRE, KAUFMANN also believes this lysin to be responsible for the production of the gall.

Reference may also be made to the reports of BERTHOLD *et al.*[82] and SMITH[1061] that the Blumenthal strain of bacteria from human breast cancer, while impotent in producing tumors in mice or rats, is capable of causing cell proliferation when inoculated into *Pelargonium zonale*, *Helianthus annuus* and pieces of *Daucus carota*. The cell-free filtrate of the organism is also reported to cause cell proliferation in plants. They have termed this filtrate as the "tumefaciens plastem".

Some workers[327] have claimed that a bacterium isolated from human uterine carcinoma, inoculated into stems of *Pelargonium zonale*, produces a smooth spherical, hard, grey coloured tumor, about the size of a hazelnut. Culture from this tumor showed small, rod-shaped, immobile, gram-negative organisms, larger and plumper than the ones originally obtained from human carcinoma. Various tests, like action on sugars, gelatin, agglutination, etc. are believed to show that the organism from this plant tumor is the one originally isolated from human carcinoma, modified by its passage through the plant.

LIESKE[693, 694] asserts that a single agent is variously recognized as filterable virus, gram-negative, monotrichous rods *(Phytomonas tumefaciens)*, variable gram-positive rods and as *Streptococcus*, and these various forms are merely cyclic expressions of one and the same organism, which he names *Polymonas tumefaciens*. The pathogen is widely distributed as a saprophyte but the causes that underlie the change of this saprophyte to pathogenic form are not clearly understood. LIESKE considers this pathogen to be responsible for animal and human cancer and also the crown-gall of plants. The pathogen is described as entering into a symbiotic relationship with the tumor cells in the shape of gonidia or symplasmas. TEUTSCH-LÄNDER *et al.* [1130] are, however, of the view that there is no universal carrier of diseases for animals and plants. Inoculations of *Pelargonium zonale*, mice and rats show in their experiments that while the plant develops tumors in many cases, none arises on animals.

3. Bacterial metabolism

The peculiarities of metabolism of the crown-gall bacterium have been studied by several workers. One of the striking features of its metabolism is the modification of the hydrogen ion concentration in one direction or the other in bacterial cultures. The addition of ammonium sulphate to the culture media as source of nitrogen moves the pH in the acid direction, as the ammonium salt is used up by the bacteria. With sodium nitrate as the source of nitrogen, the pH moves in the alkaline direction as the nitrate is used, probably because of the utilization of some protein or protein derivatives, with the production of ammonia. The oxidation-reduction potentials in aerobic cultures drop from about 500 mV to 100 or even −100 mV in a day or two, depending upon the medium and other conditions. The osmotic pressure of the medium decreases gradually perhaps with the utilization of sugar, which is converted either into the characteristic bacterial gum (with a large molecule) or into carbon dioxide, with no or little intermediate products[210, 211, 477, 774]. With the accumulation of the bacterial gum, the surface tension of the culture decreases and the culture gets a conspicuous viscous character.

In single cultures of *Phytomonas tumefaciens*, *Phytomonas savastanoi*, *Phytomonas savastanoi nerii* and *Phytomonas beticola*, PINCKARD[910] observed alkaline reactions where starch and salt of organic acids are sources of carbon. Marked acidic reactions appear, however, with other sources of carbon, but similar differences are not seen in cases of different sources of nitrogen. Acid reaction is also induced with oxamide, L-tryptophan and L-cystine. The optimal temperature is about 28 °C for growth of the bacterial culture on agar. Investigations by BERRIDGE[81] have shown that the pH of the plant sap most favourable for the growth of *Phytomonas tumefaciens* is about 5.2, which is indeed characteristic of meristematic tissues. The zoogleal mass of bacteria also appears to have the same pH value. COLLEY[207] found that in a liquid medium, consisting of monobasic potassium phosphate 0.5 g, magnesium sulphate 0.2 g, dextrose 1.0 g, asparagin 1.0 g, water 1 litre and with the initial pH between 6.0 and 8.0, the growth of *Phytomonas tumefaciens* increases as the temperature rises to 28 °C, when the oxygen supply is abundant and the quantity of dextrose is increased.

The crown-gall bacterium has the remarkable ability to utilize an unusually large number of different sources of carbon and nitrogen. Favourable conditions for fermentation of sugar have been studied by MC INTYRE, PETERSON & RIKER[774]. The observations of RIKER [457] seem to suggest that its ability to grow on so many materials, toleration of many kinds of inhibiting substances and its resistance to unfavourable substances probably account for its wide host range.

Among the metabolic products of the crown-gall bacteria, carbon dioxide is the most conspicuous in the medium itself. With well balanced media, the change of pH in the acid direction is so slight that it can be accounted for by the presence of carbon dioxide. The most common residual metabolite is the bacterial gum, the weight of which is considerably greater than indeed of the bacterial mass itself. The gum contains approximately 24 glucose molecules and is a viscous, hygroscopic, chemically inert material that neither the bacteria nor the plant has apparently an enzyme capable of attacking. No doubt the accumulation of this gum and of the bacteria in the intercellular spaces naturally interferes with the proper gaseous exchange in the plant. A polysaccharide, produced by *Phytomonas tumefaciens*[774] has a specific rotation of –9 to –10° and yields glucose on hydrolysis. The shift in rotation during hydrolysis indicates beta-linkages between the glucose units and other peculiarities of hydrolysis suggest a pyranoside-ring structure. The molecular weight is estimated at 3600 ± 200 and contains 22 anhydroglucose units per molecule. The metabolites resulting from nitrogen in the medium have perhaps received much less attention. Ammonia is one of the products that has been studied. With the use of protein derivatives as a source of carbon, the pH shifts in the alkaline direction, but when sufficient sugar is present, it seems to have a sparing action upon the source of nitrogen[210, 211]. The crown-gall organism produces, as already indicated, biotin, riboflavin, pantothenic acid and thiamin[773] and probably also other growth factors necessary for its development. LOCKE, RIKER & DUGGAR[711] have observed the production of growth substances from peptone broth containing tryptophan, probably as a precursor for indole-3-acetic acid. Various lipids, more or less toxic in strong concentrations on plants, have been demonstrated by analysis of the crown-gall bacteria[372]. BRAUN[125] indeed believes that the tumor inducing principle or something closely associated with this principle has a large complex structure, approaching that of protein. Even SMITH asserted that a bacterial endotoxin, acting on the cell nucleus of the plant, is the immediate stimulus for cell proliferation. It has also been claimed that a glucidolipid, obtained by treating the crown-gall bacterium with trichloracetic acid, is the specific endotoxin[109]. According to some workers, this endotoxin is believed to act as an antigen when injected into animals and plants and is also precipitated by the antitoxin obtained from rabbits inoculated with living cultures of *Phytomonas tumefaciens*. Injection of this glucidolipid into young branches of *Helianthus annuus* results in tumor formation in the cortex, resembling very much the tumors formed by inoculating stems with bacterial cultures. HODGSON, RIKER & PETERSON[477] have also made important contributions to our knowledge of the bacterial endotoxin, especially the wilt-inducing toxic substance produced by *Phytomonas tumefaciens*. They found in the cell-free filtrates of the crown-gall bacteria (grown in media with sucrose, urea and mineral salts), a substance capable of inducing wilting and necrosis of leaflets of *Lycopersicum esculentum* and other plants. This toxic substance is described as thermostable in neutral solution, but completely destroyed in two hours in 1/N hydrochloric acid at 100°C, non-volatile, water soluble, but relatively insoluble in most organic solvents and dialysable through a semipermeable membrane. Alcoholic fractionalizations result in a product soluble in 60% alcohol but precipitated by 90% alcohol. This product consists largely of glucosan. Several well purified preparations of glucosan induce toxic symptoms, similar to those produced by whole filtrates. According to these workers the chief toxic substance of the crown-gall bacteria thus appears to be glucosan.

RIKER[940, 941] also studied in great detail the loss of pathogenicity induced by certain amino acids, recovery of the pathogenicity of the bacterium in certain media and after ultra-violet treatment, oxygen hunger of the plant cells, changes in osmotic pressure leading to swelling of the plant cells, presence of unusual quantities of enzymes and growth-promoting substances like thiamin, riboflavin, pantothenic acid, biotin, etc., bacterial metabolites like ammonia, phospholipids and polysaccharides that irritate plant cells and the presence of food materials in abnormal amounts. Both virulent and attenuated (non-virulent) cultures of bacteria appear identical in ordinary bacteriological characters but the virulent cultures seem more viscous than the attenuated cultures on both solid and liquid media. Both cultures grow well on L-tryptophan but D-tryptophan inhibits the virulent cultures more than the attenuated culture[948]. Both strains produce comparable amounts of auxin in the presence of tryptophan[711].

The attenuation of virulent cultures of bacteria by means of certain amino acids and related compounds is of particular importance. Media with a relatively alkaline reaction, in presence of enough glycine (0.02 M) to reduce but not to stop growth, gradually abolish the capacity of cultures to induce gall formation, especially after a series of fifteen or more transfers[1202]. Restoration of virulence may be accomplished by irradiation with ultraviolet, so as to kill all bacteria but one in a thousand. The survivors usually exhibit increased virulence. If attenuation is not carried through too many transfers, restoration of virulence may be affected.

It is interesting to observe here that according to the investigations of LOCKE, RIKER & DUGGAR[711], single cultures of virulent crown-gall bacterium, attenuated crown-gall bacterium and *Bacillus radiobacter* are similar in their capacities to produce growth-promoting substances on peptone broth, but they differ widely in their capacities to induce overgrowth when inoculated into a number of plants. They failed to find evidence for the view that beta-indole-acetic acid or any other growth substance, produced in such cultures, has a direct major relation to the pathogenicity of *Phytomonas tumefaciens*.

PATEL[828, 829] has reported that overwintering strains of *Phytomonas tumefaciens* retain their virulence for at least five months of winter temperatures, ranging from –23 to 15 °C in the absence of the host plant. He is also of the view that the longevity of the bacteria in soil is influenced more by the competition of other soil microorganisms than by the nutritive content of the soil.

Heat treatment studies have shown that the crown-gall bacteria live through treatment as long as the relative humidity of the atmosphere is 50% or lower, but higher humidities enough to inhibit the development of *Phytomonas tumefaciens* favour at the same time growth of *Penicillium* at the point of inoculation of the plant. This fungus produces an antibiotic substance that destroys the crown-gall bacteria.

The bacteria grow well at 30 °C both in the plant and in soil, but at this temperature gall formation does not take place on *Lycopersicum esculentum*[937].

Exposure to temperature of 32 °C for four days immediately after inoculation prevents the formation of gall on *Vinca rosea*, but the gall develops on plants held for four days at less than 32 °C. Virulent bacteria are, however, present in the first case, because with rewounding the old inoculation site and exposing the plant to a temperature of 20 °C we find the development of typical galls. BRAUN[123] found that inoculted plants, incubated at a temperature of 32 °C for five days, then exposed to 47° for another five days so as to kill the bacteria, do not produce galls. Plants treated similarly and kept at 26 °C for five days and then exposed to 47 °C for another five days, produce galls in all cases. Changes in the cells appear to arise at temperatures below 32 °C during the four-day period immediately following inoculation by the bacteria. In the case of *Kalanchoë daigremontiana* the bacteria require incubation at a temperature of about 25 °C to be able to give rise to a gall. The size of the resulting gall is proportional to the total duration of exposure to 32 °C, alternated with 25 °C for an accumulated period of thirty hours at 25 °C. The activity of the transforming capacity of the bacteria seems to be destroyed at 32 °C at about the same rate as it is produced at 25 °C.

4. Host and tissue susceptibility

The host range of the crown-gall bacterium is unusually wide[816] and includes most Dicotyledons, with not too acid cell sap. Particularly susceptible are *Pirus malus*, *Pirus communis*, *Prunus persica*,

Prunus domestica, Rosa, Rubus, Geranium and *Chrysanthemum frutescens.* The wide host range suggests that the pathogenicity involves rather common and fundamental processes.

As early as 1911, SMITH observed that *Phytomonas tumefaciens* is capable of giving rise to crown-galls on plants of eighteen different Natural Orders, almost all of them Dicotyledonae, viz. Salicaceae, Juglandaceae, Fagaceae, Moraceae, Polygonaceae, Chenopodiaceae, Caryophyllaceae, Ranunculaceae, Lauraceae, Cruciferae, Resedaceae, Crassulaceae, Rosaceae, Leguminosae, Geraniaceae, Tropaeolaceae, Rutaceae, Euphorbiaceae, Anacardiaceae, Celastraceae, Balsaminaceae, Vitaceae, Malvaceae, Sterculiaceae, Passifloraceae, Begoniaceae, Cactaceae, Myrtaceae, Oenotheraceae, Umbelliferae, Ericaceae, Ebenaceae, Oleaceae, Apocynaceae, Polemoniaceae, Labiatae, Solanaceae, Caprifoliaceae and Compositae. STAPP[1081] has reported a crown-gall on *Asparagus,* a Monocot. DAUVERGNE & WEIL [243] have shown that *Sedum spectabile* develops tumors in all inoculations not only with virulent strains, but also with non-virulent strains on stem, leaves, flowers and equally well also with fragments of tumor tissue. SMITH[1051] recorded crown-galls on Taxaceae, Taxodiaceae and Pinaceae. Claims have also been made that *Phytomonas tumefaciens* is capable of producing tumors on animals like the brooktrout[1058–60] and on a number of Invertebrates[623, 1145, 1146]. Tumor production on *Nereis diversicolor* (Annelida) has also been reported. SORU & BRAUNER[1074] reported stimulation of cell proliferation in the bone marrow of the rabbit. Most of the experimental investigations on crown-gall problems have been carried out on *Helianthus annuus* (sunflower), *Helianthus tuberosus* (Jerusalem artichoke), *Kalanchoë daigremontiana* (kalanchoë), *Pelargonium zonale, Lycopersicum esculentum* (tomato), *Vinca rosea* (periwinkle), *Chrysanthemum frutescens* (Paris daisy), *Tagetes erecta* (marigold), *Nicotiana tabacum, Nicotiana glauca, Nicotiana langsdorfi, Beta vulgaris, Daucus carota* (carrot), *Scorzonera hispanica, Abutilon avicinnae, Geranium, Ricinus communis, Nicotiana glutinosa* and *Vitis vinifera.*

Although we mentioned in the beginning that a crown-gall tumor develops typically at the crown or the portion of the stem about the ground level, it must be pointed out that almost all organs of the susceptible plants bear these galls. In 1961 SMITH *(op. cit.)* recognized different types of crown-gall tumors, depending upon the tissues from which they are derived. He compared the ordinary crown-gall to sarcomata, as it involves connective tissue and develops most vigorously on young plants, with vessels reduced to their primitive elements. Tumors derived from epidermis, cortical and gland cells were termed by him carcinomata. Galls with mixed organs like roots, buds, leaves, etc. were similarly called embryomata. Inoculations of the bacteria penetrating only into the outer cortex result in superficial hyperplasia of limited growth and do not involve wood, cambium or phloem. They arise solely from the cells of the cortex, which grow rapidly, become smaller and more embryonic in nature than the surrounding cells and are capable of vascularizing themselves to a limited extent. SMITH concluded from this that under the action of the bacterium the cortical cells are converted into more embryonic tissue.

HILDEBRANDT[470] has described the formation of small galls, evidently from the hair cells of the epidermis, when the stem of *Lycopersicum esculentum* is stroked with a glass rod carrying the bacteria. The smallness of the resulting gall is ascribed either to the limited proliferation or to the inadequate vascularization. The participation of cambium is not necessary for the production of the crown-gall; pith, cortex and vascular parenchyma are all capable of undergoing cell proliferation under the influence of this bacterium and giving rise to gall formation. Inoculation of the cambial tissue results no doubt in much larger tumors than from non-cambial tissues, but this has not been accepted as proof that the cambium plays a decisive rôle in tumor formation. It may be interpreted as resulting from closer connection of cambial and vascular tissues. The cells of storage organs differ in their capacities to respond to the crown-gall bacterium in different plants. In *Daucus carota, Beta vulgaris* and *Helianthus tuberosus* the development of tumors closely follows lines of cambial tissue. Phloem tissue lying closest to the cambium in *Daucus carota* shows also the capacity to respond to the bacterium but the tissue of secondary xylem does not show any response. The development of crown-gall tumors in *in vitro* cultures of isolated fragments of the stem of *Helianthus annuus* shows a similar behaviour [253]. We may conclude that the more closely a cell approaches to the meristematic condition, the more readily is it transformed into a crown-gall tumor cell by *Phytomonas tumefaciens.*

It is known that the formation of callus in *in vitro* cultures on fragments of *Daucus carota* occurs almost exclusively on the lower, radical side rather than on the upper foliar side of the fragment. DE ROPP[260] found, however, no such polarity on inoculation with *Phytomonas tumefaciens* and the tumor formation on the radical side is as great as on the foliar side. In this respect the tumor tissue on *Daucus carota* differs from overgrowths induced by indole-acetic acid, where the influence of polarity of the organ is distinctly evident.

5. Host-parasite relations in the crown-gall

SMITH believed that *Phytomonas tumefaciens* is an intracellular parasite and its presence within the host cell may be demonstrated by special staining[1054]. Later, however, he doubted that the bacteria could thus be demonstrated within the host cells. The investigations of MAGROU[734, 735] PERTI[899], PINOY[912] and others seem to support

SMITH's earlier view. RIKER[934], ROBINSON & WALKDEN[965] and others have suggested that *Phytomonas tumefaciens* occurs in the form of zoogleal strands in the vessels or is also intercellular. HILDEBRANDT [470] found that injections of bacteria directly into the trichome and epidermal cells of the stem of *Lycopersicum esculentum* do not result in the formation of a crown-gall tumor and concluded therefore from this observation that the "living cell interior is an unfavourable medium for these bacteria".

These studies are apparently inconclusive in view of the fact that the crown-gall cannot be produced by a mere external application of the bacteria on the surface of uninjured tissues and that wounding is essential[935, 937, 956]. RIVERA[956] has stressed the importance of injuries in infections by *Phytomonas tumefaciens*; he failed to produce the crown-gall tumor by merely applying the crown-gall organism to the external surface of the shoot of *Ricinus communis* and *Geranium*. Cut surfaces of flowering stalks of *Brassica oleracea* and shoots of *Ricinus communis* treated with culture of *Phytomonas tumefaciens* do not produce galls. He concludes that wound or injury of the plant tissue is necessary for the crown-gall bacteria to be able to give rise to the gall. NÀBÉLEK[833] also showed that the crown-gall originates only from a wounded tissue. Intact tissue in contact with *Phytomonas tumefaciens* produces no tumor. Although the bacteria may unquestionably be sometimes demonstrated in the intercellular spaces of the host plants, it is not certain that these were actually responsible for the original transformation of the healthy cells into tumor cells. Furthermore, it is also not the number of bacteria but it is the size of the wound that is a decisive factor in determining the size of the resulting crown-gall tumor[470, 678]. The gall forming response of wounded tissue to the application of crown-gall bacteria ceases in from 5—7 days[122, 130, 254, 937]. The wound hormone proba-

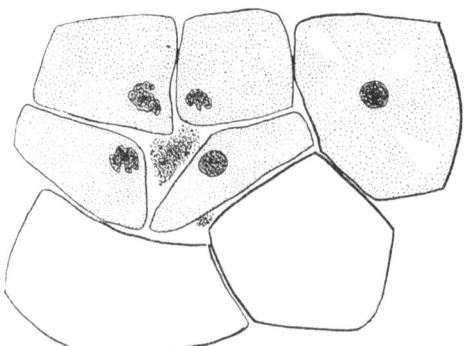

Fig. 161. Early initiation of cell division in crown-gall, with bacterial mass in between the newly divided cells.

bly plays some rôle, but the wound is more probably important in that it exposes the protoplasm direct to the action of the bacterium. All available evidence indicates that *Phytomonas tumefaciens* is incapable of exerting its tumefying influence through intact cell wall.

6. Plant response to the crown-gall bacteria

Some of the major structural changes in the plant exposed to the action of *Phytomonas tumefaciens* were briefly described by RIKER[936] at two-day intervals. The wound through which the bacteria are introduced into the plant tissue floods the intercellular spaces and provides a culture medium for the bacteria. The cells immediately around the bacteria enlarge within two days of inoculation and the adjacent cell walls turn brown and in microscopic preparation become stained more intensely than the other normal cells. The crown-gall appears within four days after inoculation (fig. 161). In the early stages of development of the crown-gall tumor the new cell walls arise more or less as in the case of normal wound healing[938]. The actual wound through which the bacteria are introduced does not, however, influence the cells that ultimately develop into crown-gall tumor. According to the summary by BANFIELD[54], the bacteria penetrate as zoogleal strands throughout the plant tissue and the middle lamella of cell is dissolved. The cells, with which the bacteria are in immediate contact, undergo more or less rapid cytolysis. In the early phase of development of the gall, the cells at some distance from the bacterial mass divide rapidly and repeatedly. Intercellular penetration by the bacteria among these cells now leads to the proliferation of still more distant cells and eventually cell division ceases, extensive intercellular penetration continues and the gall disintegrates. Two types of cellular degeneration are described by BANFIELD *(op. cit.)* viz. cytolysis and autolysis. In cytolysis all elements of cells are progressively lysed by the contiguous bacteria; the plastids disappear first, then mitochondria, nucleolus, fat globules and part or whole of the cell wall disappear. In autolysis the elements of the chondriome show pronounced tendency for agglutination, the plastids and mitochondria become fused end to end or also laterally to form reticules, solid condensation masses or open mesh-work. The chondriome elements may become vacuolated. We have an interesting review[446] of morphological anomalies, including polyploidy, induced by repeated inoculations of massive doses of *Phytomonas tumefaciens* on *Pelargonium zonale*. During ten years following the application of the bacteria on a healthy plant, gradual transformation and segregation into two distinct morphological and physiological divergent types were observed. One of these types is a tetraploid, with teratological leaves of the ascidium-type on some plants. The other type of plants is a typical diploid characterized by extreme lengthening of the stems. We are indebted to LEVINE[676] for some interesting observations on the plant responses, especially the chromosome changes; he has also described the formation of the giant cells and polyploidy under the influence of the crown-gall bacteria. According to BERRIDGE[81], the cell wall in contact with the bacterial zooglea appears to be suberized.

7. Structure of the crown-gall

In 1916 SMITH *(op. cit.)* described that the nucleus of the crown-gall cells divides both mitotically and amitotically; but nuclear division by simple cleavage is common. The nuclei are variously lobed in multinuclear cells. LEVINE[673-676] does not, however, confirm the occurrence of atypical nuclear divisions in the crown-gall tissue. Binuclear and multinuclear cells have no doubt often been described. Most crown-gall tumor cells contain diploid chromosomes. GAUTHERET[367] compared the bacteria-free crown-gall tissue of *Helianthus tuberosus* with that of its normal tissue in *in vitro* cultures and found that the structure of the normal tissue is more orderly than that of the tumor tissue. Lignified elements are associated in the normal tissue with an island of phloem, in which are normal sieve tubes with sieve plates. In the tumor tissue on the other hand the lignified elements are not associated in this manner with islands of phloem and sieve tubes are also absent. BRAUN[122] has also described the presence of lignified elements in bacteria-free crown-gall tissue cultures in isolated masses of modified scalariform cells. Most of the gall tissue is composed of thin-walled parenchyma cells, but giant cells often with 20 nuclei are also found. Occasional lagging of a chromosome is the only abnormality observed by him. The anatomical peculiarities of the crown-gall on the Himalayan *Rubus procerus*, studied by JONES[557], are interesting from the fact that the gall develops from two layers of meristematic tissues, one external and the other internal to the band of pericycle, outside the cylinder of vascular bundles. These comprise the outer and inner pericycle cambia. The gall is composed of secondary parenchyma, derived from outer pericycle cambium, carried out into the growing gall and breaking up into sections, dividing and independently giving rise to secondary parenchyma. Masses of parenchyma become converted into tracheids and short vessels by communication of tracheids. The axial cylinder becomes split, the medullary rays become meristematic and produce intrusive unlignified tissues.

NÀBÉLEK[833] has compared the normal sterile wound healing with the changes after inoculations with *Phytomonas tumefaciens* on *Pelargonium zonale*. There is no significant difference between these two in the early phase but soon a disorder of the secondary meristem arises under the influence of the bacteria, with over production of tracheids. In a third stage, which he has termed "cancerization," rapid cell proliferation produces immature, undifferentiated cells.

Interesting cytological observations are reported by WINGE[1270]. He found tetraploid tumor cells, with 36 chromosomes, in the crown-gall on *Beta vulgaris*, both in the original tumor and in the transplanted tumor. Further doubling of the chromosomes results in the formation of octoploid cells and also cells with even larger numbers of chromosomes. According to the same author, tetraploidy should give a satisfactory explanation of the increased growth activity, characteristic of the crown-gall cells. Occasionally diploid cells, also found in the tumor tissue, are considered by him to be probably derived by reduction division of the tetraploid cells.

The most important contribution perhaps on the general morphology, histology and cytology of the crown-gall on *Lycopersicum esculentum* and *Beta vulgaris* is that of KOSTOFF & KENDALL[622]. They have also made comparative studies of the chemically induced tumors and the genetic tumors in *Nicotiana* hybrids. They have described a typical precipitation reaction between plant extracts and chemicals that induce cell proliferation and they have attempted to interpret the tumor formation in the light of chromosome abnormalities observed in the tumor tissue.

RIKER[938] has described a progressive reduction in size of the cell and nucleus and in the nucleo-cytoplasmic ratio both in the gall and the wound tissue, but the minimum is reached more quickly in the wound tissue rather than in the gall tissue. The size reduction of the nucleus is accompanied by reduction in chromosome size. The first few unequal divisions of the cells of tissues inoculated with *Phytomonas tumefaciens* may perhaps be influenced by positive traumatic responses of the nuclei to the bacteria. There is, however, no evidence of amitosis in the crown-gall tissues of *Lycopersicum esculentum*. RIKER does not also find more than one nucleus. The stimulus producing wound regeneration tissue disappears quickly and is also localized, but that of the gall persists indefinitely and is at the same time diffuse.

LEVINE[675] also made an interesting comparison of the chromosome numbers of human epithelioma, Rous sarcoma of chicken, tar tumor of mouse, Jensen's rat sarcoma, mouse tumor 180 and the crown-gall tumor on *Beta vulgaris* and *Nicotiana glutinosa*. According to him, hypoploid and heteroploid cells in these cases are due to failure of chromosomes to reach to the pole and to become included in the daughter cells. The failure of the chromosomes to divide appears to be a potent factor in producing the abnormal chromosome numbers of these cells. Chromosome numbers in crown-gall tissue are more frequently diploid, but lagging chromosomes do occur. Giant cells of the crown-gall tissue show tetraploid and octoploid complexes; the giant cells appear to be due to nuclear fusions.

Uninucleate cells give rise to binucleate, trinucleate or tetranucleate cells, if nuclear divisions are not followed by cell divisions. Independent nuclear divisions in multinucleate cells were not, however, observed by him. In a later contribution, LEVINE[676] again described three types of cells in animal tumors on the basis of the size and number of chromosomes. Giant cells have 135—300 chromosomes, semi-giant cells have 96—100 chromosomes and human epithelioma cells have chromosome numbers 47—48. He has also described the stages in chromosomal disintegration of diffusion. Crown-gall tissue, obtained from inoculations in the midrib of *Nicotiana tabacum* with *Phytomonas tumefaciens*, reveals two zones of cells, viz. a peripheral zone of diploid cells and the core with tetraploid cells or also with polyploid cells. Similar organization may also be found in isolated nodules of the crown-gall on *Beta vulgaris*. The apical part of leaf galls has normal cells, viz. diploid, but at the base of the gall the cells are tetraploid. It is concluded that while cell plate formation may be initiated, it is not completed in many cases, forming binucleate cells or permitting nuclear fusion. There is, however, no evidence of cell fusions. The chromosome number in the crown-gall of *Beta vulgaris* is regularly diploid 18, tetraploid 36, octoploid 72 and large cells with even more chromosomes may also occur. In the crown-gall of *Nicotiana glutinosa* we find cells with diploid 24, tetraploid 48 and octraploid 96. LEVINE also believes that giant cells represent merely the products of nuclear divisions, without cell division. According to the interesting observations of MOTTRAM[823], endomitosis, polytene, polyploid and polynucleate cells of tumors are really secondary to an increased stiffness of cytoplasm and are no more than signs of such a change.

8. Physiology of the crown-gall

Investigations on the physiology of the crown-gall have largely been carried out in *in vitro* tissue cultures. HILDEBRANDT, RIKER & DUGGAR[466] have, for example, given an account of their studies on the *in vitro* growth of fragments of non-parasitic tissue of *Nicotiana tabacum* and secondary tumor tissue of the petiolar crown-gall on *Helianthus annuus*, at different temperatures, pH and amounts of sugars, on White's synthetic agar medium. The tissues were cultured at temperatures ranging from 4° to 37°C and the optimal temperature was found to be about 26–32°C for *Nicotiana* tissue and 24–28°C for *Helianthus* tissue. The tissues were cultured on media, with original acidities averaging respectively from pH 2 to pH 9.9 and from pH 2.5 to pH 9.9 and both tissues grew within the pH range 3.5–7.9. Best growth of tissue of *Helianthus* occurs on media with pH 5.5–5.9 original values and 6.0–6.4 final values. Cultures in basic media, with sucrose concentrations ranging from 0 g to 160 g/litre, show no growth in absence of sugar or even with 160 g/litre of sugar. Both tissues, however, grow well within 5–20 g/litre sugar, but 10 g/litre being the optimum for both plant tissues.

WHITE & BRAUN[1265] showed that the nutritional requirements of the crown-gall tumor tissue differ fundamentally from those of the stem tissue on which the gall develops. The growth of the healthy stem tissue of *Helianthus annuus* is limited and soon ceases in the medium, which, however, supports the growth of the bacteria-free tumor tissue indefinitely. A most interesting observation by DE

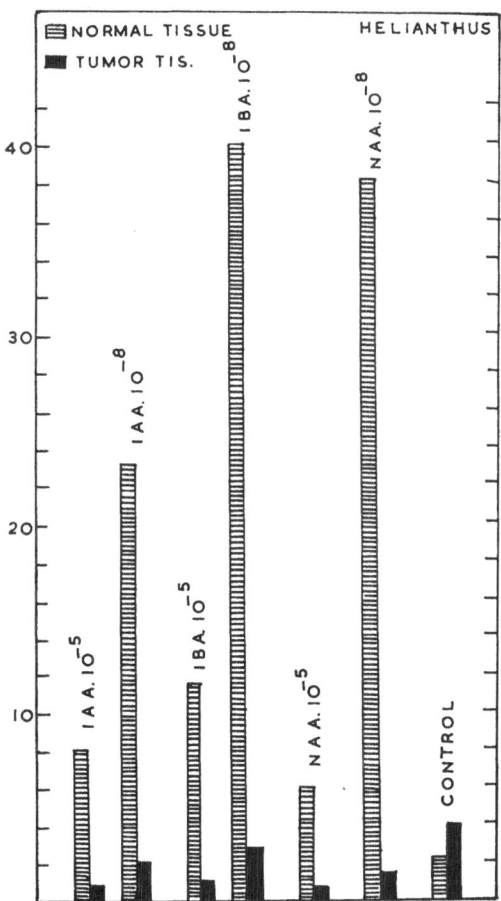

Fig. 162. Difference in vitro growth of tumor and healthy tissues of *Helianthus annuus* in presence of auxins, measured as ratio of fresh initial and final weights. IAA indole-acetic acid. IBA indole-butyric acid. NAA naphthalene-acetic acid. (Based on data of De Ropp).

Ropp[249] shows that when fragments of the stem tissue of *Helianthus annuus* are cultured in the same medium in the presence of 0.01 g/litre of indole-acetic acid, indole-butyric acid and naphthalene-acetic acid, the fragments develop adventitious roots, the growth of which increases the weight of the fragment by more than fifty times. On a medium containing 10 mg/litre of these substances, the fragments proliterate to produce loose, translucent tissue of thin-walled elongate cells. Fragments from the cambium of *Vinca rosea* also behave in a like manner under the same treatment. The growth of the bacteria-free tumor tissue shows no change, however, in these media, but is strongly inhibited by higher concentrations. Bacteria-free tumor tissues of *Helianthus tuberosus*[367], *Vitis vinifera*, *Abutilon avicennae* and *Nicotiana tabacum*[821] are also capable of growing indefinitely on a medium devoid of indole-acetic acid, but this substance is required for the continued *in vitro* growth of healthy tissues of these species. HILDEBRANDT & RIKER[464] tested the growth reactions of bacteria-free gall tissue of *Helianthus annuus* with cystine hydrochloride and nine other growth promoting substances and observed a slight increase in weight in the presence of very low concentrations, but high concentrations always inhibit the growth in each case (fig. 162). HILDEBRANDT[467-468] further reports better growth of the tumor tissue of *Helianthus annuus* in some modifications of the medium originally used by WHITE & BRAUN, with the addition of 0.8 mg/litre pyroxidine. The optimal temperature for growth is 24—28 °C, the pH is 5.0—6.4, sucrose concentration 1% [464]. Later HILDEBRANDT[465] studied the effects of different carbon compounds on the growth of bacteria-free crown-gall tumor tissues of *Helianthus annuus*, *Vinca rosea*, *Chrysanthemum frutescens* and *Tagetes erecta* and found dextrose, levulose and sucrose to be the best sources of carbon. Polysaccharides and other sugars favour some limited growth. Organic acids and alcohols mostly either inhibit or only poorly support growth. Media containing from 0.008 M to 0.064 M nitrate or urea favour excellent growth, but no growth takes place with nitrite and eleven amino acids[946]. D, L-alamine enhances growth at 0.001 M concentration, but inhibits at concentration of 0.002 M and again at 0.128 M enhances growth. None of these substances, however, cause any marked increase in the growth of tumor tissue and the original medium of WHITE & BRAUN seems to provide all the material essential for the growth of the crown-gall tissues. Only in the case of the tumor tissue of *Scorzonera* there is a doubling of the rate of growth by the addition of 50% cocoanut milk to the medium[311].

It is thus evident that the crown-gall tumor tissue shows a much simpler nutritional requirement than the tissues from which it is derived. The tumor tissue has possibly a nearer complete enzyme system than the tissue from which it is derived and tumefaction perhaps also eliminates a growth inhibitor that may inhibit or limit the growth of healthy tissue. Perhaps also the number of essential metabolites involved in the growth of the tumor tissue may be smaller than in

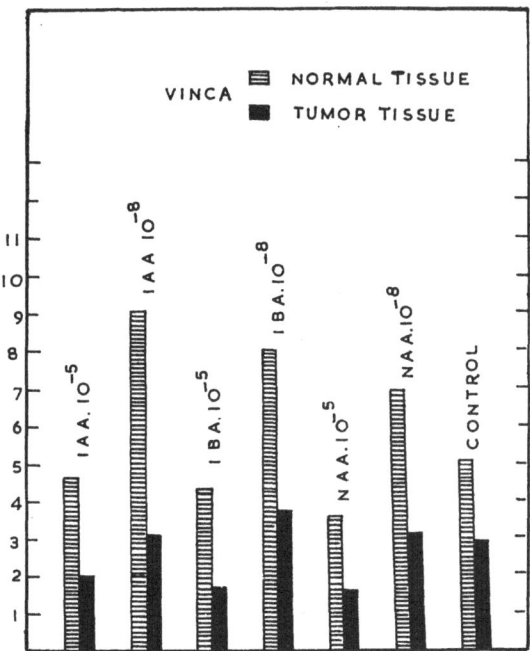

Fig. 163. Difference in vitro growth of tumor and healthy tissues of *Vinca rosea* in presence of auxins. (Based on data of De Ropp).

case of the healthy tissue, so that the tumor tissue is less exacting in its nutrient requirements. The first of these possibilities appears to be supported by the observed facts at present.

It is also known that crown-gall bacteria produce on the plant several other symptoms such as epinasty of petioles and inhibition of lateral buds, which are also produced by the action of indole-acetic acid[249, 710, 1045]. It is interesting in this connection to note that a number of workers have indeed remarked on the pronounced hyper-auxinity of crown-gall tumors[699, 700, 711]. KU-LESCHA[631] estimated the auxin content of healthy and crown-gall tumor tissues of *Helianthus tuberosus* by the *Avena*-coleoptile test and found that in the healthy tissue it is 0.3×10^{-8}, that of the crown-gall is 5×10^{-8} on the basis of fresh weights. According to KULESCHA & GAUTHERET[632], the differences in the auxin contents of healthy and crown-gall tumor tissues of *Scorzonera* are almost of the same order. The enhanced growth capacity of crown-gall tumor tissue, as compared with the normal tissue, seems therefore to depend on its ability to manufacture a growth hormone, which may be either indole-acetic acid or some other kindred substance. The tumor tissue cells differ from normal cells in being able to elaborate the auxins necessary for their proliferation. The crown-gall bacteria produce a transformation in the cell, which involves either the acquisition or the exaltation of a capacity to synthesize these auxins. The elaboration of this auxin essential for cell proliferation enables the tumor elements to multiply without limit. The continued production of this substance by the tumor cells would account for the continued unregulated proliferation of the affected cells. In an important contribution on the pronounced hyperauxinity of the crown-gall on *Lycopersicum esculentum*, LINK & EGGERS[698] have shown that the crown-gall is associated with disturbance in auxin relation (dysauxinity). The same workers again showed that crown-gall tissues are characterized by more auxin activity than normal tissues [699]. They believe that L-tryptophan is

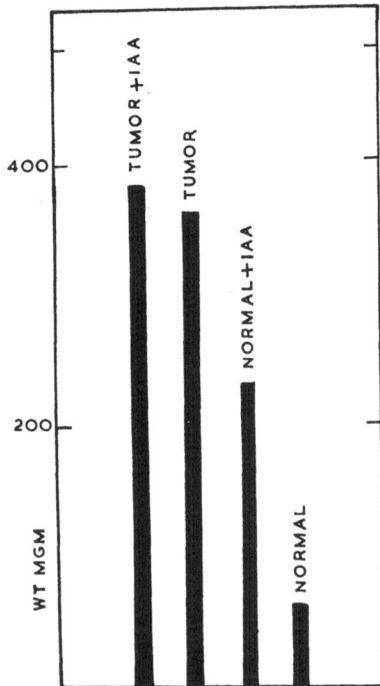

Fig. 164. In vitro growth of tumor and healthy tissues of *Daucus carota* in the presence and in the absence of IAA indole-acetic acid in 8 weeks. (Based on data of De Ropp).

converted into active substance by enzymes present in the tissues of *Lycopersicum esculentum*.

In 1930 Nèmec[863] investigated the effects of smearing fresh cultures of *Phytomonas tumefaciens* on freshly cut surfaces of the roots of *Cichorium intybus* and observed that the adventitious roots, normally formed at the cut surfaces, become suppressed. The formation of callus is also greatly increased by the bacteria. Negative results were, however, observed when the cut surfaces of *Scorzonera hispanica* are smeared with cultures of the crown-gall bacteria. Riker, Henry & Duggar[947], who investigated the growth substances in the crown-gall in relation to the time after inoculation, critical temperature and diffusion, concluded that the pathogenicity of *Phytomonas tumefaciens* is independent of the auxin production.

Crown-galls induced on the storage tissues of several plants reveal a remarkable capacity to utilize the reserve food material stored in these organs. Crown-galls on the tubers of *Daucus carota* regenerated five times in succession and transformed about 15% of the estimated dry weight of the organ into tumor tissue in about twelve weeks. The tubers of *Solanum tuberosum* on which the crown-gall was induced, became completely devoid of starch when the gall was full grown. The reserve food material stored in *Helianthus tuberosus* also disappeared completely when the crown-gall was induced on it[261].

White[1255] studied the respiratory activities by the Barcroft-Warburg method, in healthy vegetative growing points, inflorescence and internodes, the crown-gall with active colonies of *Phytomonas tumefaciens*, bacteria-free secondary tumor. tertiary graft tumor (metastatic tumor) (obtained as a result of implantation of bacteria-free tumor tissue derived from bacteria-free secondary tumor of *Helianthus annuus*), genetic tumors found on hybrid *Nicotiana glauca* × *Nicotiana langsdorfi*, tissue cultures of meristem of this hybrid plant and also tumors of hybrids arising as a result of implantation of such tumors of hybrids under the epidermis of *Nicotiana glauca*. He found no qualitative differences in the respiratory activities of the bacteria-free crown-gall of *Helianthus annuus*, healthy growing vegetative tissue, internodes and inflorecence but the respiratory level of the tumor tissue is lower than that of the healthy tissue. White has drawn significant attention to the long recognized, characteristic lowering of the respiratory levels in animal neoplasia also in this connection. The considerable excess of the oxidising enzymes suggests that in comparison to the surrounding healthy tissue, the basic metabolic activity of the crown-gall is relatively anaerobic. This condition, considered as a possible causal factor for cell stimulation, is perhaps correlated with the fact that the intercellular spaces are flooded with the bacterial zooglea and bacterial gum, so that the oxygen diffusion through normal tissue is greater than the tissue with flooded intercellular spaces. The crown-gall bacteria lower the oxidation-reduction potentials of the material in which they grow. The formation of ammonia and the consequent change of pH in the alkaline direction also lowers the oxidation-reduction potential. The bacteria and the bacterial gum block the intercellular spaces. The presence of the hygroscopic gummy material probably causes the cells to swell and to metabolize slowly. The reduced respiration levels are also associated with the effects of large amounts of auxins. A lowering of the oxygen uptake in presence of indole-3-acetic acid, at gall-inducing concentrations, has indeed been observed by MITCHELL, BURRIS and RIKER. Indole-acetic acid in a concentration of 55 mg/litre reduces the respiratory activity of stem slices of *Lycopersicum esculentum* by 55% at pH 5.0, 50% at pH 6.0. It has been found that at concentrations of 400 mg/litre chemicals with an aromatic ring and free carboxyl group are most effective in this respect. Some other structurally similar substances like benzoic, salicylic and nicotinic acids also inhibit respiration. The enzymes oxidizing cytochrome-C, ascorbic acid, catechol and glycolic acid are not affected.

9. The transformation of healthy into tumor cells

SMITH'S suggestion that an endotoxin liberated on the death of *Phytomonas tumefaciens* is responsible for the transformation of the healthy cells of the host plant into crown-gall tumor cells is not, however, supported by the results of later investigations. Although no doubt a wilt-inducing polysaccharide from *Phytomonas tumefaciens* is described by HODGSON et al. [475-477], and though more of it is formed by virulent strains than by attenuated strains of the bacteria, it does not readily follow that this is the substance concerned directly with tumor formation. In certain other investigations [688] the phosphatide fraction produces cell proliferation and the polysaccharide produces necrosis, with only limited cell proliferation. MAGROU[737] is of the view that hyperplasia is apparently not produced by the bacteria themselves, but by the toxin or chemical products of their metabolism. BOIVIN, MARBE, MESROBEANU & JUSTER[109] obtained a complex glucidolipid by treating the crown-gall bacteria with trichlor-acetic acid and claimed that this is the specific endotoxin, concerned in tumefaction. It was believed to act as antigen and also precipitate the antitoxin obtained from rabbits inoculated with bacterial cultures of *Phytomonas tumefaciens*. When injected into the young branches of *Helianthus annuus*, tumors were reported to have risen in the cortical region. The influence of the metabolites of *Phytomonas tumefaciens*, extracts of the crown-gall and of yeast on the *in vitro* growth of excised fragments of tissues of *Nicotiana tabacum* and *Helianthus annuus* was studied by HILDE-

BRANDT, RIKER & DUGGAR[468]. At low concentrations, supplements to the basal medium of the fermented cell-free, bacterial media of virulent and attenuated crown-gall bacteria have little effect, but at high concentrations the growth of these tissues is strongly inhibited. Increased concentrations of lyophilized virulent and attenuated *Phytomonas tumefaciens* generally impair the growth respectively of the cultures of *Nicotiana* and *Helianthus*. Low concentrations of unautoclaved crown-gall extract of *Lycopersicum esculentum* or autoclaved extract of the crown-gall of *Tagetes erecta* or unautoclaved extract of yeast in the basic medium generally stimulate growth but at high concentrations inhibit their growth. KAUFMANN [569] failed, however, to observe the production of tumor by treatment with bacteria-free filtrates from cultures of *Phytomonas tumefaciens* and concludes therefore that apparently the living bacteria are necessary for the formation of crown-gall tumors.

The production of indole-3-acetic acid from tryptophan by *Phytomonas tumefaciens*[82, 702] seems to suggest that perhaps the action of the bacteria on the plant tissue is due to their generating this substance. When applied in lanolin mixtures, 1—3% indole-3-acetic acid is capable of inducing the formation of tumors on a number of plants. Filtrates and ether extracts of *Phytomonas tumefaciens* also show tumor inducing power[146, 703].

The effect of synthetic auxins dissolved in lanolin, described by KELLY[572], include the inhibition of lateral buds, epinasty and formation of gall. Naphthalene-acetic acid produces thickening of the stem of *Helianthus annuus* below the second node from the decapitated stump. Concentrations that cause galls and inhibition of buds are lower than are required for epinasty. At threshold concentrations of 0.002 and 0.006% naphthalene-acetic acid is most effective in producing the inhibition of lateral buds and in causing epinasty. Indole-acetic acid in threshold concentration of 0.004% is most effective in giving rise to galls. Inhibition of lateral buds by indole-butyric acid is less than in the case of naphthalene-acetic acid, but the formation of galls is more effective. Indole-propionic acid produces large galls, phenyl-acetic acid in concentration of 0.3% produces galls, but is ineffective in producing inhibition of lateral buds and epinasty, except in higher concentrations.

Certain significant differences between the crown-gall tumors arising under the action of *Phytomonas tumefaciens* and the tumors caused by the application of indole-acetic acid, to which we shall again refer, indicate, however, the unlikely rôle of the production of this growth substance by *Phytomonas tumefaciens* in tumor induction. The transformation of normal cells to crown-gall tumor cells by *Phytomonas tumefaciens* is influenced by the temperature at which the plant tissue is incubated after inoculation. Early studies by RIKER[937] revealed that development of the crown-gall on *Lycoper-*

sicum esculentum is slight at temperatures between 28 °C and 30 °C and no tumors develop above 30 °C. Though the plant and the bacteria grow well at this temperature, the tumor formation is inhibited. Tumors form on certain species of *Nicotiana*, both above and below this critical temperature, but the tumors which appear at 31 °C are smaller and also appear about ten days later than at 26 °C[937]. RIKER *(op. cit.)* also reports that the crown-gall develops within the temperature range of 14—30 °C but the optimal development is at 22 °C. Above 30 °C the plant grows well, but the galls do not develop. At 10 °C and below the host plant shows almost no growth and the gall also does not develop.

BRAUN[121] investigated the period necessary for *Phytomonas tumefaciens* to alter the normal plant tissue cells to tumor cells. By subjecting the host-parasite complex to heat treatment of 46-47 ° it is possible to kill the causal agent of the crown-gall at any desired time, after it has become established in *Vinca rosea*. The host is not seriously affected by this heat treatment. By the application of this technique, BRAUN found that normal plant cells are converted into tumor cells as early as 36—48 hours after inoculation, but the galls remain small. Tumors initiated in four days are comparable in size to the untreated tumors. The cellular alterations are brought about within four days after inoculation of the host plant with the bacteria. After this period, the tumor development is independent of the bacterium. Later, BRAUN[123] found that the transformation of the normal tissue into the tumor tissue on *Vinca rosea* fails if the inoculated plant is maintained at a temperature of 32 °C. The plants maintained at 26 °C for one, two, three, four and five days before being subjected to the action of 32 °C differ significantly in the size of the tumors produced on them. No tumor is formed on the plant transferred from 26 °C to 32 °C in one day. The plant exposed to 26 °C for two days before being tranferred to 32 °C develops a few small-sized tumors and the tumors are large on plants kept at 26 °C for three days. The tumors on plants kept for four days or five days at 26 °C are almost as large as those kept at 26 ° continuously. As already pointed out, the high temperatures do not inhibit the growth of either the bacteria or of the plant and there is also no significant physiological disturbance in them. As a consequence of the heat treatment, there is, however, an apparent lag of a little more than one day between infection and the conversion of the normal tissue into the tumor tisse.

BRAUN[123] also discoverd that exposure of *Vinca rosea*, inoculated with the crown-gall bacterium, to temperatures of 46—47 °C destroys the bacteria, without, however, killing the plant. It was also soon found that when the bacteria are thus eliminated from the plants 36—48 hours after inoculation by this "fever" treatment, tumor still develops at the site of inoculation. It is thus evident that the crown-gall tumor is capable of continuing its growth in the absence of the inciting bacteria. It thus appears that the bacteria have to act for about four days for the plant cells to be altered to the maximum. The tranformation of the normal into tumor cell occurs at 25 °C but not at 32 °C in *Kalanchoë daigremontiana*[125] and the minimum period is 32 ± 2 hours. When the plants are exposed to 32 °C for twenty-four hours and then to 25 °C, the transformation still takes place. It is thus evident that *Phytomonas tumefaciens* requires about 20—24 hours at 25 °C for environmental adjustments and an additional 10 hours at 25 °C to elaborate the substance capable of transforming the normal cells into tumor cells. RIKER[937] and RIKER, LYENIS & LOCKE[949] are, however, of the view that the effect of temperature on the pathogenicity seems to be correlated with the plant more than with the bacteria.

These investigations show that the normal cells are converted into tumor cells at 26 °C but not above 30 °C and once this transformation has taken place at 26 °C, the tumor cells develop into neoplastic growth equally well, both above and below this critical temperature. It is thus possible for us experimentally to stop the transformation process at any desired time by simply growing the plant at a temperature above 30 °C. The relative size of the tumor developing at a temperature above 30 °C is then a measure of the degree of the transformation of the normal tissue to the tumor tissue at the lower temperature. Recently BRAUN[125] found that the optimal temperature for this transformation lies between 25 °C and 27.4 °C in the case of *Kalanchoë daigremontiana*. There is a progressive decrease in size and weight of the tumor tissue between 27.5 °C and 28.8 °C. The period of inception becomes also prolonged and no tumor is initiated at temperatures above 29 °C. BRAUN concludes from these results that the transforming substance is a thermolabile compound, which is destroyed at 29 °C as rapidly as it is synthesized by the bacteria. Applying ARRHENIUS' principle, we obtain high values above 80 000 calories per mole for the temperatures at 27.3 °C and 28.6 °C. Reactions of this magnitude are, as is well known, characteristic of denaturation of proteins. It is thus highly suggestive that the tumor-inducing substance or at least something associated with its inactivation has a complex structure.

DE ROPP[255] found that virulent crown-gall bacteria can be maintained for as long as three months at 33 °C without loss of tumefacient capacity. It is clear from this observation that although this high temperature destroys the tumor inducing substance, its production is not, however, prevented. Loss of the tumefacient power is, however, found to occur by growing bacteria in media containing 0.1—0.3% of glycerine and other amino acids[713, 1202]. The tumefacient power is sometimes restored by bringing the bacteria back to media lacking in these amino acids. DE ROPP[256] made the im-

portant discovery that tumor formation is prevented by streptomycin in concentration of 0.5 mg/litre. In 1937 KENT[577] found that
lysis of *Phytomonas tumefaciens* with a bacteriophage destroys the
tumefacient power. ISRAILSKY[541] reports the finding of a bacteriophage in the crown-gall. The lytic principle is said to withstand
70°C, the bacteriophage is destroyed by 70°C. The same author[542]
has also reported that the bacteriophage protects the plants against
infection. KAUFMANN[568, 569] obtained a bacteriophagic lysin, which
he believes to be responsible for the production of the gall. BROWN
& QUIRK[147] investigated the influence of bacteriophage on *Phytomonas tumefaciens* and potentials of juices of normal and tumor
tissues. They consider the presence of bacteriophage as established.
Accelerated growth of bacteria, with increased pathogenicity, seems
to be associated with the phenomenon of bacteriophagy, as well as
does the retarded growth of bacteria, causing decreased or delayed
or no pathogenicity.

Isolation of bacteriophage from non-viable cultures of *Phytomonas
tumefaciens*, from virulent cultures and from crown-gall on *Beta
vulgaris* is reported by MUNCIE & PATEL[830]. High concentrations
cause complete lysis of *Phytomonas tumefaciens*. The potency of the
lytic principle is greatly reduced by heating for ten minutes at
80°C and completely inactivated at 85°C. CHESTER[191] is mainly
responsible for the refinement of the technique for investigations of
bacteriophage and crown-gall. He was able to isolate the bacteriophage from the crown-gall and to show that the bacteriophage is
capable of diffusing outward from lesions into the surrounding tissue.
He found that the bacteriophage can penetrate healthy plants from
soil through the root. The weight of available evidence would appear
to suggest that the tumefacient substance is of macromolecular
nature, having the capacity to induce a self-perpetuating abnormal
pattern of cell growth in the plant. The substance may also possibly
be a virus, of which *Phytomonas* is a vector or it may be a normal
constituent of the plant cell, which under the action of the bacteria
is converted into a tumor-inducing substance.

It is not known at present whether the transformation by *Phytomonas* is confined to cytoplasm or the nucleus of the host plant cell.
The transforming substance is perhaps a mutagenic agent of considerable complexity, capable of altering permanently the growth
pattern of cells of diverse groups of plants.

Although some workers like HAMADI[430, 431] insist that the continued presence of the bacteria is necessary for the development of
the crown-gall, CENTANNI[180] and BRAUN[120] have conclusively shown
that the bacteria function only in the initial stimulation of the susceptible plant tissue and are not necessary for the continued development of the crown-gall tumor. GAUTHERET[369], in his recent review of the crown-gall problem, has stressed the fact that the effect

of *Phytomonas tumefaciens* consists simply in initiating the process of formation of tumors; the process then becomes independent of the presence of the bacteria. This is well proved by the isolation of bacteria-free secondary tumor tissue, retaining all the tumor properties of the original tumor. The modifications produced by *Phytomonas tumefaciens* persist even after the complete elimination of the bacteria, so that some permanent alteration of the plant cells seems to have been brought about.

10. Secondary tumor

We have already mentioned that in addition to the primary tumor at the site of inoculation with bacteria, secondary tumors also develop at some distance away. In 1916 SMITH considered the secondary crown-gall tumor to result not from the migration of bacteria within the plant tissue or on its surface, but a consequence of outgrowth of tumor strands. Tumor strands do often grow out as low ridges of tumor tissue, externally from the primary tumor to within some distance of the secondary tumor, but are not connected with the secondary tumor. These strands of bacterially infected tissues are sometimes large enough to be visible to the naked eye. The tumor strands consist usually of delicate threads of cells and vessels, hardly distinguishable even under the microscope. Such strands are found on *Nicotiana tabacum* in the outer cortex, but also through the medulla in the case of *Chrysanthemum frutescens*. The secondary tumors exhibit the structure of primary tumors. In 1922 SMITH[1064] discovered, however, that tumor can occur not only by tumor strand but also by the conversion of healthy cells immediately in contact with tumor tissue into tumor cells. In the course of this conversion, the healthy cells first enlarge and then divide rapidly. The stimulus causing this cell proliferation was conceived of as a physico-chemical one derived from the bacteria and acting either at distance from them on the cells in which they are not present or as due to direct transfer of the bacteria from cell to cell. Other workers have considered the secondary tumors to be produced when immature tissues are infected by *Phytomonas tumefaciens* and the infected cells are subsequently spread out by growth. Bacterial migrations in zoogleal masses over several centimetres in the intercellular spaces in *Nicotiana affinis* were thought by ROBINSON[964] to give rise to the secondary tumors. Many of the tumor strands described by SMITH are interpreted as internal tumors formed at points where vessels have ruptured[1118]. The transpiration current has been assumed to help the transport of the bacteria over considerable distances within the vessels of the infected plants. Bacteria have not doubt been detected from apparently healthy tissues at distances of 100 cm above and below the primary tumor in *Datura latula* by STAPP[1089] and from internodes of *Helianthus annuus* 5.5 cm from the point of inoculation an hour afterwards. Secondary tumors and tumor strands can be produced by attenuated strains of *Phytomonas tumefaciens*, even though no primary tumor arises at the site of inoculation in such cases. It would therefore appear that the process of formation of secondary tumor is essentially different from that concerned in primary tumors. Diffusible cell-stimulating substances formed by the bacteria confined within the vessels have also been thought to be responsible for the initiation of the secondary tumors[120]. BRAUN[120] described in 1941 the development of the bacteria-free secondary tumor on *Helianthus annuus*. According to him, the ridge-like overgrowths, which extend from the primary tumor to within a short distance of the soil level, are also secondary tumors. The tumor strands are not, as described by SMITH, direct outgrowths of the primary tumor at all, but these strands develop laterally from the region into the pith. Two distinct types of secondary tumors, differing in position and internal structure, are recognized by him. Secondary tumors and tumor strands may be produced by attenuated strains of the crown-gall bacteria even though no primary tumor is formed by these strains.

11. Bacteria-free crown-gall tissue

Even though as late as 1930 HAMADI[430] insisted that the crown-gall tumor growth is not autonomous, but depends entirely on the continued presence of the causative organism, the first indication of bacteria-free crown-gall was found by JENSEN already in 1918[548]. This observation also showed that SMITH had indeed not believed the evidence of his own observations and it was becoming increasingly clear that his ideas that bacteria are always necessary for the development of tumor may not be correct. The final doubts were roused when crown-gall tumors from cultivated varieties of *Beta vulgaris* were grafted on to other varieties[674]. The grafted tumors grew on the new host, without becoming differentiated, thus forming new tumors, corresponding entirely in appearance and structure to the mother tumor. The graft tumors, in a large number of cases, were found to be bacteria-free. JENSEN concluded from this that the cells of the tissue, under the influence of the bacterium, become altered for a series of cell generations and develop the abnormally increased proliferative power of tumor cells proper, which is independent of the continued stimulation. BRAUN & WHITE[132] inoculated *Helianthus annuus* with *Phytomonas tumefaciens* and observed primary tumors containing bacteria, tumor strands frequently containing bacteria and secondary tumors at a distance from the primary locus and completely bacteria-free. Tissue cultures of the secondary tumors, maintained over long periods, have remained entirely bacteria-free and have retained the tumor characters. As a result of a series of exhaustive tests, BRAUN & WHITE[132] were able to demonstrate that secondary tumors are free from *Phytomonas tumefaciens*. Bacteria-free secondary tumor cells exhibit many of the essential characteristics of the original crown-gall cells. Subsequent studies by numerous other investigators have shown that whereas the primary tumor is 100% infected with the bacteria, the secondary tumor on the first node above the primary tumor is only 83% infected and at higher sites only 4% infected [132]. WHITE & BRAUN[1265] have also given an account of the bacteria-free secondary tumor tissues. They found that fragments of these tissues are bacteria-free and these fragments in *in vitro* cultures maintained a rapid and disorganized type of growth. Implantation of fragments from tissue cultures of the bacteria-free secondary tumor of *Helianthus tuberosus* under the bark of *Helianthus annuus* results in the formation of rapid and disorganized growth, wholly free from bacteria. Fragments removed aseptically from such secondary tumors by WHITE & BRAUN *(op. cit.)* were placed in agar medium containing 2% suchrose, minerals, glycine and thiamine. A high proportion of these tissue fragments proved to be bacteria-free and began to grow on the medium, resulting in the formation of a white, friable mass, which did not differentiate and showed no signs of polarity in growth.

Tissues removed from chemically induced tumors (for example induced by indole-acetic acid) and cultured *in vitro* in the same medium produce a unilateral, compact, green, woody and sometimes rooted type of structures. The cultures of healthy tissues of *Helianthus annuus* increased about 250-fold in a year and those of the chemically induced tumor only by about 50-fold. The crown-gall tumor tissue, however, grew so rapidly that its increase during thirty passages was estimated at 100 000 000 000 000 times and some of the strains of tumor tissue isolated then were reported to be still growing even in 1950, demonstrating that their capacity for growth on the medium employed is unlimited. Four hundred cultures of tumor tissues were crushed in nutrient broth, incubated from one to eight hours, mixed with agar broth and poured into petri-dishes. Most of these cultures were bacteria-free, but the bacteria which did appear in a few cases, inoculated into the stem of *Lycopersicum esculentum*, proved incapable of giving rise to galls. It must therefore be concluded that these cultures of the tumor are entirely free of *Phytomonas tumefaciens*. The possibility of a virus being involved was also eliminated by grinding the tumor cultures on sterile slides and inserting the debris into a month-old plant of *Helianthus annuus* at the level of the cambium. Only one tumor developed and this also proved to be entirely bacteria-free; this isolated formation of a tumor was attributed to a few cells of the culture having remained uninjured by the grinding. No tumors arose in a second similar experiment. When bacteria-free tumor tissue is grafted back to *Helianthus annuus*, a tumor results whenever the graft takes. Similar results are also obtained with grafts of tumor tissues on the stem of *Helianthus tuberosus*. The tumors were thus believed to arise probably from pre-existing tumor cells, and the host cells are not themselves modified in their behaviour by the presence of adjacent tumor cells.

In 1945 WHITE[1258] isolated bacteria-free tumor tissue from the crown-gall on *Vinca rosea* by freeing the gall of bacteria by heat treatment. This bacteria-free gall tissue grows more rapidly than the fragment of tumor tissue from *Helianthus annuus* and produces white open-textured growths. It has been found to be capable of indefinite growth on the same medium as used in cultures of tumor tissue of *Helianthus annuus*. The fragments of *Vinca rosea* tumor tissue after twenty passages were grafted onto healthy young plants of *Vinca rosea* and in thirty-three out of forty-two grafts, tumors arose from the implant and not from the host cell. Occasionally portions of primary tumors of *Helianthus annuus* are also found to be bacteria-free and several clones of bacteria-free tumor tissue were isolated by DE ROPP[251]. Such bacteria-free crown-gall tissue has been isolated from several other plants including *Scorzonera hispanica, Helianthus tuberosus*[368], *Vitis vinifera, Opuntia monocantha, Carthamus tinctoria, Abutilon avicennae, Nicotiana tabacum, Antirrhinum majus*[821], *Chrysanthemum frutescens* and *Tagetes erecta*. DE ROPP [251] has described the isolation and behaviour of bacteria-free crown-gall tissue even from the primary tumor on *Helianthus annuus*. The interior tissue of the crown-gall is wholly free from bacteria and new strains of bacteria-free tumor tissue may readily be obtained from this source. Though often mixed with normal tissue, with prolonged culture, the tendency of the normal tissue to develop disappears.

12. Tumor induction by bacteria-free crown-gall tissue

We have referred to the formation of tumors by grafting bacteria-free secondary tumor tissue onto healthy tissues. WHITE & BRAUN [1265] and WHITE[1258] explained the tumors obtained by grafting the bacteria-free gall tissue of *Helianthus annuus* or of *Vinca rosea* onto the normal stem as arising from the implanted tumor tissue. DE ROPP[253] investigated the interaction of healthy and tumor tissues in *in vitro* grafts, in which small fragments of stem or fragments of healthy cambial tissue were used as stock and fragments of tumor tissue as scion. The healthy tissue reacts by producing a mass of hyperhydric, translucent tissue and adventitious roots, much as if stimulated by the action of indole-acetic acid in certain concentrations. The reaction thus appears to be due to the production of this auxin or some other similar substance by the tumor tissue. When the stock becomes firmly united with the tumor tissue, we observe the formation of spherical tumors, equal to or larger than the original healthy fragment. These tumors are characterized by a spherical core of xylem mixed with short tracheids and parenchyma cells. Outside the xylem is cambium and the phloem outside of this lacks sieve tubes. The outermost layer is a loose-celled cortex, containing chloroplast. The structure of the induced tumor thus differs from the tumor scion, within which there is a more or less chaotic mixture of cell types. Fragments of these induced spherical tumors placed in White's medium are capable of growing at a rate comparable to that of most strains of bacteria-free crown-gall tissue and grow as small spheres for several transfers, repeating the structure of the original tumor from which the fragments are removed. They lose the green colour and spherical form after about six months and their structure becomes increasingly chaotic, but it is still distinct from the tumor scion. The spherical tumors mostly appear to arise from the cambium of the healthy tissue in the immediate vicinity of the tumor graft. A particularly massive tumor is reported to have been observed in one instance from the end of the stem fragment opposite the graft. Different strains of tumor tissue differ in their ability to produce these induced tumors.

DE ROPP[253] failed to induce tumors by the application of extracts of tumor tissues to fragments of stems of *Helianthus annuus*. By separating the tumor scion from the healthy stock by a thin layer of agar, it was found that the tumor is unable to activate the healthy cells. Unusual growth in the healthy tissue occurs when a union between the tumor and the healthy tissue takes place. Two different growth stimulants thus appear to be produced by the bacteria-free crown-gall tissue: one of these seems to be capable of causing a generalized proliferation of tissues similar to that caused by indole-acetic acid and the other is capable of causing the formation of new tumors with a definite structure and the same nutrient requirements as the original bacteria-free tumor tissue. Both these factors appear to be capable of exerting their effects only when the tumor and the healthy tissues become united in a graft.

CAMUS & GAUTHERET[170, 171] have also shown that bacteria-free gall tissue of *Helianthus tuberosus* grafted onto healthy tissues of the same plant, produces tumors that can be cultured on the same medium as will support the growth of crown-gall tissue (but not healthy tissue). The action of the implant on healthy host was investigated by MC EWEN[771] with bacteria-free tumor tissue obtained from the primary gall of *Helianthus annuus* and cultured on White's media, implanted in healthy stems. The tumor implant is readily distinguished from the host tissue by its characteristic nodular texture and fawn colour. The host stock supporting the tumor implant exhibits varying degrees of stimulatory response, viz. abnormal swollen roots, abnormal basal callus and apical callus and overall cortical proliferations. The response is similar to the tissue of *Helianthus annuus* exposed to high concentrations of auxins. The action of a growing tumor implant on a healthy one effects a temporary stimulatory response, which does not persist in the absence of the implant and which hastens necrosis. This response could perhaps be explained on the basis of the auxin supplied by the tumor implant. However, in some cases the tumor implant is short-lived and the host stock still produces a tumor with all the characteristics of a crown-gall tumor. Though the experimental tumor produced by auxins shows intermediate response between normal and tumor tissues, the change through response to the short-lived tumor tissue is positive. LEVINE[674] has compared the behaviour of the crown-gall on *Ricinus communis, Lycopersicum esculentum*, three varieties of *Beta vulgaris* and on *Ficus elastica* and their auto-, homo- and hetero-transplants. In several cases limited growth of transplant and overgrowth of host tissue in proximity to the transplant were observed by him. He concludes that plants susceptible to form crown-galls by action of *Phytomonas tumefaciens* are capable of forming galls upon introduction of the crown-gall tissue. Tissues isolated from the stems of the hybrid *Nicotiana langsdorfi × Nicotiana glauca*, which normally produce spontaneous tumors, after being maintained in *in vitro* cultures for five years, were grafted by WHITE [1255] onto healthy plants of *Nicotiana glauca*. On this they give rise to typical tumors. The genetic tumors also thus appear to possess the property of propogating their tumor nature indefinitely.

DE ROPP[251] is of the view that a tumefacient agent is present in bacteria-free crown-gall tissue. In experiments with bacteria-free tumor-bearing stems of *Vinca rosea*, root-producing substances were observed to be generated by tumor tissue. This is evidence that bacteria-free gall tissue generates a diffusible substance capable of stimulating proliferation on normal callus tissue, but this proliferation is not a permanent change in behaviour of the callus tissue.

13. Reversion of tumor to normal tissue in crown-gall

MAGNUS[732] and SMITH[1056] observed tumors on or within which organs like roots, leaves, etc. can be recognized. SMITH indeed believed that these organs arise from stimulated totipotent cells present in the tumor. Other workers have, however, claimed that the leafy shoots may result from differentiation of gall tissue[663, 665, 667, 668]. DE ROPP[251] observed roots during the first few passages in tumor tissue isolated from a primary crown-gall tumor on *Helianthus annuus* but after prolonged culture the roots vanished. BRAUN[124] showed that strains of *Phytomonas tumefaciens* differ in their capacity to give rise to teratomas on *Kalanchoë daigremontiana*, ranging from leaves, stems or distorted root-like structures. Inoculation of decapitated plants result in a typical crown-gall, from which structurally complex growths arise. These structures, when inoculated with *Phytomonas tumefaciens*, failed in most cases to generate tumors. Plantlets originating from leaves on teratomas produce sterile gall-like overgrowths in place of roots. BRAUN[124] interpreted these observations by assuming the presence of factors determining the morphogenetic fate of cells. In undifferentiated tumors resulting from bacterial inoculation, these factors are assumed to be completely overwhelmed by the action of the tumor-inducing principle. The tumor-inducing principle was assumed to be capable of cancelling in large part the differentiation competence previously possessed by the affected cells. This conclusion seems to be justified by the fact that tissue cultures of tumors isolated from *Helianthus annuus* and from *Vinca rosea* have been kept for more than five years without showing any evidence of tendency to differentiate and produce complex tissues and organs. The balance is assumed to be in favour of varying degrees of morphogenetic determinants present in the affected cells when this same tumor-inducing principle acts on young cells of *Kalanchoë daigremontiana*. The result of such a condition would probably be the development of structural elements that show varying degrees of differentiation. Potentialities for differentiation in certain young cells appear to be so great that complete recovery of the affected cells takes place and apparently normal plants are produced. Working on the suggestion of DE ROPP[251] that the alteration of the normal into the tumor cells may be traced to certain self-propagating cytoplasmic factors, BRAUN[126, 127] succeeded in eliminating them by providing conditions that favour the more rapid multiplication of cells in relation to that of these factors. Dilution and final elimination of such cytoplasmic tumefacient factor was accomplished, with the resulting recovery of the tumor cells. He obtained tumor cells by the action of *Phytomonas tumefaciens* on the pluripotent cells of the meristem tissue of the bud — at the time of their alteration, these cells possess highly regenerative powers. The tumor cells thus obtained retain indefinitely a well developed capacity for organizing buds. Shoots derived from these adventitious tumor buds were forced into very rapid growth by a series of graftings to healthy plants. Thus made to divide rapidly, these tumor cells gradually recovered and became normal in every respect. The relative potency and regenerative competency of the cell acted upon by the tumor-inducing principle appear to be an important factor in determining the type of the resulting tumor. Cells with low potency at the time of alteration become typical tumor cells, but cells with high developmental potential at the time of alteration retain the capacity for organization.

In his studies on the recovery of tumor cells from the effects of the tumor-inducing principle, BRAUN[126, 127] refers to the peculiar behaviour of *Kalanchoë*. The altered cells appear first as an undifferentiated type, but as the tumor grows, there results, in place of the characteristic neoplasm, an overgrowth composed of not only the uncoördinated tumor cells but also of cells organized into structural complex. He deals with the question whether the structurally abnormal shoots are composed of altered cells that have acquired the capacity for differentiation and organization or these structures result from growth of normal cells stimulated to develop by the tumor. Using *Nicotiana tabacum* for tumor development by inoculation with T–37 strain of *Phytomonas tumefaciens*, BRAUN observed tumor formation to be followed by teratoma on the tumor surface. He removed fragments from the distorted but organized stem tissues developed from teratoma on the tumor and cultured them *in vitro* on White's medium. Though slower than that of tumors, the rate of growth of these bacteria-free tissue fragments was found to be greater than for normal cells. These cells differ from the tumor cells of *Nicotiana* in their capacity for differentiation and organization. The tissue cultures become covered with adventitious buds and leaves. The mass has a hyperplastic core of hypertrophied cells. BRAUN concludes from these observations that structurally distorted but organized stems developing from teratoma are composed of cells, the metabolism of which is fundamentally different from that of normal cells of *Nicotiana*. Implantation of tissue fragments from *in vitro* cultures of the teratoma overgrowth in healthy host results in teratomous growth. In their ability to grow profusely on a medium lacking growth promoting substances and in their ability to develop independently of the growth restraining influence of the host, these cells resemble tumor cells. Repeated implantations of buds, composed of altered cells into healthy plants, from which all normal buds have been removed, result in new growth, which becomes more and more normal. It is a gradual process in the direction of normal: the factors which cause tumor cells to develop abnormally seem to become diluted and eventually lost from affected cells that are forced to grow and divide sufficiently rapidly. BRAUN[127] also showed that tumors arising on *Nicotiana tabacum* as a result of inoculation with T–37 strain of *Phytomonas tumefaciens* can be induced to produce normal shoot, flower and set seeds. This complete reversion of the gall to the normal pattern of growth takes place when all normally produced buds are removed from the plants, inoculated at the cut stem surface with *Phytomonas tumefaciens*. This results in the production of a complex tumor that contains adventitious buds composed of tumor cells. The shoots arising from these buds are removed when large enough and grafted onto healthy plants, from which all normal axillary buds had been removed. The tips of the shoots are again removed and grafted to healthy plants. After this treatment the stems derived initially from the tumor tissue, appear to have reverted completely to the normal pattern of growth.

It may be remarked in this connection that tissue cultures of even normal parts of higher plants, unlike animal tissues, retain their ability to generate new organs, especially roots [1162]. Such organs arise even when the cultured tissue fragment contains no rudimentary organs. Reference may also be made to the observation of WHITE[1253] on explants of the genetic tumor of *Nicotiana glauca* × *Nicotiana langsdorfi* in simple synthetic media, growing as undifferentiated callus cells. As soon, however, as they are submerged in the liquid media of the same composition, they form abundant buds and leafy shoots. It seems that plant tissue cultures are complex systems that can change during their life time. Many observations on cultures of both normal and tumor tissue show that the occurrence of polyploid cells depends on the division of corresponding cells and also on the conversion of diploid cells. In cultures of diploid meristem tissue of pea roots there are first diploid cells, but after some time we find some tetraploids, octoploids and even higher ploid cells.

14. Recession of the crown-gall

A number of workers have reported curing the crown-gall tumor by various means. The investigations of BOEHM & KOPACZEWSKI[103] have, for example, demonstrated a biological antagonism between *Phytomonas tumefaciens* and *Streptococcus erysiplatus*. This antagonism prevents the formation of tumors caused by *Phytomonas tumefaciens* on *Pelargonium zonale* by treatment with injections of *Streptococcus erysiplatus*. Penicillin and streptomycin have also been described as having the power to destroy gall cells, without injury to the normal cells[138, 139, 433]. Streptomycin is, however, reported to be more efficient than penicillin in this respect. Penicillin G has an inhibitory action on the growth of the bacteria-free crown-gall tissue at concentration of 50 mg/ml[257]. Pteroylglutamic acid and twelve analogues were similarly tested and five of these analogues were found to completely inhibit the growth of the crown-gall at concentrations of 0.05 mg/ml and a concentration of 0.0005 mg/ml significantly reduced the amount of tumor growth. DE ROPP [257] investigated the inhibiting action of some folic acid on the growth of the crown-gall. Inhibitory properties are exhibited by 4-aminopterolyl-glutamic acid, 4-amino-N[10]-dimethylpterolyl-glutamic acid, 4-aminopterolyl-gamma-glutamic-gamma-glutamyl-glutamic acid and 4-aminopterolyl-l aspartic acid. Compound containing the methyl group as well as the 4-amino group inhibits partially at concentrations of 10 mg/litre. The growth of excised root tissues of *Lycopersicum esculentum* is inhibited more strongly than that of the tumor tissue on the same plant. Any amount of pterolylglutamic acid does not prevent inhibition. These compounds do not appear to have a specific inhibitory action on the growth of the tumor tissue and merely appear to act by preventing cell multiplication rather than cell enlargement. Mitosis is, for example, completely stopped in the root of *Allium cepa* grown in a solution of 0.1 mg/litre of 4-amino-N[10]-methlylpteroyl-glutamic acid in two hours [259]. A local application of 10 mg/litre of 4-amino-N[10]-methyl pterolyl-glutamic acid completely prevents the development of the crown-gall on the hypocotyl of *Helianthus annuus*. Galls are prevented by this treatment from growing on the fully grown hypocotyl, but severe damage is also caused to the young stem.

A number of nitrogen-mustards, cortisone and guanazolo (8-azaguanine) also inhibit the growth of the crown-gall *in vitro*. The most active of the nitrogen-mustards so far tested, methyl-bis-beta-chlorethyl amine, completely inhibits tumor growth in one hour at a concentration of 100 mg/litre and the inhibition of growth on excised roots of *Lycopersicum esculentum* is stronger than on tumor tissue. A concentration of 1.0 mg/litre of guanazolo inhibits the growth of tumor tissue *in vitro*. Cortisone inhibits at a concentration of 100.0 mg/litre [259].

We have the interesting observation of RYBAK[1001] on the bacteriostatic action in the crown-gall on *Pelargonium zonale*. Phenols, which play an important rôle in the resistance of plants to fungus diseases, is abundant in *Pelargonium zonale*, but does not prevent the formation of the crown-gall tumor on this plant. *Pelargonium zonale* produces, however, a thermostable antibiotic factor, which acts on *Phytomonas tumefaciens in vitro*.

Radiotherapy of the crown-gall tumor has also been reported by several workers. As early as 1922 LEVIN & LEVINE[664] and later LEVINE[672] found, for example, that radium emanations are capable of killing the tumor tissue of crown-gall, but the radiations affected the healthy tissues also more or less equally. RIVERA[954] has described that irradiation by X-rays of the tumor on *Pelargonium zonale* results in the formation of globose bodies or secondary tumors, surrounded by a layer of pseudocambium cells, which proliferate rapidly. Irradiation favours the formation of functioning fibro-vascular bundles. Disorganization of cytoplasm and destruction of nuclei are perhaps the chief causes of the regression of tumor by granular degeneration. This disorganization and death of the secondary tumor are due to disorganization and obliteration of the conducting system. In tests of radiotherapy on the crown-gall on *Ricinus communis*, RIVERA[957] exposed to small doses of X-ray

(3 ma, 180 kV at 40 cm without filter) to 2—4 minutes. Tumor tissues exposed for shorter periods were found to grow larger, but both the irradiated tumors had their growth arrested. Exposures of 7, 15 and 30 minutes duration result in retardation and no growth for 30 minutes. The same author also studied the effects of X-ray on tumor cells of *Ricinus communis* and *Pelargonium zonale*. Globose nodules arise on the crown-gall on *Pelargonium* and are surrounded by dead cells. The nodules arise from the tumor meristem, modified in its proliferation. By exaggerating the functions of the tumor cells, X-ray brings about the final death of these cells. Further, the comparative experiments on the effects of X-ray on the normal and tumor tissues conducted by RIVERA have shown that the X-ray sensitivity varies with the type and stage of development of the cells, so that some cells are killed but others are actually excited to growth. Owing to the death of the outer more sensitive cells and the stimulation of the inner groups of cells, the tumor becomes segregated by disorganization with intercalation of a pseudomeristem cell layer between the dead cells and the innermost tumor cells. In 1927 RIVERA[958] reports the regression of the growth of the crown-gall tumor twenty-two hours after irradiation for 35 minutes at 35 cm distance to X-rays 4 ma, 180 kV. The regression is followed by rapid increase and then decrease in tumor size. Growth increases after seven days up to 14—15 days and then the size begins to decline. Discussing the effects of strong doses of gamma rays on *Phytomonas tumefaciens*, RIVERA[959] states that even the strongest irradiation fails to kill pure cultures of this bacterium, although multiplication is arrested and growth retarded. Reference may also be made to the investigations by RIVERA regarding the influence of Lakhovsky current on the development of tumors on *Pelargonium zonale* and *Ricinus communis*, surrounded by an open circuit, attached to an oscillator, providing current wave length 2 metres, with 150 000 000 vibrations per second. The tumor on *Pelargonium zonale* dies out but that on *Ricinus communis* becomes accelerated. Applied after the tumor growth has already begun, this treatment retards the development of the tumor.

15. Concluding remarks

We may now refer to the question of the fundamental similarity of the crown-gall tumor and animal cancers. SMITH repeatedly emphasized the resemblance of the crown-gall to cancer and even went to the extent of suggesting that human and animal neoplasia, like the crown-gall, must be bacterial in origin. The fact that except in the case of Rous sarcoma and Bittner's milk-factor, neither virus nor bacteria have so far been demonstrated in cancer, is, however, no objection to the idea that the crown-gall is comparable to cancer of animals[164, 180, 383, 384, 430, 431, 614, 693, 694, 737, 771, 833, 972, 1205, 1265, 1266].

One of the main objections to the cancer analogy of the crown-gall is, however, that the structure of plants precludes metastases that characterize animal neoplasia. We have mentioned above the observations of several workers on induced tumors. These observations seem to favour the interpretation that the induced tumors are metastases of the original tumor scion. The facts which seem to indicate the contrary are that firstly the plant tissues are rigid and migration of tumor cells seems improbable and the induced tumors have a very characteristic structure, which distinguishes them from the tumor scion. If the induced tumors are not metastases, it then seems that a specific transforming principle exists in the tumor tissue itself, capable of changing the healthy tissue into tumor tissue. The transforming principle is apparently capable of passing through only a continuous layer of living tissue. It does not also seem to travel far from the point of union of the healthy-tumor tissue graft. Either the original transforming principle is transferred and perpetuated in the tumor cell or possibly some new transforming principle is developed in these tissues. On the basis of comparative studies of the crown-gall tumor, HAMADI[430, 431] believed that plant tumors and animal neoplasia are not analogous, but considered the plant tumors to correspond to malformations caused by prolonged irritation. According to him, this view is supported by his observations that the plant tumors are not autonomous, but depend on the presence of microörganisms, the physiological functions, changes of corresponding normal tissue being seen in the plant tumors, the putting out of normal twigs, etc. The so-called metastases are believed by him to be due to transportation of bacteria. McEWEN[771], however, believed that the cancerous response of bacteria-free crown-gall tissue has been fully established by the continued proliferation, free from the causal agent or other external sources of auxins.

Although some workers like LEVINE have called the crown-gall a "plant cancer", yet they do not consider it to be analogous to cancer. Further, the teratomas produced by *Phytomonas tumefaciens* on many plants like *Nicotiana tabacum*, *Bryophyllum*, *Kalanchoë daigremontiana*, etc., in addition to the regular crown-gall tumors, do not readily fit into the general conception of cancer in animals. The cells of these teratomas have unquestionably undergone the transformation induced by the crown-gall bacterium, but can change back to the normal cells and give rise to normal organs. *Phytomonas tumefaciens* seems only to have produced a temporary change in the growth pattern of these cells, comparable to that induced by indole-acetic acid and other compounds. It seems that *Phytomonas tumefaciens* can, under certain circumstances, transform some cells in certain plants in such a way that they fit the definition of cancer, as ordinarily understood and only a temporary transformation in other plants. The growth pattern of the cells exposed to the bacteria is modified in varying degrees, depending on the growth capacity of the cell itself, degree of its differentiation, duration of exposure to the influence of the bacteria and the virulence of the bacteria. DE ROPP[261] considers it advisable to avoid identifying crown-gall tumors with cancer and regards them as a condition involving increased autonomy of the affected tissues, either reversible or irreversible. Such an increased autonomy of cells is seen in some other plant tumors[90] from wound tumor virus infection and in certain hybrid tumors of *Nicotiana*[614, 1251]. Tissues of *Nicotiana*-hybrid tumors are capable of differentiating roots, stems, leaves, etc. and are thus not fully autonomous and correspond to the tumors on *Kalanchoë daigremontiana*. Root differentiation has also been described in cultures of tissues from wound tumor virus.

The discovery, however, of the bacteria-free crown-gall tissue, retaining all the tumor characters and the recognition that the crown-gall organism is necessary only in the initial stimulation, after which the tumor becomes completely autonomous like cancer, have also given fresh impetus to the discussion. Further, the production of tumors on plants by the action of diverse carcinogenic agents, described in an earlier chapter, has also strengthened the ideas on the cancer analogy of the crown-gall. Although STAPP[1081] has objected to the identification of secondary tumors with metastases, he nevertheless recognizes that there are so many points of similarities between crown-gall and cancer that a redefinition of cancer is necessary so as to be comprehensive enough to include both vegetable and animal tumors. Some of the salient features of animal cancers do not necessarily depend upon the causative factor and it seems therefore best to examine the cancer analogy of the crown-gall independent of etiological considerations[161, 322, 582, 691, 692, 706, 1231].

Cancer cells are readily distinguished from all normal and all other pathological cells by their autonomy and anaplasia. Cancer cells are autonomous in the sense that they are not subject to the restraints of the morphogenetic processes either *in vitro* or *in vivo* when grafted back into intact tissues. Cancer cells determine their own activity, quite independent of the integrating mechanism of the organism of which they form a part. Anaplasia stresses the lack of perfect form and function. Cancer cells are not also simply cells, which have acquired rapidity of multiplication, but they are in reality specifically altered cells. Although for example many healthy tissue cells are often known to proliferate far more rapidly than certain typical cancer cells, the alteration of the cancer cell is characteristically permanent, irreversible and hereditary. In cancer there is a release for further cell proliferation of tissues, the growth of which is normally inhibited by the neighbouring tissues or by the body fluids that circulate through them. The controls of normal development and growth are indeed well known[851, 852]. All living cells produce substances which affect other cells of the body[1291, 717]. Beginning with the epoch making discoveries of SPEMANN, the large series of investigations by numerous workers have shown that in the ontogenetic development of animals a series of chemical evocators come into play. It is also now known that gastrulation evocator is present not only in the embryo, but even in the adult tissues and has been shown to be related to sterols. Some oestrogenic and carcinogenic hydrocarbons like 1 : 1 : 5 : 6-dibenzanthracene have been shown by WADDINGTON & NEEDHAM[1226] to be capable of inducing neural tube formation. There is also evidence that carcinogenic substances and evocators are chemically related. Cancer cells are not subject to the action of these integrating and limiting influences, but are typically characterized by their capacity for unlimited, uncoördinated, uncontrolled proliferation. Cells of any tissue or organ not only of man but also of any animal may indeed thus become neoplastic and may then invade surrounding tissues, giving rise to secondary growths called metastases. ROUS[985] first demonstrated the filterability of cancerous tumors. Tumors are filterable, if after grinding and mixing with normal saline, they yield after Berkfeld filteration, a cell-free filtrate, which, on injection into healthy tissues, reproduces the same type of tumor. Cancer cells are frequently polyploid or polytene. Because of their high content in desoxyribose-nucleic acid[177], cancer cells are conspicuously more sensitive to the action of X-rays than normal cells. The malignancy of the cancer cell is considered by LEWIS[692] to be due to changes in their cytoplasm, rather than chromosomal or gene mutations. Cancer cells are also believed to have undergone a high degree of dedifferentiation, but with retention of specific glandular activities. Carcinomata of endocrine glands do not, for example, interfere with their endocrine functions; these glands have not indeed degenerated but have merely taken on the power of unlimited growth. Indeed in many forms of cancer, the only change known is the unlimited capacity for cell division and growth.

It is therefore concluded[717] that mitotic irregularities show more in tissue cultures of cancer cells than in those of normal tissue cells, resulting probably from colloidal changes in the cytoplasm. BURROWS[161], however, argues that there is really no reversion to the embryonic state in the neoplastic cells, but the latter are merely freed from the forces that make them function as part of the organism. Accordingly whether an embryonic heart or muscle cell is to grow, differentiate or function is dependent on the immediate environment. The change may not thus be within the cell at all but in the environment of the cells. Differentiation and functioning take place under conditions which are not suitable for growth and cell division. Growth depends on the concentration around the cells of the substances (archusia) produced by the cells themselves. Cancer may thus be partly a result of a local crowding of the cells. They migrate by liberating a lipoid substance (ergusia) absorbed by protein and fats. All cancer cells indeed contain large quantities of growth-promoting archusia, corresponding to vitamin B, and ergusia, corresponding to vitamin A. There is perhaps a local inbalance in cancer cells, in which vitamin B is largely increased and vitamin A is decreased. The chromosome irregularities of cancer cells do not, however, seem to affect either their quality or activity[453] and since the cancer cells divide but do not differentiate, the chromosomes are also physiologically inactivated. The characteristic stability of the cancer character of the cells would, however, seem to imply both nuclear and cytoplasmic mutations[427].

Abnormal functioning of oxidation enzymes has been considered to underlie cancer cells[1231]. A permanent destruction of the oxidase cytochrome respiration system of the cell is considered by some as the basis of cancer cells[322]. The cell energy is derived from glycerolysis and the oxidation of the resulting lactic acid to carbon dioxide and water. This view seems to be supported by the fact that the production of lactic acid is unusually high in tumors. If lactic acid oxidation controls glycerolysis and if there be deficiency of vitamin A or B_2, then glycerolysis will preponderate and should naturally lead to rapid cell proliferation. Neoplastic cells show an increase in lipoids and cholesterols[68]. The observations of POTTER[914, 915] that the so-called cancer virus is merely an altered protein, arising within the cell itself, emphasizes the fact that a virus can arise either under the action of a carcinogen or through other as yet unrecognized facts in the cytoplasm. WOODS & DUBUY[1274] have indeed demonstrated that virus nucleoproteins are derived from mitochondria in plants. Cancer is transmitted not from cell to cell but from organ to organ by metastasy, the cells that have escaped from the original tumor are carried by the blood stream or through the lymphatic current to distant organs, where they now settle down and give rise to new cancerous growths. This mode of dispersal shows that cancer cells possess certain special properties. Tissue culture investigations have demonstrated partly an important peculiarity of cancer cells, viz. they require less nourishment than healthy cells. It seems as if the formation of cancer is an expression of increased synthetic capacity of cells, leading to infinite increase at the cost of nutrition of the organism. The reader may refer to the works of ROUS & KIDD[989], POTTER[914, 915], HEIBERG & KEMP[453], FIRROR & GEY[332], DARLINGTON[242] and COMAN[208] for further accounts of cancerous growths in man and other animals.

In the final analysis, the fundamental character of cancer cells is not merely that their pattern of growth and division has been changed, but they transmit this alteration to their progeny. The continuing causes of malignancy are not in the carcinogenic agents, virus or bacteria, but in the cells themselves, probably in the nature of the self-duplicating (self-propagating) alteration of certain constituents, either of the cytoplasm or of the nucleus, the rôle of which is to regulate the growth and nutrition of the cell. A growth substance may perhaps be produced or abnormally increased in quantity and its production is renewed autocatalytically. The growth processes are intensified irreversibly and certain cellular, metabolic and functional processes are altered. Such growth promoting substances do not, however, seem to have definitely been demonstrated in the case of cancer cells of animals and man, but as we have already seen, the bacteria-free crown-gall tumor is remarkable for its hyperauxinity. The excess of the growth substance may result either from enhanced capacity of the tumor cell to synthesize it or from reduced capacity to destroy it.

To conclude, it may be pointed out that any cell of any tissue or organ, be it ectoderm, endoderm, skin, nerve, etc. of any animal, whether man or Invertebrate, can turn neoplastic. It is therefore quite conceivable that plant cells can also equally similarly escape from the normal restraining morphogenetic influences and become altered into neoplastic cells. As in the animal neoplasia, so also in plant neoplasia the continuing cause of malignancy is not in the carcinogenic factors at all, but in the cell itself. Plant tumor cells, such as we find in the bacteria-free crown-gall tissue, which exhibit this property of autonomous growth, are cancerous in the wide sense of the term.

BIBLIOGRAPHY

1. ADLER, H. 1877. Beiträge zur Naturgeschichte der Cynipiden. *Dtsch. ent. Z.*, **21**.
2. ADLER, H. 1881. Über Generationswechsel der Eichengallwespen. *Z. wiss. Zool.* **35**: *151*.
3. AKAI, S. 1939. On the anatomy of galls on *Crepis japonica* caused by *Protomyces inacyei. Bot. Zool.*, **7**: *875—882*.
4. AKAI, S. 1939. Studies on the pathological anatomy of the hypertrophied buds of *Camellia japonica* caused by *Exobasidium camelliae. Bot. Mag.*, **53** (627): *118—125*, fig. 1—6.
5. AKAI, S. 1940. On the pathological histology of the deformed petioles and leaves of *Camellia japonica* caused by an undescribed species of *Exobasidium. Ann. phytopath. Soc. Japan*, **10** (2/3): *104—105*, fig. 1—4.
6. ALBRECHT, W. A. & L. M. TURK, 1930. Leguminosenbakterien mit Bezug auf Licht und Lebensdauer. *Missouri agric. exp. Sta. Res. Bull.* **132**.
7. ALFIERI, A. 1921. Sopra una specie probabilmente nuovo di Afide simbionte. *Boll. Lab. Zool. Ent. Agr. Scuol. Sup. Agr. Portici*, **14**: *18—31*, fig. 1, pl. i.
8. ALICANTE, M. M. 1926. The viability of the nodule bacteria of Legume outside of the plant. *Soil Sci.*, **21**: *27—52, 93—114*.
9. ALLISON, W. F. E., C. A. LUDWING, S. R. HOOVER & F. MINOR, 1940. Biochemical nitrogen fixation studies. I. Evidence for limited oxygen supply within the nodule. II. Comparative respiration of nodules and roots, including non-legume roots. *Bot. Gaz.*, **101** (3): *513—534, 534—549*.
10. ALPANT, A. 1942. Über einige Gallen aus dem Pamuku-Tal bei Ankara. *Collegium*, 1942 (872): *417—424*, fig. 11.
11. ALTENBURG, E. 1947. Tumor formation in relation to the origin of viruses. *Amer. Anat.*, **81** (796): *72—76*.
12. ANDERS, F. 1955. Zur biologischen Charakterisierung der galleninduzierenden Substanz aus dem Speicheldrüsensekret der Reblaus *Viteus (=Phylloxera) vitifolii* Shimer. *Verh. dtsch. zool. Ges. Erlangen, 421—428*, fig. 9.
13. ANDERS, F. 1957. Über die gallenerregenden Agenzien der Reblaus *Viteus (=Phylloxera) vitifolii* Shimer. *Vitis*, 1: *121—124*.
14. ANDERS, F. 1957. Reblaus- und colchicin-induzierte Keulenbildung an der Wurzel von *Vitis*-Sämlingen. *Naturwissenschaften*, **44**: *95—96*.
15. ANDERS, F. 1958. Aminosäuren als gallenerregende Stoffe der Reblaus *Viteus (=Phylloxera) vitifolii* Shimer. *Experientia* 14: *62—63*.
16. ANDERS, F. 1958. Über die Morphogenese der Gallen von *Viteus vitifolii* Shimer (*=Phylloxera vastatrix*). *Marcellia*, **30** (Suppl.): *103—112*, fig. 4.
17. ANDERS, F. 1958. Das galleninduzierende Prinzip der Reblaus *(Viteus vitifollii* Shimer.). *Verh. dtsch. zool. Ges. Leipzig, 355—363*, fig. 10.
18. ANDERS, F. 1960. Untersuchungen über das cecidogene Prinzip der

Reblaus *Viteus (=Phylloxera) vitifolii* Shimer. I. Untersuchungen an der Reblausgalle. *Biol. Zbl.*, **79**: *47—58;* II. Biologische Untersuchungen über das galleninduzierende Sekret der Reblaus. *ibidem*, **79**: *679—700.*

19. ANDERS, F. 1961. Untersuchungen über das cecidogene Princip der Reblaus *Viteus (=Phylloxera) vitifolii* Shimer. III. Biochemische Untersuchungen über das galleninduzierende Agenz. *Biol. Zbl.*, **80** (2): *199—233,* fig. 16.

20. ANDERSON, J. A., W. H. PETERSON & E. E. FRED, 1928. The production of pyruvic acid by certain nodule bacteria of Leguminosae. *Soil Sci.*, **25**: *123—131.*

21. ANDRÉ, E. 1888. Relations des fourmis avec les pucérons et les gall insectes. *Bull. d'Insec. Agric. Ann.*, **7**: *3—7.*

22. APPEL, O. 1904. Beispiele zur mikroskopischen Untersuchung von Pflanzenkrankheiten. Berlin, J. Springer, pp. 48, fig. 53.

23. ARCIZEWSKI, W., M. BOEHM & W. KOPACZEWSKI, 1928. Studien über capillarelektrische Erscheinungen. Bakterienantagonismus und Krebsproblem. *Z. Krebsforsch.*, **27** (3): *273—289,* fig. 5.

24. ARNAUDI, C. 1926. Sull'immunita acquista nei vegetali. *Atti Soc. Ital. Sci. Nat. Milano*, **64** (3/4).

25. ARNAUDI, C. 1927. À propos de quelques expériences sur l'immunité acquise des végétaux. *Rev. Path. veg. Ent. Agric.*, **14** (2): *103—112.*

26. ARNAUDI, C. 1928. Nuova esperienze sulla vaccinazione delle piante. *Riv. Patol. Veg.*, **18** (7/8): *161—168.*

27. ARNAUDI, C. & G. VENTURELLI, 1934. L'azione del radio sui tumori vegetali. *Riv. Biol.*, **16** (1): *5—23,* pl. i.

28. ARRUDA, S. C. 1943. Obserfacões sobre algumas dõencas do eucalipto no Estado de S. Paulo. *Biologica, São Paulo*, **9** (6): *140—144,* pl.i.

29. ARZBERGER, E. G. 1910. The fungus root-tubercles of *Ceanothus americanus, Elaeagnus argentea* and *Myrica cerifera. Ann. Rep. Missouri bot. Garden, S. Louis*, **21**: *60—102,* pl. vi—xiv.

30. ASHMEAD, W. H. 1899. The largest oak gall in the world and its parasites. *Ent. News*, **10**: *193.*

31. ASHMEAD, W. H. 1904. Classification of Chalcid flies or the superfamily Chalcidoidea. *Mem. Carnegie Mus.*, **1** (4): *225—551.*

32. ASHMEAD, W. H. 1903. Classification of Cynipoidea. *Psyche*, **10**: *7, 59, 140.*

33. ATANASOFF, D. 1925. *Diplophosphora* disease of cereals. *Phytopathology*, **15**: *11—40.*

34. ATKINSON, F. T. 1889. Nematode in root-galls. *J. Elisha Mitchell Sci. Soc.*, **5**: *2, 81.*

35. AULER, H. 1924. Zur Histogenese der *Tumefaciens*-Geschwülste an der Sonnenblume. *Z. Krebsforsch.*, **21**: *354—360,* fig. 2.

36. AULER, H. 1925. Über chemische und anaerobe Tumorbildung bei Pflanzen. *Z. Krebsforsch.*, **22**: *393—403.*

37. BACCARINI, P. 1893. Sopra un curioso cecidio della *Capparis spinosa. Malpighia*, **7**: *405.*

38. BACCARINI, P. 1909. Sui micocecidii od "Ambrosiagallen." *Bull. Soc. Bot. Ital.*, **137.**

39. BACCARINI, P. 1913. Primi appunti intorno alla biologia dello *Exobasidium lauri* Gev. *Nuovo Giorn. bot. Ital.*, (NS) **20** (2): *282—301.*

40. BACHMANN, E. 1920. Über Pilzgallen auf Flechten. *Ber. dtsch. bot. Ges.*, **38**: *333—338.*

41. BACHMANN, E. 1927. Die Pilzgallen einiger Cladonien. II. *Arch. Protistenk.*, **59** (2): *373—416,* fig. 48.

42. BACHMANN, E. 1928. Die Pilzgallen einiger Cladonien. III. *Arch. Protistenk.*, **62** (2/3): *261—308,* fig. 52.

43. BACHMANN, E. 1928. Die Pilzgallen einiger Cladonien. IV. Blattgallen und beblätterte Gallen. *Arch. Protistenk.*, **64** (1/2): *109—151*.
44. BACHMANN, E. 1929. Tiergallen auf Flechten. *Arch. Protistenk.*, **66** (1): *61—103*, fig. 30.
45. BACHMANN, E. 1929. Pilz-, Tier- und Scheingallen auf Flechten. *Arch. Protistenk.*, **66** (3): *459—514*, fig. 60.
46. BACHMANN, E. 1930. Die Gallen zweier Laubflechten. *Arch. Protistenk.*, **71** (2): *323—360*, fig. 34.
47. BACHMANN, E. 1934. Pilz-, Tier- und Scheingallen auf Flechten. *Arch. Protistenk.*, **82**: *23—44*, fig. 29.
48. BACHMANN, E. 1934. Scheingallen auf *Physcia stellaris* (Linn.) Nyl. *Ber. dtsch. bot. Ges.*, **52**: *80—86*, fig. 1.
49. BAGNALL, R. S. 1928. On some new genera and species of Australian Thysanoptera (Tubulifera) with special reference to gall species. *Marcellia*, **25**: *184—203*.
50. BAIS, M. P. 1933. Dipterocécidie foliaire *(Vicia cracca* L. attaqué par *Perrisia viciae* Kieff.), ses caractères anatomiques et leur comparison avec ceux du fruit. *Rev. gén. bot. Paris*, **45**: *5—19*, fig. 1—4.
51. BAKER, A. C. 1917. On the Chinese gall. *Ent. News*, **28**: *385*, pl. xxvi.
52. BALDASSERONI, V. 1929. Osservazioni biologische surghi afidi in rapporto al tumoro batterico del pino di Aleppo. *Boll. Soc. Ital. Biol. Sperim.*, **4** (4): *393—395*.
52a. BALLY, 1912. Zytologische Studien an Chytridieen. *Jb. wiss. Bot.*, **50**: *95*.
53. BANCROFT, T. L. 1890. Hymenopterous galls found on *Acacia cunninghami* Hook. and *Acacia penninervis* Sieb. *Proc. Linn. Soc. N. S. Wales*, Sydney, (2) **5**: *680*.
54. BANFIELD, W. M. 1935. Studies in cellular pathology. I. Effects of cane gall bacteria upon gall tissue cells of the black raspberry. *Bot. Gaz.*, **97** (2): *193—238*, fig. 1, pl. i.
55. BARGAGLI-PETRUCCI, G. 1907. Cecidi della Cina. *Nuovo Giorn. bot. Ital.* (NS) **14**: *235—245*, fig. 7, pl. i.
56. BARTHEL, C. 1921. Contributions à la recherche des causes de la formation des bactéroides chez les nodosités des Legumineuses. *Ann. Inst. Pasteur*, **35**: *634—646*, fig. 6.
57. BARTLETT, A. W. 1928. *Olpidium radicicolum* de Wildeman and the hybridization on nodules of swedes. *Trans. Brit. mycol. Soc.*, **13** (3/4): *221—238*, pl. ii.
58. BARTON, E. S. 1891. On the occurrence of galls on *Rhodymenia palmata* Grev. *J. Bot.*, **26**: *65*.
59. BARTON, E. S. 1892. On malformations of *Ascophyllum* and *Desmarestia*. *Brit. Mus. mycol. Mem.*, **1**: *21*.
60. BARTON, E. S. 1901. On certain galls on *Furcellaria* and *Chondrus*. *J. Bot.*, **39**: *49*.
61. BASSETT, H. F. 1880. The structure and development of certain Hymenopterous galls. *Amer. Ent.*, **3**: *284*.
62. BASTIAN. 1865. Monograph of the Anguillulidae. *Trans. Linn. Soc. London* (Zool.) **25**: *73—184*.
63. BAUCH, R. 1942. Über Beziehungen zwischen polyploidisierenden, carcinogenen und phytohormonalen Substanzen. Auslösung von Gigas-Mutationen der Hefe durch pflanzliche Wuchsstoffe. *Naturwissenschaften*, **27**: *420—421*.
64. BAUCH, R. 1952. Das Problem der Mitosengift- und Wuchsstoffekeulen bei der Wurzel. *Ber. dtsch. bot. Ges.*, **65**: *4—6*.
65. BAUDYS, E. 1917. Massenhaftes Auftreten von Gallenerzeugern im Jahre 1910. *Z. Insektenbiol.*, **13**: *251*.
66. BAWDEN, F. C. 1913. Plant viruses and virus diseases. 2 ed. Waltham, Mass.

338

67. BAWDEN, F. C. 1951. Viruses: Molecules or organisms. *Discovery*, **12** (2): *41.*
68. BEARD, H. H. 1935. Cancer as a problem in metabolism. *Arch. intern. Med.*, **56**: *1143—1170.*
69. BECHER, E. 1917. Die fremddienstliche Zweckmäßigkeit der Pflanzengallen und die Hypothese eines überindividuellen Seelischen. Leipzig.
70. BECHHOLD, H. & L. SMITH, 1927. Das Tumefaciensplastem. *Z. Krebsforsch.*, **25** (1): *97—104,* fig. 3.
71. BECK, E. G. 1946. A study of *Solidago* gall caused by *Eurosta solidaginis. Amer. J. Bot.*, **33**: *228.*
72. BÉGUINOT, A. 1903. Studio anatomico di due cecidii del genere *Cuscuta. Marcellia*, **2**: *47—62,* pl. i.—ii.
73. BEIJERINCK, M. W. 1877. Über Pflanzengallen. *Bot. Ztg.*, **35.**
74. BEIJERINCK, M. W. 1877. Over Gallen aan Cruciferen. *Nederl. kruidk. Arch.*, (2) **2**: *164.*
75. BEIJERINCK, M. W. 1882. Die ersten Entwicklungsphasen einiger Cynipidengallen. Amsterdam.
76. BEIJERINCK, M. W. 1885. Die Galle von *Cecidomyia poae* auf *Poa nemoralis. Bot. Ztg.*, **43**: *305.*
77. BEIJERINCK, M. W. 1888. Über das Cecidium von *Nematus capreae* auf *Salix amygdalina. Bot. Ztg.*, **46**: *1.*
78. BEIJERINCK, M. W. 1895. Eucalyptusgallen. *Nederl. kruidk. Arch.*, (2) **6**: *623.*
79. BENKO, G. 1882. *Vaucheria*-gubaesok. *Bot. Jb.*, **2**: *686.*
80. BERENBLUM, I. 1944. Irritation and carcinogenesis. *Arch. Path.*, **3**: *233—244.*
81. BERRIDGE, E. M. 1930. Studies in bacteriosis. XVII. Acidic relations between crown-gall organism and its host. *Ann. appl. Biol.*, **17** (2): *280—283.*
82. BERTHELOT, A. & G. AMOUREUX, 1938. Sur la formation d'acide indole-3-acétique dans l'action de *Bacterium tumefaciens* sur le tryptophane. *C. R. Acad. Sci. Paris*, **206**: *537—540.*
83. BIGNELL, G. B. 1903. Inquiline Cynipidae. *Ent. Rec. J. Variation,* **13**: *360.*
84. BINET, L. & J. MAGROU, Glutathion, croissance et cancer des plantes. *C. R. Acad. Sci. Paris*, **192** (22): *1415—1416.*
85. BISHOP. 1911. A new root gall midge from *Smilacina. Ent. News*, **22** (8): *346.*
86. BITANCOURT, A. A. & V. ROSSETTI, 1946. La galhas pulverulentas das Lauraceae (The powdery galls of Lauraceae). *Biologico, São Paulo*, **12** (3): *55—62,* pl. xiii.
87. BITTMANN, O. 1925. Ein Beitrag zur künstlichen Erzeugung atypischer Zellenproliferation bei den Pflanzen. *Z. Krebsforsch.*, **22**: *291—296.*
88. BITTNER, J. J. 1940. Possible method of transmission of susceptibility to breast cancer in mice. *Amer. J. Cancer*, **39**: *104—113.*
89. BJÖRKENHEIM, C. G. 1904. Beiträge zur Kenntnis der Pilze in den Wurzelanschwellungen von *Alnus incana. Z. Pflanzenkr.*, **14**: *129— 133,* pl. iii.
90. BLACK, L. M. 1945. A virus tumor disease of plants. *Amer. J. Bot.*, **32** (7): *408—415,* fig. 9.
91. BLACK, L. M. 1946. Plant tumor induced by the combined action of wounds and virus. *Nature*, **158**: *56—57.*
92. BLACK, L. M. 1946. Virus tumors in plants. *Growth*, **10** (Suppl.)6 (Symposia): *79—84.*
93. BLOCH, R. 1954. Morphogenetic classification of types of abnormal growth. *C. R. Seances Rep. Comm. VIIIe Congr. Int. Bot. Paris*, **7—8**: *220—222.*

94. BLOMFIELD, J. E. & E. J. SCHWARTZ, 1910. Some observations on the tumors on *Veronica chamaedrys* caused by *Sorosphaera veronicae. Ann. Bot.*, **24**: *35*.
95. BLUM, J. L. 1953. Vascular development in three common goldenrod galls. *Papers Michigan Acad. Sci. Arts Letters*, **38**: *23—34*.
96. BLUMENTHAL, F. & H. HIRSCHFELD, 1921. Beiträge zur Kenntnis einiger durch *Bacterium tumefaciens* hervorgerufenen Pflanzenge-schwülste. *Z. Krebsforsch.*, **18**: *110—125*.
97. BLUMENTHAL, F. & P. MEYER, 1924. Über durch Acidum lactum erzeugte Tumoren auf Mohrrübenscheiben. *Z. Krebsforsch.*, **21**: *250—252*.
98. BLUNK, G. 1920. Die Anpassung der Knöllchenbakterien an Nicht-leguminosen. *Zbl. Bakt.*, **2** (51): *87—90*.
99. BLUNK, H. 1925. Thysanopteren. In: Sorauers Handbuch der Pflan-zenkrankheiten, 4. Aufl. **4**: *246*.
100. BOAS, F. 1911. Zwei neue Vorkommen von Bakterienknollen in Blät-tern von Rubiaceen. *Ber. dtsch. bot. Ges.*, **29**: *416—418*, fig. 2.
101. BODENHEIMER, F. S. 1929. Materialien zur Geschichte der Entomolo-gie bis Linné. Junk. Berlin. (vide pp. 498 in vol. **1**, 486 in vol. **2**).
102. BOEDIJN, K. B. 1937. A smut causing galls on the leaves of *Hypoly-tum. Bull. Jardin bot. Buitenzorg*, (3) **14** (3/4): *368—372*, fig. 1—2.
103. BOEHM, M. M. & W. KOPACZEWSKI, 1929. Étude sur les phénomènes électrocapillaires. IX. L'antagonismus microbiens et la thérapeuti-que du cancer. *Protoplasma*, **6**: *302—320*.
104. BÖHNER, K. 1933. Geschichte der Cecidologie. Ein Beitrag zur Ent-wicklungsgeschichte naturwissenschaftlicher Forschung und ein Führen durch die Cecidologie der Alten – Mit einer Vorgeschichte zur Cecidologie der klassischen Schriftleiter von Felix Oefele I. Teil. Arthur Neumeyer, Mitwald (Bayern), pp. 466; II Teil. Botanik und Entomologie. **6**: *1—710* (1935).
105. BÖRNER, C. 1908. Eine nonographische Studie über die Chermiden. *Arb. biol. Reichsanstalt Land.- und Forstw.*, Berlin-Dahlem, **6**: *81*.
106. BÖRNER, C. 1916. Über blutlösende Säfte im Blattlauskörper und ihr Verhalten gegenüber Pflanzensäften. *Mitt. kais. Anst. Land.-Forstw.*, **16**.
107. BÖRNER, C. 1921. Über die Umwandlung von Wurzelrebläusen zu Blattrebläusen. *Mitt. biol. Reichsanst.*, **21**: *163—166*.
108. BOGGIO, L. F. 1932. Ricerche biologiche e morfologiche su *Neuro-terus lenticularis* Oliv. *Atti Soc. Ital. Progr. Sci., 20 Riun. Milano*, **2**: *306—310*.
109. BOIVIN, A., M. MARBE, L. MESROBEANU & P. JUSTER, 1935. Sur l'exis-tence dans le *Bacillus tumefaciens* d'une endotoxin capable de pro-voquer la formation de tumeurs chez les végétaux. *C. R. Acad. Sci. Paris*, **201** (21): *984—986*.
109a. BONNER, J. 1950. The rôle of toxic substances in the interactions of higher plants. *Bot. Rev.*, **16**: *51—65*.
109b. BONNER, J. & S. G. WILDMAN, 1947. Contributions to the study of auxin physiology. VI. *Growth Symposia*, *51—68*.
110. BORCEA, I. 1912. Deformations provoquées par *Exobasidium rhodo-dendri* Cram. sur *Rhododendron myrtifolium* Sch. et Kotsch. *Ann. Sci. Univ. Jassy*, **7**: *209—210*.
111. BORELLI, N. 1919. Contributo alla conoscenza della vita nelle galle dell'Alloro. *Boll. Soc. nat. Ital.*, **51**: *1—37*, fig. 7.
112. BORGHI, B. & C. LUZZATTO, 1928. Ricerche intorno al porte oncogeno del *B. tumefaciens* e del *B. paola* Meyer nelle piante e negli animali. *Boll. Ist. Sieroterap. Milanese*, **8** (4): *243—255*.

340

113. BORM, L. 1931. Die Wurzelknöllchen von *Hippophaë rhamnoides* und *Alnus glutinosa. Bot. Arch.*, **31**: *441—448*, fig. 23.
114. BORTHWICK, H. A. *et al.* 1937. Histological and microchemical studies of the reactions of tomato plants to indole-acetic acid. *Bot. Gaz.*, **98**: *491—519*.
115. BOUDIER, E. 1893. Sur les causes de production de tubercles pileux des lames de certains agarics. *Rev. gén. Bot.*, **5**: *9*.
116. BOSCOLO, J. 1906. Tanino de las agallas de la *Duvana longifolia* forma *praecox* Crisel, Molle de incienso. *Buenos Aires Soc. nac. Farmacia*, **31**: *1—8*.
117. BOYSEN-JENSEN, P. 1948. Formation of galls of *Mikiola fagi. Physiol. Plantarum*, **1**: *95—108*.
118. BRAND, F. 1895. Über *Batrachospermum. Bot. Zbl.*, **61**: *283*.
119. BRANHOFER, K. & J. ZELLNER, 1920. Chemische Untersuchungen über Pflanzengallen. III. *Z. physiol. Chem.*, **109**: *166—177*.
120. BRAUN, A. C. 1941. Development of secondary tumors and tumor strands in the crown-gall of sunflower. *Phytopathology*, **31**: *135—149*.
121. BRAUN, A. C. 1943. Studies on tumor inception in the crown-gall disease. *Amer. J. Bot.*, **30**: *674—677*.
122. BRAUN, A. C. 1947. Recent advances in the physiology of tumor formation in the crown-gall disease of plants. Growth Symposium, *Growth*, **11**: *325—337*.
123. BRAUN, A. C. 1947. Thermal studies on the factors responsible for tumor initiation in crown-gall. *Amer. J. Bot.*, **34**: *234—240*.
124. BRAUN, A. C. 1948. Studies on the origin and development of plant teratomas incited by the crown-gall bacterium. *Amer. J. Bot.*, **35**: *511—519*.
125. BRAUN, A. C. 1950. Thermal inactivation studies on the tumor inducing principle in crown-gall. *Phytopathology*, **40**: *3*.
126. BRAUN, A. C. 1951. Recovery of crown-gall tumor cells. *Cancer Res.*, **11** (2): *839—844*.
127. BRAUN, A. C. 1951. Recovery of tumor cells from effects of the tumor-inducing principle in crown-gall. *Science*, **113** (2945): *651—653*.
128. BRAUN, A. C. & R. P. ELROD, 1946. Stages in the life-history of *Phytomonas tumefaciens. J. Bact.*, **52**: *695—702*.
129. BRAUN, A. C. & T. LASKARIS, 1942. Tumor formation by attenuated crown-gall bacteria in the presence of growth-promoting substances. *Proc. nat. Acad. Sci. America*, **28**: *468—477*.
130. BRAUN, A. C. & R. J. MANDLE, 1948. Studies on the inactivation of the tumor inducing principle in crown-gall. *Growth*, **12**: *255—269*.
131. BRAUN, A. C. & G. MOREL, 1950. A comparison of normal habituated and crown-gall tumor tissue implants in the European grape. *Amer. J. Bot.*, **37**: *499—501*.
132. BRAUN, A. C. & R. P. WHITE, 1943. Bacteriological sterility of tissues derived from secondary crown-gall tumors. *Phytopathology*, **33**: *85*.
133. BRAVO, G. A. 1927. Studio sulle galle della *Pistacia atlanta* Desf. della Libia. *Boll. R. Staz. Sperim. Industr. pelli e materie concianti, Napoli, Torino*, pp. 10, fig. 6.
134. BREIDER, H. 1939. Untersuchungen zur Vererbung der Widerstands-fähigkeit von Weinreben gegen die Reblaus *Phylloxera vastatrix* Planch. I. Das Verhalten von F_3-Generationen, die aus Selbstungen von widerstandsfähigen und anfälligen-F_2-Artbastarden gewonnen wurden. *Z. Zücht.*, (A) Pflanzenzücht., **23** (1): *145—168*.
135. BREMEKAMP. C. E. B. 1933. The bacteriophilous species of *Psychotria. J. Bot.*, **71** (850): *271—280*.
136. BRIZI, U. 1907. Ricerche su alcune singolari neoplasie del Pioppo e sul

Bacterio che le produce. *Atti Congr. nat. Ital. Milano*, 1906: *1—17*, pl. ii.
137. BROWN, J. G. 1948. Cytological effects of penicillin and streptomycin on crown-gall. *Phytopathology*, 38: *3*.
138. BROWN, J. G. & A. M. BOYLE, 1944. Penicillin treatment of crown-gall. *Science*, 100 (2696): *258*.
139. BROWN, J. G. & A. M. BOYLE, 1945. Application of penicillin to crown-gall. *Phytopathology*, 35: *521—524*, fig. 1.
140. BROWN, J. G. & M. M. EVANS, 1933. Crown-gall on a conifer. *Phytopathology*, 23 (1): *97—101*.
141. BROWN, J. G. & M. M. EVANS, 1933. The natural occurrence of crown-gall on the giant cactus *Carnegiea gigantea. Science*, 78 (2017): *167—168*.
142. BROWN, N. A. 1928. Bacterial pocket disease of sugarbeet. *J. agric. Res.*, 37 (3): *155—168*, fig. 4, pl. i.
143. BROWN, N. A. 1929. The tendency of the crown-gall organism to produce roots in conjuction with tumors. *J. agric. Res.*, 39 (10): *747—766*, fig. 10.
144. BROWN, N. A. 1939. Colchicine in the prevention, inhibition and death of plant tumors. *Phytopathology*, 29 (3): *221—231*, fig. 1—2.
145. BROWN, N. A. 1942. The effect of certain chemicals, some of which produce chromosome doubling, on plant tumors. *Phytopathology*, 32 (1): *25—45*, fig. 2.
146. BROWN, N. A. & F. E. GARDNER, 1936. Galls produced by plant hormones including a hormone extracted from *Bacterium tumefaciens. Phytopathology*, 26: *708—713*.
147. BROWN, N. A. & A. J. QUIRK, 1929. Influence of bacteriophage on *Bacterium tumefaciens* and some potential studies of filtrates. *J. agric. Res.*, 39 (7): *503—530*, fig. 1, pl. iv.
148. BROWN, W. W. & L. PEARCE, 1923. Studies on a malignant tumor of the rabbit. 1. The spontaneous tumor and associated abnormalities. *J. exp. Med.*, 37: *601—630*.
149. BUCHENAU, F. 1870. Kleinere Beiträge zur Naturgeschichte der Juncaceen. *Abh. naturw. Ver. Bremen*, 2: *390*.
150. BUCHENAU, F. 1903. Entwicklung von Staubblättern im Innern von Fruchtknoten bei *Melandrium rubrum* Garcke. *Ber. dtsch. bot. Ges.*, 21: *417*.
151. BUCHNER, P. 1921. Tier und Pflanze in intrazellularer Symbiose. Berlin, Bornträger. viii-1—462.
152. BUCKTON, G. B. 1889. Gall insects. Zoology of the Afghan Delimitation Commission, by J. E. T. Aitchison. London. *Trans. Linn. Soc. London* (Zool.) (2) 5: *141—142*, fig. 3.
153. BURDON, E. R. 1907. Influence of *Chermes* on larch canker. *Garden. Chronicle*, 17 (2): *353*.
154. BURDON, E. R. 1908. Some critical observations on the European species of *Chermes. J. econ. Ent.*, 2: *119*.
155. BURGEFF, H. 1909. Die Wurzelpilze der Orchideen. Jena.
156. BURGEFF, H. 1920. Über den Parasitismus des *Chaetocladium* und die heterocaryotische Natur der von ihm auf Mucorinen erzeugten Gallen. *Z. Bot.*, 12: *1*.
157. BURGEVIN, H. 1933. Sur la fixation de l'azote atmosphérique par les bactéries des Légumineuses. *C. R. Acad. Sci. Paris*, 196: *441*.
158. BURGERS, A. 1934. Studies on the genus *Uromycladium* (Uredineae). I. General introduction, the anatomy of the galls and the cytology of the vegetative mycelium and pycnidia of *Uromycladium tepperianum* (Sacc.) Mc Alp on *Acacia stricta* Willd. *Proc. Linn. Soc. N. S. Wales*, Sydney, 59: *212—228;* II. Notes on dikaryon of *Uromycladium tepperianum. ibid.*, *94—96*.

342

159. BURNET, F. M. 1946. Virus as organism. Cambridge, Mass.
160. BURRILL, T. J. & R. HAUSEN, 1917. La symbiose entre les bactéries des nodosités des légumineuses et les racines des plantes non légumineuses est elle possible? *Bull. Univ. Ill. agric. exp. Sta.*, **202**: *115—181*, pl. xvii.
161. BURROWS, M. T. 1926. Studies on the nature of the growth stimulus in cancer. *J. Cancer Res.*, **10**: *239—251*.
162. BÜSGEN, M. 1890. Beobachtungen über das Verhalten des Gerbstoffes in den Pflanzen. *Jenaische Z. Naturw.*, **24**: *11*.
163. BÜSGEN, M. 1895. Zur Biologie der Galle von *Hormomyia fagi* Htg. *Forstl. naturw. Z.*, **4**: *9*.
164. BUSCALIONI, L., P. SCARAMELLA & L. BERNARDI, 1935/1937. I tumori maligni degli animali superiori ed i loro rapporti morfologici, anatomici e citologici con talune neoformazioni delle piante. *Malpighia*, **34**: *343—446*, pl. vii—x.
165. BUTLER, E. J. 1930. On some aspects of the morbid anatomy of plants. *Ann. appl. Biol.*, **17** (2): *175—212*, fig. 13, pl. viii—xii.
166. BUVAT, R. 1942. Sur l'action d'hydrocarbures cancérigènes sur le tissue libérien de carotte cultivé in vitro. *C. R. Acad. Sci. Paris*, **214**: *129—130*.
167. BUVAT, R. 1944. Recherches sur la dédifférentiation des cellules végétales. *Ann. Sci. nat. Bot.*, (11) **5**: *1—130*.
168. CAMERON, P. 1883. On the origin of the form of galls. *Trans. nat. Hist. Soc. Glasgow.*
169. CAMUS, G. & R. J. GAUTHERET, 1948. Sur le caractère tumoral des tissus de Scorzonère avant subi le phénomène de accoutumance aux hétéro-auxines. *C. R. Acad. Sci. Paris*, **226**: *774—745*.
170. CAMUS, G. & R. J. GAUTHERET, 1948. Nouvelles recherches sur le greffage des tissus normaux et tuméraux sur des fragments des racines Scorzonère cultivés in vitro. *C. R. Acad. Sci. Paris*, **142**: *769—771*.
171. CAMUS, G. & R. J. GAUTHERET, 1948. Sur le repiquage des proliférations induits sur des fragments des racines de Scorzonère par des tissus de crown-gall et des tissus avant subi le phénomène d'accoutumance aux hétéro-auxines. *C. R. Acad. Sci. Paris*, **142**: *771—773*.
172. CAMUS, G. & R. J. GAUTHERET, 1948. Sur transmission par greffage des propriétés tumorales des tissus de crown-gall. *C. R. Acad. Sci. Paris*, **142**: *15—16*.
173. CANTACUZENE, A. 1928. Structure anatomique des tumeurs bactériennes de *Saccorhiza bulbosa*. *C. R. Soc. Biol. Paris*, **99** (25): *1715—1717*.
174. CANTACUZENE, A. 1928. Tumeurs bactériennes des thalles de *Saccorhiza bulbosa*. *C. R. Soc. Biol. Paris*, **99** (25): 565—566.
175. CAPPALLETTI, C. 1928. I tubercoli radicali delle Leguminidae considerati nei loro rapporti immunitari e morfologici. *Ann. Bot.*, **17** (5): *211—297*.
176. CASELLO, D. 1933. Un tumore prodotto da *Bacterium tumefaciens* Sm. & Towns. su arantia ovale e la selezione gemmaria. *Ann. R. Staz. sperim. Agrumicol. Frutticolt. Acireole*, **1**: *43—45*.
177. CASPERSSON, T., CL. NYSTRÖM & L. SANTESSON, 1946. Zytoplasmatische Nucleotide in Tumorzellen. *Naturwissenschaften*, **29**: *29—30*.
178. CAVADAS, D. S. 1922. Étude morphologique, histologique, et cytologique d'une mycocécidie provoquée chez l'*Urtica dioica* (Linn.) par la *Puccinia caricis* (Schaum.) Reb. *Mem. fac. Sci. Nancy*, *1—14*, pl. iii.
179. CELASKOVSKY, L. 1885. Neue Beiträge zur Foliartheorie des Ovulums. *Abh. böhm. wiss. math.-natur. Klasse*, **6**: *12*.
180. CENTANNI, E. 1929. Sopra alcuni rapporti fra tumori vegetali e tumor animali. *Tumori*, **15** (1): *17—26*.

181. CHADWICK, G. H. 1907. A catalogue of the Phytoptid galls of North America. *N. Y. St. Mus. Bull.*, **124.**
182. CHAMPION, G. C. & T. A. CHAPMAN, 1903. Some notes on the habits of *Nanophyes durieui* Lucas as observed in Central Spain, with a description of the larvae and pupa. *Trans. ent. Soc. London*, **1**: *87—91*, pl. i.
183. CHARGAFF, E. & M. LEVINE, 1936. Chemical composition of *Bacterium tumefaciens. Proc. Soc. exp. Biol. Med.*, **34**: *675—677*.
184. CHAUDHURI, H. 1951. Recherches sur la bactérie des nodosites radiculaires du *Casuarina equisetifolia* (Fort.). *Bull. Soc. Bot. France*, **78** (7/8): *447—452*.
185. CHEMIN, E. 1931. Sur la présence de galles chez quelques Floridées. *Rev. Algol.*, **5** (3/4); *315—325*.
186. CHEMIN, E. 1932. Sur l'existence des galles chez *Ceramium rubrum. C. R. Soc. Biol. Paris*, **109** (3): *135—157*.
187. CHEMIN, E. 1937. Rôle des bactéries dans la formation des galles chez les Floridales. *Ann. Sci. nat. Bot.*, (10) **19**: *61—71*, pl. i.
188. CHEN, H. K. 1938. Production of growth substance by clo ver nodule bacteria. *Nature*, **142**: *753—754*.
189. CHEN, H. K. & H. G. THORNTON, 1940. The structure of infective nodules and its influence on nitrogen fixation. *Proc. R. Soc. London*, (B) **129**: *208—229*, fig. 17, pl. ii.
190. CHESTER, K. S. 1933. Studies on bacteriophage relation to phytopathogenic bacteria. *Zbl. Bakt.*, (2) **89** (1/4): *1—30*.
191. CHI, P. 1915. Some inhabitants of the round gall of goldenrod. *J. ent. Zool.*, **7** (3): *161—177*, pl. ii.
192. CHILD, C. M. 1941. Patterns and Problems of Development. Chicago University Press, Illinois.
193. CHITWOOD, I. N. & E. BUHRER, 1946. Further studies on the life-history of *Heterodera rostochiensis. Proc. helminthol. Soc., Washington*, **13.**
194. CHITWOOD, I. N. & E. BUHRER, 1946. Life-history of *Heterodera rostochiensis* under Long Island conditions. *Phytopathology*, **36.**
195. CHOLODKOVSKY, N. 1903. Aphidologische Mitteilungen 18. *Chermes*-Gallen auf einer Weisstanne. 19. Zur Biologie von *Chermes pini* Koch. *Zool. Anz.*, **26**: *259—263*, pl. i.
196. CHRISTIE, J. R. 1936. The development of root-knot Nematode galls. *Phytopathology*, **26**: *1—22*, fig. 8.
197. CHRISTIE, J. R. 1949. Host-parasite relationships of the root-knot Nematode *Meloidogyne* spp. III. The Nematode resistance in plants to root-knot. *Proc. helminthol. Soc. Washington*, **16** (2): *104—108*.
198. CHRISTIE, J. R. & F. ALBIN, 1944. Host parasite relations of root-knot Nematode. I. The question of races. *Proc. helminthol. Soc. Washington*, **11.**
199. CHRISTIE, J. R. & G. COBB, 1941. Notes on the life-history of *Heterodera marioni. Proc. helminthol. Soc. Washington*, **8.**
200. CHRISTMANN, C. 1934. La galle *Perrisia carpini* (F. Löw) sur *Carpinus betulus* (Linn.). *Rev. gén. Bot.*, **46**: *470—484*, fig. 6.
201. CHRISTY, T. 1881. Specimens of horn-shaped galls from a branch of *Pistacia atlanta. Proc. Linn. Soc. London*, 1880/1882: *6.*
202. CHUPP, C. 1917. Studies on bulbroots of Cruciferous plants. *Cornell Univ. agric. exp. Sta. Bull.*, **382**: *421—452*, fig. 13, pl. ii.
203. COBB, N. A. 1901. Root galls. *Agric. Gaz. N. S. Wales*, Sydney, **12**: *1041—1052*, fig. 1—8.
204. COCKERELL, T. D. A. 1890. The evolution of galls. *Entomologist*, **23** (322): *73—76*.
205. COCKERELL, T. D. A. 1900. A new genus of Coccidae, injuring the roots of the grape-vine in South Africa. *Entomologist, 173—174.*

206. COLE, C. S. & H. W. HOWARD, 1958. Observations on giant cells in potato root infected with *Heterodera rostochiensis*. *J. Helminthol.*, **32** (3): *135—144*.
207. COLLEY, M. W. 1931. Culture experiments with *Pseudomonas tumefaciens*. *Amer. J. Bot.*, **18** (3): *211—214*.
208. COMAN, D. R. 1947. Mechanism of invasiveness of cancer. *Science*, **105**: *347—348*.
209. CONKLIN, M. E. 1936. Studies of the root nodule organism of certain wild legumes. *Soil. Sci.*, *167—185*.
210. CONNER, H. A., W. H. PETERSON & A. F. RIKER, 1937. The nitrogen metabolism of the crown-gall and hairy root bacteria. *J. agric. Res.*, **54**: *621—628*.
211. CONNER, H. A., A. J. RIKER & W. H. PETERSON, 1937. The carbon metabolism of the crown-gall and hairy root organism. *J. Bact.*, **34** (2): *221—236*, fig. 1—7.
212. COOK, M. T. 1902/1904. Galls and insects producing them. Parts I-IX and Appendix. *Ohio Naturalist*, **2** (6) (15): *263—278*, pl. xviii-xxi; **3** (7) (20): *419—436*, pl. xiii-xviii; **4** (8) (13): *125—147*, pl. ix-xv.
213. COOK, M. T. 1909. The development of insect galls as illustrated by the genus *Amphibolips. Proc. Indiana Acad. Sci.*, 25 *Ann. Meet.*, **5**.
214. COOK, M. T. 1923. The origin and structure of plant galls. *Science*, **26**: *6*.
215. COOK, M. T. 1923. Early stages of crown-gall. *Phytopathology*, **13**: *475—482*, fig. 14.
216. COOK, W. R. 1933. A monograph of Plasmodiophorales. *Arch. Protistenk.*, **179**, pl. xiv.
217. CORDER, M. 1933. Observations on the length of dormancy in certain Nematodes infesting plants. *J. Parasitol.*, **20**.
218. CORNU, M. 1874. Altération des radicelles de la vigne sous l'influence du *Phylloxera vastatrix* Planchon. *Bull. Soc. Bot. France*, **22**: *290*.
219. CORNU, M. 1878. Études sur le *Phylloxera vastatrix*. *Mém. Pres. Div. Sav. Acad. Nat. France*, **26**: *1*.
220. CORTI, A. 1903. Di una nuova galla d'*Apion pubescens* Kirby e dei Coleotterocecidii in genere. *Riv. Coleottereologica Ital.*, **1**: *179—182*.
221. COSENS, A. 1912. A contribution to the morphology and biology of insect galls. *Trans. Canad. Inst.*, **9** (3): *297—387*, fig. 9, pl. i—xiii.
222. COSENS, A. 1912. Insect galls. *Canad. Ent.*, **45**: *380*.
223. COTTE, J. 1911. Remarques au sujet des zoocécidies et leur origine. *C. R. Soc. Biol. Marseille*, **71**: *737—739*.
224. COTTE, J. 1911. Un ennemi des cécidies: *Polydrusus murinus* Gyllh. *Bull. Soc., Linn. Provance*, *146—148*.
225. COTTE, J. 1913. Un oiseau cécidophage. La mésange bleu. *Le Feuille des Jeunes Nat.*, **43** (506): *21—24*.
226. COTTE, J. 1915. Observations sur quelques cécidozoaires. *Ann. Mus. Hist. nat. Marseille*, **15**: *14*.
227. COTTE, J. 1925. Considerations sur l'hétérogonie chez les Cynipides cécidogènes. *Marcellia*, **22**: *89—119*.
228. COURCHET, L. 1880/1881. Études sur les galles causées par des Aphidiens. *Mem. Acad. Montpellier*, **10**: *1*.
229. CRANE, M. B. 1945. Origin of virus. *Nature*, **155**: *115*.
230. CRAWFORD, D. L. 1914. A monograph of the jumping plant lice or Psyllidae of the New World. *U.S. Nat. Mus. Bull.*, **85**: *186*.
231. CRAWFORD, D. L. Psyllidae of South America. *Brotéria*, (Zool.) **22**: *56—74*, pl. v.
232. CRISTINZIO, M. 1932. Nota priliminare su di una galla di *Bromus sterilis* Linn. nuova per l'Italia. *Marcellia*, **28**: *31—32*, fig. 5.

233. CROSBY, C. R. 1909. Chalcid-flies from galls from Zumbo, E. Africa. *Brotéria*, **8**: 77—90.
234. CURTIS, K. M. 1921. The lite history and cytology of *Synchytrium endobioticum* (Schil.) Perc. the cause of wart disease in potato. *Trans. R. phytopathol. Soc. London*, (B) **5**: 210, 409—475.
235. DAGUILLON, A. 1904. Sur une Acrocécidie de *Veronica chamaedrys* Linn. *Rev. gén. Bot.*, **16** (187): 257—264, fig. 6.
236. DAGUILLON, A. 1907. Les cécidies de *Rhopalomyia tanaceticola* Karsch. *Rev. gén. Bot.*, **19** (219): 112—115.
237. DALLA TORRE, W. K. & J. J. KIEFFER, 1902. Genera Insectorum. Cynipidae. **4**: 1—84, pl. iii.
238. DALLA TORRE, W. K. & J. J. KIEFFER, 1910. Cynipidae. Das Tierreich, **24**: xxxv—891, fig. 422.
239. DAME, F. 1928. *Pseudomonas tumefaciens* (Sm. & Towns.) Stev. der Erreger des Wurzelkropfes, in seiner Beziehung zur Wirtspflanze. *Zbl. Bakt.*, (2) **98** (21/24): 385—429, fig. 1—18.
240. DANGEARD, P. A. 1908. Note sur les zoocécidies recontrés chez un ascomycète l'*Ascobolus furguraceus*. *Bull. Soc. Bot. France*, **55**: 54.
241. DANGEARD, P. A. 1926. Recherches sur les tubercules radicaux des Legumineuses. *Le Botaniste*, **16**: 1—266, pl. xxviii.
242. DANGEARD, P. A. & M. T. LECHTOVA, 1929. Sur les phénomènes de symbiose chez le *Myrica gale*. *C. R. Acad. Sci. Paris*, **188**: 1584; *Le Botaniste*, (21) **5—6**: 345—350, fig. 1.
242a. DARLINGTON, C. D. 1949. The plasmogene theory of the origin of cancer. *Brit. J. Cancer*.
243. DAUVERGNE, J. & L. WEIL, 1927. Le cancer experimental du *Sedum spectabile*. *C. R. Soc. Biol.*, **97** (25): 815—816.
244. DEAN, H. L. 1937. Gall formation in host plants following haustorial invasions by *Cuscuta*. *Amer. J. Bot.*, **64**: 229—233, fig. 1.
245. DEL GUERICO, G. 1905. Intorno a tre species rare di Mizozilini Italiani e alle diverse gall prodotte da vaiii Afidi sul *Populus nigra*. *Redia*, **3** (2): 360—385, fig. 31.
246. DE MAN, J. C. 1892. Über eine in Gallen einer Meeresalge lebende Art der Gattung *Tylenchus* Bast. *Festschr. Leuckart, 191*.
247. DENIZOT, G. 1911. Sur une galle du chêne provoquée par *Andricus radicis*. *Rev. gén. Bot.*, **23**: 165—175, fig. 5.
248. DERMEN, H. & N. A. BROWN, 1940. Cytological basis of killing plant tumors by colchicine. *J. Hered.* **31** (4): 187—199, fig. 2.
249. DE ROPP, R. S. 1947. The response of normal plant tissues and of crown-gall tumor tissues to synthetic growth substances. *Amer. J. Bot.*, **34**: 53—56.
250. DE ROPP, R. S. 1947. The growth promoting and tumefacient factors of bacteria-free crown-gall tumor tissue. *Amer. J. Bot.*, **34** (5): 248—261.
251. DE ROPP, R. S. 1947. The isolation and behaviour of bacteria-free crown-gall tissue from primary galls of *Helianthus annuus*. *Phytopathology*, **37**: 201—206.
252. DE ROPP, R. S. 1948. The growth promoting action of bacteria-free-crown-gall tissue. *Bull. Torrey bot. Club*, **75**: 45—50,
253. DE ROPP, R. S. 1948. The interaction of normal and crown-gall tumor tissue *in vitro* grafts. *Amer. J. Bot.*, **35**: 372—377.
254. DE ROPP, R. S. 1948. Tumor formation in stem fragments *in vitro*. *Cancer Res.*, **8** (11): 519—530, fig. 4.
255. DE ROPP, R. S. 1948. The movement of crown-gall bacteria in isolated stem fragments of sunflower. *Phytopathology*, **38** (12): 993—998.
256. DE ROPP, R. S. 1948. Action of streptomycin on plant tumors. *Nature*, **162**: 459.

257. DE ROPP, R. S. 1949. The action ot antibacterial substances on the growth of *Phytomonas tumefaciens* and of crown-gall tumor tissue. *Phytopathology*, **39** (10): *822—828*.
258. DE ROPP, R. S. 1949. The inhibiting action of some analogues of folic acid on the growth of plant tumors. *Nature*, **164**: *954*.
259. DE ROPP, R. S. 1950. Some new plant growth inhibitors. *Science*, **112**: *500—501*.
260. DE ROPP, R. S. 1950. The comparative growth-promoting action of 3-indole-acetic acid and *Agrobacterium tumefaciens*. *Amer. J. Bot.*, **37**: *352—363*.
261. DE ROPP, R. S. 1951. The crown-gall problem. *Bot. Rev.*, **17** (9): *629—670*.
262. DE ROSSI, G. 1907. Über die Mikroörganismen, welche die Wurzel-knöllchen der Leguminosen erzeugen. *Zbl. Bakt.*, (2) **18**: *289—314, 481—489*, pl. ii.
263. DE STEFANI PEREZ, T. 1907. Contributo all conoscenza degli zoocecidii della Colonia Eritrea. *Marcellia*, **6**: *46—61*, fig. 16.
263a. DE VRIES. 1889. Intrazellulare Pangenesis. **117**.
264. DEWITZ, J. 1899. Die Lebensfähigkeit von Nematoden ausserhalb des Wirtes. *Zool. Anz.*, **22** (580): *91*.
265. DEWITZ, J. 1915. Über die Einwirkung der Pflanzenschmarotzer auf die Wirtspflanze. *Naturw. Z. Forst.- und Landw.*, **13**: *288—389*.
266. DEWITZ, J. 1917. Über Hämolsine (Aphidolysine) bei Pflanzenläusen. *Zool. Anz.*, **13**: *389*.
267. D'HERELLE, F. & E. PEYRE, 1927. Contribution à l'étude des tumeurs expérimentales. *C. R. Acad. Sci. Paris*, **185**: *227—230*.
268. DICKMANN, H. 1911. Einige Bemerkungen über die Galle von *Cecidoses eremita*. *Dtsch. ent. Nations-bibliothek*, **2** (17/21): *156—164*.
269. DICKMANN, H. 1913. Der Harzgallenwickler und sein Bau. *Natur und Kultur*, **10**: *326*.
270. DIELS, L. 1912. Der Formbildungsprozeß bei der Blütececidie von *Lonicera* Untergattung *Periclymenum*. *Flora*: *Alg. bot. Ztg.*, **105**: *184—223*, fig. 26, pl. ii.
271. DIEUEIDE, R. 1928. Contribution à l'étude des neoplasmas végétaux. Le rôle des pucérons en phytopathologie. *Act. Soc. Linn. Bordeaux*, **81**: *1—241*, fig. 64.
272. DITTRICH, R. 1924. Die Tenthredinidocecidien. In: Die Zoocecidien Deutschlands und ihre Bewohner (Rübsaamen & Hedicke), **4** (1). *Zoologica*, **61**.
273. DIXON, H. N. 1905. Nematode galls on mosses. *J. Bot.*, **43**: *251*.
274. DOLK, H. E. & K. V. THIMANN, 1932. Studies on the growth hormones of plants. *Proc. nat. Acad. Sci. U.S.A.*, **18**: *30—46*.
275. DONTCHO, K. 1935. Heritable tumors in plants experimentally produced. *Genetics*, **17**: *367—376*.
276. DOCTERS VAN LEEUWEN, W. 1909. Een gal op de bladstelen en de bladnerven van de Dadap door een vlieg, *Agromyza erythrinae* de Meij., gevormd. *Meded. Alg. Proefsta. Salatiga*, (2) **19** (Cultuurgids 2) **6**: *227—240*, fig. 1—8, pl. i.
277. DOCTERS VAN LEEUWEN, W. 1919. Über eine Galle an *Kibessia azura* DC, irrtümlich angesehen für eine Frucht einer anderen *Kibessia*-Art: *Kibessia sessilis* Bl. *Bull. Jardin bot. Buitenzorg*, (3) **1**: *131—135*, fig. 4.
278. DOCTERS VAN LEEUWEN, W. 1920. A mite gall on *Broussaisia arguta* Gaud, occurring in the Sandwich-Islands. *Marcellia*, **19**: *58—62*, fig. 6.
279. DOCTERS VAN LEEUWEN, W. 1929. Ein neuer Typus eines Thysano-pterocecidiums. *Marcellia*, **26**: *3—5*, fig. 1—2.

280. DOCTERS VAN LEEUWEN, W. M. 1924. Some Australian zoocecidia. *Marcellia*, **21**: *138—163*, fig. 37.
281. DOCTERS VAN LEEUWEN, W. M. 1925. Second contribution to the knowledge of the zoocecidia of Siam. *Marcellia*, **22**: *25—32*, fig. 12.
282. DOCTERS VAN LEEUWEN, W. M. 1928. Over een gal op *Symplocos fasciculata* Zoll., veroorzaakt door een Galmug *Asphondylia bursaria* Felt, welke samenleeft met een Schimmel. *Nederlandsch-Indisch Natuurw. Congr.*, 1928: *416—420*.
283. DOCTERS VAN LEEUWEN, W. M. 1933. Biology of plants and animals occurring in the higher parts of Mount Pangrongo-gedeh in West Java. *Verh. kon. Acad. Wet. Amsterdam*, (2) **31**: *1—278*, fig. 67, pl. xxx.
284. DOCTERS VAN LEEUWEN, W. M. 1935. New and noteworthy zoocecidia from the Netherlands. *Marcellia*, **29**: *73—86*.
285. DOCTERS VAN LEEUWEN, W. M. 1938. Ambrosia galls. *Chron. Bot.*, **4** (1): *13*.
286. DOCTERS VAN LEEUWEN, W. M. 1939. An ambrosia gall on *Symplocos fasciculata* Zoll. *Ann. Jardin bot. Buitenzorg*, **49** (1): *27—42*, pl. i-ii.
287. DOCTERS VAN LEEUWEN, W. M. 1939. Een mijtgal op de vrouwelijke kegels van *Podocarpus neriifolia* Don. *Natuurw. Tijds.*, **21** (7): *333—338*, fig. 2.
288. DOCTERS VAN LEEUWEN, W. M. & H. KARNY, 1924. Two new thrips galls and their inhabitants from New South Wales. *Proc. Linn. Soc. N.S. Wales*, Sydney, **49** (3): *4*, fig. 3.
289. DOCTERS VAN LEEUWEN W. & J. DOCTERS VAN LEEUWEN-REIJN-VAAN, 1904. Die Entwicklung der Galle von *Lipara lucens*. *Rec. Trav. bot. Neerl.*, **2**: *245*.
290. DOCTERS VAN LEEUWEN, W. & J. DOCTERS VAN LEEUWEN-REIJN-VAAN, 1907. Die Galle von *Eriophyes psilaspis* auf *Taxus baccata* und der normale Vegetationspunkt dieser Pflanze. *Beih. bot. Cbl.*, **23** (2): *14*, fig. 11.
291. DOCTERS VAN LEEUWEN, W. & J. DOCTERS VAN LEEUWEN-REIJN-VAAN, 1907. Über die Anatomie und die Entwicklung einiger *Isosoma*-Gallen auf *Triticum repens* und *junceum* und über die Biologie der Gallformer. *Marcellia*, **7**: *68—101*, fig. 36, pl. i.
292. DOCTERS VAN LEEUWEN, W. & J. DOCTERS VAN LEEUWEN-REIJN-VAAN, 1909. Beiträge zur Kenntnis der Gallen von Java. Über die Anatomie und Entwicklung der Galle von *Erythrina lithosperma* Miq. von einer Fliege *Agromyza erythrinae* de Meij. gebildet. *Rec. Trav. bot. Neerl.* **6**: *67—98*, fig. 16—21, pl. iv.
293. DOCTERS VAN LEEUWEN, W. & J. DOCTERS VAN LEEUWEN-REIJN-VAAN, 1909. Einige Gallen aus Java. *Marcellia*, **8**: *21—35*, fig. 1—17.
294. DOCTERS VAN LEEUWEN, W. & J. DOCTERS VAN LEEUWEN-REIJN-VAAN, 1909. Einige Gallen aus Java. Zweiter Beitrag. *Marcellia*, **8**: *85—122*, fig. 18—48.
295. DOCTERS VAN LEEUWEN, W. & J. DOCTERS VAN LEEUWEN-REIJN-VAAN, 1909. Kleinere zezidologische Mitteilungen. I. Eine von der Sesiide *Aegeria unifomis* Snellen an *Commelina communis* L. verursachte Stengelgalle. *Ber. dtsch. bot. Ges.*, **27**: *572—581*, fig. 1—6.
296. DOCTERS VAN LEEUWEN, W. & J. DOCTERS VAN LEEUWEN-REIJN-VAAN, 1910. Beiträge zur Kenntnis der Gallen aus Java. II. Über die Entwicklung einiger Milbengallen. *Ann. Jardin Bot. Buitenzorg*, (2) **8**: *119—183*, pl. xxiv-xxxi.
297. DOCTERS VAN LEEUWEN, W. & J. DOCTERS VAN LEEUWEN-REIJN-VAAN, 1910. Kleinere cecidologische Notizen: Über die Anatomie der Luftwurzeln von *Ficus pilosa* Reinw. und *Ficus nitida* var. *retusa* King und der von Chalcididen auf denselben gebildeten Gallen. *Ber. dtsch. bot. Ges.*, **28**: *169*.

298. DOCTERS VAN LEEUWEN, W. & J. DOCTERS VAN LEEUWEN-REIJN-
VAAN, 1911. Beiträge zur Kenntnis der Gallen von Java. 3. Über die
Entwicklung und Anatomie einiger Stengel-, Markgallen und über
Kallus. *Rec. Trav. bot. Neerl.*, **8**: *1—56*, fig. 1—6, pl. i.

299. DOCTERS VAN LEEUWEN, W. & J. DOCTERS VAN LEEUWEN-REIJN-
VAAN, 1911. Einige Gallen aus Java. Vter Beitrag. *Marcellia*, **10**:
65—93, fig. 103.

300. DOCTERS VAN LEEUWEN, W. & J. DOCTERS VAN LEEUWEN-REIJN-
VAAN, 1911. Kleinere cecidologische Mitteilungen. III. Über die
unter Einfluss eines Cocciden entstandene Umbildung oberirdischer
Triebe von *Psilotum triquetrum* Sw. in dem Rhizom ähnlich gebau-
ten Wucherungen. *Ber. dtsch. bot. Ges.*, **29**: *166—175*.

301. DOCTERS VAN LEEUWEN, W. & J. DOCTERS VAN LEEUWEN-REIJN-
VAAN, 1912. Einige Gallen aus Java. 6ter Beitrag. *Marcellia*, **11**:
49—100, fig. 104—155.

302. DOCTERS VAN LEEUWEN, W. & J. DOCTERS VAN LEEUWEN-REIJN-
VAAN, 1912. Beiträge zur Kenntnis der Gallen aus Java. 4. Über
einige von Cecidomyiden an Gräsern gebildete Blattscheidegallen.
Rec. Trav. bot. Neerl., **9**: *382—399*, pl. vi.

303. DOCTERS VAN LEEUWEN, W. & J. DOCTERS VAN LEEUWEN-REIJN-
VAAN, 1914. Kleinere cecidologische Mitteilungen. IV Über die von
Gynaikothrips pallipes Karny an *Piper sarmentosum* Roxb. (=*Piper
zollingerianum* Bl.) verursachte Blattgalle. *Marcellia*, **13**: *127—135*,
fig. 11.

304. DOCTERS VAN LEEUWEN, W. & J. DOCTERS VAN LEEUWEN-REIJN-
VAAN, 1916. Beiträge zur Kenntnis der Gallen von Java. 7. Über die
Morphologie und die Entwicklung der Galle von *Eriophyes sesbaniae*
Nal. an den Blättern und Blumen von *Sesbania sericea* DC. gebildet.
Rec. Trav. bot. Neerl., **13** (1): *30—43*, fig. 10.

305. DOCTERS VAN LEEUWEN, W. & J. DOCTERS VAN LEEUWEN-REIJN-
VAAN, 1918. Niederländisch-Ost-indische Gallen. Nr. 10. Einige
Gallen aus Java. VIIIter Beitrag. *Bull. Jardin. bot. Buitenzorg*, (3)
1: *17—76*, fig. 1—99.

306. DOCTERS VAN LEEUWEN, W. & J. DOCTERS VAN LEEUWEN-REIJN-
VAAN, 1925. Over de aetiologie van de Gallen. *Handl. derde Nederl.
Indische Natuurw. Congr.* Buitenzorg, 25—28.

307. DOCTERS VAN LEEUWEN, W. & J. DOCTERS VAN LEEUWEN-REIJN--
VAAN, 1926. The Zoocecidia of the Netherlands East Indies. Batavia,
pp. 601, fig. 1088, pl. vii.

308. DOCTERS VAN LEEUWEN-REIJNVAAN, J. & W. M. DOCTERS VAN
LEEUWEN, 1928. Über ein von *Gynaikothrips devriesii* Karny aus
einer Gallmückengalle gebildetes Thysanopterocecidium. *Rec.
Trav. bot. Neerl.*, **25**a: *99—114*, fig. 9.

309. DOCTERS VAN LEEUWEN, W. & J. & H. KARNY, 1913. Beiträge zur
Kenntnis der Gallen von Java. 5. Über die javanischen Thysanopterocecidien. *Bull. Jardin bot. Buitenzorg*, (2) **10**: *1—126*, fig. 86.

310. DUBUY, H. G. & M. W. WOODS, 1943. Evidence for the evolution of
of phytopathogenic viruses from mitochondria and their derivatives
II. Chemical evidence. *Phytopathology*, **33**: *766—777*.

311. DUHAMET, L. 1950. Action du lai de coco sur le croissance des tissus de
crown-gall de Scorzonère cultivées in vitro. *C. R. Acad. Sci. Paris*,
230: *770—771*.

312. DUNN, J. A. 1961. The formation of galls by some species of *Pemphigus* (Homoptera: Aphididae) *Marcellia*, **30** (Suppl.): *155—167*,
pl. i-ii.

313. EBERHARDT, A. 1904. Contribution à l'étude de *Cystopus candida* Lev.
Zbl. Bakt., (2) **12**: *235*.

314. ECKHARDT, M., M. L. BALDWIN & E. B. FRED, 1931. Studies of the root nodule organism of *Lupinus*. *J. Bact.*, **21**: *273—285*.
315. EFFLATOUN, H. C. 1924. A new species of the galligenous genus *Euaresta* (Diptera: Trypetidae). *Bull. Soc. R. ent. d'Egypte*, **1923**: *152—156*.
316. EHRHORN, E. M. 1908/1912. Gall making coccids. *Proc. Hawaii. ent. Soc.*, **2**: *179*.
317. ELCOCK, H. A. 1928. The anatomy of the overgrowth on sugarbeets caused by *Bacterium beticola*. *Papers Michigan Acad. Sci.*, **9**: *111—115*.
318. ENGEL, H. & M. ROBERG, 1938. Die Stickstoffausscheidung der Wurzelknöllchen. *Ber. dtsch. bot. Ges.*, **56**: *337—352*, pl. ii.
319. ENSLIN, E. 1914. Die Blatt- und Holzwespen (Tenthrediniden). In: Die Insekten Mitteleuropas. (Schröder) 2 (3).
320. ESTEE, L. M. 1913. Fungus galls on *Cytoseira* and *Halidrys*. *Univ. California Pub. Bot.*, **4** (17): *305—316*. pl. xxxv.
321. EWING, J. 1940. Neoplastic disease. 4th ed. W. B. Saunders, Philadelphia.
322. FAEBER, E. E. 1942. A suggestion as to the nature of cancer and abnormal growth. *South. Afr. J. Sci.*, **38**: *278*.
323. FAGAN, M. M. 1918. The uses of insect galls. *Amer. Naturalist*, **52**: *155—176*.
324. FAVORSKY, W. 1910. Nouvelles recherches sur le développement et la cytologie du *Plasmodiophora brassicae* WOI. *Mem. Soc. nat. Kieff*, **20**: *149—184*.
325. FEBIGER, J. 1935. Untersuchungen über eine Nematode und deren Fähigkeit papillomatöse und carcinomatöse Geschwulstbildungen in Magen der Ratte hervorzurufen. *Z. Krebsforsch.*, **13**: *217—280*.
326. FEHER, D. & R. BOKOR, 1926. Untersuchungen über die bakterielle Wurzelsymbiose einiger Leguminosenhölzer. *Planta*, **2**: *406—413*, fig. 4.
327. FEJGIN, B., T. EPSTEIN & C. FUNK, 1926. Sur une tumeur végétale provoquée par une bactérie isolée d'un carcinome humain. *C. R. Soc. Biol. Paris*, **94** (14): *1097—1098*.
328. FELT, E. P. 1928. Key to gall midges — A resumé of studies I-VII. *N. Y. St. Mus. Bull.*, **257**: *1—239*.
329. FELT, E. P. 1936. The relation of insects and plants in gall production. *Ann. ent. Soc. Amer.*, **29**: *694—700*.
330. FELT, E. P. 1940. Plant galls and gall makers. Ithaca: New York: Comstock Publishing Company Inc., pp. *1—364*, fig. 344, pl. xli.
331. FILIPJEW, J. N. & SCHUURMANS STEKHOVEN J. K. 1941. Manual of agricultural Helminthology. E. J. Brill, Leiden, pp. 1000, fig. 400.
332. FIROR, W. M. & O. G. GEY, 1945. Observations on the conversion of normal into malignant cells. *Ann. Surg.*, **121**: *700—703*.
333. FISCHER, E. 1907. Über durch parasitische Pilze (besonders Uredineen) hervorgerufene Missbildungen. *Verh. schweiz. naturf. Ges.*, **89**: *170—177*.
334. FISCHER, E. & E. GÄUMANN, 1929. Biologie der Pflanzenbewohnenden parasitischen Pilze. Jena.
335. FLINT, L. H. & C. F. MORELAND, 1945. Note on gall formation in decapitated young bean plants. *Plant Physiol.*, **20** (3): *453—456*, fig. 1.
336. FOCKEU, H. 1889. Contributions à l'histoire des galles. Étude anatomique de quelques espèces. Imprimérie & Librairie Camille Robbe. Lille.
337. FOCKEU, H. 1849. Note pour servir à l'histoire de la mycocécidie des *Rhododendrons*. *Rev. biol. Nord. France*, **6**: *355*.

350

338. FOCKEU, H. 1896. Recherches anatomiques sur les Galles. Étude de quelques Dipterocécidies et Acarocécidies. Imprimérie & Lithographique de Bigot Frères.

339. FORESSELL, 1883. Studier öfver Cephalodierna. *Bih. Svenska vetensk.-Akad. Handl.*, **8.**

340. FORMANEK, R. & L. MELICHER, 1916. Die Rüsslergattung *Nanophyes* und ihre Arten. *Wien. ent. Z.*, **35**: *65—79.*

341. FRANK, A. B. 1885. Über das Wurzelälchen und die durch dasselbe verursachten Beschädigungen der Pflanzen. *Ber. dtsch. bot. Ges.*, **2**: *145.*

342. FRANK, A. B. 1896. Die Krankheiten der Pflanzen. 2te Auflage. vol. 2. Die pilzparasitären Krankheiten. vol 3. Die tierparasitären Krankheiten. Breslau.

343. FRANKLIN, M. 1930. Experiments with cysts of *Heterodera schachtii*. *J. Helminthol.*, **16.**

344. FRANKLIN, M. 1937. The survival of free larvae of *Heterodera schachtii* in soil. On the survival of *Heterodera marioni* infection out-of-doors in England. *J. Helminthol.*, **16.**

345. FRAZIER, W. C. & E. B. FRED, 1922. Movement of Legume bacteria in soil. *Soil Sci.*, **14**: *29—36.*

346. FRIEDERICHS, K. 1909. Die Schaumzikade als Erzeuger von Gallbildungen. *Z. wiss. Insektenbiol.*, **5**: *175.*

347. FRIEDMAN, B. A. & T. FRANCIS, Jr., 1942. Gall formation by *Phytomonas tumefaciens* extract and indole-3-acetic acid in cultures of tomato roots. *Phytopathology*, **32**: *762—772.*

348. FROGGATT, W. W. 1890. Two large apple-shaped galls nearly two inches in diameter found on *Eucalyptus* sp. *Proc. Linn. Soc. N. S. Wales*, Sydney, (2) **5**: *413.*

349. FROGGATT, W. W. 1892. Gall-making Buprestids. *Proc. Linn. Soc. N. S. Wales*, Sydney, (2) **7**: *323—326.*

350. FROGGATT, W. W. 1898. The growth of vegetable galls. *Agric. Gaz. N. S. Wales*, Sydney, **9**: *385—391, 488—499,* pl. iv; *Misc. Pub.*, **221**: *1—19,* pl. iv.

351. FROGATT, W. W. 1901. Galls on myall *(Acacia pendula)* from Tamworth. *Proc. Linn. Soc. N. S. Wales*, Sydney, **26**: *146.*

352. FROGGATT, W. W. 1892/1898. Notes on the familiy Brachyscelidae, with some account of their parasites and descriptions of new species. I. *Proc. Linn. Soc. N. S. Wales*, Sydney, (2) **7**: *353—372,* pl. vi-vii (1892); II. **8**: *209—214,* pl. viii (1893); V. **23**: *370—379,* pl. viii-x (1898).

353. FULLER, C. 1913. Root-knot gall-worm and eelworms. *Agric. J. Union S. Africa*, **6**: *440—443, 792—802.*

354. FRÜHAUF, E. 1924. Legeapparat und Eiablage bei Gallenwespen (Cynipiden). *Z. wiss. Zool.*, **121**: *656—723.*

355. GABRIEL, C. 1922. Cécidie de *Vaucheria aversa* produites par *Notommata werneckii. C. R. Soc. Biol. Marseille*, **86**: *453.*

356. GAHAN, A. B. 1922. A list of phytophagous Chalcidoidea, with descriptions of two new species. *Proc. ent. Soc. Washington*, **24**: *33—58.*

357. GAHAN, A. B. & CH. FERRIÈRE, 1947. Notes on some gall inhabiting Chalcidoidea (Hymenoptera). *Ann. ent. Soc. America*, **11** (2): *271.*

358. GAINOR, C. & F. D. CRISLEY, 1961. Proteolytic activity of crude stem extracts from normal and tumor tissues of plants. *Nature*, **190**: *1031—1032.*

359. GALLOWAY, B. T. 1919. Giant crown-galls from Florida everglades. *Phytopathology*, **9**: *207—208.* pl. x.

360. GAMBIER, MME. 1924. Recherches sur quelques cécidies florales. *Marcellia*, **21**: *10—30,* fig. 57.

361. GARRIGUES, R. 1947. Aperçu sur les modifications anatomiques et cytologiques observées dans les zoocécidies. *Bull. Soc. Bot. France*, **94**: *115—116*.

362. GARRIGUES, R. 1950. Sur un type particulier de noyau trouvé dans les Hyménoptérocécidies. *C. R. Acad. Sci. Paris*, **231**: *984—986*.

363. GARRIGUES, R. 1954. De l'existence d'un gradient chimique agent d'action cécidogène. *C. R. Seances Rep. Comm. VIIIe Congr. int. Bot. Paris*, **7/8**: *222—226*, fig. 1.

364. GAUTHERET, R. J. 1939. Sur la possibilité de réaliser la culture indéfinie des tissus de tubercules de carotte. *C. R. Acad. Sci. Paris*, **208**: *118—120*.

364a. GAUTHERET, R. J. 1942. Hétéro-auxin et cultures de tissus végétaux. *Bull. Soc. Chim. biol.*, **24**: *13—47*.

364b. GAUTHERET, R. J. 1944. Recherches sur la polarité des tissus végétaux. *Rev. Cytol. Cytophys. Veg.*, **7**: *45—185*.

365. GAUTHERET, R. J. 1946. Comparison entre l'action de l'acide indole acétique et celle du *Phytomonas tumefaciens* sur la croissance des tissus végétaux. *C. R. Soc. Biol.* **140**: *169—171*.

366. GAUTHERET, R. J. 1947. Comparison entre la structure de cultures de tissus normaux et des cultures des tissus de crown-gall de Topinambour. *C. R. Soc. Biol.*, **141**: *598*.

367. GAUTHERET, R. J. 1947. Action de l'acide indole acétique sur le développement des tissus normaux et des tissus de crown-gall de Topinambour cultivés in vitro. *C. R. Acad. Sci. Paris*, **224**: *1728—1730*.

368. GAUTHERET, R. J. 1948. Sur la culture de trois types des tissus de Scorsonère: tissus normaux, tissus de crown-gall et tissus accoutumés à l'hétéro-auxin. *C. R. Acad. Sci. Paris*, **226**: *270—271,*

369. GAUTHERET, R. J. 1950. Pflanzenkrebs. *Endeavour*, (German ed.) **9** (33): *21*.

370. GAVAUDAN, P. & N. GAVAUDAN, 1939. Tuméfaction des racines par les substances modificatrices de la caryocinèse. *C. R. Biol. Paris*, **131** (16): *168—171*, fig. 10.

371. GAVAUDAN, P. & N. GAVAUDAN, 1939. Mise en évidence sur les méristèmes radiculaires de *Triticum* vulgare de l'existence d'une propriété mitoinhibitrice commune aux divers apiols. *C. R. Soc. Biol. Paris*, **131** (21): *998—1000*.

372. GEIGER, W. B. Jr. & R. J. ANDERSON, 1939. The chemistry of *Phytomonas tumefaciens*. I. The lipids of *Phytomonas tumefaciens*. The composition of phosphatides. *J. biol. Chem.*, **129**: *258—264*.

373. GEISENHEYNER, L. 1903. Über einige Monstruositäten an Laubblättern. *Ber. dtsch. bot. Ges.*, **21**: *440*.

374. GEISSLER, G. 1950. Über die Wirkung von Mitosengiften, Wuchs- und Keimungshemmstoffen auf die Wurzelzellen von *Allium cepa*. *Naturwissenschaften*, **37**: *141*.

375. GEITLER, L. 1953. Endomitose und endomitotische Polyploidisierung. *Protoplasmatologie*, **6**: *1—89*.

376. GÉNEAU DE LAMARLIERE, L. 1905. Sur les mycocécidies des Gymnosporangium. *Ann. Sci. Natur.*, (9) **2**: *313*.

377. GENEVES, L. 1946. Recherches sur la formation de deux categories cellulaires dans les tumeurs corticales de la tomate. *Rev. gén. Bot.*, **53** (633): *381—411*, fig. 10, pl. i.

378. GEORGEVITCH, P. 1910. De la morphologie des microbes des nodosites des Legumineuses. *C. R. Soc. Biol.*, **69** (29): *276—278*, fig. 10.

379. GERHARDT, K. 1922. Über die Entwicklung der Spirallockengalle von *Pemphigus spirothecae* an der Pyramidenpappel. *Z. Pflanzenkr.*, **32**: *177—189*.

352

380. GERTZ, O. 1917. Studier öfver Klyningarnus morfologi. *Lund Univ. Arskr.* (NF 2) **15** (7): *1—85*, fig. 182.
381. GERTZ, O. 1918. Über einige durch schmarotzende *Cuscuta* hervorgerufene Gewebeveränderungen bei Wirtspflanzen. *Ber. dtsch. bot. Ges.*, **36**: *62*.
382. GERTZ, O. 1928. Linne sasom cecidology en studie till cecidologiens aldre historia. *Acta Univ. Lund*, (NF) **24** (5): *115*.
383. GHEORGHIU, I. 1933. Le cancer des plantes et l'immunité anticancéreuse. *Ann. Inst. Pasteur Paris*, **51** (4): *535—544*, fig. 4.
384. GHEORGHIU, I. 1938. Les tumeurs expérimentales des plantes causées par divers agentes chimiques, essai d'explication. *Ann. Inst. Pasteur Paris*, **60** (5): *549—558*, fig. 1—4.
385. GIARD, A. 1869. Sur l'hermaphroditisme de *Melandrium album* anfeste par *Ustilago antherarum. Bull. Soc. Bot. France*, *16*: *215*.
386. GIARD, A. 1888. Sur la castration parasitaire du *Lychnis dioica* Linn. par l'*Ustilago antherarum. C. R. Acad. Sci. Paris*, **107**: *757*.
387. GIARD, A. 1889. Sur la castration parasitaire de *l'Hypericum perforatum* Linn. par la *Cecidomyia hyperici* Bremi et par *l'Erysiphe martii* Lev. *C. R. Acad. Sci. Paris*, **109**: *324*.
388. GIBBS, M. 1905. Jumping gall. *Bull. U.S. Dep. agric. Ent.*, **54**: *81*.
389. GIESENHAGEN, K. 1899. Über einige Pilzgallen an Farnen. *Flora*, **86**: *100*.
390. GIESENHAGEN, K. 1917. Entwicklung einer Milbengalle an *Nephrolepis biseriata* Schott. *Pringsh. Jb. Wiss. Bot.*, Leipzig, **58**: *66—104*, fig. 3, pl. ii—iii.
391. GIOELLI, F. 1940. Filtrati de *Bacterium tumefaciens* su culture in vitro di tessuti. *Riv. Pat. Veg.*, **30**: *117—130*.
392. GODAN, D. 1955. Beitrag zur stofflichen Beeinflussung des Gallegewebes durch Gallmückenlarven. *Mitt. dtsch. ent. Ges.*, **14**: *8—11*.
393. GODAN, D. 1956. Beiträge zur Autökologie der Veilchenmücke *Dasyneura affinis* Kieff. *Z. angew. Ent.*, **39**: *1—19*.
394. GODFREY, G. H. 1929. Effect of some environmental factors on the root-knot Nematodes. *Phytopathology*, **21**.
395. GODFREY, G. H. 1940. Ecological specialization in the stem- and bulbinfesting Nematode *Ditylenchus dispasci* var. *amsinckiae. Phytopathology*, **30** (1): *41—53*, fig. 6.
396. GODFREY, G. H. & J. OLIVERIA, 1932. Development of root-knot Nematode. *Phytopathology*, **22**.
397. GÖBEL, K. 1928. Organographie der Pflanzen. 3. Aufl. 1.
398. GOFFART, H. 1930. Die Aphelenchen der Kulturpflanzen. In: Monographie zum Pflanzenschutz. 4. Berlin.
399. GOFFART, H. 1930. Rassenstudien an *Heterodera schachtii. Arb. biol. Reichsanst. Land.-Forstw.*, **18** (1): *83—100*, fig. 7.
400. GOIDANICH, A. 1940. Interpretazione simbiotica di una associazione micoentomatice gallare. *Atti R. Acad. Sci. Torino*, **76**: *1—14*, fig. 3.
401. GOODEY, J. B. 1939. The structure of the leaf gall of *Flantago lanceolata* Linn. induced by *Anguillulina dispasci* (Kühn) Ger. & v. Ben. *J. Helminthol.*, **17** (4): *183—190*.
402. GOODEY, J. B. 1948. The galls caused by *Anguillulina balsamophila* (Thorne) Goodey on the leaves of *Wythia amplexicaulis* Nutt. and *Balsamorhiza sevithoa* Nutt. *J. Helminthol.*, **27** (2): *109—116*.
403. GOODEY, T. 1923. Review of plant parasitic members of the genus *Aphelenchus*. Quiescence and revivascence in Nematodes. *J. Helminthol.*, **1**.
404. GOODEY, T. 1929. On some details of comparative anatomy of *Aphelenchus, Tylenchus* and *Heterodera. J. Helminthol.*, **7**.
405. GOODEY, T. 1930. *Tylenchus agrostis* (Steinh.). *J. Helminthol.*, **8** (4): *197—210*, fig. 9, pl. i.

406. GOODEY, T. 1931. Biological races in Nematodes and their significance in evolution. *Ann. appl. Biol.*, **18** (3)): *414—419*.
407. GOODEY, T. 1932. Observations on the biology of the root-knot Nematode. *J. Helminthol.*, **10**.
408. GOODEY, T. 1932. Some observations on the biology of the root gall Nematode *Anguillulina radicicola* (Greeff) 1872. *J. Helminthol.*, **10**: *33*.
409. GOODEY, T. 1934. *Anguillulina cecidoplastes*, n. sp. a Nematode causing galls on the grass *Andropogon pertusus* Willd. *J. Helminthol.*, **12** (4): *225—236*.
410. GOODEY, T. 1934. Gall formation due to *Anguillulina graminis*. *J. Helminthol.*, **12** (4).
411. GOODEY, T. 1935. The pathology and aetiology of plant lesions caused by parasitic Nematodes. *Imp. Bur. agric. Parasitol.*, St. Albans, pp. 34.
412. GOODEY, T. 1938. Observations an *Anguillulina millefolii* (Löw) Goodey, from galls on the leaves of Yarrow *Achillea millefolium* Linn. *J. Helminthol.*, **16** (2): *93—108*, fig. 8, pl. i.
413. GOODEY, T. 1945. *Anguillulina brenari*, sp. nov. a Nematode causing gall on the moss *Pottia bryoides* Mitt. *J. Helminthol.*, **21** (2/3): *105—110*, fig. 7.
414. GOSSET, A., A. TEHAKIRIAN & J. MAGROU, 1939. Sur la composition chimique des tumeurs bactériennes de *Pelargonium zonale* et des tissus aux dépenses desquelles elles se développent. *C. R. Acad. Sci. Paris*, **208**: 424.
415. GRANDI, G. 1935. Agaonidi. *Boll. Lab. Ent. Ist. Sup. Agr. Bologna*, **7**.
416. GREEFF, R. 1872. Über Nematoden in Wurzelanschwellungen (Gallen) verschiedener Pflanzen. *S.B. Ges. Beford. Naturw. Marburg*, **172**.
417. GREVILLIUS, A. Y. 1909. Ein Thysanopterocecidium auf *Vicia cracca* Linn. *Marcellia*, **8**: *37*.
418. GREVILLIUS, A. Y. 1910. Notizen über Thysanopterocecidien auf *Stellaria media* usw. *Marcellia*, **9**: *161*.
419. GREVILLIUS, A. Y. & J. NIESSEN, 1906/1912. Begleitwort zur Sammlung von Tiergallen und Gallentieren, insbesondere aus dem Rheinlande, Heft 1—6, Köln & Kempen.
420. GRIEVE, B. J. 1941. Studies in the physiology of host-parasite relations. II. Adventitious root formation. *Proc. R. Soc. Victoria*, **53** (2): *323—341*, fig. 2, pl. i.
421. GRIEVE, B. J. 1943. Mechanism of abnormal and pathological growth: A review. *Proc. R. Soc. Victoria*, **55**: *109—132*.
422. GROSSBARD, E. 1951. Antibiotica und mikrobieller Antagonismus und ihre Bedeutung für den Pflanzenschutz. *Endeavour*, **10** (39): *145—150*.
423. GUÉGEN & F. HEIM, 1901. Variations florales tératologiques, d'origine parasitaire chez le florale du *Lonicera perichimentum* Linn., produite par *Rhopalomyia xylostei* Schr. *C. R. Ass. Franç. pour l'Avance Sci.*, **30** (10): *130—131*.
424. GUTTENBERG, H. 1905. Beiträge zur physiologischen Anatomie der Pilzgallen. Leipzig.
425. GUTTENBERG, H. 1909. Cytologische Studien an *Synchytrium*-Gallen. *Jb. wiss. Bot.*, **46**: *453*.
426. HABERLANDT, G. 1923. Wundhormone als Erzeuger von Zellteilungen. *Beitr. allg. Bot.*, **2**: *1*.
427. HADDOW, A. 1944. Transformation of cells and viruses. *Nature*, **154**: *194—199*.
428. HADDOW, A., G. A. R. KON *et al.* 1947. Chemical carcinogens. *Brit. med. Bull.*, **4**: *309—426*.

354

429. HAGERUP, O. 1928. En hygrofil Baegplante *(Aeschynomene aspera* Linn). med Bakterie knolde paa staengelen. *Dansk. bot. Ark.*, **5** (14): *1—9*, fig. 9.
430. HAMADI, H. 1930. Über die Histogenese, Bau und Natur des sog. Pflanzenkrebs und dessen Metastasen. *Z. Krebsforsch.*, **30** (6): *547—552*.
431. HAMADI, H. 1932. Über den sog. Pflanzenkrebs und seine Metastasen und ihre Vergleichung mit den Tiergewächsen. *Virchows Arch. path. Anat. u. Physiol.*, **287** (1): *29—33*.
432. HAMMERSCHMIDT, C. E. 1838. Anatomischpathologische Untersuchungen über die Natur und Entwicklung der Pflanzenauswüchse. *Allg. österr. Z. Landw. Gärtn.*, **10**: *35*.
433. HAMPTON, J. E. 1948. Cure of crown-gall with antibiotics. *Phytopathology*, **38**: *11*.
434. HAMNER, K. C. & E. J. KRAUS, 1937. Histological reactions of bean plants to growth-promoting substances. *Bot. Gaz.*, **98**: *735—807*.
435. HANSEN, H. N. & R. E. SMITH, 1937. A bacterial gall disease of Douglas fir *Pseudotsuga taxifolia*. *Hilgardia*, **10**: *569—577*.
436. HANSEN, R. & E. W. TANNER, 1931. The nodule bacteria of the Leguminosae with special reference to the mechanism of inoculation. *Zbl. Bakt.*, (2): **85**: *129—152*.
437. HARMS, H. 1922. Knöllchenförmige Pilzgallen an der Wurzel von *Myrica gale*. *Verh. bot. Ver. Prov. Brandenburg*, **64**: *158—159*.
438. HARRISON, F. C. & B. BARLOW, 1907. The nodule organism of the Leguminosae, its isolation, cultivation, identification and commercial application. *Zbl. Bakt.*, (2) **19**: *264—272, 426—441*, pl. ix.
439. HARRISON, J. W. HESLOP, 1924. Sex in Salicaceae and its modification by Eriophyid mites and other influences. *Brit. J. exp. Biol.*, **1**: *445—472*, fig. 4.
440. HARTIG, TH. 1840/1843. Über die Familie der Gallwespen. *Z. Ent.*, **2**: *176* (1840); Nachträge, *ibidem*, **3**: *322* (1843).
441. HARTWICH, C. 1883. Übersicht der technisch und pharmaceutisch verwendeten Gallen. *Arch. Pharm. Berlin*, (3) **21** (221): *820—840, 881—911*, fig. 58.
442. HARTWICH, C. 1885. Über Gerbstoffe und Ligninkörper in der Nahrungsgeschichte der Infectoria-Gallen. *Ber. dtsch. bot. Ges.*, **3**: *146*.
443. HAUMESSER, J. 1933. Dipterocécidie florale produite par *Asphondylia dufori* (Kieff.) sur *Verbascum floccosum* (W. & K.). *Rev. gén. Bot.*, **45**: *71—87*, fig. 8.
444. HAVAS, L. J. 1939. Growth of induced plant tumors. *Nature*, **143** (3628): *789—891*.
445. HAVAS, L. J. 1937. Colchicine "phytocarcinomata" and plant hormones. *Nature*, **140**: *191—192*.
446. HAVAS, L. J. 1942. L'évolution graduelle d'anomalies morphologiques, y compris la polyploidie, induites chez le *Pelargonium zonale* par inoculation repetées du *Bacterium tumefaciens*. *Acad. R. Belg. Bull.*, Cl. Sci., **28** (4/6): *318—340*.
447. HEALD, F. O. Aerial galls of the mesquite. *Mycologia*, **6** (1): *37—38*, fig. 2, pl. cxvii.
448. HEDGCOCK, G. G. 1910. Field studies of the crown-gall and hairy root gall of apple tree. *Bull. Dep. Agric. Washington*, **186**: *1—108*, pl. x.
449. HEDGES, F. & L. S. TENNY, 1912. A knot of citrus trees caused by *Sphaeropsis tumefaciens*. *U.S. Dep. Agric. Bur. Plant Industries Bull.* **247**: *1—74*, pl. x.
450. HEDICKE, H. 1914. Zur Kenntnis abnormaler Gallenbildungen. *S.B. Ges. naturf. Freunde Berlin*, *424*.

355

451. HEDICKE, H. 1920. Beiträge zu einer Monographie der paläarktischen Isosominen. *Arch. Naturg.*, **86**: *1.*
452. HEDICKE, H. 1924. Die Isthmosominocecidien, durch Isthmosominen verursachte Pflanzengallen und ihre Erreger. In: Die Zoocecidien Deutschlands und ihre Bewohner. 4. 2. Stuttgart.
453. HEIBERG, A. & T. KEMP, 1929. Über die Zahl der Chromosomen in Carcinomzellen beim Menschen. *Virchows Arch. path. Anat.*, **2013**: *693—700.*
454. HEIMHOFFEN. 1858. Beobachtungen über die Menge und das Vorkommen der Pflanzengallen und ihre spezielle Verteilung auf die verschiedenen Pflanzengattungen und Arten. *Verh. zool.-bot. Ges.*, **8**: *285.*
455. HEINRICHER, C. 1929. Allmähliches Immunwerden gegen Mistelbefall. *Planta*, **7**: *165—173.*
456. HENCKEL, A. 1923. Zur Entwicklungsgeschichte der Kohlhernie *(Plasmodiophora brassicae* WOR.). *Bull. Inst. rec. biol. Univ. Perm.*, **2.**
457. HENDRICKSON, A. A., I. L. BALDWIN & A. J. RIKER, 1934. Studies on certain physiological characters of *Phytomonas tumefaciens, Phytomonas rhizogenes* and *Bacterium radiobacter.* Part. II. *J. Bacter.*, **28**:*597—618.*
458. HENRY, B. W., A. J. RIKER & B. M. DUGGAR, 1942. The relation of vitamin B₁ to crown-gall development. *Phytopathology*, **32** (1): *8* (Abstract).
459. HENRY, B. W., A. J. RIKER & B. M. DUGGAR, 1943. Thiamine in crown-gall as measured with *Phycomces* assay. *J. agric. Res.*, **67** (3): *89—110*, fig. 1.
460. HERING, M. 1926/1927. Beiträge zur Histologie der Pflanzengallen. *Mikrokosmos*, **20**: *228.*
461. HEYN, A. N. J. 1931. Der Mechanismus der Zellstreckung. *Rec. Trav. bot. Néerl.*, **28**: *113—244.*
462. HIERONYMUS, G. 1890. Beiträge zur Kenntnis der europäischen Zoocecidien und Verbreitung derselben. *Erg. 68 Jahrb. Schles. Ges. Vater. Kultur*, **49**: *272.*
463. HILDEBRANDT, A. C. 1950. Some important galls and wilts of plants and the inciting bacteria. *Bact. Rev.*, **14** (3): *259—272.*
464. HILDEBRANDT, A. C. & A. J. RIKER, 1947. Influence of some growth-regulating substances on sunflower and tobacco tissue in vitro. *Amer. J. Bot.*, **34** (8): *421—427.*
465. HILDEBRANDT, A. C. & A. J. RIKER, 1948. The influence of various carbon compounds on the growth of marigold, Paris-daisy, periwinckle, sunflower and tobacco tissue in vitro. *Amer. J. Bot.*, **36** (1): *74—85.*
466. HILDEBRANDT, A. C., A. J. RIKER & B. M. DUGGAR, 1945. Growth in vitro of excised tobacco and sunflower tissue with different temperatures, hydrogen ion concentrations and amounts of sugar. *Amer. J. Bot.*, **32** (7): *357—361*, fig. 1.
467. HILDEBRANDT, A. C., A. J. RIKER & B. M. DUGGAR, 1946. The influence of composition of the medium on the growth in vitro of excised tobacco and sunflower tissue cultures. *Amer. J. Bot.*, **33**: *591—598.*
468. HILDEBRANDT, A. C., A. J. RIKER & B. M. DUGGAR, 1946. Influence of crown-gall bacterial products, crown-gall tissue extracts and yeast extract on growth in vitro of excised tobacco and sunflower tissue. *CancerRes.*, **6** (7): *368—377·*
469. HILDEBRANDT, E. M. 1940. Cane gall of brambles caused by *Phytomonas rubi*, sp. nov. *J. agric. Res.*, **61** (9): *685—696*, fig. 3.

356

470. HILDEBRANDT, E. M. 1942. A micrurgical study of crown-gall infection in tobacco. *J. agric. Res.*, **85**: *45—59*.
471. HILDEBRANDT, E. M. & L. M. MASSEY, 1942. Crown-gall on the weed *Malva rotundifolia*. *U.S. Dep. Agric. Plant Diseases Rep.*, **26** (1): *23*.
472. HILL, J. B. 1928. The migration of *Bacterium tumefaciens* in the tissues of the tomato plant. *Phytopathology*, **18**: *553—563*.
473. HILTNER, L. & K. STRÖMER, 1903. Neue Untersuchungen über die Wurzelknöllchen der Leguminosen und deren Erreger. *Arb. biol. Abt. Land-forstw. Kais. Gesundheits*, Berlin; 3 (3): *151—307* fig. 5 pl. iv.
474. HOCQUETTE, M. 1930. Evolution du noyau dans les cellules bactérifères des nodosites d'*Ornithopus perpusillus* pendant les phénomènes d'infection et de digestion intracellulaire. *C. R. Acad. Sci. Paris*, **191** (25): *1363—1365*.
475. HODGSON, R. *et al.* 1945. Polysaccharide production by virulent and attenuated crown-gall bacteria. *J. biol. Chem.*, **158**: *89—100*.
476. HODGSON, R., W. H. PETERSON & A. J. RIKER, 1949. The toxicity of polysaccharides and other large molecules to tomato cuttings. *Phytopathology*, **39**: *47—62*.
477. HODGSON, R., A. J. RIKER & W. H. PETERSON, 1947. A wilt-inducing toxic substance from crown-gall. *Phytopathology*, **37** (5): *301—318*, fig. 3.
478. HODGSON, W. E. H. 1931. The stem and bulb eelworm *Tylenchus dispasci* (Kühn) Bastian: A further contribution to our knowledge of the biologic strains of the nematode. *Ann. appl. Biol.*, **18**: *83—97*.
479. HOFMEISTER, W. 1867. Allgemeine Morphologie der Gewächse.
480. HOLMES, B. & A. PIRIE, 1937. Biochemistry and pathogenic viruses. Perspectives in Chemistry. Cambridge.
481. HOPKINS, E. W., W. H. PETERSON & E. B. FRED, 1930. Compositions of the gum produced by root nodule bacteria. *J. Amer. chem. Soc.*, **52**: *3659—3668*.
482. HONIG, F. 1931. Der Kohlkopferreger *(Plasmodiophora brassicae* Wor.*)*. Eine Monographie. *Gartenbauwissenschaft*, **5**: *116—225*, fig. 11.
483. HOPP, H. H. 1955. Wirkungen von Blattreblauspeichel auf Pflanzengewebe. *Weinbau Wiss. Beihefte*. 9: *9—23*.
484. HOUARD, C. 1901. Quelques notes sur les zoocécidies de l'*Artemisia herba-alba* Asso. *Bull. Soc. ent. France*, *92—93*, *1—3*.
485. HOUARD, C. 1903. Recherches anatomiques sur les galles des tiges: Pleurocécidies. *Bull. Sci. France & Belgique*, **38**: *140—419*, fig. 394.
486. HOUARD, C. 1903. Recherches sur la nutrition des tissus dans les galles des tiges. *C. R. Acad. Sci. Paris*, **136**: *1489*.
487. HOUARD, C. 1904. Recherches anatomiques sur les galles des tiges: Acrocécidies. *Ann. Soc. Nat.*, (8) **20**: *289—384*, fig. 189.
488. HOUARD, C. 1904. Les galles laterales des tiges. *Marcellia*, 3: *126—146*.
489. HOUARD, C. 1904. Caractères morphologiques des acrocécidies caulinaires. *C. R. Acad. Sci. Paris*, 98 (1) (2/3): *102*.
490. HOUARD, C. 1905. Variation des caractères histologiques des feuilles dans les galles du *Juniperus oxycedrus*. *C. R. Acad. Sci. Paris*, **140**: *1412—1414*.
491. HOUARD, C. 1905. Sur l'accentuation des caractères alpines des feuilles dans les galles des Génévriers. *C. R. Acad. Sci. Paris*, 2: *3*.
492. HOUARD, C. 1905. Caractères morphologiques et anatomiques des Dipterocécidies des Génévriers. *Rev. gén. Bot.*, **17**: *198—222*, fig. 1—46.
493. HOUARD, C. 1905. Recherches anatomiques sur les Dipterocécidies des Génévriers. *Ann. Sci. nat., Bot.*, (9) **1**: *67—100*, fig. 1—59, pl. i.

494. HOUARD, C. 1905. Les galles de l'Afrique Occidentale Française. II. *Marcellia*, **4**: *106*.
495. HOUARD, C. 1906. Les galles de l'Afrique Occidentale Française. III. Cécidies du *Dialium nitidum* Guill. et Perr. IV. Cécidies de *Khaja*, de *Parinarium* et de deux Graminées. *Marcellia*, **5**: *3—22*, fig. 1—18 1—23.
496. HOUARD, C. 1906. Sur les caractères histologiques d'une cécidie de *Cissus discolor* produite par l'*Heterodera radicicola* Greeff. *C. R. Ass. France Avance Sci., Congr. Lyons, 447—453*, fig. 7.
497. HOUARD, C. 1906. Sur une Coléopterocécidie du Maroc. *Marcellia*, **5**: *32—38*, fig. 1—8.
498. HOUARD, C. 1907. Sur les zoocécidies des Muscinées. *Rev. bret. bot. Rennes*, **2**: *61*.
499. HOUARD, C. 1908/1913. Les Zoocécidies des plantes d'Europe et du Bassin de la Méditerranée. **1**: *1—570*, fig. 824, pl. i (1908); **2**: *571—1248*, fig. 825—1365, pl. ii (1909); **3**: *1249—1560*, fig. 1366—1566, pl. iii (1913).
500. HOUARD, C. 1910. Les Zoocécidies des Salsolacées de la Tunisie. *C. R. Ass. Franç. Avance Sci. Paris*, Resumés, pl. iii.
501. HOUARD, C. 1911. Action de cécidozoaires externes, appartenant au genre *Asterolecanium*, sur les tissus de quelques tiges. *Marcellia*, **10**: *3—25*, fig. 21.
502. HOUARD, C. 1911. Les Zoocécidies des Crucifères de la Tunisie. *C. R. Ass. franç. Sci.*, **108**: *495—499*, fig. 1—12.
503. HOUARD, C. 1911. Les Zoocécidies de la Tunisie. *Marcellia*, **10**: *160—184*.
504. HOUARD, C. 1912. Les Zoocécidies du Nord de l'Afrique. *Ann. Soc. Ent. France*, **81**: *1—236*, fig. 427, pl. ii.
505. HOUARD, C. 1912. Les galles de l'Afrique Occidentale Française. V. Cécidies nouvelles. *Marcellia*, **11**: *176—201*, fig. 121.
506. HOUARD, C. 1912. Sur les zoocécidies des Cryptogames. *Bull. Soc. Linn. Normandie*, (6) **4**: *107—118*, fig. 6, pl. i.
507. HOUARD, C. 1913. Recherches anatomiques sur les cécidies foliaires marginales. *Marcellia*, **12**: *124—144*, fig. 1—15.
508. HOUARD, C. 1913. Les collections cécidologiques du laboratoire d'Entomologie du Museum d'Histoire Naturelle de Paris: Galles de Burseraceae. *Marcellia*, **12**: *57—75*, fig. 86.
509. HOUARD, C. 1913. Les galles de l'Afrique Occidentale Française. VI. Cécidies du Haut-Sénégal-Niger. *Marcellia*, **12**: *76—101*, fig. 88.
510. HOUARD, C. 1913. Les collections cécidologiques du Laboratoire d'Entomologie du Museum d'Histoire Naturelle de Paris: Galles d'Afrique et d'Asie. *Marcellia*, **12**: *102—117*, fig. 38.
511. HOUARD, C. 1914. Les collections cécidologiques du Laboratoire d'Entomologie du Museum d'Histoire Naturelle de Paris: Galles de Nouvelle-Calédonie (Premier Mem.). *Marcellia*, **14**: *143—182*, fig. 142; (2 Mem.) *ibidem*, **16**: *3—66*, fig. 143—177 (1917).
512. HOUARD, C. 1915. Les collections cécidologiques du Laboratoire d'Entomologie du Museum d'Histoire Naturelle de Paris: Galles du Congo Française. *Marcellia*, **14**: *14—71*, fig. 171.
513. HOUARD, C. 1916. Caractères morphologiques et anatomiques des zoocécidies des Bruyeres. *Marcellia*, **15**: *3—57*, fig. 76.
514. HOUARD, C. 1917. Les collections cécidologiques du Laboratoire d'Entomologie du Museum d'Histoire Naturelle de Paris: Galles de l'Ancien Continent extra Européenne. *Marcellia*. **16**: *79—102*.
515. HOUARD, C. 1917/1918. Galles d'Europe. *Marcellia*, **16**: *108—125*, fig. 14 (1917); **17**: *93—113*, fig. 16 (1918).

516. HOUARD, C. 1918. Les collections cécidologiques du Laboratoire d'Entomologie du Museum d'Histoire Naturelle de Paris: L'Herbier de Galles du Dr Giraud. *Marcellia*, **17**: *3—56*, fig. 47.

517. HOUARD, C. 1918. Les collections cécidologiques du Laboratoire d'Entomologie du Museum d'Histoire Naturelle de Paris: Galles du Nord de l'Afrique. *Marcellia*, **17**: *114—148*, fig. 33.

518. HOUARD, C. 1919. Les collections cécidologiques du Laboratoire d'Entomologie du Museum d'Histoire Naturelle de Paris: L'Herbier de Galles de C. Houard. *Marcellia*, **18**: *3—189*.

519. HOUARD, C. 1920. Les collections cécidologiques du Laboratoire d'Entomologie du Museum d'Histoire Naturelle de Paris: Galles du Madagascar. *Marcellia*, **19**: *34—46*, fig. 27.

520. HOUARD, C. 1922/1923. Les Zoocécidies des Plantes d'Afrique, d'Asie et d'Océanie. 1: *1—498*, fig. 1—1049, pl. i; 2: *499—1058*, fig. 1050—1909.

521. HOUARD, C. 1923. Les collections cécidologiques du Laboratoire d'Entomologie du Museum d'Histoire Naturelle de Paris: Galles de la Guyane Française (I Mem.) *Marcellia*, **20**: *3—24*, fig. 65.

522. HOUARD, C. 1923. Les collections cécidologiques du Laboratoire d'Entomologie du Museum d'Histoire Naturelle de Paris: Galles du Maroc et de l'Algérie. *Marcellia*, **20**: *122—162*, fig. 58.

523. HOUARD, C. 1924. Les collections cécidologiques du Laboratoire d'Entomologie du Museum d'Histoire Naturelle de Paris: Galles de la Guyane Française (II Mem.) *Marcellia*, **21**: *97—127*.

524. HOUARD, C. 1924. Les collections cécidologiques du Laboratoire d'Entomologie du Museum d'Histoire Naturelle de Paris: Galles de Nouvelle-Calédonie. *Marcellia*, **21**: *59—93*.

525. HOUARD, C. 1925. Les collections cécidologiques du Laboratoire d'Entomologie du Museum d'Histoire Naturelle de Paris: Galles de l' Afrique Equatoriale Française (Gabon et Caméroun). *Marcellia*, **22**: *33—49*.

526. HOUARD, C. 1925. Les collections cécidologiques du Laboratoire d'Entomologie du Museum d'Histoire Naturelle de Paris: Galles de l'Europe Occidentale. *Marcellia*, **22**: *3—24*, fig. 16.

527. HOUARD, C. 1926. Les collections cécidologiques du Laboratoire d'Entomologie du Museum d'Histoire Naturelle de Paris: Galles de l'Asie Orientale (Inde, Indo-Chine, Chine). *Marcellia*, **23**: *3—82*, fig. 182.

528. HOUARD, C. 1926. Les collections cécidologiques du Laboratoire d'Entomologie du Museum d'Histoire Naturelle de Paris: Galles de l'Amérique Tropicale. *Marcellia*, **23**: *95—124*, fig. 67.

529. HOUARD, C. 1927. Les collections cécidologiques du Laboratoire d'Entomologie du Museum d'Histoire Naturelle de Paris: Galles des Etats-Unis. *Marcellia*, **24**: *99—141*, fig. 161.

530. HOUARD, C. 1927. Les collections cécidologiques du Laboratoire d'Entomologie du Museum d'Histoire Naturelle de Paris: Galles du Méxique. *Marcellia*, **24**: *30—81*, fig. 186.

531. HOUARD, C. 1929. Les collections cécidologiques du Laboratoire d'Entomologie du Museum d'Histoire Naturelle de Paris: Galles de Nouvelle-Calédonie. *Marcellia*, **26**: *17—30*, fig. 46.

532. HOUARD, C. 1929. Les collections cécidologiques du Laboratoire d'Entomologie du Museum d'Histoire Naturelle de Paris: Galles de l'Afrique du Nord. *Marcellia*, **26**: *37—48*, fig. 6; *49—71*, fig. 29.

533. HOUARD, C. 1933. Les Zoocécidies des plantes de l'Amérique de sud et l'Amérique Centrale. Paris: Hermann & Co. *1—519*,fig. 1027.

534. HOUARD, C. 1933. Les collections cécidologiques du Laboratoire d'entomologie du Museum d'Histoire Naturelle de Paris: Galles des Etats-Unis. *Marcellia*, **28**: *33—96*, fig. 140 .

535. HOUARD, C. 1933. Les collections cécidologiques du Laboratoire d'Entomologie du Museum d'Histoire Naturelle de Paris: Galles des Etats-Unis. *Marcellia*, **28**: *97—192*, fig. 225.
536. HOUGH, J. S. 1954. The future of gall-induction studies. *C. R. Seances Rep. Comm. VIIIe Congr. int. Bot. Paris*, **7**/8: *217—220*.
537. HUMM, H. J. 1944. Bacterial leaf nodules. *J. N. Y. bot. Gard.* **45** (537): *193—199*, fig. 3.
538. HYDE, K. C. 1922. Anatomy of a gall on *Populus trichocarpa*. *Bot. Gaz.*, **74**: *185—196*, pl. iv.
539. INOUYE, M. 1953. Monographische Studie über die japanischen Koniferengallenläuse (Adelgidae). *Bull. Sapporo Branch Govt. Forest exp. Sta.*, Sapporo, Hokkaido, **15**: *1—91*, fig. 90.
540. ISRAILSKII, V. P. 1926. d'Herelle phenomenon and plant cancer (In Russian with German summary). *Rec. Bact. Agron. Sta. Moscow*, **1926** (24): *143—157*.
541. ISRAILSKII, V. P. 1926. Bacteriophage und Pflanzenkrebs. *Zbl. Bakt.*, (2) **67** (8/15): *236—242*, pl. i.
542. ISRAILSKII, V. P. 1927. Bacteriophage und Pflanzenkrebs. *Zbl. Bakt.*, (2) **71** (8/14): *302—311*.
543. ITANO, A. & A. MATSUMARA, 1936. Studies on the nodule bacteria. VI. Influence of different parts of the plant on the growth of nodule bacteria. *Ber. Chara Inst. Landw. Forsch.*, Kurashiki **7**: *379—401, 459—577*, pl. i.
544. ITANO, A. & A. MATSUMARA, 1937. Studies on the nodule bacteria. VIII. Influence of ash-content of the nodule on the growth of the nodule bacteria with special reference to the titanium salts. *Ber. Chara Inst. Landw. Forsch.* Kurashiki **7**: *501—1515;* IX. On the electrical properties of the accessory substance. *ibidem*, **7**: *517—527*, fig. 1.
545. IVANOFF, S. S. & A. J. RIKER, 1930. Studies on the movement of the crown-gall organism within the stems of tomato plants. *Phytopathology*, **20**: *817—829*.
546. IVIMEY COOK, W. R. 1931. The life history of *Sorodiscus radicicolus*, sp. n. *Ann. Mycol.*, **29** (5/6): *313—324*, fig. 2, pl. ii.
547. JAMES, H. C. 1927. The life-history and biology of a British phytophagous Chalcidoid of the genus *Harmolita (Isosoma)*. *Ann. appl. Biol.*, **14** (1): *132—149*, fig. 12.
548. JENSEN, C. O. 1918. Undersogelser verdrorend nogle svulstlignende Dannelser hos plantes. *Meded. k. Vet. og Landbogiskoles Serumlb.* (Investigations upon certain tumor-like formations in plants) *Mede. kong. Veterin. og Landbojs. Arsskr.*, **45**: *91—143*, pl. i.
549. JENSEN, J. 1933. Leaf enations resulting from tobacco mosaic infection in certain species of *Nicotiana*. *Contr. Boyce Thompson Inst.*, **5**: *129—142*.
550. JIMBO, P. 1927. On the budding of nucleoli in the root nodule of *Wistaria*. *Bot. Mag.*, Tokyo **489**: *551—553*.
551. JIMBO, T. 1928. On the root gall of *Trachelospermum jasminoides* Lam., caused by a gall gnat. *Bot. Mag.*, Tokyo, **42**: *325—327*.
552. JOANNIS, J. DE 1922. Revision critique des espèces de Lepidoptères cécidogènes d'Europe et du Bassin de la Méditerranée. *Ann. Soc. ent. France*, **91**: *73—155*.
553. JÖRGENSEN, P. 1916. Zoocecidios argentinos. *Bol. Soc. Physis Buenos Aires*, **2**: *349—365*.
554. JÖRGENSEN, P. 1917. Zoocecidios argentinos. *Bol. Soc. Physis Buenos Aires*, **3**: *1—29*, pl. i—iii.
555. JOHNSON, J. 1942. Studies on the viroplasm hypothesis. *J. agric. Res.*, **64**: *443—454*.

556. JONES, K. L. 1944. Root nodules on *Zinnia* produced by nematodes. *Pap. Michigan Acad. Sci. Arts Letters*, **30**: *67—70*, pl. i.
557. JONES, S. G. 1947. An anatomical study of crown-gall tumors on Himalaya blackberry *(Rubus procerus)*. *Phytopathology*, **37** (9): *613—624*, fig. 8.
558. JUEL, H. O. 1912. Zur Kenntnis der Gattungen *Taphrina* und *Exobasidium. Svensk. Bot. Tidskr.*, **6** (28).
559. KARLING, J. S. 1928. A parasitic Chytrid causing cell hypertrophy in *Chara. Amer. J. Bot.*, **15** (8): *485—496*.
560. KARNY, H. 1911. Über Thripsgallen und Gallenthrips. *Zbl. Bakt.*, (2) **30**: *556—572*, fig. *1—30*.
561. KARNY, H. 1912. Gallenbewohnende Thysanopteren aus Java. *Marcellia*, **11**: *115—169*, fig. 5.
562. KARNY, H. 1913. Über gallenbewohnende Thysanopteren. *Verh. zool.-bot. Ges. Wien*, **63**: *5—12*.
563. KARNY, H. 1923. Gallenbewohnende Thysanopteren von Celebes und den Inseln südlich davon. *Treubia*. **3**: *300—326*.
564. KARNY, H. 1923. On two Tubulifera inhabiting *Acacia* galls in Egypt. *Bull. Soc. R. ent. Egypt*, **15**: *127—132* (1922).
565. KARNY, H. 1926. Phylogenetic considerations. In: DOCTERS VAN LEEUWEN, W. & J. DOCTERS VAN LEEUWEN-REIJNVAAN, The Zoocecidia of the Netherlands East-Indies, Batavia, *37—47*.
566. KARNY, H., W. & J. DOCTERS VAN LEEUWEN 1914. Beiträge zur Kenntnis der Gallen aus Java. 2te Mitteilung. Über die javanischen Thysanopterocecidien und deren Bewohner. *Z. wiss. Insektenbiol.*, **10** (6/12): *288—291, 355—369*.
567. KARSCH, JUN. F. 1880. Neue Zoocecidien und Cecidozoen. *Z. ges. Naturw.*, **53**: *286*.
568. KAUFMANN, F. 1928. Zur Tumefaciensfrage. *Z. Krebsforsch.*, **25** (2): *109—120*.
569. KAUFMANN, F. 1929. Über Veränderlichkeit von Tumefaciensbacillen. *Z. Krebsforsch.*, **26** (4): *330—332*.
570. KAULE, A. 1931. Die Cephalodien der Flechten. *Flora*, **26**: *1—44*, fig. 16.
571. KELLERMAN, W. A. & W. T. SWINGLE, 1889. New species of Kansas Fungi. *J. Mycol.*, **5**: *1*.
572. KELLY, S. M. 1944. Effect of different concentrations of synthetic auxins on decapitated sunflower stems. *Bull. Torrey bot. Club*, **71** (5): *549—554*.
573. KELLY, S. M. & L. M. BLACK, 1949. The origin, development and cell structure of a virus tumor in plants. *Amer. J. Bot.*, **36**: *65—73*.
574. KENDALL, J. 1930. The structure and development of certain Eriophyid galls. *Z. Parasitenk.*, **2** (4): *477—501*.
575. KENNEDY, J. S. 1951. A biological approach to plant viruses. *Nature*, **168** (4282): *890—894*.
576. KENT, G. C. 1937. Some physical, chemical and biological properties of specific bacteriophage of *Pseudomonas tumefaciens. Phytopathology*, **27**: *871—902*.
577. KENT, N. L. 1941. The influence of lithium salts on certain cultivated plants and their parasitic diseases. *Ann. appl. Biol.*, **28** (3): *189—209*.
578. KERNER, A. 1913. Pflanzenleben. 2 (3te Aufl.).
579. KESSLER, H. F. 1877/1878. Die Lebensgeschichte der auf *Ulmus campestris* Linn. vorkommenden Aphidenarten und die Entstehung der durch dieselben bewirkten Missbildungen an den Blättern. *Progr. höh. Bürgerschule Cassel*, 24/25 Ber. Ver. Naturk., Cassel, 1878. 1.
580. KESSLER, H. F. 1880. Neue Beobachtungen und Entdeckungen an den auf *Ulmus campestris* Linn. vorkommenden Aphiden-Arten. 26/27 Ber. Ver. Natur. Cassel, **57**.

581. KESSLER, H. F. 1881. Die auf *Populus nigra* Linn. und *Populus dilatata* Ait. vorkommenden Aphiden-Arten und die von denselben bewirkten Missbildungen. 28 *Ber. Ver. Naturk.* Cassel, **38**.
582. KIDD, J. G. 1946. Distinctive constituents of tumor cells and their possible relations to the phenomena of autonomy, anaplasia and cancer causation. *Cold Spring Harbour Symposia Quant. Biol.*, **11**: *94—112.*
583. KIEFFER, J. J. 1894. Sur la rôle de la spatule sternale chez les larves de Cécidomyes. *Ann. Soc. ent. France*, **63**: *37—44.*
584. KIEFFER, J. J. 1903. Zur Lebensweise einiger *Synergus*-Arten. *Allg. Z. Ent.*, **8**: *122—123.*
585. KIEFFER, J. J. 1905. Étude sur de nouveaux insectes et phytoptides gallicoles du Bengale. *Ann. Soc. Sci. Bruxelles*, **29** (2): *143—200*, fig. 15, pl. i—iii.
586. KIEFFER, J. J. 1908. Descriptions de quelques galles et d'insectes gallicoles du Columbie. *Marcellia*, **7**: *140—142*, fig. 1.
587. KIEFFER, J. J. 1908. Descriptions de galles et d'insectes gallicoles d' Asie. *Marcellia*, **7**: *149—167*, fig. 1—4, pl. ii—iv.
588. KIEFFER, J. J. 1913. Cecidomyiidae. Genera Insectorum, fascicle **152**: *1—346.*
589. KIEFFER, J. J. & G. CECCONI, 1906. Un nuovo Dittero galligeno su fogli di *Mangifera indica*. *Marcellia*, **5**: *135—136*, fig. 1—3.
590. KIEFFER, J. J. & P. HERBST, 1906. Descriptions de galles et d'insectes gallicoles du Chile. *Ann. Soc. Sci. Bruxelles*, **30**: *223—236*, fig. 1—5, pl. i.
591. KIEFFER, J. J. & P. HERBST, 1909. Über einige neue Gallen und Gallenerzeuger aus Chile. *Zbl. Bakt.*, (2) **23**: *119—146*, fig. 1—7.
592. KIEFFER, J. J. & P. HERBST, 1911. Über Gallen und Gallentiere aus Chile. *Zbl. Bakt.*, (2) **29**: *696—703*, fig. 1—8.
593. KIEFFER, J. J. & P. JÖRGENSEN, 1910. Gallen und Gallentiere aus Argentinien. *Zbl. Bakt.*, (2) **27**: *326—441*, fig. 1—61.
594. KINSEY, A. C. 1920. Life histories of American Cynipidae. *Bull. Amer. Mus. Nat. Hist.*, **42**: *319—357*, pl. xxviii—xxxi.
595. KINSEY, A. C. 1920. Phylogeny of Cynipid genera and biological characteristics. *Bull. Amer. Mus. Nat. Hist.*, **42**: *357—402.*
596. KINSEY, A. C. 1920. New species and synonyms of American Cynipidae *Amer. Mus. nat. Hist. Bull.*, **42**: *293—317*, pl. xx—xxvii.
597. KINSEY, A. C. 1920. The life-history of American Cynipidae. *Amer. Mus. nat. Hist. Bull.*, **42**: *319—357*, pl. xxviii—xxxi.
598. KISSER, J. 1939. Über die Wirkungen carcinogener Substanzen bei Pflanzen. *Ber. dtsch. bot. Ges.*, **57** (10): *506—515.*
599. KLEIN, G., KISSNER E. & W. LIESE, 1932. Beiträge zum Chemismus pflanzlicher Tumoren II. Wasserstoffionenkonzentration in pflanzlichen Tumoren. III. Der Katalasegehalt von pflanzlichen Tumoren im Vergleich zum Katalasegehalt gesunden Pflanzengewebes. *Biochem. Z.*, **254**: *251—285*, fig. 18.
600. KLEIN, G. KISSNER E. & W. LIESE, 1933. Beiträge zum Chemismus pflanzlicher Tumoren. IV. Über Peroxydase in pflanzlichen Tumoren. *Biochem. Z.*, **267**: *22—25*, pl. iii.
601. KLEIN, R. M. & A. C. BRAUN, 1960. On the presumed sterile induction of plant tumors. *Science*, **131** (3413): *1612.*
602. KLICA, J. 1924. Histologische Bemerkungen zu einigen Gallen. Beiträge zur Kenntnis der Gallen. *Vestnik Kral. cesk. spolecn. nauk. Prag.* (1923) 2: *1—13*, fig. 17.
603. KLOFT, W. 1956/1957. Untersuchungen über pflanzensaugende Insekten und Reaktionen des Wirtspflanzengewebes. *Ber. physik.-medizin. Ges. Würzburg*, (NF) **68**: *64—72.*

362

604. KLOFT, W. 1960. Wechselwirkungen zwischen pflanzensaugenden Insekten und den von ihnen besogenen Pflanzengeweben. *Z. angew. Ent.*, **45**: *337—381*; 46: 42—70.
605. KNY, L. 1877. Über künstliche Verdoppelung des Leitbündelkreises im Stamm der Dikotylen. *S.B. Ges. naturf. Freunde Berlin*, **189**.
606. KÖHLER, E. 1936. Untersuchungen über *Synchytrium endobioticum*. *Z. Pflanzenk.*, **46** (3/4): *214—223*.
607. KOLLER, P. C. 1947. Abnormal mitosis in tumors. *Brit. J. Cancer*, 1: *38—47*.
608. KOMURO, H. 1924. The cells of *Vicia faba* modified by Röntgen rays and their resemblance to malignant tumor cells with cytological observations of tumors. *Jap. J. Bot.*, **2**: *133—156*.
609. KOMURO, H. 1928. Über den Ort der in dem Wurzelspitzengewebe von *Vicia faba* gebildeten Röntgengeschwulst. *Gann*, Tokyo, 22 (1): *4—14*, pl. ii (In Japanese, with German summary).
610. KOMURO, H. 1931. Betrachtungen über die zytologischen Veränderungen in den in Kohlenteerlösungen getauchten Wurzelspitzen junger Pflanzen. *Proc. Imp. Acad. Tokyo*, 7: *110—113*.
611. KONDO, T. 1931. Zur Kenntnis des N-Gehaltes des Mykorrhiza-Knöllchen von *Podocarpus macrophyla* D. Don. *Bot. Mag. Tokyo*, **45**: *495—501*.
612. KORSAKOVA, M. P. & G. W. LOPATINA, 1934. Wechselwirkung der Knöllchenbakterien und Leguminosen. I. Stickstoffaufnahme durch die Knöllchenbakterien. *Mikrobiologie*, 3: *204—220*, pl. iii.
613. KOSTOFF, D. 1930. Cytology of Nematode gall on *Nicotiana* roots. *Zbl. Bakt.*, (2) **81**: *86—91*.
614. KOSTOFF, D. 1930. Tumors and other malformations on certain *Nicotiana* hybrids. *Zbl. Bakt.*, (2) **81**: *244—260*.
615. KOSTOFF, D. 1930. Protoplasmic viscosity in plants. IV. Cytoplasmic viscosity in tumors of *Nicotiana* hybrids. *Protoplasma*, 11: *193 —195*, fig. 1.
616. KOSTOFF, D. 1933. Tumor problem in the light of researches on plant tumors and galls and its relation to the problem of mutation. A critical review from biophysical, biochemical and cytogenetical point of view. *Protoplasma*, **20** (3): *440—546*.
617. KOSTOFF, D. 1935. Heritable tumors in plants experimentally produced. *Genetics*, **17**: *367—376*, fig. 3.
618. KOSTOFF, D. & J. KENDALL, 1920. Irregular meiosis in *Lycium halmifolium* Mill., produced by gall mites *(Eriophyes)*. *J. Genet.*, **21**: *111—115*.
619. KOSTOFF, D. & J. KENDALL, 1929. Studies on the structure and development of certain cynipid galls. *Biol. Bull.*, **56**: *402—459*.
620. KOSTOFF, D. & J. KENDALL, 1930. Cytology of Nematode galls on *Nicotiana* roots. *(vide* 613).
621. KOSTOFF, D. & J. KENDALL. 1930. Protoplasmic viscosity in plants. III. Cytoplasmic viscosity in cynipid galls. *Protoplasma*, 11: *190— 192*, fig. 2.
622. KOSTOFF, D. & J. KENDALL, 1933. Studies on plant tumors and polyploidy produced by bacteria and other agents. *Arch. Mikrobiol.*, 4: *487—508*, fig. 15.
623. KOSTRITSKY, M. *et al.* 1924. *Bacterium tumefaciens* chez la chenille de *Galleria melonella*. *C. R. Acad. Sci. Paris*, **179**: *225—227*.
624. KRAUS, E. J., N. A. BROWN & K. C. HAMNER, 1936. Histological reactions of bean plants to indole-acetic acid. *Bot. Gaz.*, **98**: *370—420*.
625. KRAUSSE, A. 1916. Zur Systematik und Naturgeschichte der Psylliden (Springläuse) und speziell von *Psyllopsis fraxini* Linn. *Zbl. Bakt.*. (2) **46**: *80*.

626. KREBBER, O. 1932. Untersuchungen über die Wurzelknöllchen der Erle. *Arch. Mikrobiol.*, **3**: *588—608*, fig. 2.

627. KRÜGER, R. 1913. Beiträge zur Artenfrage der Knöllchenbakterien einiger Leguminosen. Leipzig. *56.*

628. KULESCHA, Z. 1947. Production de tumeurs par l'inoculation de *Phytomonas tumefaciens* dans des fragments de parenchyme vasculaire de topinambour et de liber de carotte cultivés in vitro. *C. R. Soc. Biol. Paris*,**141**: *24—25.*

629. KULESCHA, Z. 1947. Comparison entre la structure anatomique des néoformations provoquées par l'action de l'acide indole-acétique et du *Phytomonas tumefaciens* sur les fragments de parenchyme vasculaire de topinambour cultivés in vitro. *C. R. Soc. Biol. Paris,* **141**: *358—360.*

630. KULESCHA, Z. 1947. Comparison entre l'action du *Phytomonas tumefaciens* et celle de l'acide indole-acétique sur des fragments de parenchyme vasculaire de topinambour cultivés in vitro. *C. R. Soc. Biol. Paris*, **141**: *232—234.*

631. KULESCHA, Z. 1949. Relation entre le pouvoir de proliferation spontanée des tissus de topinambour et leur teneur en substance de croissance. *C. R. Soc. Biol. Paris*, **143**: *354—355.*

632. KULESCHA, Z. & R. GAUTHERET, 1948. Sur l'élaboration des substances de croissance par 3 types des culture de scorsonère: cultures normales, cultures de crown-gall et cultures accoutumées à l'hétéro-auxine. *C. R. Acad. Sci. Paris*, **227**: *292—294.*

633. KUNKEL, L. O. 1918. Tissue invasion by *Plasmodiophora brassicae. J. agric. Res.*, **14** (12): *543—572*, pl. lxi—lxxx.

634. KÜSTENMACHER, M. 1895. Beiträge zur Kenntnis der Gallenbildungen mit Berücksichtigung der Gerbstoffe. *Jb. wiss. Bot.*, **26**: *82.*

635. KÜSTER, E. 1903. Pathologische Pflanzenanatomie. Gustav Fischer, Jena. pp. vii-312, fig. 121.

636. KÜSTER, E. Zur Morphologie der von *Eriophyes dispar* erzeugten Galle. *Marcellia*, **3**: *60—63.*

637. KÜSTER, E. 1904. Vergleichende Betrachtung über die abnormalen Gewebe der Tiere und Pflanzen. *Münchn. med. Wschr.*, **46**: *110.*

638. KÜSTER, E. 1906. Histologische und experimentale Untersuchungen über Intumeszenzen. *Flora*, **96.**

639. KÜSTER, E. 1911. Die Gallen der Pflanzen. Ein Lehrbuch der Botaniker und Entomologen. Leipzig. pp. 437, fig. 158.

640. KÜSTER, E. 1913. Neue Resultate und Streitfragen der allgemeinen Cecidologie. In: Abderhaldens Fortschritte Naturw. Forschung, **8**: *115.*

641. KÜSTER, E. 1925. Pathologische Pflanzenanatomie in ihren Grundzügen. Gustav Fischer Jena. 3te Auflage, pp. 558, fig. 285. (Gallen on pp. *184—242, 529—532*; Histogenese der pathologischen Gewebe pp. *245—404*; Entwicklungsmechanik der pathologischen Gewebe pp. *405—510*; Intumeszenzen pp. *60—72*; Wundgewebe und Regeneration pp. *76—183).*

642. KÜSTER, E. 1926. Regenerationserscheinungen an Bakteriengallen. *Flora*, (NF) **20**: *179—197.*

643. KÜSTER, E. 1930. Anatomie der Gallen. In: Schröders Handbuch der Entomologie, **1**: *1—197*, fig. 108.

644. KÜSTER, E. 1930. Über verirrte Gallen. *Biol. Zbl.*, *685—703*, fig. 13.

645. KÜSTER, E. 1930. Cecidologische Beobachtungen auf der Insel Sylt. *Marcellia*, **27**: *3—8.*

646. KÜSTER, E. 1937. On the histological structure of some Australian galls. *Proc. Linn. Soc. N. S. Wales*, Sydney, **52** (1/2): *57—64*, fig. 14.

646a. KUSANO, 1907. On the cytology of *Synchytrium. Zbl. Bakt.*, (2) **19**: *538*.

647. LABOULBENE, A. 1892. Essai d'une théorie sur la production des diverses galles végétales. *C. R. Acad. Sci. Paris*, **114**: *720*.

648. LACAZE-DUTHIERS, De H. 1853. Recherches pour servir à l'histoire des galles. *Ann. Sci. nat. Bot.*, (3) **19**: *273*.

649. LACEY, MARGARET S. 1939. Studies on bacteriosis. XXIV. Studies on a bacterium associated with leafy galls, fasciation and cauliflower disease of various plants. Part III. Further isolations, inoculations experiments and cultural studies. *Ann. appl. Biol.*, **26** (2): *252—278*, pl. xvii—xviii.

650. LAGERHEIM, G. 1900. Mycologische Notizen. III. *Bih. Svensk. Vet. Akad. Handl.*, **26**: III, 4.

651. LAIRD, D. G. 1932. Bacteriophage and the root nodule bacteria. *Arch. Mikrobiol.*, **3**: *152—193*, fig. 9.

652. LA RUE, C. D. 1936. Intumescence on poplar leaves. III. The role of plant growth hormones in their production. *Amer. J. Bot.*, **23** (8): *520*.

653. LEACH, F. H. 1923. Jumping "seeds", plant growths that hop about like fleas. *J. Amer. Mus. nat. Hist.*, **23**: *295—300*.

654. LEACH, R. & C. SMEE. 1933. Gnarled stem canker of tea caused by the capsid bug *Helopeltis bergrothi. Ann. appl. Biol.*, **20** (4): *691—706*.

655. LE CERF, F. 1914. Sur une chenille de Lycaenide élevées dans des galles d'*Acacia* par des fourmis du genre *Cremastogaster. C. R. Acad. Sci. Paris*, **158**: *1127—1129*.

656. LECHTOVA, T. M. 1931. Étude sur les bactéries des Legumineuses et observations sur quelques Champignons parasites des nodosites. *Le Botaniste*, **23** (5/6): *301—530*, fig. 33, pl. xxiv—lxiv.

657. LECONTE, H. 1915. Le tubercule des Balanophoracees. *Bull. Soc. bot. France*, **62**: *216—225*, fig. 1—4.

658. LEMESLE, R. 1935. Mycocécidie florale produite par le *Fusarium moniliforme* Sh. *anthophilum* (A. Br.) Wor. sur le *Scabiosa succisa* Linn. *Rev. gén. Bot. Paris*, **47** (558): *337—362*, fig. 1—13, pl. i—iii.

659. LEMESLE, R. 1937. Mycocécidie florale produite par le *Fusarium moniliforme* Sh. *anthophilum* (A. Br.) Wor. sur le *Scabiosa succisa* Linn. *Ann. Sci. Nat. Bot. Paris*, (10) **19**: *341—350*, fig. 1—5, pl. i; *Ass. France Avance Sci., Chambery Bull. Soc. bot. France*, **84**: *141*.

660. LERA, F. B. 1941. Una interresante struttura nelle galle di *Cynips kollari* Htg. ed affini. *Marcellia*, **30**: *153—184*.

661. LEVIN, I., MICHAELIS & H. HIBBERT, 1943. Studies on plant tumors. IV. Oxidase in normal and tumor beet root tissue. *Arch. Biochem.*, **3** (2): *167—174*.

662. LEVIN, I. & M. LEVINE, 1918. Malignancy of the crown-gall and its analogy to animal cancer. *Proc. Soc. exp. Biol. Med.*, **16**: *21—22*.

663. LEVIN, I. & M. LEVINE, 1920. Malignancy of the crown-gall and its analogy to animal cancer. *J. Cancer Res.*, **5**: *243—260*, fig. 1—15.

664. LEVIN, I. & M. LEVINE, 1922. The action of buried tubes of radium emanation on neoplasia of plants. *J. Cancer Res.*, **7**: *163—170*.

665. LEVINE, M. 1919. Studies on plant cancers. I. The mechanism of the formation of the leafy crown-gall. *Bull. Torrey bot. Club*, **46**: *447 —452*.

665a. LEVINE, M. 1921. Studies on Plant cancer. II. The behaviour of crown-gall on the rubber plant *(Ficus elastica). Mycologia*, **13** (1): *1—11*. pl. i—ii.

666. LEVINE, M. 1921. Studies in plant cancers. III. The nature of the soil as a determining factor in health of the beet *Beta vulgaris* and its relation to the size and weight of the crown-gall produced by inoculation with *Bacterium tumefaciens. Amer. J. Bot.*, **7** (10): *507—525*, fig. 9.

667. LEVINE, M. 1922. Studies on plant cancers. IV. Effects of inoculating various quantities of different dilutions of *Bacterium tumefaciens* into the tobacco plant. *Phytopathology*, **12**: *56*.
668. LEVINE, M. 1923. Studies on plant cancers. V. Leafy crown gall on tobacco plants resulting from *Bacterium tumefaciens* inoculations. *Phytopathology*, **13**: *107—116*.
669. LEVINE, M. 1924. Crown-gall on *Bryophyllum calycinum*. *Bull. Torrey bot. Club*, **51**: *449—456*.
670. LEVINE, M. 1925. The socalled strands and secondary tumors in crown-gall disease. *Phytopathology*, **15**: *435—451*, pl. v.
671. LEVINE, M. 1925. Morphological changes in *Bacterium tumefaciens*. *Science*, **62**: *424*.
672. LEVINE, M. The effects of radium emanations on the crown-gall tissue. *Amer. J. Roentgenol.*, **14**: *221—233*.
673. LEVINE, M. 1929. The chromosome number in crown-gall and cancer tissue. *Phytopathology*, **19**: *97*.
674. LEVINE, M. 1929. A comparison of the behaviour of crown-gall and cancer transplants. *Bull. Torrey bot. Club*, **56**: *299—314*.
675. LEVINE, M. 1930. Chromosome numbers in cancer tissue of man, of rodent, or bird and in crown-gall tissues of plants. *J. Cancer Res.*, **14**: *400—425*.
676. LEVINE, M. 1931. Studies in the cytology of cancer. *Amer. J. Cancer*, **15**: *1410—1494*, fig. 63.
677. LEVINE, M. 1934. A preliminary report on plants treated with carcinogenic agents of animals. *Bull. Torrey bot. Club*, **61**: *103—118*.
678, LEVINE, M. 1936. Plant tumors and their relation to cancer. *Bot. Rev.*, Lancaster Pa., 2 (9): *439—455*.
679. LEVINE, M. 1936. Studies on *Bacterium tumefaciens* in culture media. *Amer. J. Bot.*, **23**: *191—198*.
680. LEVINE, M. 1936. The response of plants to localised application of various chemical agents. *Bull. Torrey bot. Club*, **63**: *177—199*.
681. LEVINE, M. 1940. Plant responses to carcinogenic agents and growth substances: Their relation to crown-gall and cancer. *Bull. Torrey bot. Club*, **67**: *199—226*.
682. LEVINE, M. 1942. Formative influences of carcinogenic substances. *Cold Spring Harbour Symposia on quant. Biol.*, **10**: *70—78*.
683. LEVINE, M. 1947. Crown-gall disease on rubarb. *Bull. Torrey bot. Club*, **74** (2): *115—120*.
684. LEVINE, M. 1950. The growth of normal plant tissue in vitro as affected by chemical carcinogens and plant growth substances. I. The culture of carrot tap root meristem. *Amer. J. Bot.*, **37**: *445—458*.
685. LEVINE, M. 1950. The growth of normal plant tissues in vitro as affected by chemical carcinogens and plant growth substances. II. The cytology of the carrot root tissue. *J. nat. Cancer Inst.*, **10**: *1005—1043*.
686. LEVINE, M. 1950. The growth of normal plant tissues in vitro as affected by chemical carcinogens and plant growth substances. III. The culture of sunflower and tobacco stem segments. *Bull. Torrey bot. Club*, **77**: *110—132*.
687. LEVINE, M. 1951. The effect of growth substances and chemical carcinogens on fibrous roots or carrot tissue grown in vitro. *Amer. J. Bot.*, **38**: *132—138*.
688. LEVINE, M. & E. CHARGAFF, 1937. The response of plants to chemical fractions of *Bacterium tumefaciens*. *Amer. J. Bot.*, **24**: *461—472*.
689. LEWIS, I. F. & L. WALTON, 1947. Initiation of the cone gall of witch hazel. *Science*, **106** (2757): *419—420*.

366

690. LEWIS, K. H. & E. MC COY, 1933. Root nodule formation on the garden bean, studied by a technique of tissue culture. *Bot. Gaz.*, **95** (2): *316—329*.

691. LEWIS, W. H. 1935. Normal and malignant cells. *Science*, **81** (2110): *345—355*.

692. LEWIS, W. H. 1937. The cultivation and cytology of cancer cells. *Occasional Papers AAAS*, **4**: *119—120*.

693. LIESKE, R. 1928. Untersuchungen über die Krebskrankheiten bei Pflanzen, Tieren und Menschen. *Zbl. Bakt.*, (1) **108**: *118—146*, pl. i—iii.

694. LIESKE, R. 1929. Untersuchungen über die Krebskrankheiten bei Pflanzen, Tieren und Menschen. II. Teil. *Zbl. Bakt.*, (1) **111** (6/8): *419—425*.

695. LINDFORD, M. B. 1941. Parasitism of the root-knot Nematode in leaves and stems. *Phytopathology*, **31** (7): *634—648*, fig. 9.

696. LINDINGER. L. 1912. Eine weitverbreitete Gallenerzeugende Schildlaus. *Marcellia*, **11**: *3—6*.

697. LINGELSHEIM, A. 1916. Durch Hemipteren verursachte Missbildungen einiger Pflanzen. *Z. Pflanzenkrankh.*, **26**: *378—383*, fig. 3.

698. LINK, GEORGE K. K. & V. EGGERS, 1941. Hyperauxinity in crown-gall of tomato. *Bot. Gaz.*, **103** (1): *87—106*, fig. 1.

699. LINK, GEORGE K. K. & V. EGGERS, 1943. Enhanced auxinic activity of tomato tissues in presence of l-tryptophan. *Bot. Gaz.*, **105** (2): *282—284*.

700. LINK, GEORGE K. K. & H. W. WILCOX, 1936. Gall production in high and low carbohydrate tomato plants. *Phytopathology*, **26**: *100*.

701. LINK, GEORGE K. K. & H. W. WILCOX, 1937. Tumor production by hormones from *Phytomonas tumefaciens*. *Science*, **86**: *126—127*.

702. LINK, GEORGE K. K., H. W. WILCOX & ADEL DE S. LINK, 1937. Responses of bean and tomato to *Phytomonas tumefaciens*, *Phytomonas tumefaciens* extract, indole-acetic acid and wounding. *Bot. Gaz.*, **98** (4): *816—867*, fig. 22.

703. LINK, GEORGE K. K. & J. E. MOULTON, 1941. Use of frozen vacuum-dried material in auxin and other chemical analysis of plant organs: Its extraction with dry ether. *Bot. Gaz.*, **102**: *590—601*.

704. LIPETZ, J. 1959. A possible role of indole-acetic acid oxidase in crown-gall tumor induction. *Nature*, **184**: *1076—1077*.

705. LISO, A. 1928. Sul significator biologici di gemmazioni normotypiche dal meristema di tumore sperimentale. *Pathologica*, **20** (436): *61—65*.

706. LITTLE, C. C. & L. C. STRONG, 1924. Genetic studies on the transplantation of two adenocarcinomata. *J. exp. Zool.*, **41**: *93—114*.

707. LIZER, C. 1916. Sobre una nuova hemipterocecidia Argentina. *Prim. Renn. Soc. Arg. Cienc. Nat. Tucuman* (Zool.): *383—388*, fig. 6.

708. LOCKE, S. B. 1939. Production of growth substances on peptone broth by crown-gall bacteria and related non-gall-forming organisms. *J. agric. Res.*, **59**: *519—525*.

709. LOCKE, S. B., A. J. RIKER & B. M. DUGGAR, 1937. A growth hormone in the development of crown-gall. *Phytopathology*, **27**: *134* (Abstract).

710. LOCKE, S. B., A. J. RIKER & B. M. DUGGAR, 1938. Growth substance and the development of crown-gall. *J. agric. Res.*, **57**: *21—39*.

711. LOCKE, S. B., A. J. RIKER & B. M. DUGGAR, 1939. The nature of growth substances originating in crown-gall tissue. *J. agric. Res.*, **59**: *535—539*.

712. LÖHNIS, M. P. 1930. Investigations upon the ineffectiveness of root nodules on Leguminosae. *Zbl. Bakt.*, (2) **80**: *342—368*, fig. 2, pl. iii.

713. LONGLEY, B. J., T. O. BERGE, J. M. VAN LANEN & I. L. BALDWIN, 1937. Changes in the infective ability of *Rhizobia* and *Phytomonas tumefaciens* induced by culturing on media containing glycine. *J. Bact.*, 33: 29.
714. LÖW, E. 1874. *Tylenchus millefolii*, n. sp. eine neue gallenerzeugende Anguillulidae. *Verh. zool.-bot. Ges. Wien*, 24: 17—24.
715. LÖW, E. 1885. Beiträge zur Kenntnis der Helminthocecidien. *Verh. zool.-bot. Ges. Wien*, 35: 471.
716. LÖW, F. 1885. Beiträge zur Naturgeschichte der gallenerzeugenden Cecidomyiden. *Verh. zool.-bot. Ges. Wien*, 35.
717. LUDFORD, J. R. 1940. Interaction in vitro of fibroblasts and sarcoma cells with leucocytes and macrophages. *Brit. med. J.*, 1: 201—205.
718. LUDWIG, F. & J. VON RIES, 1933. Wachstumsvorgänge und Hochfrequenz (Versuche an Pflanzen und Tumoren). *Z. Krebsforsch.*, 40 (2): 117—121.
719. LUND, E. J. 1928. Relation between continuous bio-electric currents and respiration. II. 1. A theory of continuous bio-electric currents and electric polarity of cells. 2. A theory of cell correlation. *J. exp. Zool.*, 51: 26—290.
720. LUTZ, F. E. & M. BROWN, 1928. A new species of Bacteria and the gall of an aphid. *Amer. Mus. Nov.*, 305: 1—5.
721. MACCAGNO, T. 1928. Il fenomeno gallare. Considerazioni etiologiche. *Ricerche teleologiche. Riv. Fis. Mat. Sci. Nat.*, (2) 3: 67—78,141—148.
722. MACHARADZE, N. G. 1930. Verschiedene Widerstandsfähigkeit einiger Rebsarten gegenüber *Phylloxera* und deren Abhängigkeit vom anatomischen Aufbau des Wurzelsystems. *Bull. Inst. exp. Georgia. Tiflis*, 19, fig. 6.
723. MÄGDEFRAU, K. 1942. Paläontologie der Pflanzen. Jena, Gustav Fischer. pp. 396, fig. 305. (fossil galls on *40*, *308* and *311*).
724. MAGNIN, A. 1888. Sur l'hermaphroditisme du *Lychnis dioica* atteint d'*Ustilago. C. R. Acad. Sci. Paris*, 107: 663.
725. MAGNIN, A. 1892. Nouvelles observations sur la sexualité et la castration parasitaire. *C. R. Acad. Sci. Paris*, 115: 675.
726. MAGNUS, P. 1876. Eine von *Anguillula* herrührende Galle an den Blättern von *Agrostis canina. Verh. bot. Ver. Prov. Brandenburg*, 18: 61.
727. MAGNUS, P. 1893. Über *Synchytrium papillatum* Farl. *Ber. dtsch. bot. Ges.*, 11: 538.
728. MAGNUS, P. 1897. On some species of the genus *Urophlyctis. Ann. Bot.*, 41: 87.
729. MAGNUS, P. 1901. Über eine neue unterirdisch lebende Art der Gattung *Urophlyctis. Ber. bot. Ges.*, 19: 145.
730. MAGNUS, W. 1903. Experimentelle morphologische Untersuchungen (Vorläufige Mitteilung. II). Zur Ätiologie der Gallbildungen. *Ber. dtsch. bot. Ges.*, 21 (2): 131—132.
731. MAGNUS, W. 1914. Die Entstehung der Pflanzengallen verursacht durch Hymenopteren. Jena, 1—160, fig. 32, pl. iv.
732. MAGNUS, W. 1915. Durch Bakterien hervorgerufene Neubildungen bei Pflanzen. *S. B. Ges. naturf. Freunde Berlin*, 1915 (7): 263—277, pl. v.
733. MAGNUS, W. 1918. Wound-Callus und Bakterientumore. *Ber. dtsch. bot. Ges.*, 36: 20—29.
734. MAGROU, J. 1927. Recherches anatomiques et bactériologiques sur le cancer des plantes. *Ann. Inst. Pasteur*, 41: 758—801, fig. 1—9.
735. MAGROU, J. 1927. Remarques sur la bactériologie et l'anatomie du crown-gall. *Rev. Path. veg. Ent. agric.*, 14 (1): 46—50.
736. MAGROU, J. 1929. Études sur les galles produites par le *Bacterium tumefaciens. Ann. Sci. natur.*, (10) 10: 546.

737. MAGROU, J. 1931. Réactions d'immunité des plantes vis-à-vis du *Bacterium tumefaciens*. *C. R. Acad. Sci. Paris*, **200**: *256—259*.
738. MAGROU, J. 1935. Immunité et hypersensibilité de *Pelargonium zonale* vis-à-vis des réinfections par le *Bacterium tumefaciens*. *C. R. Acad. Sci. Paris*, **201**: *986—988*.
739. MAGROU, T. & M. MAGROU, 1927. Radiations émises par le *Bacterium tumefaciens*. *Rev. Path. veg. Ent. agric.*, **14**: *244—246*.
740. MAGROU, J. & P. MANIGAULT, 1946. Action du champ magnétique sur le développement des tumeurs expérimentales chez *Pelargonium zonale*. *C. R. Acad. Sci. Paris*, **223**: *8—11*.
741. MAIRE, R. 1917. Les mycocécidies des feuilles du Tirza *(Rhus oxycantha)*. *Bull. Sta. Rec. forest. Nord de l'Afrique*, 1: *183—186*.
742. MANGENOT, G. 1947. Sur les galles de *Thoningia coccinea*. *C. R. Acad. Sci. Paris*, **224** (9): *665—666*.
743. MANGENOT, G., H. ALIBERT & A. BASSET, 1946. Sur les lésions caractéristique du "swollen shoot" en Côte d'Ivoire. *C. R. Acad. Sci. Paris*, **222** (13): *749—751*.
744. MANI, M. S. 1946. Studies on Indian Itonididae (Cecidomyiidae: Diptera). Part VIII. Key to Oriental gall midges. *Indian J. Ent.*, **7** (1/2): *189—235*, fig. 1—117.
745. MANI, M. S. 1960. Cecidotheca indica. *Agra Univ. J. Res.* (Sci.) **8** (2): *91—280*, p. xxxii.
746. MANSON, G. F. 1931. Aphid galls as a noctuid feeding ground. *Canad. Ent.*, **63** (7): *171—172*.
747. MARCINOWSKI, K. 1909. Parasitisch und semiparasitisch an Pflanzen lebende Nematoden. *Arb. biol. Anst. Land.- und Forstw.*, **7**: *1—192*, fig. 76, pl. i.
748. MARCHAL, E. 1906. Une déformation causée par un Nématode. *Rev. Bryol.*, **33**: *106*.
749. MARELLI, CARLOS A. 1925. La *Agromyza* produttora de agallas nodicilas en la lagunila y sus parasites e hiperparasitos. *Rev. Soc. ent. Argentina*, 1 (2): *13—21*, fig. 1.
750. MARESQUELLE, H. J. 1937. Les processus fondamentaux de la cécidogenèse. *Ann. Soc. Nat. Bot. Zool.*, **19**: *379—392*.
751. MARESQUELLE, H. J. & R. SCHNELL, 1936. Étude expérimentale des phases de l'action cécidogenèse dans une galle. *C. R. Acad. Sci. Paris*, **203**: *270*.
752. MARSDEN-JONE, E. M. 1953. A study of the life cycle of *Adleria kollari* Htg. The marble or Devonshire gall. *Trans. R. ent. Soc. London*, **104** (7): *195—222*, fig. 9, pl. i—ii.
752a. MARSH, G. 1928. Relation between continuous bio-electric currents and cell respiration. IV. The origin of electric polarity in the onion root. *J. exp. Zool.*, **51**: *309—325*.
753. MARTIN, A. C. 1924. An ontogenetic study of the gall of *Phylloxera carvoseptem* Shim. *Proc. 23 Ann. Meet. N. Carolina Acad. Sci., J. Elisha-Mitchell Sci. Soc.*, **40**: *108—109*.
754. MARTIN, E. M. 1936. Morphological and cultural studies of *Taphrina potentillae*. *Bot. Gaz.*, **98** (2): *339—347*, fig. 11.
755. MARTIN, J. P. 1942. Stem galls of sugarcane induced with insect extracts. *Science*, **96** (2480): *39*.
756. MASSALONGO, C. 1897. Intorno all'acarocecidia della *Stipa pennata* Linn., causato dal *Tarsonemus caestrini* N. *Giorn. Bot. Ital.*, (NS) **4**: *103*.
757. MASSALONGO, C. 1898. Nuova elmintocecidio scoperto sulla *Zieria julacea* Schimp. *Riv. patol. Veg.*, **7**: *87*.
758. MASSALONGO, C. 1898. Le galle nell' Anatome Plantarum di M. Malpighi. *Malpighia*, **12**: *20*.

759. MASSALONGO, C. 1903. Di un nuovo elmintocecidio de *Ranunculus bulbosus* Linn. *Marcellia,* **2**: *139—140,* fig. 1.
760. MASSALONGO, C. 1904. Di un nuovo micocecidio dell' *Amaranthus, sylvestris* Desf. *Boll. Soc. bot. Ital., 354—356.*
761. MASSALONGO, C. 1920. Osservazione considerazioni intorno al cecidio della *Lonchaea lasiophthalma* Macq. *Atti R. Ist. Ven. Sci. Lett. Art,* **79** (2): *531—539.*
762. MATOUSCHEK, F. 1904. Über Nematodengallen bei Laubmoosen. *Hedwigia,* **43** (5): *343—345.*
763. MATTEI, G. E. Osservazioni biologiche intorno ad una galla. *Boll. Orto bot. Napoli,* **1** (4): *1—13,* pl. i.
764. MAYR, G. 1870/1871. Die mitteleuropäischen Eichengallen in Wort und Bild. Wien.
765. MAYR, G. 1872. Die Einmiethler der mitteleuropäischen Eichengallen. *Verh. zool.-bot. Ges. Wien,* **22**: *669.*
766. MAYR, G. 1876. Die europäische Cynipidengallen mit Ausschluss der auf den Eichen vorkommenden Arten. Wien.
767. MC BURNEY, C. A., W. B. BOLLEN & R. J. WILLIAMS, 1935. Pantothenic acid and the nodule bacteria legume symbiosis. *Proc. Nat. Acad. Sci. Washington* **21**: *301—304.*
768. MC CALLA, T. 1937. Behaviour of legume Bacteria *(Rhizobium)* in relation to exchangeable calcium and hydrogen ion concentration of the colloidal fraction of the soil. *Univ. Missouri agric. exp. Sta. Res. Bull., 256*: *1—14,* fig. 14.
769a. MC CALLUM, W. B. 1905. Regeneration in plants. *Bot. Gaz.,* **40**: *97—120,* 241—263.
769. MC CARTY, M. 1946. Chemical nature and biological specificity of the substance inducing transformation of pneumococcal types. *Bact. Res.,* **10**: *63—71.*
770. MC COY, E. E. 1929. A cytological and histological study of the root nodules of the bean *Phaseolus vulgaris* Linn. *Zbl. Bakt.,* (2) **79**: *394—412,* pl. iii.
771. MC EWEN, D. M. 1952. Cancerous response in plants. *Nature,* **169** (4307): *839.*
772. MC FARLANE, A. S. 1939. The chemistry of plant viruses. *Biol. Rev.,* **14**: *223—242.*
773. MC INTYRE, F. C., W. H. PETERSON & A. J. RIKER, 1941. The rôle of certain vitamins and metallic elements in the nutrition of the crown-gall organism. *J. Bact.,* **42**: *1—13.*
774. MC INTYRE, F. C., W. H. PETERSON & A. J. RIKER, 1942. A polysaccharide produced by the crown-gall organism. *J. biol. Chem.,* **143** (2): *491—496.*
775. MC WHORTHER, F. P. 1922. The nature of the organism found in Fiji galls of sugarcane. *Philip. Agric.,* **11**: *103; Zbl. Bakt.,* (2) **63** : *144.*
776. MEESS, A. 1922. Die cecidogenen und cecidocolen Lepidopteren. Gallenerzeugende und Gallenbewohnende Schmetterlinge und ihre Cecidien. In: Die Zoocecidien Deutschlands und ihre Bewohner. 3. *Zoologica,* Stuttgart. 61.
777. MEYER, J. 1940. Développement des galles de *Perrisia urticae* Perris sur pétiole et tige d'*Urtica dioica* Linn. *Marcellia,* **30**: *185—202,* fig. 11, pl. i.
778. MEYER, J. 1940. Développement des galles de *Perrisia urticae* Perris sur les deux faces du limbe foliaire d'*Urtica dioica* Linn. *Marcellia,* **30**: *90—112,* fig. 13.
779. MEYER, J. 1944. Observations sur les premiers effets cytologiques de la cécidogénèse chez deux Cécidomyides. *Rev. gén. Bot.,* **55**: *256—283,* fig. 6.

370

780. MEYER, J. 1950. Gigantisme nucléolaire et cécidogenesis. *C. R. Acad. Sci. Paris*, **2131**: *1333—1335*.
781. MEYER, J. 1951. Observations cytologiques sur la succion de la fondatrice d'*Adelges abietis* Kalt. sur *Picea excelsa* Lk. et la différenciation d'un tissu nourricier primaire par métaplasie. *C R. Acad. Sci. Paris*, **233**: *631—633*.
782. MEYER, J. 1951. Origine maternelle des principaux tissus et du plastème nourricier des galles larvaires d'*Adelges abietis* Kalt. sur *Picea excelsa* Lk. *C. R. Acad. Sci. Paris*, **233**: *886—888*.
783. MEYER, J. 1952. Dedifférenciation cellulaire et clivage des chondriocontes lors de l'évolution des cellules nourricières des galles de *Diastrophus rubi* Htg. sur la Ronce. *C. R. Acad. Sci. Paris*, **234**: *463—464*.
784. MEYER, J. 1952. Édification de galles multiples par une même fondatrice et peuplement des galles d'*Eriophyes macrorrhynchus* Nal. sur *Acer pseudoplatanus* L. *C. R. Acad. Sci. Paris*, **235**: *1428—1430*.
785. MEYER, J. 1952. Rapports entre l'évolution cytologique du tissu nourricier d'*Eriophyes macrorrhynchus* Nal. et la biologie de l'Acarien. *C. R. Acad. Sci. Paris*, **235**: *1545—1547*.
786. MEYER, J. 1952. Cécidogenèse de la galle de *Lasioptera rubi* Heeger et rôle nourricier d'un mycélium symbiotique. *C. R. Acad. Sci. Paris*, **234**: *2556—2558*.
787. MEYER, J. 1952. Réduction plastidale et dedifférenciation au cors de l'hyperplasie du tissu palissadique de la feuille de Hêtre sous l'action cécidogène d' *Oligotrophus annulipes* Htg. *C. R. Acad. Sci Paris*, **235**: *391—339*.
788. MEYER, J. 1954. Sur l'accumulation et l'origine de lipides dans certains tissus nourriciers de galles. *C. R. Acad. Sci. Paris*, **238**: *1066—1069*.
789. MEYER, J. 1954. Les étapes cécidogènes de *Neuroterus quercusbaccarum* Linn. ♀ ♂ et les relations entre le parasite et l'hôte dans les tout premiers stades. *C. R. Acad. Sci. Paris*, **238**: *1922—1924*.
790. MEYER, J. 1954. Sur quelques types de tissus nourriciers de cécidies d'Arthropodes et leur évolution. *C. R. Seances Rep. Comm. VIIIe Congr. int. Bot. Paris*, **7/8**: *227—229*.
791. MEYER, J. 1958. Cécidogénèse comparée. I. Les primiers cloisonnements cécidogènes de la galle internode de *Perrisia urticae* Perris sur *Urtica dioica* Linn., étude comparée avec les cas de la galle correspondante de le nervure. II. Morphogénèse comparée des galles de *Phylloxera vastatrix* Planch. sur feuilles vrilles et tiges de vigne. *Marcellia*, **30**: *69—91*, fig. 19.
792. MICHAELIS, M., I. LEVIN & H. HIBBERT, 1943. Differential inhibition between normal and tumor (crown-gall) tissue in beet roots. *Science*, **98** (2534): *89—90*.
792a. MIEHE. 1905. Wachstum, Regeneration und Polarität isolierter Zellen. *Ber. dtsch. bot. Ges.*, **23**: *257*.
793. MILLER, P. R. 1936. Morphological aspects of *Gymnosporangium* galls. *Phytopathology*, **26**: *795*.
794. MILOVIDOV, P. F. 1926. Über einige neue Beobachtungen an den Lupinenknöllchen. *Zbl. Bakt.*, (2) **68**: *333—345*, fig. 1, pl. ii.
795. MILOVIDOV, P. F. 1928. Recherches sur les tubercles du lupin. *Rev. gén. Bot.*, **40**: *192—205*, fig. 2.
796. MILOVIDOV, P. F. 1930. Zur Cytologie der Pflanzentumoren. *Protoplasma*, **10**: *294—296*.
797. MILOVIDOV, P. F. 1931. Cytologische Untersuchungen über *Plasmodiophora brassicae* Wor. *Arch. Protistenk.*, **73**: *1—46*, fig. 6, pl. iii.

798. MIX, A. J. 1935. The life history of *Taphrina deformis*. *Phytopathology*, 23 (19): *41—66*.
799. MOLLIARD, M. 1893. Sur deux cas de castration parasitaire observés chez *Knautia arvensis* Coultier. *C. R. Acad. Sci. Paris*, 116: *1306*.
800. MOLLIARD, M. 1895. Recherches sur les cécidies florales. *Ann. Sci. Nat. Bot.*, (8) 1: *67*.
801. MOLLIARD, M. 1897. Hypertrophie pathologique des cellules végétales. *Rev. gén. Bot.*, 9: *33*.
802. MOLLIARD, M. 1899. Sur la galle de l'*Aulax papaveris* Pers. *Rev. gén. Bot.*, 11: *209*.
803. MOLLIARD, M. 1899. Sur les modifications histologiques produites dans les tiges par l'action des *Phytoptus*. *C. R. Acad. Sci. Paris*, 129: *841*.
804. MOLLIARD, M. 1900. Cas de virescence et de fasciation d'origine parasitaire. *Rev. gén. Bot.*, 12: *323*.
805. MOLLIARD, M. 1900. Sur quelques caractères histologiques des cécidies produites par l'*Heterodera radicicola* Greeff. *Rev. gén. Bot.*, 12: *157—165*.
806. MOLLIARD, M. 1902. La galle du *Cécidomyia cattleyae*, n. sp. *Marcellia*, 1: *165*.
807. MOLLIARD, M. 1902. Caractères anatomiques de deux Phytoptocécidies caulinaires internes. *Marcellia*, 1: *22*.
808. MOLLIARD, M. 1904. Structure de quelques Tylenchocécidiens foliaires. *Bull. Soc. bot. Paris*, (4) 4: *51*, ci-cxii, fig. 1—5.
809. MOLLIARD, M. 1904. Une coleoptérocécidie nouvelle sur *Salix caprea* type de cécidies facultatives. *Rev. gén. Bot.*, 16: *91—95*, fig. 16—18.
810. MOLLIARD, M. 1904. Virescence et proliferations florale produites par des parasites agisant à distance. *C. R. Acad. Sci. Paris*, 139: *130*.
811. MOLLIARD, M. 1909. Une nouvelle Plasmodiophorée, parasite du *Triglochin palustra*. *Bull. Soc. bot. France*, 56: *23*.
812. MOLLIARD, M. 1910. Remarques physiologiques relatives au déterminisme des galles. *Bull. Soc. bot. France*, 57: *24*.
813. MOLLIARD, M. 1912. Action hypertrophiante de produits élaborés par le *Rhizobium radicicola* Beyer. *C. R. Acad. Sci. Paris*, 155: *1531*.
814. MOLLIARD, M. 1913. Recherches physiologiques sur les galles. *Rev. gén. Bot.*, 25 (294/296): *225—252, 285—307, 341—370*, fig. 4, pl. vii—ix.
815. MOLLIARD, M. 1926. Dimorphisme déterminé chez la galle de *Mikiola fagi* Htg. par un parasite secondaire. *C. R. Acad. Sci. Paris*, 183 (16): *624—626*.
816. MONTEMATINI, L. 1938. Il *Bacterium tumefaciens*. *Boll. Ist. Sieroterapio Milanese*, 17: *551—588*.
817. MORDWILKO, A. 1907. Beiträge zur Biologie der Pflanzenläuse (Aphididae): Die zyklische Fortpflanzung der Pflanzenläuse. *Biol. Zbl.*, 27: *747—816*.
818. MORDWILKO, A. 1928. Nouvelle contribution à l'étude de l'anholocyclie chez les Pemphigiens des Pistachiers. *C. R. Acad. Sci. Paris*, 186.
819. MOREL, G. 1946. Remarques sur l'action de l'acide naphthyle acétique sur le développement des tissus de vigne Vierge. *C. R. Soc. Biol. Paris*, 140: *269*.
820. MOREL, G. 1947. Transformation des cultures des tissus de vigne produit par l'hétéro-auxine. *C. R. Soc. Biol. Paris*, 141: *280—282*.
821. MOREL, G. 1948. Recherches sur la culture associée des parasites obligatoires et des tissus végétaux. *Ann. Epiphyt.*, 24: *1—234*.
822. MORGAN, W. L. 1933. Flies and Nematodes associated in flower bud gall of spotted gum. *Agric. Gaz. N. S. Wales*, Sydney, 46, (2): *125—127*, fig. 2.

823. MOTTRAM, J. C. 1944. Production entomitosis in bean roots and its bearing on the genesis of tumors. *Nature*, **154** (3922): *828*, fig. 1.
824. MOYSE, A. 1937. Sur des cas de duplicature florale produite par des *Eriophyes*. *Rev. gén. Bot.*, **579**: *182—202*.
825. MÜHLDORF, A. 1926. Über den Ablösungsmodus der Gallen von ihren Wirtspflanzen, nebst einer kritischen Übersicht über die Trennungs-erscheinungen im Pflanzenreiche. *Beih. bot. Zbl.*, **42**: *1—110*, pl. ii.
826. MÜLLER, C. 1877. Eine neue Milbengalle auf *Lysimachia vulgaris* Linn. *Bot. Ver. Prov. Brandenburg, Sitzb.* **19**: *105*.
827. MÜLLER, C. 1880. Einige Bemerkungen über die durch Anguillulen auf *Achillea* erzeugten Gallen. *Bot. Zbl.*, **1**: *187—188*.
828. MUMFORD, E. P. 1931. On the fauna of the diseased big-bud of the black currant, *Ribes nigrum* Linn., with a note on some fungous parasites of the gall-mite *Eriophyes ribis* (Westw.) Nal. *Marcellia*, **27**: *29—62*.
829. MUNCIE, J. H. 1926. A study of crown-gall caused by *Pseudomonas tumefaciens* on rosaceous hosts. *Iowa State Coll. J. Sci.*, **1** (1): *67—110*, fig. 7, pl. iv.
830. MUNCIE, J. H. & M. K. PATEL, 1930. Studies upon a bacteriophage for *Pseudomonas tumefaciens*. *Phytopathology*, **20** (4): *289—305*.
831. MUNRO, H. K. 1939. On certain South African gall forming Trypetidae (Diptera), with description of new species. *J. ent. Soc. S. Africa*, **2**: *154—164*.
832. MURRAY, M. A. & A. G. WHITING, 1946. A comparison of histological responses of bean plants to tryptophane and to low concentrations of indole acetic acid. *Bot. Gaz.*, **108** (1): *74—100*, fig. 17.
833. NÀBÉLEK, V. 1930. Das Krebsproblem der Pflanze. Eine phytopatho-logische Studie über die Einwirkung der Bakterien auf die Heilungs-prozesse der Pflanze. Praga, pp. *55*, pl. viii. (Problem rakoviny u rostlin Fytophathologiska studie o pusbeni bakterii na hojive pro-cessy u rostlin. *Prace Ucene spolec. Safarikovv v Bratislave*. **2**: *1—55*, pl. viii.
834. NAGAKURA, K. 1930. Über den Bau und die Lebensgeschichte der *Heterodera radicicola*. *Jap. J. Zool.*, **3**.
835. NAGY, R., W. H. PETERSON & A. J. RIKER, 1937. Comparison of enzymes in crown-gall and non-inoculated plant tissue. *Phytopa-thology*, **27**: *136*.
836. NAGY, R., A. J. RIKER & W. H. PETERSON, 1938. Some physiological studies of crown-gall and contiguous tissue. *J. agric. Res.*, **57**: *545—556*, fig. 4.
837. NALEPA, A. 1898. Eriophyidae. Das Tierreich, **4**: *1—74*.
838. NALEPA, A. 1909. Eine Gallmilbe als Erzeugerin der Blattgallen von *Cinnamomum zylanicum* Breyn. *Marcellia*, **8**: *3—6*.
839. NALEPA, A. 1910. Eriophyiden. In: Die Zoocecidien Deutschlands und ihre Bewohner. Stuttgart. 1.
840. NALEPA, A. 1917. Die Systematik der Eriophyidae, ihre Aufgabe und Arbeitsmethode nebst Bemerkungen über die Umbildung der Arten. *Verh. zool.-bot. Ges. Wien*, **67**: *12*.
841. NALEPA, A. 1920. Zur Kenntnis der Milbengallen einiger Ahornarten und ihrer Erzeuger. *Marcellia*, **19**: *3—33*.
842. NALEPA, A. 1923. Index Nominum quae ab A 1886 Eriophyidarum generibus, speciebus et subspeciebus imposite sunt, conscriptus ab. Eine alphabetische Liste der Gattungen, Art- und Unterartnamen der Familie Eriophyidae, nebst Angabe des Autors, des Jahres und der Schrifstelle der ersten Veröffentlichung. *Marcellia*, **20**: *25—66*.
843. NALEPA, A. 1924. Die systematische Abgrenzung der Species. Subspe-cies und Varietäten der Eriophyiden. *Marcellia*, **21**: *129—139*.

844. NALEPA, A. 1926. Beobachtungen über die Verbreitung der Gall-milben. *Marcellia*, 23: *89—94*.
845. NALEPA, A. 1927. Zur Phänologie und Entwicklungsgeschichte der Milbengallen. *Marcellia*, 24: *87—98*.
846. NALEPA, A. 1928. Problem der Eriophyiden-Systematik. *Marcellia*, 24: *3—29*.
847. NALEPA, A. 1929. Neuer Katalog dei bisher beschriebenen Gallmilben, ihrer Gallen und Wirtpflanzen. *Marcellia*, 25: *67—183*.
848. NALEPA, A. 1929. Untersuchungen über die Variabilität einiger Eriophyiden-Arten. *Marcellia*, 25: *44—60*.
849. NAREK, J. 1961. Die Wirkung von Aphidenstichen auf pflanzliche Zellen. *Ent. exp. appl.*, 4: *20—34*.
849a. NAWASCHIN, S. 1899. Beobachtungen übei den feineren Bau und Umwallungen von *Plasmodiophora brassicae* Wor. *Flora*, 86: *406*.
850. NAYAR, K. K. 1948. Descriptions of plant galls from Travancore. *J. Bombay nat. Hist. Soc.*, 47 (4): *668—675*.
851. NEEDHAM, J. 1936. New advances in the chemistry and biology of organized growth. *Proc. R. Soc. Med.*, 29: *1577—1626*, fig. 42.
852. NEEDHAM, J. 1942. Biochemistry and morphogenesis. Cambridge University Press, pp. *785*, fig. 328.
853. NEGER, F. W. 1908. Ambrosia-Pilze. *Ber. dtsch. bot. Ges.*, 26: *735*.
854. NEGER, F. W. 1910. Ambrosia-Pilze. III. Weitere Beobachtungen an Ambrosiagallen. *Ber. dtsch. bot. Ges.*, 23: *455—480*, fig. 4.
855. NEGER, F. W. 1913. Ambrosia-Pilze. In: Biologie der Pflanzen. Stuttgart, p. *490*.
856. NEGER, F. W. 1914. Zur Frage der systematischen Stellung der so-genannten Ambrosia-Pilze. *Zbl. Bakt.*, (2) 42 (1/4): *45—49*.
857. NEISH, A. C. & H. HIBBERT, 1940. Effects of crown-gall formation on the chemical composition of beets. *Canad. J. Res.*, (C) Bot. Sci., 18 (2): *613—623*.
858. NEISH, A. C. & H. HIBBERT,1943. Studies on plant tumors. I. Car-bohydrate metabolism of normal and tumor tissues of beet root. *Arch. Biochem.*, 3 (2): *141—157*; III. Nitrogen metabolism of normal and tumor tissues of beet root. *ibidem*, 3 (2): *159—166*.
859. NÈMEC, B. 1911. Über die Nematodenkrankheit der Zuckerrüben. *Z. Pflanzenkr.*, 21: *1*.
860. NÈMEC, B. 1913. Pflanzengeschwülste und ihre Beziehung zu den tierischen Geschwülsten. *Lékařské Rozledy Prag odděl imm. Serol.*, 481; *Bot. Zbl.*, 123: *409*; *Zbl. Bakt.*, (2) 58: *164*.
861. NÈMEC, B. 1924. Untersuchungen über Eriophyidengallen. *Stud. Plant physiol. Lab. Charles Univ. Praha*, 2: *47—94*.
862. NÈMEC, B. 1928. Über Pflanzentumoren. *Arch. exp. Zellforsch.*, 6: *172—177*.
863. NÈMEC, B. 1930. Bakterielle Wuchsstoffe. *Ber. dtsch. bot. Ges.*, 48 (3): *72—74*.
864. NÈMEC, B. 1932. Die Wurzelbildung an den bakteriellen Pflanzentumo-ren. *Stud. Plant physiol. Lab. Charles Univ. Praha*, 4 (2): *1—6*, fig. 1.
865. NÈMEC, B. 1932. Über die Gallen von *Heterodera schachtii* auf Zucker-rübe. *Studies Plant physiol. Lab. Charles Univ. Praha*, 4 (2): *1—14*.
866. NIBLETT, N. 1931. Some gall causing Trypetidae. *Nature, 139*.
867. NIELSEN, J. C. 1912. Untersuchungen über die Lebensweise und Ent-wicklung einiger Arten der Gattung *Synergus*. *Allg. Z. Ent.*, 8: *35—37*
868. NIERENSTEIN, M. 1930. Interrelation between gall producer and galls. *Nature*, 125 (3149): *348—349*.
869. NIERENSTEIN, M. & K. VON STOCKERT, 1944. Chemical differentiation of galls. *Analyst*, 69 (822): *272—273*.

374

870. Nobécourt, P. 1937. Les tumeurs causées chez les végétaux par le *Bacterium tumefaciens* et leurs relations avec le cancer. *Rev. Med. France, 17.*

871. Nobécourt, P. 1939. Sur la pérennité et l'augmentation de volume des cultures de tissus végétaux. *C. R. Soc. Biol.,* **130**: *1270.*

872. Nobécourt, B. & A. Dusseau, 1938. Sur la prolifération in vitro de fragments des végétaux et la formation de tumeurs aseptiques. *Science* (rev. fr.), **20**: *53—56.*

873. Noble, N. S. 1933. The Citrus gall wasp *(Eurytoma felis* Gir.). *Agric. Gaz. N. S. Wales,* Sydney **44**: *465—469.*

874. Noble, N. S. 1936. The Citrus gall wasp *Eurytoma felis* Gir. *Dept. Agric. N. S. Wales, Sydney, Sci. Bull.,* **53**: *1—41,* fig. 18.

875. Noble, N. S. 1938. *Tepperella trilineata* Cam., a wasp causing galling of the flower buds of *Acacia decurrens. Proc. Linn. Soc. N. S. Wales,* Sydney, 63 (5/6): *389—411,* fig. 9, pl. xx.

876. Noble, N. S. 1940. *Trichilogaster acaciaefoliae* (Froggatt) (Hymenoptera: Chalcidoidea), a wasp causing galls on the flower buds of *Acacia longifolia* Willd., *Acacia floribunda* Sieber and *Acacia sophora* R. Br. *Trans. ent. Soc. London,* 90 (2): *13—38,* fig. 8, pl. i—ii.

877. Nolte, H. W. 1960. Der Einfluss der Wirtspflanze auf die Entwicklung von Galleninsekten. Ontogeny of insects. Acta symposia de evolutione insectorum, Praha, **1959**: *309—313.*

878. Nourteva, P. 1955. On the nature of plant injuring salivary toxins of insects. *Ann. Ent. Fenn.,* **21**: *33—38.*

879. Nourteva, P. 1956. Notes on the anatomy of the salivary glands and the occurrence of protease in these organs in some leaf-hoppers (Homoptera: Auchenorrhyncha). *Ann. Ent. Fenn.,* **22**: *103—108.*

880. Nourteva, P. 1956. Studies on the effect of the salivary secretions of some Heteroptera and Homoptera on plant growth. *Ann. Ent. Fenn.,* **22**: *117—119.*

881. Nourteva, P. 1958. Die Rolle der Speichelsekrete im Wechselwirkung zwischen Tier und Nahrungspflanze bei Homopteren und Heteropteren. *Ent. exp. appl.,* 1: *41—49.*

882. Nüsslin-Rhumbler. 1927. Forstinsektenkunde. IV. Auflage. Berlin.

883. Nysterakis, F. 1946. Nouvelle interprétation de la formation des cécidies. *C. R. Acad. Sci. Paris,* **222**: *1133—1134.*

884. Nysterakis, F. 1947. Zoocécidies et substance de croissance. *C. R. Soc. Biol. Paris,* **141**: *1218—1219.*

885. Nysterakis, F. 1948. Phytohormone et inhibition de la croissance des organes végétaux attaqués par des Aphides. *C. R. Acad. Sci. Paris,* **226**: *746—757.*

886. Nysterakis, F. 1948. Autres preuves sur la secrétion d'auxines par certain insects. Un nouveau test très sensible pour le dosage des substances de croissance. *C. R. Acad. Sci. Paris,* **226**: *1917—1919.*

887. Palm, B. T. 1931. A disease of *Hibiscus sabdariffa* caused by *Rhodochytrium. Phytopathology,* 21 (2): *1201—1202.*

888. Pape, H. 1925. Beitrag zur Frage der Übertragbarkeit des Veilchenbrandes *(Urophlyctis violae)* durch die Samen. *Zbl. Bakt.,* (2) **65**: *301.*

889. Paquet, E. 1939. Sur les nodosités radicales de *Datisca cannabina* Linn.: Leur endophyte et les réactions cellulaires qui l'accompagnent. *C. R. Acad. Sci. Paris,* **209** (5): *330—332,* fig. 4.

890. Paris, G. & A. Trotter, 1911. Sui composti azotati nelle galle dei *Neuroterus baccarum. Marcellia,* **10**: *150—159.*

891. Parker, R. N. 1932. *Casuarina* root nodules. *Indian Forester,* 8 (7): *361—364.*

892. Paszlavsky, J. 1882. Über die Bildung des Bedeguars. *Termeszetrajzie Fiizetek,* **5**: *277.*

893. PATEL, M. K. 1928. A study of pathogenic and non-pathogenic strains of *Pseudomonas tumefaciens* (Sm. & Towns.) *Phytopathology*, **18** (4): *331—343*.
894. PATEL, M. K. 1929. Biological studies of *Pseudomonas tumefaciens* (Sm. & Towns.) and fifteen related non-pathogenic organisms. *Iowa State Coll. J. Sci.*, **3** (3): *271—298*.
895. PATTON, W. H. 1897. A gall inhabiting ant. *Amer. Nat.*, *126—127*.
896. PEGALION, V. 1892. Studi anatomici di alcune ipertrofie indotte dal *Cystopus candidus* in alcuni organi di *Raphanus raphanistrum*. *Riv. Pat. Veg.*, **1**: *265*.
897. PERTI, G. 1924. Sulle consequenze prodotte dall' *Eriophyes avellanae* Nal. nel Noccielo. *Marcellia*, **21**: *3—9*, fig. 5.
898. PERTI, L. 1907. Osservazioni sulle galle foliari di *Azalea indica* prodotte dall' *Exobasidium discoideum* Ellis. *Ann. Mycol.*, **5**: *341—347*, fig. 8.
898a. PERTI, L. 1923. Sulla produzione sperimentali di iperplasi nelle piante. *Rend. Acad. Lincei*, (5) **22** (2): *509*; *Zbl. Bakt.*, (2) **58**: *164*.
899. PERTI, L. 1924. I tumori batterici del Pino d'Aleppo. *Ann. R. Ist. Sup. Naz.*, **9**: *1—43*.
900. PETERS-POSTUPOLSKA ZOFIA. 1932. Recherches anatomiques sur les zoocécidies de *Mecinus lineariae* (Curculionidae). *C. R. Soc. Sci. L. Varsovie*, (4) **24** (1931): *29—43*, fig. 5, pl. ii (In Polish with French summary).
901. PEYERITSCH, J. 1878. Über Placentarsprosse. *S.B. Akad. Wiss.*, (1) **78**: *1*.
902. PEYERITSCH, J. 1882. Zur Ätiologie der Chloranthien einiger *Arabis*-Arten. *Jb. wiss. Bot.*, **13**: *1*.
903. PEYERITSCH, J. 1888. Über künstliche Erzeugung von gefüllten Blüten und anderen Bildungsabweichungen. *S.B. Akad. Wiss., math.-naturw. Klasse, Wien*, **97** (1): *597*.
904. PEYERITSCH, J. 1896. Über Bildungsabweichungen bei Umbelliferen. *S.B. Akad. Wiss. math.-wiss. Klasse, Wien*, **60** (1): *899*.
905. PFEIFFER, H. 1928. Die pflanzlichen Trennungsgewebe. Die Trennungsgewebe der Pflanzlichen Gallbildungen. In: Handbuch der Pflanzenanatomie. K. Libauer.
906. PHILLIPS, W. J. & F. F. DICKE, 1935. Morphology and biology of the wheat jointworm gall. *J. agric. Res.*, **50** (1): *359—386*, fig. 1—13.
907. PIERRE, A. 1902. Sur le ponte d'un Neuroptère cécidozoen, *Lestes viridis* van der Lind. *Rev. Sci. Bourbonnais et Centre France*, **15**: *181*; *Marcellia*, **1**: *186*.
908. PIERRE, A. 1913. La cryptocécidie de *Balanius nucum* Linn. *Rev. Sci. Bourbonnais et Centre France*, **26**: *3—10*, fig. 2.
909. PIETZ, J. 1938. Beiträge zur Physiologie des Wurzelknöllchenbakteriums. *Zbl. Bakt.*, (2) **99**: *1—32*, fig. 7.
910. PINCARD, J. A. 1935. Physiological studies of several pathological bacteria that induce cell stimulation in plants. *J. agric. Res.*, **50**: *933 —952*.
911. PING, Chi. 1920. Some inhabitants of the round gall of golden-rod. *J. Ent. Zool.* **7** (3): *161—177*.
912. PINOY, E. 1925. Apropos du cancer des plantes ou crown-gall. *C. R. Acad. Sci. Paris*, **180**: *311—313*.
913. PIRONE, P. P. 1945. Control of the gall disease of *Gypsophila* caused by *Phytomonas gypsophilae* (Brown) Magrou. *Phytopathology*, **35** (5): *368 —369*.
913a. POPOFF. 1924. Zellstimulationsforschung. **1**: *107*.
914. POTTER, V. R. 1943. An enzyme-virus theory regarding carcinomata. *Cancer Res.*, **3**: *358—361*.

376

915. POTTER, V. R. 1945. The genetic aspects of the enzyme-virus theory of cancer. *Science*, **101**: *609—610*.
916. PRAT, H. 1955. Relations existant entre les notions de gradient chimique et de rayon d'activité cécidogénétique. *Marcellia*, **30**: *51— 53*, fig. 1.
917. PRILLIEUX, E. 1876. Étude sur la formation et le développement de quelques galles. *Ann. Sci. Natur*, (6) **3**: *113*.
918. PUZANOWA, E. W. 1934. *Asphondylia prunorum* Wachtl. (Diptera: Cecidomyiidae) und deren Pilzgallen auf dem Pflaumenbaum. *Z. angew. Ent.*, **21**: *443—463*, fig. 16.
919. QUINTARET, G. 1911. Observations sur deux Rhizocécidies nouvelles ou peu connues de la Provence. *Ann. Fac. Sci. Marseille*, **20** (Suppl.) (1): *1—4*, fig. 2.
920. QUINTARET, G. 1911. Études anatomiques d'une Rhizocécidie du *Linaria striata* DC racoltée en Provence. *Bull. Soc. Linn. de Provence*, *133—138*, fig. 3.
921. RABAUD, E. 1914. Sur la significance de la cryptocécidie de *Balanius nucum* Linn. *Rev. Sci. Bourbonnais et du Centre France*, **1914**: *1—13*.
922. RACK, G. 1958. Eriophyiden als Bewohner der Wirrzöpfe zweier Weidenarten. *Mitt. Hamburg. zool. Inst.*, **56**: *31—80*.
923. RAINER, A. 1911. Einige Bemerkungen über die Familie der Gallwespen im allgemeinen, über die äussere Gestalt, den Bau und die Lebensweise, der seltenen und wenig bekannten *Ibalia cultellator* im besonderen. *Österr. Monatschr. grundleg. naturf. Unterricht*, **7**: *283—290*.
924. RAMACHANDRA RAO, Y. 1924. A gall forming thrips on *Calycopteryx floribunda: Austrothrips cochinchinensis*. *Agric. J. India*, **19** (4): *435—437*.
925. RANFIELD, W. M. 1935. Studies in cellular pathology I. Effects of cane gall bacteria upon gall tissue cells of the black raspberry.
926. RAPOSO, H. 1943. A galha da azalea *Rhododendron indicum* Sweet, provocado pelo fungo *Exobasidium discoideum* Ellis. *Bol. Soc. brasil. Agron.*, **6** (1): *61—70*, fig. 3.
927. RECHINGER, C. 1902. Über ein seltenes Phytoptocecidium auf *Artemisia campestris* Linn. und seine Ähnlichkeit mit *Filago arvensis* Linn. *Verh. zool.-bot. Ges. Wien*, **52**: *152*.
928. REHWALD, VON, CHR. 1927. Über pflanzliche Tumoren als vermeintliche Wirkung chemischer Reizung. *Z. Pflanzenkr.*, **37**: *65—86*.
929. RENNER, O. 1906. Über Wirrzöpfe an Weiden. *Flora*, **96**: *322—328*.
930. REUTER, E. 1903/1904. Gallbildning hos *Achillea millefolium* förorsakad af *Tylenchus millefolii* Löw. *Meded. Soc. Fauna Flora Fenn.*, **30**: *25—26*.
931. REWALD, B. & W. RIEDE. 1932. Knöllchenbakterien und Phosphatbildung bei *Soja hispida*. *Biochem. Z.*, *424—428*.
932. RIEDEL, M. 1896. Gallen und Gallwespen. Stuttgart. pp. 96, pl. ii. (2. Auflage 1910).
933. RIEDEL, M. 1900. Insekten auf *Polyporus*. *Illustr. Z. Ent.*, **5**: *9*.
934. RIKER, A. J. 1922. Studies of crown-gall. *Phytopathology*, **12**: *55*.
935. RIKER, A. J. 1923. Some relations of the crown-gall organism to its host tissue. *J. agric. Res.*, **25**: *119—132*, pl. v.
936. RIKER, A. J. 1923. Some pathological responses of the host tissue to the crown-gall organism. *J. agric. Res.*, **26**: *425—435*, pl. vi.
937. RIKER, A. J. 1926. Studies on the influence of some environmental factors on the development of crown-gall. *J. agric. Res.*, **32**: *83—96*.
938. RIKER, A. J. 1927. Cytological studies of crown-gall tissue. *Amer. J. Bot.*, **14** (1): *25—37*, fig. 5.

939. RIKER, A. J. 1939. Physiological relations between host and parasite in crown-gall: An example of basic biological research with plant material. *Amer. J. Bot.*, **26**: *159—162*, fig. 3.
940. RIKER, A. J. 1941. Attenuation of bacteria pathogenic on plants. *Chron. Bot.*, **6**: *392—393*.
941. RIKER, A. J. 1942. The relation of some chemical and physiological factors to the initiation of pathological plant growth. *Growth*, **6** (Suppl.) *105—117*.
942. RIKER, A. J., W. M. BANFIELD, W. H. WRIGHT & G. W. KEITT, 1928. The relation of certain bacteria to the environment of roots. *Science*, **68**: *357—359*.
943. RIKER, A. J. & T. O. BERGE, 1935. Atypical and pathological multiplication of cells approached through studies on crown-gall. *Amer. J. Cancer*, **25**: *310—357*.
944. RIKER, A. J. & B. M. DUGGAR, 1938. Growth substance and the development of crown-gall. *J. agric. Res.*, **57**: *21—39*.
945. RIKER, A. J. & B. M. DUGGAR, 1939. The nature of growth substance originating in crown-gall tissue. *J. agric. Res.*, **59**: *535—539*.
946. RIKER, A. J. & A. E. GUTSCHE, 1948. The growth of sunflower tissue in vitro on synthetic media with various organic and inorganic sources of nitrogen. *Amer. J. Bot.*, **35**: *227—238*.
947. RIKER, A. J., B. HENRY & B. M. DUGGAR, 1941. Growth substance in crown-gall as related to the time after inoculation and diffusion. *J. agric. Res.*, **63** (7): *395—405*.
948. RIKER, A. J. & A. C. HILDEBRANDT, 1951. Pathological plant growth. *Ann. Rev. Microbiol.*, **5**: *223—240*.
949. RIKER, A. J., M. LYENIS & S. R. LOCKE, 1941. Comparative physiology of crown-gall, attenuated crown-gall, Radiobacter and hairy-root bacteria. *Phytopatholgy*, **31**: *964—977*, fig. 3.
950. RIKER, A. J., E. SPOERL & ALICE E. GUTSCHE, 1946. Some comparisons of bacterial plant galls and their causal agents. *Bot. Rev.*, **12**: *57—82*.
951. RILEY, C. V. 1883. Jumping seeds and galls. *Proc. U S. Nat. Mus.*, **5**: *632*; *Amer. Mag. nat. Biol.*, (5) **12**: *140*.
952. RISCHKOV, V. L. 1943. The nature of ultra viruses and their biological activity. *Phytopathology*, **33**: *950—955*.
953. RIVERA, V. 1928. Degeneration granulaire di tumori vegetali irradiati. *Boll. Accad. Pugliese Sci.*, **1** (4): *3—7*, fig. 3.
954. RIVERA, V. 1926. Seggi di radioterapia vegetale. *Boll. R. Staz. Pat. Veg.*, **6** (4): *337—345*.
955. RIVERA, V. 1926. Transformazione indotte dei raggi X in tessuti tumorali. *Riv. Biol.*, **8** (1): *1—15*, fig. 4.
956. RIVERA, V. 1926. E necessaria la ferita del tessuta per la produzione di tumori da *Bacterium tumefaciens* su vegetali? *Boll. Accad. Pugliese Sci.*, **1** (6): *1—5*.
957. RIVERA, V. 1926. Effets des rayons X sur les tissus végétaux normaux et pathologiques. *Ann. Inst. Pasteur*, **40** (7): *614—658*, fig. 12.
958. RIVERA, V. 1927. Dipressioni ed esaltazione dell'accrescimento in neoplasmi vegetali sperimentali irradiati. *Riv. Biol.*, **9** (1): *3—10*.
959. RIVERA, V. 1928. Azione di forti dosi di raggi gamma sopra il *Bacterium tumefaciens* Sm. Towns. *Lab. Patol. Veg. R. Ist. Sup. Agr. Perugia Mem.*, **2**: *876—869*, pl. i.
960. RIVERA, V. 1928. Influenza dei circuiti aperti di Lakhovsky sullo sviluppo di tumori nel vegetali. *Boll. R. Staz. Patol. Veg.* (Firenze) **8** (4): *357—373*.
961. RIVERA, V. 1931. Fattori eccitativi dell'accrescimento di neoplasmi vegetali da *Bacterium tumefaciens*. *Atti R. Accad. Naz. Lincei R. Cl. Sci.Fis. Mat. et Nat.*, **13** (8): *621—627*, fig. 2.

962. ROBERG, M. 1934. Über den Erreger der Wurzelknöllchen von *Alnus* und den Elaeagnaceen, *Elaeagnus* und *Hippophoë*. *Jb. wiss. Bot.* 79: *472—492*.

963. ROBERTSON, R. A. 1906. On the histology of plant galls. I. *Xestophanes tormentillae*. *Proc. Scott. micr. Soc.*, 4 (2/3): *136—141*.

964. ROBINSON, W. 1936. Some features of crown-gall in plants in reference to comparions with cancer. *Proc. R. Soc. Med.*, 20: *1507—1509*.

965. ROBINSON, W. & H. WALKDEN, 1923. A critical study of crown-gall. *Ann. Bot.*, 37: *299—324*.

966. RODELLA, G. 1907. Die Knöllchenbakterien der Leguminosen. *Zbl. Bakt.*, (2) 18: *455—461*.

967. RODELLA, G. 1907. I batteri radicali delle Leguminose, studio critico-sperimentale d'alcuni problemi di batteriologia agraria e di fisio-patologia umana. Padova, Prosperini, pp. 87, fig. 6, pl. iii.

968. ROMANES, G. J. 1890. Galls. *Nature*, 41: *80, 173, 369*.

969. ROMEO, A. 1935. Sugli zoocecidii a fungaia di *Coronella emerus* Linn. var. *emeroides* (Boiss. & Spr.). *Ann. R. Ist. Sup. Agr. Portici*, 7: *1—42*, fig. 15 (pagination in reprint).

970. ROMEO, A. 1939. Zoocecidii a fungaia di origine fiorale su *Satureia gracea* Linn. *Ann. Fac. Agr. R. Univ. Napoli*, (3) 10 (1938): *42—59*, fig. 5.

971. RONCALI, F. 1904. Contributo allo studio della composizione chimica delle galle. *Marcellia*, 3: *54—59*.

972. ROSE, M. 1939. Recherches expérimentales sur la cécidogénèse et les néoplasies chez les végétaux. *Bull. Biol.*, Paris, 73: *336—366*, fig. 5.

972a. ROSE, S. M. 1957. Cellular interaction during differentiation. *Biol. Rev.*, 32 (4): *351—382*.

973. ROSEN, H. R. 1924. Ist die Saugtätigkeit der anfängliche Reiz bei Hemipterengallen? *Z. Pflanzenkunde Gallenkunde*, 34: *344*.

974. ROSEN, H. G. 1936. Morphological notes together with some ultra-filtration experiments on the crown-gall pathogen *Bacterium tumefaciens*. *Mycologia*, 18: *193—205*.

975. ROSENSTOCK, E. 1907. Beiträge zur Pterydophytenflora Südbrasiliens. *Hedwigia*, (Dresden) 46: *57—167*.

976. ROSS, H. 1910. Beiträge zur Kenntnis der Anatomie und Biologie deutscher Gallenbildungen. I. *Ber. dtsch. bot. Ges.*, 28: *228*.

977. ROSS, H. 1911. Die Pflanzengallen (Cecidien) Mittel- und Nordeuropas, ihre Erreger und Biologie und Bestimmungstabellen. Gustav Fischer, Jena, pp. 350, fig. 24, pl. x.

978. ROSS, H. 1912. Adventivblättern auf Melastomaceenblättern, verursacht durch parasitisch lebende Älchen. *Ber. dtsch. bot. Ges.*, 30: *346—361*, fig. 8.

979. ROSS, H. 1914. Über verpilzte Tiergallen. *Ber. dtsch. bot. Ges.*, 32: *574—597*.

980. ROSS, H. 1922. Weitere Beiträge zur Kenntnis der verpilzen Mückengallen. *Z. Pflanzenkr.*, 32 (1/2): *83—93*.

981. ROSS, H. 1932. Praktikum der Gallenkunde (Cecidiologie). Entstehung, Entwicklung, Bau der durch Tiere, und Pflanze hervorgerufenen Gallbildungen sowie Ökologie der Gallenerreger. Biologische Studienbücher. W. Schönichen, Berlin. Julius Springer, pp. 312, fig. 181.

982. ROSS, H. & H. HEDICKE, 1927. Die Pflanzengallen Mittel- und Nordeuropas, ihre Erreger und Biologie und Bestimmungstabelle. Gustav Fischer, Jena, pp. 348, fig. 33, pl. i-x.

983. RÖSSIG, H. 1904. Von welchen Organ der Gallwespenlarven geht der Reiz zur Bildung der Pflanzengallen aus? Untersuchung der Drüsenorgane der Gallwespenlarven, zugleich ein Beitrag zur postembryonalen Entwicklung derselben. *Zool. Jb.*, (Syst.) 20 (1): *19—90*, pl. iv.

984. ROTHERT, W. 1896. Über die Gallen der Rotatorie *Notommata werneckii* auf *Vaucheria walzi*, sp. nov. *Jb. wiss. Bot.*, 29: 524—594.
985. ROUS, P. 1910. A transmissible avian neoplasm (sarcoma of the common fowl). *J. exp. Med.*, 12: 696—705.
986. ROUS, P. 1912. An avian tumor in its relation to the tumor problem. *Proc. Amer. phil. Soc.*, 51: 201—205.
987. ROUS, P. 1936. Virus tumors and the tumor problem. *Amer. J. Cancer*, 28: 233—272.
988. ROUS, P. & J. G. KIDD, 1938. The carcinogenetic effect of a papilloma virus on the tarred skin of rabbits. *J. exp. Med.*, 67: 399—428.
989. ROUS, P. & J. G. KIDD, 1941. Conditional neoplasms and subthreshold neoplastic status. A study of the tar tumor of rabbits. *J. exp. Med.*, 73: 365—390.
990. ROZE. 1888. L'*Ustilago caricis* Fuckel, aux environs de Paris. *Bull. Soc. bot. France*, 35: 277.
991. RÜBSAAMEN, E. H. 1898. Über Gallen, das Sammeln und Konservieren und die Zucht der Gallenerzeuger. *Illustr. Z. Ent.*, 3: 67.
992. RÜBSAAMEN, E. H. 1899. Mitteilungen über neue und bekannte Gallen aus Europa, Asien, Afrika und Amerika. *Ent. Nachr.*, 25: 234.
992a. RÜBSAAMEN, E. H. 1900. Über Zoocecidien von der Balkanhalbinsel. *Illustr. Z. Ent.*, 5: 177.
993. RÜBSAAMEN, E. H. 1907. Beiträge zur Kenntnis aussereuropäischer Zoocecidien. III. Beitrag. Gallen aus Brasilien und Peru. *Marcellia*, 6: 110—173.
994. RÜBSAAMEN, E. H. 1910. Beiträge zur Kenntnis aussereuropäischer Zoocecidien. IV. Afrikanische Gallen. *Marcellia*, 9: 3—36.
995. RÜBSAAMEN, E. H. 1910. Über deutsche Gallmücken und Gallen. *Z. wiss. Insektenbiol.*, 6: 125.
996. RÜBSAAMEN, E. H. 1910. Die Zoocecidien, durch Tiere erzeugte Pflanzengallen Deutschlands und ihre Bewohner. *Zoologica*, 61: 1—293, fig. 3, pl. vi.
997. RÜBSAAMEN, E. H. 1911. Beiträge zur Kenntnis aussereuropäischer Zoocecidien. V. Beitrag. Gallen aus Afrika und Asien. *Marcellia*, 10: 100—132, fig. 43.
998. RÜBSAAMEN, E. H. & H. HEDICKE, 1923/1925. Die Zoocecidien, durch Tiere erzeugte Pflanzengallen Deutschlands und ihre Bewohner. *Zoologica*, Stuttgart, 77.
999. RÜBSAAMEN, E. H. & H. HEDICKE, 1926. Die Cecidomyiden (Gallmücken) und ihre Gallen. In: Die Zoocecidien Deutschlands und ihre Bewohner.
1000. RUGGIERI, G. 1935. Osservazioni istologiche sopra le galle della *Viola odorata* Linn. prodotte *Dasyneura affinis* Kieff. *Boll. R. Staz. Pat. Veg. Roma*, 15 (2): 301—312, fig. 4.
1001. RYBAK, B. 1946. Actions bactériostatiques chez *Pelargonium zonale* et crown-gall. *C. R. Acad. Sci. Paris*, 223 (16): 586—587.
1002. RYTZ, W. 1907. Beiträge zur Kenntnis der Gattung *Synchytrium*. *Zbl. Bakt.*, (2) 18: 635—655, 799—825, fig. 10, pl. i.
1003. RYTZ, W. 1917. Beiträge zur Kenntnis der Gattung *Synchytrium*. I. Fortsetzung: Die cytologische Verhältnisse bei *Synchytrium taraxaci*. De B. & Wor. *Beih. bot. Zbl.*, 34: 343—372, pl. ii-iv.
1004. SADEBECK, R. 1903. Einige kritische Bemerkungen über Exoascaceen. *Ber. dtsch. bot. Ges.*, 21 (10): 539.
1005. SAIRD, D. G. 1932. Bacteriophage and the root nodule bacteria. *Arch. Mikrobiol.*, 3: 159—193, fig. 3.
1006. SAJO, K. 1907. Über die Linsengallen der Eichenblätter und über Gallwespen überhaupt. *Prometheus*, 18: 433—455.

380

1007. SALAMAN, R. N. 1938. A discussion on new aspects of virus diseases. A recent development of plant virus research. *Proc. R. Soc. London*, (B) **125**: *291—310*.
1008. SASAKI, C. 1911. A new aphid gall on *Styrax japonicus* Sieb. et Zuck. *Mem. I. Congr. int. Ent.*, (1910) **2**: *449—456*, pl. xxv-xxvi.
1009. SAUNDERS, W. W. 1847. On the gall formed by *Diphucrania auriflua* Hope, a species of Buprestidae. *Trans. ent. Soc. London*, (1847/ 1849) **5**: *27—28*, fig. 5—9, pl. ii.
1010. SAUVAGEAU, C. 1892. Sur quelques algues phéosporées parasites. *J. Bot.*, **6**: *57*.
1011. SCHAEDE, R. 1933. Über die Symbionten in den Knöllchen der Erle und des Sanddorns und die cytologischen Verhältnisse in ihnen. *Planta*, **19**: *389—416*, fig. 19.
1012. SCHAEDE, R. 1941. Untersuchungen an den Wurzelknöllchen von *Vicia faba* und *Pisum sativum*. *Beitr. Biol. Pflanzen*, **27**: *165—188*, fig. 15, pl. i.
1013. SCHAEDE, R. 1943. Über die Symbiose in dem Wurzelknöllchen der Podocarpeen. *Ber. dtsch. bot. Ges.*, **61** (2): *39—41*.
1014. SCHÄLLER, G. 1959. Untersuchungen über die Gallenbildung und Nekrosereaktion der Rebensorten unter Berücksichtigung der Rassendifferenzierung der Reblaus. *Phytopathol. Z.*, **36**: *67—83*.
1014a. SCHÄLLER, G. 1960. Untersuchungen über den Aminosäuregehalt des Speicheldrüsensekretes der Reblaus *(Viteus vitifolii* Shimer) (Homoptera). *Ent. exp. appl.*, **3**: *128—136*, fig. 1.
1015. SCHÄLLER, G. 1961. Aminosäuren im Speichel und Honigtau der grünen Apfelblattlaus *Aphis pomi* Deg. (Homoptera). *Ent. exp. appl.*, **4**: *73*.
1016. SCHEERLINCK, H. 1936. Les nodosites radicales des Légumineuses. *Ann. Soc. Sci. Bruxelles*, **56** (B): *250—303*, fig. 1, pl. vi.
1017. SCHELLENBERG, H. C. 1911. Über Speicherung von Reservestoffen in Pilzgallen. *Verh. schweiz. naturf. Ges.*, 94 Jahresversammlung, **1**: *277*.
1018. SCHIFFNER, V. 1905. Beobachtungen über Nematodengallen bei Laubmoosen. *Hedwigia*, **44**: *218*.
1019. SCHIFFNER, V. 1906. Neue Mitteilungen über Nematodengallen auf Laubmoosen. *Hedwigia*, **45**: *189*.
1020. SCHILBERSZKY, K. 1935. Beiträge zur Biologie von *Pseudomonas tumefaciens*. *Z. Pflanzenk.*, **45** (3): *146—160*.
1021. SCHLECHTENDAL, D. H. R. VON, 1880. Pflanzenmissbildungen: die Vergrünung der Blüten von *Daucus carota* Linn. *Jb. Ver. Naturk.* Zwickau, 70.
1022. SCHLECHTENDAL, D. H. R. VON, 1882. Übersicht der zurzeit bekannten mitteleuropäischen Phytoptocecidien und ihrer Literatur. *Z. ges. Naturw.*, **55**: *480*.
1023. SCHLECHTENDAL, D. H. R. VON, 1888. Über Zoocecidien. Beiträge zur Kenntnis der Acarocecidien. *Z. Naturw.*, **56**: *93*.
1024. SCHLECHTENDAL, D. H. R. VON, 1903. Beiträge zur Kenntnis der durch Eriophyiden verursachten Krankheitserscheinungen der Pflanzen. *Marcellia*, **2**: *117*.
1025. SCHLECHTENDAL, D. H. R. VON, 1916. Eriophyidocecidien, die durch Gallmilben verursachten Pflanzengallen. In: RÜBSAAMEN: Die Zoocecidien Deutschlands und ihre Bewohner. *Zoologica*, **61** (2): *1—204*, fig. 34, pl. xviii.
1026. SCHMIDT, H. 1910. Notizen zur Biologie unserer gallenbildenden Rüsselkäfer. *Ent. Rundschau*, **27**: *111, 137—138*.
1027. SCHMIDT, H. 1912. Biologische Bemerkungen zu einigen gallenerzeugenden Schmetterlingen III. Ein Beitrag zur Mikrolepidopterenfauna Niederschlesiens. *Soc. ent.*, **27**: *25—26*.

1028. SCHMIDT, H. 1914. Einige Notizen über das Zusammenleben von Gallinsekten und Pilzen an einheimischen Pflanzen. *Fühlings landw. Ztg.*, **63** (4): *143—146*.
1029. SCHMIEDEKNECHT, O. 1909. Chalcidoidea. Genera Insectorum, fascicle 9.
1030. SCHMITZ, F. 1892. Über knöllchenartigen Auswüchse an der Sprossen einiger Florideen. *Bot. Ztg.*, **50**: *624*.
1031. SCHNEIDER-ORELLI, O. 1909. Die Miniergänge von *Lyonetia clerkella* und die Stoffwanderung in Apfelblättern. *Zbl. Bakt.*, (2) **24**: *158*.
1032. SCHNEIDER-ORELLI, O., ROOS & R. WEISMANN, 1938. Untersuchungen über die Generationsverhältnisse der Fichtengallenlaus, *Sacchiphantes (Chermes) abietis* (Linn.). *Vierteljahresschr. naturf. Ges. Zürich*, **83**: *29—107*, fig. 1—20.
1033. SCHOUTENDEN, H. 1903. Les aphidocécidies palaeartiques. *Ann. Soc. Ent. belg.* **47**: *167—193*.
1034. SCHRADER, H. L. 1862. Observations on certain gall-making Coccidae of Australia. *Trans. ent. Soc. N. S. Wales*, Sydney, (1862/1863) **1**: *1—6*, pl. i-ii; *6—8*, pl. iii.
1035. SCHRADER, H. L. 1863. Über gallenbildende Insekten in Australien. *Verh. zool.-bot. Ges. Wien*, **13**: *189*.
1036. SCHRENK, H. VON, 1905. Intumescenes formed as a result of chemical stimulation. *Rep. Missouri bot. Garden*, **16**: *125—148*.
1037. SCHULZE, P. 1916. Die Galle von *Rhopalomyia ptarmicae* Vallot. *S.B. Ges. naturf. Freunde Berlin*, **8**: *277—241*, fig. 20; **10**: *381—385*, fig. 5.
1038. SCHULZE, P. 1917. Mischgallen und behaarte Hörnchengallen bei unseren Linden. *S.B. Ges. naturf. Freunde Berlin*, **1917** (8/10): *519—527*, fig. 7.
1039. SCHWARTZ, M. 1911. Die Aphelenchen der Veilchengallen und der Blattflecken an Farnen und Chrysanthemum. *Arb. kais. biol. Anst. Land.- und Forstw.*, **8** (2): *303—334*.
1040. SCHWAZBACH, E. 1959. Regeneration bei Gallen von *Pemphigus spirothecae* Pass. *Naturwissenschaften*, **46**: *337*, fig. 3.
1041. SCHWEIZER, J. 1932. Über das Verhalten der Bakterienknöllchen bei einigen chlorophyllfreien Leguminosen. *Verh. schweiz. naturf. Ges.*, **113**: *376—377*.
1042. SEVERINI, G. 1907. Ricerche bacteriologiche sui tubercoli dell' *Hedysarum coronarium* Linn. *Atti R. Accad. Lincei Roma, Cl. Sci. Fis. Mat. Nat.*, **16** (3): *219—226*.
1043. SEVERINI, 1920. Sui tubercoli radicali di *Datisca cannabina*. *Annali Bot.*, **15**: (1): *29—52*, pl. ii.
1044. SHOPE, R. E. 1933. Infectitious papillomatosis of rabbits. *J. exp. Med.*, **58**: *607—624*.
1045. SIEGLER, E. A. 1928. Studies on the etiology of apple crown-gall. *J. agric. Res.*, **37**: *301—331*.
1046. SIEGLER, E. A. 1929. The woolly-knot type of crown-gall. *J. agric. Res.*, **39** (6): *427—450*, fig. 8.
1047. SIEGLER, E. A. & R. E. PIPER, 1929. Aerial crown-gall of apple. *J. agric. Res.*, **39** (4): *249—262*.
1047a. SILBERBERG, B. 1909. Stimulation of storage tissues of higher plants by zinc sulphate. *Bull. Torrey bot. Club*, **36**: *489*.
1048. SKRZIPIETZ, P. 1900. Die Aulaxgallen auf *Hieracium*-Arten. Rostock. pp. 52, pl. ii.
1048a. SINNOT, E. W. 1962. Plant morphogenesis. Mc Graw-Hill Book Company, New York, pp. 550.
1049. SJÖSTEDT, Y. 1910. Hymenoptera, 4. Akaziagallen und Ameisen auf den Ostafrikanischen Steppen. Biologische Studien. Miss. Ergebnis

schwedischen zool. Exped. nach Kilimandjaro, dem Meru und den Umgebenden Massaisteppen Deutsch-Ostafrika 1905–1906 unter Leitung von Prof. Yngve Sjöstedt. 2 (8/14): *844*, pl. i-xiv.

1050. SMITH, C. O. 1939. Susceptibility of species of Cupressaceae to crown-gall as determined by artificial inoculation. *J. agric. Res.*, 59 (12): *919—925*.

1051. SMITH, C. O. 1940. Gall on *Pseudotsuga macrocarpa* induced by *Bacterium pseudotsugae*. *Phytopathology*, 30 (7): *624*, fig. 1.

1052. SMITH, C. O. 1942. Crown-gall on species of Taxaceae, Taxadiaceae and and Pinaceae as determined by artificial inoculation. *Phytopathology*, 32 (11): *1005—1009*, fig. 3.

1053. SMITH, E. F. 1911. Bacteria in relation to plant Diseases.

1054. SMITH, E. F. 1912. On some resemblances of crown-gall to human cancer. *Science*, 35: *161—172*.

1055. SMITH, E. F. 1912. The staining of *Bacterium tumefaciens* in tissue. *Phytopathology*, 2: *127—128*.

1056. SMITH, E. F. 1916. Studies on the crown-gall of plants. Its relation to human cancer. *J. Cancer Res.*, 1: *231—258*.

1057. SMITH, E. F. 1916. Crown-gall studies showing changes in plant structures to a changed stimulus. *J. agric. Res.*, 6: *179—182*.

1058. SMITH, E. F. 1916. Crown-gall and cancer. *J. Amer. med. Ass.*, 67: *1318*.

1059. SMITH, E. F. 1916. Further evidence that crown-gall of plants is cancer. *Science*, 43: *871—889*.

1060. SMITH, E. F. 1916. Further evidence on the relation between crown-gall and cancer. *Proc. nat. Acad. Sci.*, 2: *444—448*.

1061. SMITH, E. F. 1917. Mechanism of tumor growth in crown-gall. *J. agric. Res.*, 8: *165—188*; *Proc. Amer. phil. Soc.*, 56: *437—444*.

1062. SMITH, E. F. 1917. Embryomas in plants. *Bull. Hopkins Hosp.*, 28: *319*.

1063. SMITH, E. F. 1921. Effect of crown-gall inoculation. *J. agric. Res.*, 21: *593—598*.

1064. SMITH, E. F. 1922. Appositional growth in crown-gall tumors and in cancers. *J. Cancer Res.*, 7: *1—49*, fig. 4, pl. xxviii.

1065. SMITH, E. F., N. A. BROWN, L. M. MC CULLOCH, 1912. The structure and development of crown-gall. *U. S. Dep. Agric. Bull.*, 255: *1—61*.

1066. SMITH, E. F. & C. O. TOWNSEND, 1907. A plant tumor of bacterial origin. *Science*, 25: *671—673*.

1067. SMITH, E. F. & C. O. TOWNSEND, 1911. Crown-gall of plants: Its cause and remedy. *U. S. Dep. Bur. Plant Ind. Bull.*, 213: *1—215*.

1068. SMITH, K. M. 1945. Transmission by insects of a plant virus. *Nature*, 155: *174*.

1069. SMITH, W. G. 1894. Untersuchungen über die Morphologie und Anatomie der durch Exoasceen verursachten Spross- und Blattdeformationen. *Forstl. Naturw. Z.*, 3: *433*.

1070. SOLACOLU, T. & D. CONSTANTINESCU, 1937. Tumeurs à caractères néoplastiques formées sur les plantes par l'action de l'acide indole acétique. *C. R. Acad. Sci. Paris*, 204: *290—292*.

1071. SOLEREDER, H. 1905. Über Hexenbesen usw. *Naturw. Z. Forst.-Landw.*, 5: *17*.

1072. SOLOWIOW, P. 1907. Microlepidoptera Gallarum. *Z. wiss. Insektenbiol.*, 3 (7): *222*.

1073. SORHAGEN, L. 1898. Gallenbewohnende Schmetterlingslarven. *Illustr. Z. Ent.*, 3: *114—117*.

1074. SORU, E. & R. BRAUNER, 1933. Action à distance de *Bacillus tumefaciens* sur la moëlle asseuse du lupin. *C. R. Soc. Biol. Paris*, 112: *623—625*.

383

1075. SPRATT, E. R. 1912. The morphology of the root tubercles of *Alnus* and *Elaeagnus* and the polymorphism of the organism causing their formation. *Ann. Bot.*, **26**: *119—128*.
1076. SPRATT, E. R. 1912. The formation and physiological significance of root nodules on the Podocarpinae. *Ann. Bot.*, **26**: *801—844*, pl. iv.
1077. SPRATT, E. R. 1915. The root nodules of Cycadaceae, *Bacillus radicicola*. *Ann. Bot.*, **29**: *616—626*, pl. i.
1078. SPRENGEL, F. 1936. Über die Kropfkrankheit an Eiche, Kiefer und Fichte. *Phytopathol. Z.*, **9** (6): *583—635*, fig. 53.
1079. STÄMPFLI, R. 1910. Untersuchungen über die Deformationen welche bei einigen Pflanzen durch Uredineen hervorgerufen werden. *Hedwigia*, **49**: *230*.
1080. STANER, P. 1935. L'*Acacia* à galles du Congo. *Ann. Soc. Sci. Bruxelles*, **55**: *310—314*.
1081. STAPP, C. 1927. Der bakterielle Pflanzenkrebs und seine Beziehungen zum tierischen und menschlichen Krebs. *Ber. dtsch. bot. Ges.*, **45**: *480—504*.
1082. STAPP, C. 1942. Der Pflanzenkrebs und sein Erreger *Pseudomonas tumefaciens*. X. Die Virulenzsteigerung von *Pseudomonas tumefaciens* durch Titan. *Zbl. Bakt.*, (2) **104** (23/24): *395—400*, fig. 2.
1083. STAPP, C. 1942. Der Pflanzenkrebs und sein Erreger *Pseudomonas tumefaciens*. XI. Mitt. Zytologische Untersuchungen des bakteriellen Erregers. *Zbl. Bakt.*, (2) **105** (1/4): *1—14*, pl. i-iii.
1084. STAPP, C. 1944. Der Pflanzenkrebs und sein Erreger *Pseudomonas tumefaciens*. XIII. Über die Bedeutung des Colchicine als polyploidisierendes Mittel für den Erreger und als angebliche Bekämpfungsmittel gegen den Wurzelkropf. *Zbl. Bakt.*, (2) **104** (16/19): *338—350*, fig. 5.
1085. STAPP, C. 1947. Der bakterielle Pflanzenkrebs und seine Bedeutung im Lichte allgemeiner Krebsforschung. *Naturwissenschaften*, **34**: *81—88*.
1086. STAPP, C. & H. BORTELS, 1931. Der Pflanzenkrebs und sein Erreger *Pseudomonas tumefaciens*. I. Konstitution und Tumorbildung der Wirtspflanze. *Z. wiss. Biol.*, **3** (4): *654—663*.
1087. STAPP, C. & H. BORTELS, 1931. Der Pflanzenkrebs und sein Erreger *Pseudomonas tumefaciens*. II. Über den Lebenskreislauf von *Pseudomonas tumefaciens*. *Z. Parasitenk.*, **4**: *101—125*.
1088. STAPP, C. *et al.* 1938. Der Pflanzenkrebs und sein Erreger *Pseudomonas tumefaciens*. VII. Untersuchungen über die Möglichkeit einer wirksamen Bekämpfung an Kernobstholzen. *Zbl. Bakt.*, (2) **99**: *210—276*.
1089. STAPP, C. & E. PFEILL, 1939. Der Pflanzenkrebs und sein Erreger *Pseudomonas tumefaciens*. VIII. Zur Biologie der Krebsgewebes. *Zbl. Bakt.*, (2) **101** (14/17): *261—268*.
1090. STARNACH, K. 1930. Narosle bakteryine na niektoych slodkowodnych gatunkach rodzaju *Chantransia* Fr. (The bacterial galls on freshwater *Chantransia*). (With German summary). *Acta Soc. Bot. Polon.*, **7** (4:) *435—460*.
1091. STEC, W. 1927. Über das Vorkommen von Bakteriocecidien an Kartoffelantheren. *Bull. int. Acad. polon. Sci. Lett. Cl. Sci. math. Nat.*, (B) Sci. Nat. (Bot) **1** (7B): *705—712*.
1092. STEFANI-PEREZ, T. DE, 1903. Alterazioni tardive d'alcune piante per influesso di insetti. *Marcellia*, **2**: *44*.
1093. STEFANI-PEREZ, T. DE, 1904. Mimismo di una galla. *Marcellia*, **3**: *66—70*.
1094. STEFANI-PEREZ, T. DE, 1906. Contributo all'entomofauna dei cecidii. *Marcellia*, **5**: *113*.
1095. STEFANI-PEREZ, T. DE, 1912. Notizi su alcuni zoocecidii della Libia. *Boll. Orto bot. Giardino Colo. Palermo*, **11**: *144—151*.

1096. STEFANI-PEREZ, T. DE, 1914. Aggiunte ai zoocecidii della Tripolitania. *Boll. Studi Inform. Giardino Colon. Palermo*, **1**: *177—179*.
1097. STEGAGNO, G. 1904. I locatari dei cecidozoi sin qui noti in Italia. *Marcellia*, **3**: *18—53*.
1098. STEIN, E. Über Gewebeentartung in Pflanzen als Folge von Radiumbestrahlung (Zur Radiomorphose von *Antirrhinum*). *Biol. Zbl.*, **49** (2): *112—126*.
1099. STEIN, E. 1930. Über Karzinomähnliche erbliche Gewebeentartungen in *Antirrhinum*, dem Soma durch Radiumbestrahlung induziert. *Strahlentherapie*, **37** (1): *137—141*.
1100. STEIN, E. 1930. Weitere Mitteilung über die durch Radiumbestrahlung induzierten Gewebeentartungen in *Antirrhinum* (Phytocarcinomata) und ihr erbliches Verhalten (Somatische Induktion und Erblichkeit). *Biol. Zbl.*, **50** (3): *129—158*, fig. 27.
1101. STEINER, G. 1919. The problem of host selection and host specialization in certain plant-infesting Nemas. *Phytopathology*, 15.
1102. STEINER, G. & E. M. BUHRER, 1933. Recent observations on diseases caused by Nematodes. *Plant Disease Rep.* 17 (14): *1—172*.
1103. STEINER, G., E. M. BUHRER & A. S. RHODES, 1934. Giant galls caused by the root-knot Nematode. *Phytopathology*, 24 (2): *161— 163*, fig. 1.
1104. STEINER, G. & F. E. ALBIN, 1946. Resusciation of the Nematode *Tylenchus polyphypnus*, n. sp. after almost 30 years dormancy. *J. Washington Acad. Sci.*, 36 (3): *97—99*, fig. 1.
1105. STEWART, A. 1914. Notes on the anatomy of the *punctatus* gall. *Amer. J. Bot.*, **1**: *531—546*, pl. i—iii.
1106. STEWART, A. 1915. An anatomical study of *Gymnosporangium*-galls. *Amer. J. Bot.*, **2**: *402—417*, pl. ii.
1107. STEWART, A. 1916. Concerning certain peculiar tissue strands in *Protomyces*-gall on *Ambrosia trifolia* (Ref.). *Science*, (NS) **43**: *365 —366*.
1108. STEWART, A. 1916. Notes on the anatomy of *Peridermium* galls. I. *Amer. J. Bot.*, **3**: *12*.
1109. STICHEL, W. 1916. Massenhaftes Auftreten von Gallen. *Z. Insektenbiol.*, **12**: *213, 250*.
1110. STOCKERT, K. R. & J. ZELLNER, 1914, Chemische Untersuchungen über Pflanzengallen. *Z. physiol. Chem.*, **90**: *495—501*.
1111. STRASBURGER, E. 1873. Einige Bemerkungen über Lycopodiaceen. *Bot. Ztg.*, **31**: *81—103*.
1112. STRONG, L. C. 1926. Changes in the reactional potential of transplantable tumor. *J. exp. Med.*, **43**: *713—742*, pl. i.
1113. STRUBBEL, A. 1888. Untersuchungen über den Bau und die Entwicklung des Rübennematoden. *Bibliotheca Zool.*, 1, 2.
1114. STRUCKMEYER, B. E., A. C. HILDEBRANDT & A. J. RIKER, 1949. Histological effects of growth-regulating substances on sunflower tissue of crown-gall origin grown in vitro. *Amer. J. Bot.* 36:(7): *491—495*.
1115. SÜCHTING, H. 1904. Kritische Studien über die Knöllchenbakterien. *Zbl. Bakt.*, (2) **11** (12/13): *377—388, 417—441*.
1116. SUESSENGUT, K. & R. BEVERLE, 1935. Über Bakterienknöllchen am Spross von *Aschynomene paniculata* Willd. *Hedwigia*, **75** (4): *234 —237*, fig. 1.
1117. SUIRE, J. 1934. Une espèce nouvelle de *Coleophora* de l'*Atriplex halinus*, et ses premiers états. (Lep. Tineidae). *Bull. Soc. ent. France*, **39** (14): *202—204*, pl. i.
1118. SUIT, R. F. & E. A. EARDLEY, 1935. Secondary tumor formation on herbaceous hosts induced by *Pseudomonas tumefaciens* Sm. & Towns. *Sci. Agric.*, **15** (5): *345—358*.

1119. SYLWESTER, E. P. & MARY C. COUNTRYMAN, 1933. A comparative histological. study of crown-gall and wound callus on apple. *Amer. J. Bot.*, **20** (5): *328—340*, pl. xiii-xiv.
1120. SZAFER, W. 1915. Anatomische Studien über javanische Pilzgallen. *Bull. Acad. Sci. Cracovie, Cl. Sci. Mat. Nat.*, (B): *80—85*, pl. iv.
1121. TAKAHASHI, R. 1934. Association of different species of thrips in their galls. *Bot. & Zool., Tokyo*, **2** (11): *1827—1836*, fig. 4.
1122. TANKA, MME. L. 1931. Étude sur les bactéries des Légumineuses et observations sur quelques champignons parasites des nodosités. *Les Botaniste*, **23**: *301—530*, fig. 33, pl. xx.
1123. TAVARES, DA SILVA J. 1902. Descripção de seis Coelopterocecidias novas. *Broteria*, **1**: *172—184*.
1124. TAVARES, DA SILVA J. 1903. Bewegungen der Galle des Käfers *Nanophyes pallidus* Oliv. *Insektenbörse*, *60—61*.
1125. TAVARES, DA SILVA J. 1917. As cecidias o Brazil que se criam nas plantes da familia das Melastomataceae. *Broteria*, **15**: *18—44*, fig. 1—8, pl. i-v.
1126. TAVARES, DA SILVA J. 1917. Cecidias brazeleiras que se criam em plantes das familias das Compositae, Rubiaceae, Tiliaceae, Lythraceae e Artocarpaceae. *Broteria*, **15**: *113—181*, fig. A-C, 4, pl. vi-xi.
1127. TEPPER, J. G. O. 1893. Descriptions of South Australian Brachyscelid galls. *Trans. R. Soc. S. A., Adelaide*, **17**: *265—280*, pl. iii-v.
1128. TEPPER, J. G. O. 1893. Südaustralische Brachysceliden. *Greiz. Abh. Ver. Naturf.*, *1—16*, pl. i-iii.
1129. TERBY, J. 1925. Études cytologiques sur les nodosités radicales les Legumineuses. *Acad. R. belg. Sci. Mem.*, (8) **8** (8): *1—32*, pl. ii.
1130. TEUTSCHLÄNDER & F. KRONENBERGER, 1926. Über Versuche mit *Bacterium tumefaciens*. *Z. Krebsf.*, **23** (2): *177—182*.
1131. THIMANN, K. V. 1939. Auxins and the inhibition of plant growth. *Biol. Rev.*, **14**: *314—337*.
1131a. THIMANN, K. V. & J. BONNER, 1949. Experiments on the growth and inhibition of isolated plants. II. The action of several enzyme inhibitors on the growth of *Avena*-coleoptile and on *Pisum*-internodes. *Amer. J. Bot.*, **36**: *214—222*.
1132. THIMANN, K. V. & B. M. SWEENEY, 1937. The effect of auxins upon the protoplasmic streaming. *J. gen. Physiol.*, **21**: *123—135*.
1133. THOM, CH. 1903. A gall upon a mushroom. *Bot. Gaz.*, **36**: *223—225*, fig. 6.
1134. THOMAS, B. F. 1945. Tissue responses to physiologically active substances. *Bot. Rev.*, **11** (10): *593—610*.
1135. THOMAS, F. 1872. Zur Entstehung der Milbengallen und verwandten Pflanzenauswüchse. *Bot. Ztg.*, **30**: *284*.
1136. THOMAS, F. 1873. Beiträge zur Kenntnis der Milbengallen und Gallmilben. Die Stellung der Blattgallen an den Holzgewächsen und die Lebensweise von *Phytoptes. Z. ges. Naturw.*, **42**: *513—537*.
1137. THOMAS, F. 1877. Ältere und neue Beobachtungen über Phytoptocecidien. *Z. ges. Naturw.*, **49**: *329*.
1138. THOMAS, F. 1877. Über Einteilung der Phytoptocecidien. (Milbengallen). *Verh. bot. Ver. Prov. Brandenburg*, **19**: *76*.
1139. THOMAS, F. 1880. *Synchytrium* und *Anguillula* auf Dryas. *Bot. Zbl.*, *761*.
1140. THOMAS, F. 1889. Über das Heteropterocecidium von *Teucrium capitatum* und anderen *Teucrium*-Arten. *Verh. bot. Ver. Prov. Brandenburg*, **31**: *103*.
1141. THOMAS, F. 1891. Die Blattflohkrankheit der Lorbeerbäume. *Gartenflora*, **40**: *42*.
1142. THOMAS, F. 1893. Cecidologische Notizen. I. *Ent. Nachr.*, **19**: *289*.

1143. THOMAS, F. 1902. Die Dipterocecidien von *Vaccinium ulginosum* mit Bemerkungen über Blattgrübchen und terminologische Fragen. *Marcellia*, 1: *146.*

1144. THOMAS, J. A. 1931. Production des tumeurs d'appérance sarcomateuse chez l'annélide *Nereis diversicolor* (O.F.M.) par l'inoculation de *Bacterium tumefaciens* (Smith). *C. R. Acad. Sci. Paris*, 193: *1045—1047.*

1145. THOMAS, J. A. 1931. Sur les réactions de la tunique d'*Ascidia mentula* Mull., à l'inoculation de *Bacterium tumefaciens* Smith.*C. R. Soc. Biol. Paris*, 108: *772—774.*

1146. THOMAS, J. A. 1930. Contribution à l'étude des réactions de quelques invertébrés à l'inoculation des substances à propriétés cancérigènes et du *Bacterium tumefaciens. Ann. Inst. Pasteur*, 49: *234—274.*

1147. THOMAS, J. A. & A. J. RIKER, 1948. The effects of representative growth substances upon the attenuated bacterial crown-galls. *Phytopathology*, 38: *25* (Abstract).

1148. THOMAS, P. T. 1945. Experimental imitation of tumor condition. *Nature*, 156: *738—740.*

1149. THOMPSON, J. 1926. Studies in irregular nutrition. I. The parasitism of *Cuscuta reflexa* (Roxb.). *Trans. R. Soc. Edinburgh*, 54 (2): *343—356*, pl. viii.

1150. THORNE, G. 1926. *Tylenchus balsamophilus*, a new plant parasitic Nematode. *J. Parasitol.*, 12: *141—145.*

1151. THORNTON, B. H. & N. GANGUELLE, 1926. The life cycle of the nodule organism *Bacillus radicicola* Beij. in soil and its relation to the infection of the host plant. *Proc. R. Soc.*, (B) (699) 99: *427.*

1152. THORNTON, H. G. 1930. The study of development of the root nodule of lucerne *(Medicago sativa). Ann. Bot.*, 14: *385—392*, pl. ii.

1153. THORNTON, H. G. & E. F. MAC COY, 1931. The relation of the nodule organism *(Bacterium redicicola)* to its host plant. *Rep. Proc. V int. bot. Congr., Cambridge*, 1930: *44—45.*

1154. THRIFFT, M. 1928. Morphology of *Heterodera schachtii. J. Helminthol.*, 6.

1155. THRIFFT, M. 1930. Observations on the life-history of *Heterodera schachtii. J. Helminthol.*, 8.

1156. THUNG, T. H. 1929. Experimenten met *Bacterium tumefaciens* Sm. & Towns. *Tijdsch. Plantenziekten*, 35 (10): *265—269.*

1157. TISCHLER, G. 1901. Über Heterodera-Gallen an den Wurzeln von *Circaea luteiana* Linn. *Ber. dtsch. bot. Ges.*, 19: *95.*

1158. TISCHLER, G. 1912. Untersuchungen über die Beeinflussung von *Euphorbia cyparissus* durch *Uromyces pisi. Flora*, 104: *1.*

1159. TOBLER, G. 1913. Die Synchytrien. Studien zu einer Monographie der Gattung. *Arch. Protistenk.*, 3: *1—98*, pl. iv.

1160. TÖTH, L. 1939. Über die Biologie der Blattlaus *Pemphigus spirothecae* Pass. *Z. angew. Ent.*, 26 (2): *297—311.*

1161. TOMASZEWSKII, W. 1931. Cecidomyiden (Gallmücken) als Grasschädlinge. *Arb. biol. Reichsanst. Land.- Forstw. Berlin-Dahlem*, 19 (1): *1—715.*

1162. TORRY, J. G. 1959. Experimental modification of development in the root in "Cell, Organism and Milieu". The Ronald Press.

1163. TREUB, M. 1882. Abnormal gezwollen ovarien van *Liparis latifolia* Lindl. *Nederl. kruidk. Arch.*, (2) 3 (4): *404.*

1164. TROTTER, A. 1901. La cecidogenesi nelle alghe. *Nuova Notarisia*, 12.

1165. TROTTER, A. 1903. Contributo all conoscenza del sistema secretore in alcuni tessuti prosoplasmatici. *Ann. Bot.*, 1 (3): *123—133*, fig. 5.

1166. TROTTER, A. 1903. Studi cecidologici. III. Le galle de i cecidozoi fossili. *Riv. Ital. Palaeontol.*, 9 (1/2): *12—21.*

1167. TROTTER, A. 1903. Di una forte infezione di *Anguillula radicicola* in piante de Garofano *(Dianthus caryophyllus)*. *Boll. Soc. bot. Ital.*, *156—157.*

1168. TROTTER, A. 1904. Galla della Colonia Eritrea (Africa). *Marcellia*, **3**: *95—112.*

1169. TROTTER, A. 1905. Sulla strutura istologica di un micocecidio prosoplastico. *Malpighia*, 19.

1170. TROTTER, A. 1907. *Cynips fortii*, sp. n. Descrizione ed istologia di una nuova galla d'Asia Minore. *Marcellia*, **6**: *12—23*, fig. 5.

1171. TROTTER, A. 1910. Le cognizioni cecidologiche e teratologiche di Ulisse Aldrovandi e della sua scuola. *Marcellia*, **9**: *114.*

1172. TROTTER, A. 1911. Contributo alla conoscenza delle galle dell'America del Nord. *Marcellia*, **10**: *28—61*, fig. 21, pl. i.

1173. TROTTER, A. 1914. Nuovo contributo alla conoscenza delle galle della Tripolitania. *Marcellia*, **13**: *3—23*, pl. i-ii.

1174. TROTTER, A. 1915. Atrofia parasitaria della corolla e virescenze nel *Trifolium angustifolium* Linn. *Marcellia*, **14**: *136—142.*

1175. TROTTER, A. 1961. Osservazioni e ricerche istologiche sopra alcune morfosi vegetali determinate da funghi. *Marcellia*, **15**: *58—111*, fig. 14, pl. i-iii.

1176. TROTTER, A. 1920/1923. Intorno all'evoluzione morfologica delle galle. *Marcellia*, **19**: *120—147*, fig. 2; **20**: *67—86* (1923).

1177. TROTTER, A. 1929. Contributo alla illustrazione cecidologica dell' Anteille. *Marcellia*, **26**: *78—114*, fig. 11.

1178. TROTTER, A. 1931. Nuovo contributo alla conoscenza delle galle Colonia Eritrea (Africa Or.). *Marcellia*, **27**: *63—105*, fig. 21, pl. ii-iii.

1179. TROTTER, A. 1932. Nuovo contributo all cecidologia della Libia. *Marcellia*, **28**: *14—30*, fig. 9.

1180. TROTTER, A. 1932. Una rara deformazione parasitaria del *Pinus mugus*. *Marcellia*, **28**: *4—7*, fig. 4.

1181. TROTTER, A. 1934. Manipolo di galle dell'Isola di Formosa. *Marcellia*, **29**: *87—101.*

1182. TROTTER, A. 1934. Galle su *Genista andreana*. *Il Giardino Fiorito*, 9.

1183. TROTTER, A. 1934. Osservazioni ricerche istologische su vari zoocecidi. *Marcellia*, **29**: *111—183*, fig. 50, pl. i-v.

1184. TROTTER, A. 1939. Galle in "Missione Biologica nel paese dei Borana racolte botaniche". *Pub. R. Accad. d'Ital.*, **1939**: *421—423.*

1185. TROTTER, A. 1940. Galle dell'Africa Orientale Italiana. *Marcellia*, **30**: *113—152*, fig. 17.

1186. TROTTER, A. 1940. Contributo all conoscenza delle galle del Sahara Tripolitano. *Marcellia*, **30**: *79—89*, fig. 8.

1187. TROTTER, A. 1940. Galle dell'Eritrea. Ulteriori osservazioni e galle nuove. *Marcellia*, **30**: *203—245*, fig. 14.

1188. TSCHIRCH, A. 1890. Über durch *Astegopteryx*, eine neue Aphidengattung, erzeugte Zoocecidien auf *Styrax benzoin* Dryand. *Ber. dtsch. bot. Ges.*, **8**: *48—53*, pl. iv.

1189. TUBEUF, K. VON, 1896. Die Haarbildungen bei den *Chermes*-Gallen des Fichten. *Forst. naturw. Z.*, **5**: *121.*

1190. TUBEUF, K. VON, 1898. Die Zweiggallen der Kiefer. *Forst. naturw. Z.*, **7**: *252—331.*

1191. TUBEUF, K. VON, 1904. Wirrzöpfe und Holzkröpfe der Weiden. *Naturw. Z. Land.-Forstw.*, **2**: *330.*

1192. TUBEUF, K. VON, 1912. Nicht-Parasitische Hexenbesen. *Wiss. Z. Forst.-Landw.*, **10**: *62.*

1193. TUBEUF, K. VON, 1930. Das Problem der Knollenkäfer. *Z. Pflanzenkr.*, **40** (5): *225—251*, fig. 25.

388

1194. TUBEUF, K. VON, 1933. Das Problem der Hexenbesen. *Z. Pflanzenkr.*, 43 (5): *194—242*, fig. 60.
1195. TUBEUF, K. VON, 1936. Tuberkulose, Krebs und Rindengrind der Echsen *(Fraxinus)* Arten und die sie veranlassenden Bakterien, Nektriapilze und Borkenkäfer. *Z. Pflanzenkr.*, 46 (10): *449—483*.
1196. TYLER, J. 1933. Reproduction without males of the root-knot Nematode. *Hilgardia*, 7 (10): *391—415*.
1197. TYZZER, E. E. 1916. Tumor immunity. *J. Cancer Res.*, 1: *125—153*.
1198. UICHANCO, L. B., 1919. A biological and systematic study of Philippine plant galls. *Philip. J. Sci.*, 14: *527—554*, pl. i-xv.
1199. ULBRICH, E. 1939. Eine bisher unbekannte Gallenbildung des Weiden-Holzschwämmer *Fomes salicinus* (Pers.) Fr. und über die Gallen am Flachenporling *Ganoderma aplanatum* (Pers.) Pat. *Ber. dtsch. bot. Ges.*, 57 (8): *397—402*.
1200. VALIANTE, 1883. Sopra un Ectocarpia parasita dell *Cystoseira opuntioides - Streblonemopsis irritans*. *Boll. zool. Sta. Naples*, 6: *489*.
1201. VALLEAU, W. D. 1947. A wound tumor-like graft transmitted disease. *Phytopathology*, 37 (8): *580—582*, fig. 1.
1202. VAN LANEN, I. L. BALDWIN & J. A. RIKER, 1940. Attenuation of cell stimulating bacteria by specific amino acids. *Science*, 92: *512—513*.
1203. VAN SLOGTEREN, E. 1931. Les helmintoses des Plantes. *II. Congr. int. Path. Compar. Paris*, 1: *432—447*.
1204. VAN TIEGHEM, P. & H. DOULIOT, 1888. Origine, structure et nature morphologique des tubercules radicaux des Legumineuses. *Bull. Soc. bot. France*, 35: *105—109*.
1205. VASILIU, I. 1927. Étude sur les tumeurs des végétaux et leur analogie avec les tumeurs animales. *Bull. Ass. Franc. étude Cancer, Paris*, 16 (4): *256—277*, fig. 16.
1206. VENKATARAYAN, S. J. 1932. *Tylenchus* sp. forming leaf galls on *Andropogon pertusus* Willd. *J. Indian bot. Soc.*, 11: *243—247*, pl. ii.
1207. VERONA, O. 1942. Nutrizione e virulenza in *Bacterium tume faciens*. *Riv. Patol. Veg.*, 32 (2/10): *173—179*, fig. 1.
1208. VERRIER, MARIE-LUISE. 1928. Études anatomiques et cytologiques d'une cécidie sur *Senecio cacaliaster* Lamarck. *Ann. Soc. Ent. France*, 97: *19—26*, fig. 10.
1209. VERRIER, MARIE-LUISE. 1929. Contribution à l'étude de la cécidie de *Livia juncorum* Latr. sur *Juncus conglomeratus* Linn. *Bull. Soc. Ent. France*, 1929 (4): *77—80*, fig. 4.
1210. VERRIER, MARIE-LUISE. 1928. Sur les particularités de l'appareil mitochondrial de quelques cécidies. *C. R. Acad. Sci. Paris*, 187: *611*.
1211. VERRIER, MARIE-LUISE. 1930. Etude biologique de quelques galles des capitules de Composées. *Bull. biol. France Belg.*, 64 (2): *191*.
1212. VERRIER, MARIE-LUISE. & F. LÖW, 1938. Recherches sur la composition des galles *Pemphigus* sur *Pistacia terebinthus*. *C. R. Soc. Biol. Paris*, 127: *1401—1403*.
1213. VIENNOT-BOURGIN, G. 1937. Les déformations parasitaires provoquées par les Ustilaginées. Paris, *1—189*.
1214. VIERMANN, H. 1929. Die Wurzelknöllchen der Lupine. *Bot. Arch.*, 25 (1/2): *45—86*, fig. 29, pl. i.
1215. VIRTANEN, A. J., S. VON HAUSEN & H. KARSTRÖM, 1933. Untersuchungen über die Leguminosebakterien und Pflanzen. XII. Die Ausnutzung der aus den Wurzelknöllchen der Leguminosen herandiffusierten Stickstoffverbindungen durch Nichtleguminosen. *Biochem. Z.*, 258: *105—117*, fig. 3, pl. ix.
1216. VIRTANEN, A. J. & T. LAINE. 1936. Investigations on the root-nodule bacteria of Leguminous plants. XVIII. Break-down of proteins by the root nodule bacteria. *Biochem. Z.*, 30: *377*.

389

1217. VÖCHTING, H. 1892. Über Transplantation am Pflanzenkörper. Tübingen, *1—162*.
1218. VÖCHTING, H. 1900. Zur Physiologie der Knollengewächse. *Jb. wiss. Bot.*, **34**: *1*.
1219. VÖCHTING, H. 1918. Untersuchungen zur experimentelle Anatomie und Pathologie des Pflanzenkörpers. Tübingen.
1220. VOGLER, P. 1899. Insekten auf *Polyporus*. *Illustr. Z. Ent.*, **4**: *345*.
1221. VOIGT, G. 1932. Galle oder Blattmine? Beitrag zur Histologie der vergallten Mine und anderer *Sedum*-Minen von *Apion sedi* Germ. *Anz. Schädlingsk.*, **8**: *135—143*, fig. 6.
1222. VORMS, MME. 1933. Caractères anatomiques résultant de l'arrêt du développement chez les galles. *C. R. Acad. Sci. Paris*, **196** (8): *558—560*.
1223. VOSSELER, J. 1906. Eine Psyllide als Erzeuger von Gallen am Maulebaum. *Z. wiss. Insektenbiol.*, **2**: *276, 308*.
1224. VUILLEMIN, P. 1895. Sur une maladie des agarics produite par une association parasitaire. *Bull. Soc. mycol. France*, **11**: *16*.
1225. VUILLEMIN, P. 1904. Les castrations femelles et l'androgénie parasitaire du *Lonicera periclymenum*. *Bull. mens. Soc. Sci. Nancy*.
1226. WADDINGTON, C. H. & J. NEEDHAM, 1935. Induction by synthetic polycyclic hydrocarbons. *Proc. R. Soc. London*, **177** (B): *310—317*.
1227. WAGNER, W. 1905/1907. Über die Gallen *Lipara lucens* Meig. *Verh. Ver. naturw. Unterhalt. Hamburg*, **13**: *120—135*, fig. 10.
1228. WAKKER, J. H. 1892. Untersuchungen über den Einfluss parasitischer Pilze auf ihre Nährpflanze. *Jb. Bot.*, **24**: *499*.
1229. WALLER, A. D. 1900. Four observations concerning the electrical effects of light upon green leaves. *J. Physiol.*, **25**: *18—22*.
1230. WALSH, B. D. 1864/1866. On the insects, Coleopterous, Hymenopterous and Dipterous, inhabiting the galls of certain species of willows. *Proc. ent. Soc. Philadelphia*, **3**: *543—641*; **6**: *223—288*.
1231. WARD, H. B. 1937. Some fundamental aspects of the cancer problem. New York. Science Press, *248*.
1232. WARNSTOFF, C. 1906. Die ersten von mir an einem Lebermoos beobachteten Nematodengallen. *Allg. bot. Z.*, **12**: *194*, fig. 3.
1233. WEBER VAN BOSSE. Études sur des algues de l'Archipel Malasien II. *Phytophysa treubii*. *Ann. Jardin bot. Buitenzorg*, **8**: *165*.
1234. WEIDEL, F. 1911. Beiträge zur Entwicklungsgeschichte und vergleichenden Anatomie der Cynipidengallen der Eiche. *Flora* (NF) **2**: *279—334*, fig. 49, pl. i.
1235. WEIDNER, H. 1957. Neuere Anschauungen über die Entstehung der Gallen durch die Einwirkung von Insekten. *Z. Pflanzenkr.*, **64**: *287 —309*.
1236. WEIMER, J. L. 1917. The origin and development of the galls produced by two cedar rust fungi. *Amer. J. Bot.*, **4**: *241—251*. fig. 1, pl. iv.
1237. WEIN, K. 1834. Beiträge zur Geschichte der Cecidologie mit besonderen Ausblicken auf die Entwicklung in Thüringen. *Marcellia*, **29**: *7—72*.
1238. WEISSE, A. 1902. Über die Blattstellung an einigen Triebspitzengallen. *Jb. wiss. Bot.*, **37**: *593*.
1239. WELD, L. H. 1921. American gallflies of the family Cynipidae, producing subterranean galls on oaks. *Proc. U. S. Nat. Mus.*, **59**: *187 —246*, pl. xxviii-xxxvii.
1240. WELD, L. H. 1922. Notes on American gallflies of the family Cynipidae, producing galls on acorns with descriptions of new species. *Proc. U. S. Nat. Mus.*, **61** (19): *1—32*, pl. v.
1241. WELD, L. H. 1926. Field notes on gall inhabiting cynipid wasps with descriptions of new species. *Proc. U. S. Nat. Mus.*, **68** (10): *1—131*, pl. viii.

1242. WELD, L. H. 1952. Cynipoidea (Hymenoptera) 1905—1950, being a supplement to the Dalla Torre and Kieffer Monogiaph - the Cynipidae in Das Tierreich, Lief. **24,** 1910 and bringing the systematic literature of the world up to date, including keys to families and subfamilies and lists of new genera and specific and variety names. Ann Arbor. Michigan.

1243. WELLS, B. W. 1916. The comparative morphology of the zoocecidia of *Celtis occidentalis. Ohio J. Sci.*, **16** (7): *249—290,* pl. xii-xi x.

1244. WELLS, B. W. 1920. Early stages of the development of certain *Pachypsylla*-galls on *Celtis. Amer. J. Bot.*, **7** (7): *275—285,* pl. i.

1245. WELLS, B. W. 1921. Evolution of zoocecidia. *Bot. Gaz.*, **71:** *358—377,* pl. xxi-xxii.

1246. WENDEL, E. 1918. Physiologische Anatomie der Wurzelknöllchen einiger Leguminosen. *Beitr. allg. Bot.*, **1:** *151.*

1247. WENT, F. W. 1932. Eine botanische Polaritätstheorie. *Jb. wiss. Bot.*, **76:** *528—557.*

1248. WERTH, E. 1932. Die Galle des *Pemphigus cornicularius* Pass. an *Pistacia terebinthus* Linn. *Ber. dtsch. bot. Ges.*, **50:** *529,* fig. 2, pl. i.

1249. WESTWOOD, J. O. 1885. Galls on the roots of orchids. *Gardners Chronicle* (NS) **24:** *24.*

1250. WHEELER, W. M. 1910. Ants. 663 pp.

1251. WHITAKER, T. W. The occurrence of tumors on certain *Nicotiana*-hybrids. *J. Arn. Arb.*, **15:** *144—153.*

1252. WHITE, O. E. 1945. The biology of fasciation and its relation to abnormal growth. *J. Heredity,* **36** (1): *11—22,* fig. 11.

1253. WHITE, P. R. 1939. Controlled differentiation in plant tissue culture. *Bull. Torrey bot. Club,* **66:** *507—513.*

1254. WHITE, P. R. 1943. A Handbook of plant tissue culture. The Jaques Cattell Press, Lancaster, Pa., pp. *277.*

1255. WHITE, P. R. 1944. Transplantation of tumors of genetic origin. *Cancer Res.,* **4** (12): *791—794,* fig. 2.

1256. WHITE, P. R. 1945. Do plants too have cancer? *Plants and Garden,* **1** (3): *184—188,* fig. 4.

1257. WHITE, P. R. 1945. Respiratory behaviour of bacteria-free crowngall tissue. *Cancer Res.,* **5** (5): *302—311.*

1258. WHITE, P. R. 1945. Metastatic (graft) tumors of bacteria-free crowngall on *Vinca rosea. Amer. J. Bot.,* **32:** *237—241.*

1259. WHITE, P. R. 1948. A plant physiologist looks at the cancer problem. *Sci. Monthly,* **47** (3): *187—192.*

1260. WHITE, P. R. 1950. Les tumeurs végétales et le problème de l'étiologie cancereuse. *Année Biol.,* **26** (12): *745—761* (Colloque International du Centre National de la Recherche Scientifique sur la Morphogénèse, Strasbourg, juillet 1949).

1261. WHITE, P. R. 1951. Neoplastic growth in plants. *Quart. Rev. Biol.,* **26** (1): *1—16.*

1262. WHITE, P. R. 1951. Nutritional requirements of isolated plant tissues and organs. *Ann. Rev. Plant Physiol.,* **2:** *231—244.*

1263. WHITE, P. R. 1954. Morphological ecological evidence on the etiology of a localized epiphytic tumor of *Picea glauca. VIIIe Congr. int. Bot. Paris, C. R. Seances Rep. Comm.,* *7/8: 214—217.*

1264. WHITE, P. R. & A. C. BRAUN, 1941. Crown-gall production by bacteria-free tumor tissue. *Science,* **94** (2436): *239—241.*

1265. WHITE, P. R. & A. C. BRAUN, 1942. A cancerous neoplasm of plants. Autonomous bacteria-free crown-gall tissue. *J. Cancer Res.,* **2:** *597—617.*

1266. WHITE, P. R. & A. C. BRAUN, 1943. A cancerous neoplasm of plants produced by autonomous bacteria-free crown-gall tissue. *Proc. Amer. phil. Soc.*, **86**: 467—469.
1267. WILLE, J. 1926. *Cecidoses eremita* Curt. und ihre Galle an *Schinus dependens* Ortega. *Z. Morphol. Ökol. Tiere*, **7** (1/2): 1—101, fig. 49.
1268. WILLFORD, B. H. 1937. The spruce gall aphid *(Adelges abietis* Linn.) in southern Michigan. *Univ. Michigan School Forest & Conserv. Circ.*, **2**: 1—35, fig. 98.
1269. WILSON, E. E. 1935. The olive-knot disease, its inception, development and control. *Hilgardia*, **9**: 233—265.
1270. WINGE, Ö. 1927. Zytologische Untersuchungen über die Natur maligner Tumoren. I. "Crown-gall" der Zuckerrübe. *Z. wiss. Biol.*, (B), *Z. Zellforsch. Mikroskop. Anat.*, **6** (3): 397—423, fig. 13.
1271. WOLFF, M. 1921. Notizen zur Biologie, besonders auch zur Frage nach des Verbreitungsmodus der Eriophyiden. *Z. Forst.- Jagdwesen*, **53**: 162.
1272. WOLL, E. 1954. Austritt von Nucleolarsubstanz im Nährzellen von Gallen. *8 int. Bot. Congr.*, 283—284.
1273. WOLL, E. 1954. Untersuchungen über die zytologische Differenzierung einiger Pflanzengallen. *Planta*, **43**: 477—494.
1274. WOODS, M. W. & H. G. DUBUY, 1943. Evidence for the evolution of pathogenic viruses from mitochondria and their derivatives. *Phytopathology*, **33**: 637—655; 766—777, fig. 11.
1275. WÖRNLE. 1894. Anatomische Untersuchung der durch *Gymnosporangium*-Arten hervorgerufene Missbildungen. *Forstl. Naturw. Z.*, **3**: 68.
1276. WORONIN, M. 1867. Observations sur certaines excroissance que présentent les racines de l'Aune et du Lupin de jardins. *Ann. Soc. Nat. Bot.*, (5) **7**: 73.
1277. WORONIN, M. 1878. *Plasmodiophora brassicae*, Urheber der Kohlpflanzenhernie. *Jb. wiss. Bot.*, **11**: 548—574.
1278. YENDO, Y. & H. TAKASE, 1933. On the root-tubercles of *Elaeagnus*. *Alumni Assoc. Uyeda Coll. Agric. and Silk Industr.*, **4**: 114—130, fig. 3, pl. ii.
1279. ZACH. 1909. Über den in den Wurzelknöllchen von *Elaeagnus angustifolia* und *Alnus glutinosa* lebenden Fadenpilz. *S.B. Akad. Wiss. Wien, math.-natur. Klasse* 117 (1): 973.
1280. ZACHER, F. 1916. Die Literatur über die Blattflöhe und die von ihnen verursachten Gallen, nebst einem Verzeichnis der Nährpflanzen und Nachträgen zum Psyllidarum Catalogus. *Zbl. Bakt.*, (2) **46**: 111.
1281. ZELLER, A. 1937. Über Nematodengallen an Wasserpflanzen. *Ber. dtsch. bot. Ges.*, **55**: 473—484, fig. 11.
1282. ZELLNER, J. 1913. Über die durch *Exobasidium vaccinii* Woron. auf *Rhododendron ferrugineum* Linn. erzeugte Galle. *Akad. Wiss. math.-naturw. Klasse, 24 Oktober, Öster. bot. Z.*, **63**: 45.
1283. ZIEGENSPECK, H. 1929. Die cytologischen Vorgänge in den Knöllchen von *Hippophaë rhamnoides* (Sanddorn) und *Alnus glutinosa* (Erle). *Ber. dtsch. bot. Ges.*, **47**: 50—58, pl. i.
1284. ZIMMERMANN, A. 1900. Über einige javanische Thysanoptera. *Bull. Inst. bot. Buitenzorg*, **17**: 6—19, fig. 9.
1285. ZIMMERMANN, A. 1902. Über Bakterienknoten in den Blättern einiger Rubiaceen. *Jb. wiss. Bot.*, **37**: 1.
1286. ZIPFEL, H. 1911. Beiträge zur Morphologie und Biologie der Knöllchenbakterien den Leguminosen. *Zbl. Bakt.*, (2) **32**: 97—139.
1287. ZOPF, W. 1887. Die Pilztiere oder Schleimpilze. *Handb. Bot.*, **3**: 2, 127.
1288. ZOPF, W. 1897. Untersuchungen über die durch parasitische Pilze hervorgerufenen Krankheiten der Flechten. *Nova Acta Leop.-Carol. Akad.*, 70.

392

1289. Zopf, W. 1907. Biologische und morphologische Beobachtungen an Flechten. III. Durch tierische Eingriffe hervorgerufene Gallenbildungen an Vertreten der Gattung *Ramalina*. *Ber. dtsch. bot. Ges.*, **25**: *233—237*, pl. i.

1290. Zorin, F. M. 1925. Sur une excroissance singulière des bourgeons de poirier les faisant resembler à des fruits. *Sovietsk. Bot.*, **6**: *95—99*, fig. 5.

1291. Zuckermann, S. 1936. The endocrine control of prostate. *Proc. R. Soc. Med.*, **29**: *1557—1568*.

1292. Zweigelt, F. 1914. Beiträge zur Kenntnis des Saugphänomens der Blattläuse und der Reaktion der Pflanzenzellen. *Zbl. Bakt.*, (2) **42** (10/14): *265—336*, fig. 7, pl. ii.

1293. Zweigelt, F. 1917. Blattlausgallen unter besonderer Berücksichtigung der Anatomie und Ätiologie. *Zbl. Bakt.*, (2) **47**: *408—535*, fig. 32.

1294. Zweigelt, F. 1918. Biologische Studien an Blattläusen und ihren Wirtspflanzen. *Verh. zool.-bot. Ges. Wien*, **68**: *121*.

1295. Zweigelt, F. 1929. Gallenbildung und Spezialisation. *Verh. Ges. angew. Ent.*, 7 Mitgliederversammlung München.

1296. Zweigelt, F. 1930. Anpassung und Spezialisation. Rassenbildung und Immunität. Festschritt anlässlich des 70 jährigen Bestandes der höheren Bundeslehransalt und Bundesversuchstations für Wein,- Obst- und Gartenbau in Klosterneuburg.

1297. Zweigelt, F. 1931. Blattlausgallen. Histogenetische und biologische Studien an *Tetraneura*- und *Schizoneura*-gallen. Die Blattlausgallen im Dienste prinzipieller Gallenforschung. *Monogr. angew. Ent.*, 11 (Beiheft zu *Z. angew. Ent.*, **27**: *1—684*, fig. 155, pl. v.

1298. Zweigelt, F. 1941. Immunität und Gallenproblem. *Z. angew. Ent.*, **28** (2/3): *194—210*.

1299. Zweigelt, F. 1942. Beiträge zur Kenntnis der Blattlausgallen. *Biol. Gen.*, **16** (4): *554—572*.

1300. Zweigelt, F. 1947. Problematik der Gallenforschung. *Mikroscopie*, **1** (5/6): *159—173*, fig. 18.

INDEX

406

412

426

RUGGERI 35
rugosus, Sphaerococcus 179
Rumex 252
— *nervosa usambarensis* 102, 103, 159
rupestre, Phagnolon 64
Rutaceae 220, 302
RYBAK 328
RYTZ 244

sabdariffa, Hibiscus 33
Sabia campanulata 74, 75, 193, 276
Sabiaceae 221
Sabicea venosa 67, 85
sabinae, Gymnosporangium 248
Saccorhiza bulbosa 13, 258
Sactogaster curvicauda 230
Sageretia oppositifolia 85
Sagittaria montevidiensis immaculata 180
sagittata, Balsamorrhiza 154, 155
sagittifolia glabrescens, Nesaea 180
SAJO 126
Salicaceae 16, 302
salicifolia, Baccharis 234
salicifolium, Helianthemum 193
salicinus, Fomes 14
salicis, Rhabdophaga 71, 80, 224
saliva, action on plant cell 270
—, cecidogenetic action 284
—, cecidozoa 201
salivary chemical 268
Salix 1, 17, 28, 41, 45, 48, 80, 84, 104, 145, 162, 180, 183, 193, 202, 204, 209, 224, 225, 236, 239, 262, 268
— *caprea* 71, 269
— *cinerea* 42, 43
— *purpurea* 2
—, stomata in gall 46
— *triandra* 2, 191
— *viminalis* 110, 194, 197
saltans, Neuroterus 212
Salvadora 80
— *oleoides* 80
— *persica* 114, 126, *Plate IV, 1*
salvadorae, Eriophyes 114
—, *Thomasiniana* 80
sambucifolia, Scrophularia 139
Sambucus 126
— *nigra* 267
samia, Phlomis 49
sanguinea, Cornus 131
—, *Pelesseria* 13
sanguineum, Geranium 67, 159
sap drain 224
Sapholitus 236
Saphonecrus 236
Sapindaceae 16, 132, 220

Sapotaceae 221
Saprolegnia 13
saprolegniae, Olpidiopsis 13
Sarcinastrum urosporae 13
Sarcococca brevifolia 114
sarcoma, Rous 7, 329
sarcomata 302
sarmentosum, Piper 168
sarothamni, Asphondylia 30, 191, 232, 234
Sarothamnus 88, 233
— *scoparius* 71, 72, 191, 232
SASAKI 177
sativa, Medicago 138, 202
—, *Onobrychis* 138
—, *Oryza* 191
sativum, Petroselinum 224
SAUVAGEAU 13
savastanoi, Phytomonas 258, 298
— *nerii, Phytomonas* 298
Scardia boleti 14
Scelionidae 230
schachti, Heterodera 22
SCHAEDE 59, 60, 246
SCHAEDEL 59
SCHÄLLER 268, 271, 273, 275
scharlach-red 267
SCHEERLINK 59
Schefflera divaricata 169
— *hexapetalum* 169
— *odorata* 169
— *polybotrya* 169
scheppigi, Schizomyia 90
SCHILBERSZKY 295
Schinus 91, 211
— *dependens* 30, 181, 212
Schinzia 246
Schizomyia 31, 133, 142, *Plate VII, 3, Plate IX, 2*
— *acaciae* 195
— *cheriani* 195
— *cocculi* 142, 195
— *galiorum* 142, 234
— *meruae* 106, 107, 194, *Plate III, 4, 7*
— *pimpenellae* 140
— *scheppigi* 90
Schizoneura lanigerum 177
— *lanuginosa* 230
— *ulmi* 208
SCHLECHTENDAL 138, 157
Schlechtendalia chinensis 98
SCHMIEDEKNECHT 30
SCHMIDT 28, 30, 234
Schmidtiella gemmarum 16
SCHMITZ 12
scholaris, Alstonia 171, 173
Schoutenia ovata 38, 167